Communications
in Computer and Information Science   206

T0074409

Geuk Lee   Daniel Howard
Dominik Ślęzak (Eds.)

Convergence and Hybrid Information Technology

5th International Conference, ICHIT 2011
Daejeon, Korea, September 22-24, 2011
Proceedings

 Springer

Volume Editors

Geuk Lee
Hannam University, Computer Engineering Department
70 Hannamro, Daedeuk-gu, Daejeon, Korea
E-mail: leegeuk@hnu.kr

Daniel Howard
QinetiQ Company Fellow, Howard Science Limited
24 Sunrise, Malvern, WR14 2NJ, UK
E-mail: dr.daniel.howard@gmail.com

Dominik Ślęzak
University of Warsaw, Institute of Mathematics
ul. Banacha 2, 02-097 Warsaw, Poland
E-mail: slezak@mimuw.edu.pl

ISSN 1865-0929   e-ISSN 1865-0937
ISBN 978-3-642-24105-5   e-ISBN 978-3-642-24106-2
DOI 10.1007/978-3-642-24106-2
Springer Heidelberg Dordrecht London New York

Library of Congress Control Number: Applied for

CR Subject Classification (1998): C.2, H.4, H.3, D.2, I.2, H.5, D.2

Typesetting: Camera-ready by author, data conversion by Scientific Publishing Services, Chennai, India

Printed on acid-free paper

Springer is part of Springer Science+Business Media (www.springer.com)

Preface

This volume contains 85 out of 179 papers presented at the 5th International Conference on Convergence and Hybrid Information Technology (ICHIT) held during September 22–24, 2011, at the Legend Hotel (Yuseong) in Daejeon, Korea. ICHIT is a series of scientific events spanning the last six years, placing a special emphasis on new theoretical inspirations and hybrid solutions to complex problems with real-world application. It investigates the meeting points between a number of areas related to IT research, such as communications and networking, security and safety systems, pattern recognition and information retrieval, data mining and knowledge discovery, as well as soft computing and intelligent systems, to mention a few domains represented this year.

We received 467 paper submissions from over 20 countries. Each paper was reviewed by at least two chairs or Program Committee members. Author contributions were considered for inclusion into one of two volumes of proceedings: volume one published by Springer within the *Lecture Notes in Computer Science* series and volume two also published by Springer within the *Communications in Computer and Information Science* series. The LNCS volume with 323 papers submitted, and an acceptance ratio below 30%, contains the more scientifically oriented papers, while the CCIS volume with 144 papers submitted, and an acceptance ratio below 60%, contains the more technically oriented papers.

We would like to thank all authors and reviewers for their work and excellent contributions. We are very grateful to Miyagi Hayao for accepting our invitation to deliver the keynote talk. We would also like to acknowledge the following organizations and sponsoring institutions: SERC (Security Engineering Research Center), IWIT (Institute of Webcasting, Internet Television and Telecommunication), KIAS (Korea Information Assurance Society), and KIISE (Korea Institute of Information Science and Engineering).

July 2011

Geuk Lee
Daniel Howard
Dominik Ślęzak

Organization

General Chairs

Geuk Lee   Hannam University, Korea
Daniel Howard   QinetiQ, UK

Program Chairs

Dominik Ślęzak   University of Warsaw, Poland
You-Sik Hong   Sangji University, Korea
Chung-Huang Yang   National Kaohsiung Normal University,
Taiwan

Publicity Chairs

Wai Chi Fang   National Chiao Tung University, Taiwan
Osvaldo Gervasi   University of Perugia, Italy
Dae Yeol Kim   National Institute for Mathematical Science,
Korea

Publication Chairs

Dhananjay Singh   National Institute for Mathematical Sciences,
India
Andrzej Skowron   Warsaw University, Poland
Eun Ser Lee   ANU, Korea

Workshop Chairs

Jeong Jin Kang   Dong Seoul University, Korea
SangSuk Lee   Sangji University, Korea

Organization Chairs

Hanku Lee   Konkuk University, Korea
Keun Ho Ryu   Chungbuk National University, Korea
BongHwa Hong   KyungHee Cyber University, Korea

Local Arrangements Chairs

Jeom Goo Kim	NSU, Korea
Phil Kyu Rhee	Inha University, Korea
Tae Nam Ahn	SERC, Korea

Advisory Board

Aboul Ella Hassanien	University of Cairo, Egypt
Adrian Stoica	JPL NASA, Pasadena, USA
Cheon Hee, Yi	Cheong Ju University, Korea
Hojjat Adeli	Ohio State University, USA
James F. Peters	University of Manitoba, Canada
Juan-Carlos Cubero	University of Granada, Spain
Jun Liu	Harvard University, USA
Kouichi Sakurai	Kyushu University, Japan
Lotfi A. Zadeh	University of California, USA
Nicholas Cercone	Dalhousie University, Canada
Patrick Doherty	Linkoping University, Sweden
Sankar Kumar Pal	Indian Statistical Institute, India
Zbigniew Michalewicz	University of Adelaide, Australia

Program Committee

Ajith Abraham	Norway University of Science and Technology, Norway
Akingbehin Kiumi	University of Michigan-Dearborn, USA
Andrew Kusiak	The University of Iowa, USA
Antonio Lagana	University of Perugia, Italy
Bing Chen	Memorial University, Canada
CheonShik Kim	Anyang University, Korea
Chunnian Liu	Beijing Polytechnic University, China
Conor Ryan	University of Limerick, Ireland
Dieter Kranzlmueller	Johannes Kepler University Linz, Austria
Edward David Moreno	Euripides Foundation of Marilia, Brazil
Elena Zudilova-Seinstra	University of Amsterdam, The Netherlands
Frank Klawonn	Fachhochschule Braunschweig, Germany
Gary B. Fogel	NSI, USA
Gongzhu Hu	Central Michigan University, USA
Guenther Gediga	University of Münster, Germany
Gustavo Olague	CICESE Research Center, USA
Hai Jin	Huazhong University of Science and Technology, China
Hans-Dieter Burkhard	Humboldt Universität Berlin, Germany

Hassan Diab	American University of Beirut, Lebanon
Hideyuki Sawada	Kagawa University, Japan
Hideyuki Suzuki	The University of Tokyo, Japan
Hisao Ishibuchi	Osaka Prefecture University, Japan
Igor Kotenko	St. Petersburg Institute for Informatics and Automation, Russia
Injoo Jeong Kim	East-West University, USA
J. A. Rod Blais	University of Calgary, Canada
Jawed Siddiqi	Sheffield Hallam University, UK
Jianbing Li	University of Northern British Columbia, Canada
Jiman Hong	Kwangwon University, Korea
Jiming Liu	Hong Kong Baptist University, China
JingTao Yao	University of Regina, Canada
Jiong Yang	Case Western Reserve University, USA
Jongmoo Choi	Dankook University, Korea
Jose Negrete-Martinez	Universidad Nacional Autonoma de Mexico, Mexico
Joseph Kolibal	University of Southern Mississippi, USA
Karin Kailing	IBM Research, USA
Kuan-Ching Li	Providence University, Taiwan
Kuntinee Maneeratana	Chulalongkorn University, Thailand
Lei Liu	Dalhousie University, Canada
Ling Zhang	Anhui University, China
Mike Nachtegael	Ghent University, Belgium
Min-Ling Zhang	Hohai University, China
Min Wook Kil	Mun Kyung College, Korea
Mokhtar Beldjehem	St. Anne's University, Canada
Pabitra Mitra	Indian Institute of Technology, India
Pablo Moscato	The University of Newcastle, Australia
Pawan Lingras	Saint Mary's University, Canada
Phil Kyu Rhee	Inha University, Korea
Rainer Unland	University of Duisburg-Essen, Germany
Rajkumar Buyya	University of Melbourne, Australia
Rene Mayorga	University of Regina, Canada
Robert C. Meurant	Institute of Traditional Studies, New Zealand
Roman Slowinski	Poznan University of Technology, Poland
SeokSoo Kim	University of Hannam, Korea
Sergei O. Kuznetsov	National State University Higher School of Economics, Moscow, Russia
Shoji Hirano	Shimane University, Japan
Shuiyuan Cheng	Beijing Polytechnic University, China
Shusaku Tsumoto	Shimane University School of Medicine, Japan
Stefano Cagnoni	University of Parma, Italy
Sushmita Mitra	Indian Statistical Institute, India

Thomas M. Gatton	National University, USA
Torab Torabi	La Trobe University, Australia
Tsau Young Lin	San Jose State University, USA
Umberto Straccia	ISTI-CNR, Italy
Vijay Raghavan	University of Louisiana, USA
Witold Pedrycz	University of Alberta, Canada
Wojciech Ziarko	University of Regina, Canada
Xia Jun	Chinese Academy of Sciences, China
Xiaohua Hu	Drexel University, USA
Xiao-Lin Li	Nanjing University, China
Xin Geng	Deakin University, Australia
Xue-wen Chen	The University of Kansas, USA
Yiming Li	National Chiao Tung University, Taiwan
Yiyu Yao	University of Regina, Canada
Young-Jun Song	Chungbuk National University, Korea
Yong-Kee Jun	Gyeongsang National University, Korea

Table of Contents

Security Systems

Cloud, RFID and Robotics

Industrial Application of Software Systems

Hardware and Software Enginering

Healthcare, EEG and E-Learning

HCI and Data Mining

Software System and Its Applications

Study on Architectural Model and Design of P2P-Based Streaming Client

Wook Hyun[1], SungHei Kim[1], ChangKyu Lee[2], and ShinGak Kang[1]

[1] Electronics and Telecommunications Research Institute, 138 Gajeongno,
Yuseong-gu, Daejeon, Republic of Korea
[2] University of science & technology, 176 Gajungro Yuseong-Gu,
Daejeon 305-350, Republic of Korea
{whyun,shkim,echkyu,sgkang}@etri.re.kr

Abstract. P2P technology is widely used for distributing contents. The P2P applications increase the load of ISP on handling massive and unpredictable P2P traffic. By the way, P2P itself provides very good performance on receiver-side when it receive large-sized files from multi-points. Furthermore, the usage of P2P technology is enhanced its area on multimedia streaming. Nowadays, the volume of multimedia is tends to be massive and need high performance of network. In the case of server-client model, there are definitely limitations on accommodating a number of users. In this paper, we present architecture for providing P2P-based streaming services and framework model for client agent.

Keywords: P2P, streaming, architecture, client framework.

1 Introduction

Since P2P technology has been emerged, the volume of P2P traffic has increased up to more than 40% of total Internet traffic. According to the *Internet Study 2008/2009* *[1]* that is published by *ipoque*, it said that web and streaming traffic is taking over P2P traffic for video contents. It also says that P2P data is still major traffic in all regions *[2]*. P2P traffic burden ISP on handling massive and unpredictable traffic and contents are shared without copyright protections. The most part of P2P traffic is for video including movie, anime and TV. P2P is efficient on distributing massive volume of data for large number of participants. The throughput of P2P network increases proportionally to the number of peers. Nowadays, the volume of contents tends to be bigger because the quality of video gets higher and the size of contents are larger. Hence, we think that P2P-based streaming solution is good for providing high density multimedia streaming for large number of subscribers.

In this paper, we present some related works that provide streaming services using P2P technologies. In chapter 3 and 4, we describe our architecture of P2P-based streaming services and framework model for P2P-based streaming client. We also issue future works to be done for enhance performances and new features with some conclusion remarks in chapter 5.

G. Lee, D. Howard, and D. Ślęzak (Eds.): ICHIT 2011, CCIS 206, pp. 1–7, 2011.
© Springer-Verlag Berlin Heidelberg 2011

2 Related Works

There are several attempts to provide streaming services over P2P overlay network such as *coolstreaming[3]*, *goalbit[4][5]* and *afreeca[6]*. There are two main classification based on the architecture of overlay network creation; Tree-based model, mesh-based model and hybrid model.

2.1 CoolStreaming

CoolStreaming[3] divides MPEG stream into N sub-streams and transmit them as a unit of MOB. In the point of streaming, it constructs virtual parent-child relationship among peers. In the case of poor performance with specific peers, it releases the relation and makes another partnership with other peer.

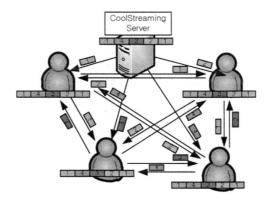

Fig. 1. P2P Service model of CoolStreaming

Different from *BitTorrents[7]*, this approach maintain some kind of partnership with the other peer for receiving and sending sub-stream. Each peer receives all sub-streams from different peers and each sub-stream data is transmitted within parent-child relationship.

2.2 GoalBit

GoalBit[4] has a similar characteristics with BitTorrent on distributing contents data to other peers. This is differ from on providing enhanced QoS on using PSQA(Pseudo Subjective Quality Assessment) for monitoring QoS parameters on providing HD live media contents. GoalBit does not create any hierarchical relationship on constructing overlay network and this mean that all peers have same position within the overlay network.

2.3 Afreeca

Afreeca[5] is one of most popular application in korea that provide multimedia streaming services using tree-type hierarchy. It has unique structures that service

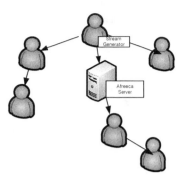

Fig. 2. P2P Service model of Afreeca

provider's server is located within the hierarchy. We think this is for monitoring the stream contents and for providing network stability.

When a node gets out from its parent node, it seems to seek another parent by requesting to service provider's server.

3   Architectural Model for P2P-Based Streaming Service

We designed architectural model for providing P2P-based streaming service as show figure 3. In our architecture, we modeled after the BitTorrent structure. All participating peers are organized with mesh-structure and have no hierarchical relationship with the other peer.

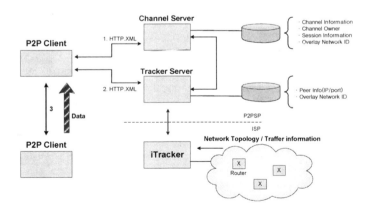

Fig. 3. Architectural model for P2P-based streaming service

The tracker server maintains whole participating peer's information that includes the address of peer. The channel server maintains information for channel descriptions such as channel owner, session information and overlay network-ID.

In this architecture, all multimedia stream exchanged among peers is treated as just data. Therefore, we don't care what kinds of codecs are used for encoding the contents. This may leads to lack of considerations on multimedia characteristics, but this gives flexibilities on codec dependencies.

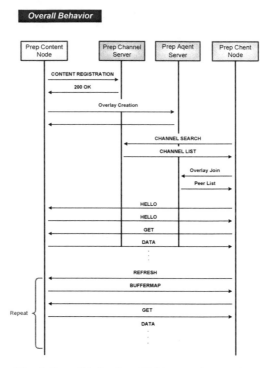

Fig. 4. Overall behavior of P2P streaming service

When a P2P client wants to provide streaming session, it register to channel server by let it know the information about the session. On receiving the request, the channel server allocate overlay network id to the client and notify to the tracker server about newly allocate overlay network id. After that, P2P client that wants to initiate streaming session access to tracker server in order to create P2P overlay network. In this procedure, tracker server checks some credentials whether the client has authority on creating the overlay network.

When a P2P client wants to watch some specific channel session after selecting from channel server, it access the overlay network by using overlay network id that is contained from the channel server. When channel server lists up channels, it also provide overlay network id for each channel.

On receiving overlay network join request from peers, the tracker server responds with peer list for the specified overlay network. By using this peer-list, a client access every peer for retrieving buffermap that indicates what fragmented data it has. If there are any fragment that it interests, it request that. Different from file sharing services, the buffermap negotiations are processed with one peer repeatedly, because of the characteristics of streaming service.

4 Framework for PREP (P2p-Based stREaming Player) Client

The P2P-based streaming client is capable of transmitting media data that is received from other peer at the same time. Following figure 5 shows the framework of our P2P-based streaming client.

Multimedia Player			
PREP Agent			
PREP Buffer	PREP Node	TSComm	CSComm
PREP Proto		XML/Utility library	
TCP/UDP			

Fig. 5. Framework model for PREP client

4.1 Channel Server and Tracker Server Communication

Since the client should be able to create streaming session, it should have functionalities for communicating with channel server. It also should be able to create and access overlay network through interaction with tracker server. On designing interface between client and channel/tracker server, XML is used for describing information and HTTP is used for delivering such information.

4.2 Channel Server and Tracker Server Communication

PREP Agent functions takes role of invocation of overlay network activity such as join and leave. This functions control other underlying functions to organize the behavior of P2P-based media distribution. Because one client can join multiple overlay networks, there can be multiple agent instances. In other words, one agent instance takes charge of one overlay network. When Prep agent invoked, it access tracker server in order to retrieve the list of peers. After receiving the list of peers, it begins to access each peers for querying fragments that it needs.

4.3 Buffer and Buffermap Managements

Different from file sharing using P2P, the size of sharing data is relatively small and there's no need to save all received stream. Hence, we adapted ring type buffermap over temporary file system.

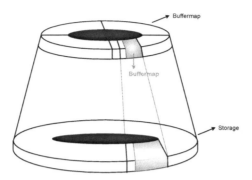

On using ring-type buffermap and buffer storage, it is crucial to keep synchronization and data validity. If there is some fragment to be negotiated to exchange but overwritten, it will break the protocol and make it more complex. In order to avoid such situation, we have reserved enough empty buffer space. When it notifies it's buffermap to the other peer, it does not provide its current full buffermap state but provide subset of them begin with latest one.

4.4 Interface between GUI and PREP Agent

When a client is acting as a streaming server that generates multimedia stream, prep agent receives media data from player through UDP. If the size of received stream exceeds the pre-configured fragment size, it notifies to the buffer manager and set the flag of the bitmap. In order to make it easy to modify user interface, we have divided GUI and prep agent, and they does not share any memory spaces but interact with local socket interface.

4.5 PREP Peer-to-Peer Communication

On retrieving fragmented data from other peers, there are two ways; pull-type and push-type. In case of pull-type, client peer requests buffermap of server peer and then request fragmented data that it has not. On the other hand, in case of push-type, each peer makes some kind of relationship such as parent-child model. In the latter case, it can receive fragmented data without the requesting process. Both approaches have each pros and cons. In order to simplify the behavior of peer, we implemented a pull-type communications among peers as a first step. When a peer is connected to the other peer, it request buffermap and check its availability. If it has interest on specific fragmented data, it sends requests to retrieve them. Unlikely from P2P file sharing, buffermap of peers keeps changing because it receives streaming data from other peer at the same time. Hence, when a peer received all fragments what it wants, it re-request to the other peer for refreshing peer's buffermap. At the same time, it connects to another peer for retrieving another fragment. If it decides that currently connected peer does not have what it wants, it then disconnects and release used resources.

5 Conclusions

In this paper, we have presented architecture for multimedia streaming through using P2P technology and framework model for P2P-based streaming client. In the next steps, we will extend the protocol in order to add partnership concept for exchanging media data. The pull-type approach provide simplicity for implementing streaming services, but need some enhancements for providing stable streaming. To improve performance, we have planned to make it share internal memory spaces on media data that is exchanged with other peers.

Acknowledgments. This research was supported by the ICT Standardization program of MKE (The Ministry of Knowledge Economy)/KCC (Korea Communications Commission).

References

1. Schulze, H., Mochalski, K.: Internet Study 2008/2009, Report, Ipoque (2009)
2. P4P(Proactive network Provider Participation for P2P),
   http://www.pandonetworks.com/p4p
3. Xie, S., et al.: Coolstreaming: Design, Theory, and Practice. IEEE Transaction on Multimedia 9(8) (December 2007)
4. Elisa, M., et al.: GoalBit: The First Free and Open Source Peer-to-Peer Streaming Network. In: LANC 2009 (2009)
5. Project page, http://goalbit.sourceforge.net/
6. Afreeca, http://afreeca.com/
7. BitTorrent WiKi, http://en.wikipedia.org/wiki/BitTorrent

Time Slot Assignment Algorithm
in WSNs to Reduce Collisions

Jookyoung Kim[1], Joonki Min[2], and Youngmi Kwon[3]

[1,3] Dept. of Information Communications Engineering,
Chungnam National University, Yuseong-gu, Deajeon, Korea
{iliwhoth,ymkwon}@cnu.ac.kr
[2] Agency for Defense Development
Taean P.O.Box 1, Chungnam, Korea
minopia@gmail.com

Abstract. Existing hybrid MAC in wireless sensor networks (WSNs) is to combine TDMA and CSMA MAC techniques to adapt sensor nodes into the environment from low contention state to high contention state. It operates in pure CSMA mode when the traffic requirement is comparably low in WSN. On the other hand, as the sensor data is generated in high rate, hybrid MAC is shifted to TDMA-based operation. The main advantage of pure TDMA is no collision among nodes with pre-allocation of each time slot. Every time slot is allocated to nodes in need and it is owned only by pre-allocated node. Non-owner node may not use any other time slots that are not allocated to it. We propose new time slot assignment method to assign every empty slot to nodes which are not in need currently. Even though they are not in need now, the sensor nodes may get more data to send in some time later. In that case, if more time slots were allocated to nodes in advance, the owner nodes can get empty slots to send without contention resulting of reduction in collision and increase in overall traffic throughput.

Keywords: Sensor Network, MAC protocol, Time slot.

1 Introduction

CSMA/CA is widely used multiple access method in Wireless Sensor Networks (WSNs). It is popular because of its simplicity (doesn't need specific hardware) and cheap implementation cost. But, network performance can be reduced when the data traffic has been increased.

To improve weakness of CSMA/CA, research on hybrid MAC has been proposed. They combine CSMA with other multiple access method like TDMA or FDMA. A hybrid MAC that combines CSMA with TDMA needs time slot assignment algorithm to assign time slot to the nodes in a network. But, existing slot assignment algorithms have several problems to be applied instantly to WSN.

G. Lee, D. Howard, and D. Ślęzak (Eds.): ICHIT 2011, CCIS 206, pp. 8–14, 2011.
© Springer-Verlag Berlin Heidelberg 2011

2 Related Work

In this section we briefly overview the MAC layer protocols that have been proposed for WSNs. S-MAC[1] and T-MAC[2] are hybrid CSMA and TDMA MAC protocol each other. They employ local sleep-wake schedules to coordinate packet exchanges and reduce idle listening. The use of RTS/CTS control packet makes up for the hidden terminal problem and failures of the clock synchronization. However, This RTS/CTS method INIT high overhead, because RTS/CTS packet size is bigger than WSN's packet size.

B-MAC[3] achieves LPL(Low Power Listening)[4]. B-MAC periodically wakes up after a sampling interval and check whether channel is active or not. If channel is detected to be active, the node stays awake to receive the data. In other case, the node turn in sleep mode. This MAC protocol has scalability problem.

Z-MAC[5] is a hybrid MAC of TDMA and CSMA MAC protocols. This MAC uses DRAND[6] time slot assignment algorithm. Z-MAC gives transmission priority to nodes. Transmission priority determined by current slot. Slot owner has shorter back-off period than non-owner. This transmission priority can reduce the collision among nodes in a network. And, To solve hidden terminal problem, Z-MAC use ECN message. This message prevent data transmission between two hop distance neighbor at the same time.

HyMAC[7] is a hybrid MAC of TDMA / FDMA MAC protocols. This MAC protocol need specific hardware that can support multi-channel transmission (like CC2420 Radio chip [8]). The nodes in a network get different frequency and time slot to avoid collision. This MAC protocol can transmits data at the same time through separate frequencies. So, HyMAC can get high throughput and low end to end delay.

Existing method needs specific hardware (can support multi-channel transmission) and reduces the network performance under high contention state.

To alleviate this weakness, we proposed a new time slot assignment algorithm. This algorithm doesn't need any kind of specific hardware and can improve network performances under high contention state.

3 Algorithm Design

3.1 Non-priority Mechanism

Nodes in a network have their neighbor information. This information is used as neighbor list to get responses from all the neighbor nodes.

Time slot assignment algorithm consists of 3 states as shown in figure 1: INIT, WAIT_ALL_RESP and NORMAL. Each node changes the state following according to message flows. Figure 2 describe message flow used in our algorithm.

1)   INIT state

Node state transition starts from INIT state. Nodes in INIT state try to get timeslot for exclusive usage for themselves. Nodes calculate their own candidate timeslot position in a distributed manner to avoid excessive collision of SLOT_REQ messages emitted from many nodes at the same time, nodes in INIT state issue SLOT_REQ message with random delays each other.

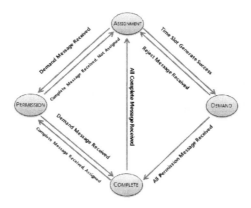

Fig. 1. Slot Assignment State Transition Diagram

2)   WAIT_ALL_RESP state

After transmitting SLOT_REQ message, a node goes to the WAIT_ALL_RESP state. Only when a node receives the positive response messages (SLOT_RESP_POS messages) from all the neighbors, the candidate timeslot is changed to the fixed timeslot for the requesting node. Nodes maintain a neighbor list so to know if all the neighbors responded to their SLOT_REQ message positively. If SLOT_RESP_NEG (negative response) message is received, nodes go to the INIT state. If all the neighbors grant the SLOT_REQ message with positive response, the node goes to the NORMAL state.

3)   NORMAL state

The NORMAL state means that a node took its own fixed timeslot for transmission. Nodes start to sense, collect and transmit the data using the allocated timeslot. Even when other nodes are not in the NORMAL state yet, because they operate in CSMA/CA manner until they get their own fixed timeslot, all the nodes can send/receive data.

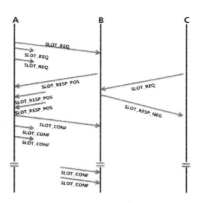

Fig. 2. Message Flow

The above state description is same as in the existing hybrid algorithms. Our algorithm is different from the existing ones in that all the timeslots in an active period are allocated to the nodes in a network. It means that even though nodes are not in need now, the sensor nodes hold extra timeslots in advance preparing for more heavy data transmission afterwards. Figure 3 shows the difference in the allocation of timeslots

3.2 Priority Mechanism

Many slots was assigned nearby sink when apply priority to slot assignment. Few slot is assigned if faraway sink node. Nearby sink node occurs bottle neck effect if many data is entered. So limited slot assign nearby sink node. It make reduce degradation of network by bottleneck effect.

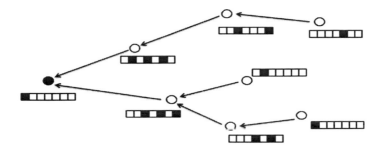

Fig. 3. Slot Assignment Example

3.2.1 Allocation Rule
Sink node broadcast *search message* in the network. Use this message, sink node obtain network path information and allocate time slot to node in network. Sink node calculate maximum allocation slot follow path information and [Equation 1].

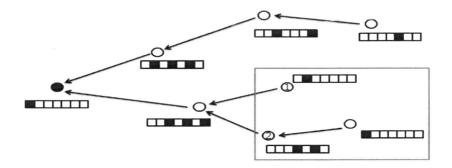

Fig. 4. Slot Assignment Example

$$N = \begin{cases} \dfrac{Totalslot}{2} & , \ Slot \geq \dfrac{Totalslot}{2} \\ Slot & , \ 0 < Slot < \dfrac{Totalslot}{2} \\ 1 & , \quad Slot \leq 0 \end{cases} \tag{1}$$

In the case of lack of time slots in frame, a node who have more childe node has more priority. So, high priority node has more time slot than low priority node.

For example, In [Figure 4] node 1 and 2 need two time slots. But total number of slots what can allocated are 3. So, follow allocation rule, node 2 has two time slots and node 1 has one time slot.

3.3 Summary

The above state description is same as in the existing hybrid algorithms. Our algorithm is different from the existing ones in that all the timeslots in an active period are allocated to the nodes in a network. It means that even though nodes are not in need now, the sensor nodes hold extra timeslots in advance preparing for more heavy data transmission afterwards. [Figure 3] shows the difference in the allocation of timeslots

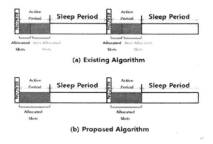

Fig. 5. Algorithm Compare

4   Simulation Results

In this chapter, we evaluate priority and non-priority slot assignment method through simulation. We compare throughput and end-to-end delay. [Table 1] shows simulation parameters.

Table 1. Simulation Parameters

Node Number	25
Packet size	1 Kbyte
Area	100x100m
Time Slot size	30msec

Table 2. Throughput Result

ddddd	dd	dd	dd	dd	dd	dd	dd	dd
dd d ddd	555	555	555	555	555	555	555	555
dddd dd	555	5555	555	555	5555	5555	5555	5555
d ddddd dd	555	555	555	5555	5555	5555	5555	5555

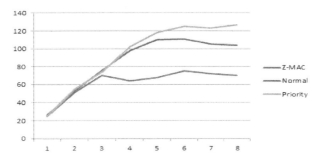

Fig. 6. Throughput Result

Figure 6 shows the comparison of network throughput, A Packet size used in the simulation is 1 Kbyte. A PPS (Packets per Second) value was changed from 1 PPS to 8 PPS. Up to 3 PPS, the proposed algorithm doesn't show much improvement in the words of network throughput. But the traffic is getting heavy, it shows evident better throughput. Table 1 shows the improvement of proposed algorithm.

Table 3. End-to-End Delay Result

ddddd	dd	dd	dd	dd	dd	dd	dd	dd
dd d ddd	55555	55555	5555	5555	5555	555	5	5
dddd dd	55555	55555	55555	5555	5555	5555	5	5
d dddddd d	55555	55555	55555	5555	5555	5555	5	5

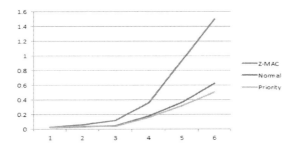

Fig. 7. End-to-End Delay

Figure 7 represent end-to-end delay. The traffic is getting heavy, it shows evident lower end-to-end delay. Table 2 shows this evidence.

5 Conclusion

In this paper, we proposed a new time slot assignment algorithm that assigns multiple time slots to one node in a frame. This algorithm leads high throughput, low end-to-end delay and small average queue size under high contention state.

The data rate in wireless sensor networks is comparatively lower than in the traditional other networks. But more and more, the applications requiring high data rate are being merged into the WSN. With high throughput and low end-to-end delay performances, the new proposed hybrid MAC can be one option to adopt those future applications in the WSN.

References

1. Ye, W., Heidemann, J., Estring, D.: An Energy-Efficient MAC Protocol for Wireless Sensor Networks. In: IEEE INFOCOM 2002, pp. 1567–1576 (2002)
2. Dam, T., Langendoen, K.: An Adaptive Energy-Sufficient MAC Protocol for Wireless Sensor Networks. In: ACM SenSys 2003 (November 2003)
3. Polastre, J., Hill, J., Culler, D.: Versatile Low Power Media Access for Wireless Sensor Networks. In: ACM SenSys 2004 (November 2004)
4. El-Houydi, A.: Spatial TDMA and CSMA with Preamble Sampling for Low Power Ad Hoc Wireless Sensor Networks. In: ISCC 2002, pp. 685–692 (July 2002)
5. Rhee, I., Warrier, A., Aia, M., Min, J.: Z-MAC: a Hybrid MAC for Wireless Sensor Networks. In: Proc. of 3rd ACM Conference on Embedded Networked Sensor Systems(SenSys 2005) (November 2005)
6. Rhee, I., Warrier, A., Xu, L.: Randomized dining philosophers to TDMA scheduling in wireless sensor networks. Technical report, Computer Science Department, North Carolina State University, Raleigh, NC (2004)
7. Salajegheh, M., Soroush, H., Kalis, A.: HyMAC: Hybrid TDMA/FDMA Medium Access Control Protocol for Wireless Sensor Networks. In: IEEE International Symposium on Personal, Indoor and Mobile Radio Communications, PIMRC 2007 (2007)
8. CC2420 2.4 GHz IEEE 802.15.4 / ZigBee ready RF Transceiver,
   http://www.chipcon.com

A Management for the Deployment of PoC Using Dynamic Routing Algorithm in the IMS Nodes

Jae-Hyoung Cho and Jae-Oh Lee

Information Telecommunication Lab,
Dept. of Electrical and Electronics Engineering
Korea University of Technology and Education
Cheon Ahn, Korea

Abstract. Nowadays, by increasing the network traffic in the IMS, the role of Network Management System (NMS) is very important because of limited network resource. It can perform two kinds of routing ways with the capability of static or dynamic routing. In the IMS network, network traffic is very fickle, therefore, a dynamic routing way is more efficient than static routing one because it can make the flow of traffic changeful among nodes in the IMS. In this paper, we suggest a management function of NMS, applying a dynamic routing algorithm in order to manage IMS nodes. Finally, we apply the algorithm by deploying a PoC service, one of the prominent application services in the IMS.

Keywords: PoC, IMS, SIP, NMS, SNMP.

1 Introduction

By increasing network traffic, we need to manage this traffic because of limited resource. The IP Multimedia Subsystem (IMS) is an architectural framework for delivering IP multimedia service. As proposed by the 3GPP, the IMS is not intended to standardize applications but rather to aid the access of multimedia and voice applications from wireless and wire-line terminals. The IMS is able to manage network traffic by controlling IMS nodes. Additionally, network traffic of the IMS increases steadily, so the IMS need to manage network traffic related the IMS nodes. The IMS nodes are called Call Session Control Function (CSCF) that controls session establishment [1]. If a CSCF is overloaded, it might have some problems such as traffic jam and system down. In order to solve these problems, the IMS need to apply the routing algorithm that is able to control CSCFs. Because the traffic of the IMS is very fickle, it is desirable to make the NMS control the traffic flow among IMS nodes. Therefore we study a dynamic routing algorithm which is much more efficient than a static routing algorithm to manage the IMS nodes. In this paper, we suggest a dynamic routing algorithm to manage traffic among CSCFs. Also, we show the scenario that uses Push to talk over Cellular (PoC), one of application services, by applying the proposed algorithm. In section 2, we explain related works such as the

G. Lee, D. Howard, and D. Ślęzak (Eds.): ICHIT 2011, CCIS 206, pp. 15–22, 2011.
© Springer-Verlag Berlin Heidelberg 2011

structure of the IMS, PoC and routing way in the IMS. And in section 3, we show the process of dynamic routing algorithm by deploying the service of PoC.

2 Related Works

2.1 The Structure of PoC Service and the IMS

PoC is a walkie-talkie type of service. Users press (and hold) a button when they want to say something, but they do not start speaking until their terminal tells them to do so. There are several incompatible PoC specifications at present. Most of them are not based on the IMS, but consist of proprietary solutions implemented by a single vendor. As a result these PoC solutions generally cannot interoperate with equipment from other vendors. This situation prompted Open Mobile Alliance (OMA) to create the PoC working group to start working on the OMA PoC service. OMA's PoC service is based on the IMS.

The reason for PoC service deployment in the IMS is that PoC server itself does not need to handle such as SIP signaling routing, discovery and address resolution services, Session Initiation Protocol (SIP) compression, Authentication Authorization Accounting (AAA), QoS control, and so on.

The fig 1 describes the functional entities and reference interfaces that are involved in the support of the PoC services. And the PoC architecture also uses the XML Document Manager (XDM) to manage XML documents like group list, access control lists, presence lists and URI lists which are created by the user and stored by the service provider by means of specific XML documents [5].

The IMS specifies a SIP-based common interface so called ISC by which applications hosted on the SIP, Parlay/OSA and CAMLE application servers are able to interact with the S-CSCF node in the IMS core network. The ISC interface implements filters for subscribers that have been stored in the AAA. It compares the filters with the incoming the SIP message and determines which application services should be invoked [3]. Trigger points in initial Filter Criteria (iFC) are matched with information in the initial the SIP request (a specific the SIP header, to decision which application server are to be included in the SIP chain). There is an another advantage of PoC service by embedding the PoC client into both the mobile phones and PC that users could not only talk by cellular access but also by a fixed connection with wild coverage area. Another importance of PoC service needed to be mentioned about is media control for session. Session control and other signaling are based on SIP, while voice traffic is carried through a RTP/RTCP based streaming bearer. Among them SIP is used for session set up and RTP/RTCP are used for voice data transfer and flow control, respectively. As PoC is a half-duplex communication, its under-control implementation is owed to Tagged Binary Core Protocol (TBCP) or Media Burst Acknowledgement Protocol (MBCP) used to arbitrate request from the PoC clients for the right to send media.

The IMS has been developed to realize real-time or non-real-time multimedia service in the mobile environment. The service object of the IMS provides synthetic multimedia services using IP protocol such as voice, video and data services. Additionally, the IMS has advantages that are easy in service developments and

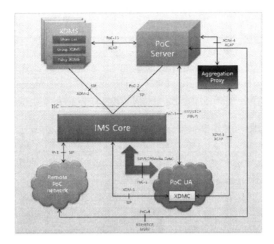

Fig. 1. Related PoC Architecture in OMA 2.0

modifications. The IMS can also is easy to interlock with various 3rd party applications that are based on efficient session management. It enables mobile operator to support the structure based on Session Initiation Protocol (SIP) such as session management, security, mobility, Quality of Service (QoS) and charging. Fig 2 shows the simplification of the IMS structure. Protocols for the definition of the IMS interfaces are Diameter and SIP.

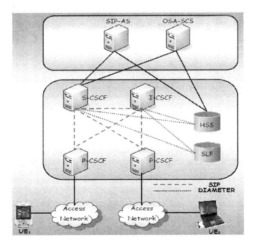

Fig. 2. The Simplification of the IMS Structure

In the IMS network, CSCFs process their related signal. The CSCFs based on SIP implement basic functions to control multimedia session with infra-system, and are divided into P(Proxy)-CSCF, I(Interrogating)-CSCF, S(Serving)-CSCF by depending on their role. The P-CSCF is the first point of contact between the UE (User

Equipment) and the IMS network. From the SIP point of view the P-CSCF is acting as an outbound/inbound SIP Proxy server [2]. The P-CSCF performs the role of either Proxy or User Agents that is defined in the IETF RFC 3261. The I-CSCF is a SIP proxy located at the edge of an administrative domain. The address of the I-CSCF is listed in the Domain Name System (DNS) records of the domain. When the P-CSCF follows SIP procedures to find the next SIP hop for a particular message the CSCF obtains the address of an I-CSCF of the destination domain. In order to discover P-CSCF, 3GPP is presenting method using Dynamic Host Configuration Protocol (DHCP) or gets address through PDP context. The I-CSCF retrieves user location information and routes the SIP request to the appropriate S-CSCF. The S-CSCF is the central node of the signaling plane. The S-CSCF is essentially a SIP server, but it performs session control as well. In addition to the S-CSCF also acts as a SIP registrar. Also, one of the S-CSCF's main functions is to provide SIP routing services. And profile information, authentication and related location information data of a subscriber are stored in Home Subscriber Server (HSS).

Network Management System (NMS) is used to monitor and administer network resources. Effective planning for a NMS requires that a number of network management tasks be performed. NMS should discover the network inventory, monitor the health and status of devices and provide alerts to conditions that impact system performance. NMS systems make use of various protocols for the purpose they serve. SNMP protocol allows them to simply gather the information from the various devices down the network hierarchy. NMS software is responsible for identification of a problem, the exact source(s) of the problem, and solving them. NMS systems not only are responsible for the detection of faults, but also for collecting device statistics over a period of time. A NMS may include a library of previous network statistics along with problems and solutions that were successful in the past—useful if faults recur. NMS software can then search its library for the best possible method to resolve a particular problem. Network elements or devices are managed by the NMS, so these devices are used to be called as managed devices. Device management includes Faults, Accounting, Configuration, Performance, and Security (FCAPS) management.

2.2 A Dynamic Routing Algorithm in the IMS

In typical SNMP use, one or more administrative computers called managers have the task of monitoring or managing a group of hosts or devices on a computer network. Each managed system executes, at all times, a software component called an agent which reports information via SNMP to the manager. Essentially, SNMP agents expose management data on the managed systems as variables. The protocol also permits active management tasks, such as modifying and applying a new configuration through remote modification of these variables. The variables accessible via SNMP are organized in hierarchies. These hierarchies, and other metadata, are described by Management Information Base (MIB) [4].

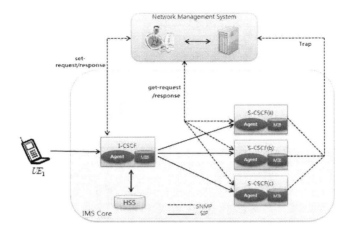

Fig. 3. The Structure of the NMS in the IMS

Fig 3 shows dynamic SIP message flow using network management system (NMS). When there are many S-CSCF (a, b, c) in the NMS, NMS notifies routing address to I-CSCF and selects suitable S-CSCF (a or b or c). The NMS includes information such as performance of CSCF, routing algorithm and so on. There are many kinds of the way about information of the performance but in this paper, we consider the following information:

-System down
-CPU Utilization (Tc)
-Memory Utilization (Tm)
-Throughput: coefficient of utilization of process
-Response Time

Using the above mentioned information, the manager can select more efficient and better CSCF using the dynamic routing algorithm.

3   An Implementation for Dynamic Routing in a Test Bed

Fig 4 shows the dynamic routing algorithm, using information of MIB such as Tc, Tm and response time and so on. A NMS periodically gets the information of performance from CSCF agent. Basically SNMP based manager can get the performance related information from the agent periodically to perform the monitoring function. So the period value for monitoring can depend on the performance of CSCF's agent. So, NMS administrator should condignly adjust monitoring per.

When there are numerous CSCFs, it is possible to change the routing information in CSCF's agent using SNMP Set message by analyzing the monitoring information related to CSCFs. Additionally, when the utilization(CPU and memory) is exceeded to limited value (Tm, Tc), CSCF's agent must notify manager of *Trap* message. And then the manager can store the received *Trap* message in *Log_DB*. In order to analyze the performance information of CSCFs, the manager must perform the following four steps [6]:

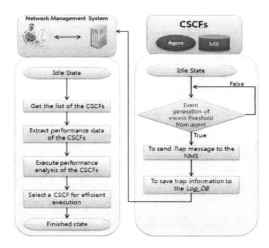

Fig. 4. Dynamic Routing Algorithm of CSCF

• **Step 1:** The manager obtains a list of CSCFs by using the management information from each CSCF and initializes/updates the agent information after performing SNMP *Get-Operation* to agents. The information includes CSCF ID, Type, IP address and state information of CSCF which are shown in Table 1. The *CSCF_ID* field indicates the ID of CSCF and the *CSCF_Class* field denotes the type of CSCF (e.g., P-CSCF, I-CSCF, and S-CSCF). The *CSCF_Name* field is the name of CSCF and the IP field indicates the IP address of CSCF. The *Port_Num* field means the port number of CSCF and the *Status* field represents the operation status of CSCF. The *SysUpTime* field is represented as the time that goes by an activated state of CSCF. The *Description* field means the characteristics of CSCF.

Table 1. MIB of the CSCF

CSCF_ID	CSCF_Class	CSCF_Name	IP	Port_Num	Status	SysUpTime	Description	....

• **Step2:** The manager extracts the performance information from the agent. In this processing, the manager should go through authentication. If the manager successes with authentication, it can get the performance information from agent. But if the manager fails to authentication, it receives the *Trap* message from the agent. The manager looks up the information about a present running process by using *hrSWRunPerfCPU* and *hrSWRunPerfMem* field value. However, the manager needs whole of CPU utilization rate and memory utilization rate of CSCFs for the dynamic routing of SIP message. Therefore, the structure of performance information in the database of manager is shown in Table 2. The *PerfCPU* field is the information about *hrSWRunPerfCPU* and the *PerfMem* field indicates the information about *hrSWRunPerfMem(mib refer)*. The *CPU_Util* field means a CPU utilization rate of CSCF and the *Mem_Util* field denotes a memory utilization rate of CSCF. The

Throughput field indicates the information about a processing rate and the *Delaytime* field means the information about transmission delay time.

Table 2. Performance Information

CSCF_ID	CSCF_Class	PerfCPU	PerfMem	CPU_Util	Mem_Util	Throughput	Delaytime	·········

• **Step3:** In this step, the manager analyzes the performance of each CSCF. After finishing the step 2, the manager can preserve the performance information about CSCFs in the management database. The manager decides the priority order for SIP routing among CSCFs. And then, it renews and modifies the priority information in the management database which is depicted in Table 3. The *Priority* field indicates the information about routing priority order and the *OperStatus* field means state information (e.g., active, standby).

Table 3. Priority-based Routing Information

CSCF_ID	CSCF_Class	IP	Priority	OperStatus	··········

• **Step4:** The manager selects the best CSCF for performing a dynamic routing among CSCFs. After finishing the step 3, the manager decides a CSCF with the highest priority field and configures the best CSCF using *Set-Request* message. Additionally, if the CSCF exceeds the defined value of agent's threshold (*Tm, Tc*), it notifies the manager of the *Trap* message that is stored into *Log_Db*. The *Log_ID* field means ID information of the *Trap* message and the *TrapType* field indicates the type information of *Trap* message. The *TrapInfo* field denotes the detail information about *Trap* message and *Timestamp* field means the information about the creation date and the creation time of *Trap* message.

Fig 5 presents the GUI of NMS manager and S-CSCF including PoC enabler. Every CSCF is implemented in Java language using JAIN SIP 1.2, and the experiment is tested in the environment of 100Mbps LAN. To get the performance information, the administrator should add the IP address of the CSCF's agent. SNMP OID is the unique identifier for collecting performance information. **CPU Utilization** shows a coefficient of utilization about CPU, and **Memory Utilization** shows a coefficient of utilization about memory. CSCF Information includes various information such as CSCF Agent's IP, *sysName, sysDesc* (H/W and S/W information), *sysUpDown*(server operation), *hrSystemUptime*(recent initiation time), *CPUUtilization, Memory_ Utilization* and so on. Fig 6 shows the GUI S-CSCF including presence enabler. Additionally, S-CSCF includes information of registered users. Users can access to S-CSCF (or presence enabler) by set dynamic routing algorithm after session establishment.

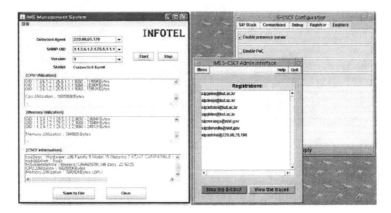

Fig. 5. The GUI of NMS and S-CSCF including PoC Enabler

4 Conclusions

In this paper, we suggest the dynamic routing algorithm to manage network traffic among CSCFs using SNMP. We show the process of dynamic routing algorithm by deploying the service of presence service. Later, we will create the enhanced algorithm which is combined in the management functions such as Accounting, Configuration, and Security. Additionally, we should evaluate the performance in order to optimize the dynamic routing algorithm.

Acknowledgments. This paper was (partially) supported by the Education and Research Promotion Program of KUT.

References

1. Camarillo, G., Garcia-Martin, M.A.: The 3G IP Multimedia Subsystem (IMS) merging the internet and the cellular worlds (2006)
2. RFC 3859, A Presence Event Package for Session Initiation Protocol (SIP) (August 2004)
3. Lee, J.-O., et al.: JNMWare: Java-based Network Management Platform. In: APNOMS (December 2000)
4. Jain SIP API 1.2 specification (December 2006)
5. Cho, J.-H., Lee, J.-O.: An Implementation of the SDP Using Common Service Enablers. In: Hong, C.S., Tonouchi, T., Ma, Y., Chao, C.-S. (eds.) APNOMS 2009. LNCS, vol. 5787, pp. 448–452. Springer, Heidelberg (2009)
6. Cho, J.-H., Lee, J.-O.: A Dynamic Routing Algorithm for Management of the IMS Nodes Using SNMP. In: KICS, vol. 36 (March 2011)

Network-Oriented Intelligent Agent Infrastructure for Internetworking with Guaranteed Quality of Service in High-Speed Network

Joonhyeon Jeon[*], Donghyeok Kim, and Gunmin Kim

Department of Information and Communications Engineering, Dongguk University 26, 3ga, Pil-dong, Chung-gu, Seoul 100-715, Korea
{memory,khb007,tggun}@dgu.edu

Abstract. This paper describes an Agent-based InterNetworking Platform (AINP) which has a hierarchical infrastructure for internetworking with guaranteed Quality of Service in a high-speed network. The AINP system intelligently controls and manages the AINP network through network-oriented agents due to its policy which is periodically updated using the user/network-state information. In this approach, for implementing network-oriented agents and agent-based communications in intelligent AINP infrastructure, a Common Information Model is used which is the international standardization of object-oriented modeling. Extensible Markup Language is used for encapsulation and the HyperText Transfer Protocol is used for transference of user/network-state information. We concluded that the AINP was robust and suitable for an intelligent network management system in a high-speed network.

Keywords: xDSL, Network Management, Quality of Service, Agent-based Network, Network Convergence.

1 Introduction

For the purpose of guaranteeing the service quality of high-speed networks such as x-Digital Subscriber Lines (xDSL), a fair amount of attention has been focused on Service Level Agreements (SLAs) which ensure Quality of Service (QoS) to guarantee the quality of a user-access network [1, 2, 3]. Especially, user demand for various higher-quality services requires intelligent network management systems controlling QoS and collecting network information automatically. Hence the access network takes on an important role in providing the user with network connectivity [4, 5, 6]. One effective user-help service is an agent service that prepares setup information or executes exchanges on behalf of user applications in Internet Protocol (IP) networks [7, 8]. Earlier agent-based management systems [9~12] play a simple role in monitoring the user-network environment through user agents on client-server internetworking. Hence most of these techniques are not suitable for internetworking of guaranteed QoS in a high-speed network having a large number of subscribers

---

[*] Corresponding author.

G. Lee, D. Howard, and D. Ślęzak (Eds.): ICHIT 2011, CCIS 206, pp. 23–30, 2011.
© Springer-Verlag Berlin Heidelberg 2011

[7~12]. In this paper, to supply users with various advanced services effectively and automatically in the environment of a high-speed access network, we proposed an Agent-based InterNetworking Platform (AINP) for internetworking with guaranteed QoS between a user-access network and a public network. The proposed AINP system uses an internetworking platform which intelligently controls and manages network-oriented agents. Conclusively, the AINP management system can provide users with efficient network connectivity of guaranteed QoS for supporting value-added services in a high-speed network.

2   Agent Based InterNetworking Platform

2.1   Agent Based InterNetworking Platform Infrastructure

Fig. 1 shows that the proposed AINP consists of layered management servers which hierarchically control and manage the AINP network through their own network-oriented agents. AINP management servers consist of a Central Management Server (CMS) at the top and Local Management Servers (LMS) at each layered point. Agents are differentiated into user agents, LMS agents, and CMS agents, and are used for either collecting user/network-state information from users and LMSs or transferring a new policy up to each LMS at a layered point. The agent-based data flow seignior in the AINP management system is summarized as follows.

First, each end-point LMS collects user-state information through each user agent with which the LMS connects locally. The user/network-state information is transferred up to a CMS through each LMS agent after LMSs located at each layered network add their network information to user-state information. The CMS at the top point builds up and updates the policy of AINP network management, which is transferred to each LMS at the nearest lower point thorough a CMS agent. Finally, this policy is propagated to each end-point LMS through LMSs and their agents located at each layer of the network. In this process, each LMS at each network layer reflects the CMS's policy of AINP internetworking for guaranteeing the QoS of each corresponding network. Finally, each end-point LMS updates and allocates the user-network environment again due to the policy transferred from the corresponding LMS at the upper level. The internetworking in AINP is periodically accomplished since the state information is transferred and collected periodically through the user or LMS agents. Each end-point LMS periodically controls and manages user-access networks belonging to the local area due to user information. There are two kinds of user information: one is user-profile information for user-access authentication and admission, and the other is user-state information, distinguished from static data and dynamic data, for guaranteeing the QoS of the user-access network. Each end-point LMS provides a user's network with differential QoS due to the grade of the user's class recorded in user-state information itself. If a user connects with the corresponding end-point LMS through its agent, the LMS assigns itself network resources to the user-access network after computing the priority grade of the user. Table 1 shows weighted priority due to the user class. Note that each end-point LMS provides each user with different services due to a different user class. This is to guarantee constant bandwidth for each user and allocate the network resources of

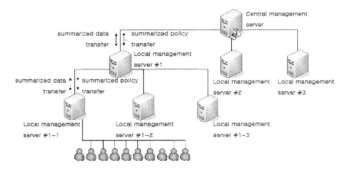

Fig. 1. Hierarchical management server

Table 1. Example of priority grade due to user class / service weight

User	Class	Class Weight	Used Service	Service Weight	Service Quality[1]	Weighted Service Quality	Weighted priority order
100	Business 1st class	1	streaming	1	70	70	2
101	Business 2nd class	0.9	FTP	0.7	70	77.7	3
102	Personal 1st (premium)	0.8	mail	0.9	60	75	1
103	Personnal 2nd (economy)	0.7	HTTP	0.8	80	114	4

such network bandwidth and network delay. Thus each LMS reflects and allocates the policy of CMS such as routing path, surrounding network and bandwidth information, traffic blocking, limited number of users, and so on.

For guaranteed QoS in an AINP system with the hierarchical structure, the transfer process of user state information can be accomplished in the following 5 steps.

1) Step 1: Once a user logs on, a user requests an authentication through its agent in order to connect to its own access network which belongs to a corresponding LMS. Thus the LMS recognizes the user agent as a managing target and permits the network-connectivity request to the user through its own agent. For the internetworking of the user-access network between the user agent and the LMS agent, user profile data and user static data are used. Tables 2 and 3 show the attributes used for the user-access service.

2) Step 2: Since a lot of user-state information is transferred to the corresponding LMS either periodically or aperiodically, this might be caused by data loss due to network delay. In addition, the load of LMSs on the AINP network should also increase. Therefore, to reduce the loads on the AINP network and LMSs, user data is collected and transferred in a circle rotation by sequential order which is distinguished with aria code allocated according to the localization of LMSs. For example, let us assume that the update period is 3-minute, and the area codes with the local interval 1 min between each area are A(01), B(02) and C(03). The update period of each end-point LMS is 4 minutes for area A, 5 minutes for area B, and 6 minutes for area C, since each update period is obtained by adding the local interval to the update period. In this case, this load balance by time division

---

[1] Refer to Service of Quality in Table 2.

multiplexing (TDM) is propagated from LMSs at upper level to the CMS at the top level. As a result, the bottleneck state of each LMS can be eliminated using the TDM method. However, the synchronization of each LMS for data load balancing by TDM should be done first.

Table 2. Attributes of user profiles

Attribute	Data type[2]	Description
UniqueIdentifier	string	User ID
Password	string	Password of user account
ClassType	unit32	The service-type regulation (Family Premium, Family Economy, Single Premium, Single Economy)
NumberOfAccount	unit32	Maximum number of accounts issuable including one's own (if single, 1)
IsMaster	unit32	In the case of Family-type service, whether another lower account can be created or not
ServiceLocation	string	Classified into home, office, and so on.
DependentMaster	string	Indicates Master account of Family-type service. If one is not a Master, service profile is not modifiable (Use SELF if it is Master, while use Master account ID if lower account)

Table 3. Static attributes of user agent

Attributes	Data Type[2]	Description
SessionID	string	Give personal ID to check effectiveness
AreaCode	uint32	Collect and process information in sequence by area code and local code number that are given.
CurrentMode	uint32	① Active : Regardless of server's intention, it transfers information periodically. After synchronization between time and server is executed, the information is transferred periodically according to the property received from the server agent. The data reliability is low, but it can be transferred without any requests from the server. ② Passive : Transfer information only at the request of server. High reliability of data. Efficient. Easy to determine the state of user with accuracy.
AvailablePort	uint32	Port number for connection to local management server.
UpdateInterval	uint32	Transfer user information to local management server, once values of TimeStamp and UpdateInterval are added.
BusinessCategory	uint32	User class(Business, Ordinary). Standard for priority in allocating bandwidth or process.
ServiceCategory	uint32	Service class(Premium, Economy). Standard for service quality.
PreferredServiceLoc	string	Item to provide same service, in spite of the modification of location, with user's own IP address which represent the current location of service connection.
ServiceProfile	string	Service profile for personalized service of user (reference value of user service profile).

[2] Refer to Table 4.

Table 4. Data type

Built in data type	Description	Built in data type	Description
uint8	Unsigned 8-bit integer	string	UCS-2 string
sint8	Signed 8-bit integer	boolean	Boolean
uint16	Unsigned 16-bit integer	real32	IEEE 4-byte floating-point
sint16	Signed 16-bit integer	real64	IEEE 8-byte floating-point
uint32	Unsigned 32-bit integer	datetime	A string containing a date-time
sint32	Signed 32-bit integer	<classname> ref	Strongly typed reference
uint64	Unsigned 64-bit integer	char16	16-bit UCS-2 character
sint64	Signed 64-bit integer		

Table 5. Dyanmic attributes of user agent

Attributes	Data Type[2]	Description
TimeStamp	datetime	Point of time data is transferred by user agent (units in seconds)
UniqueIdentifier	string	Identifier to make a service profile of user
Bandwidth[BW] (units in kbps)	uint32	BW allocated by network management server for current user
Download-Speed	uint32	Current download speed is determined by the delay of the first byte and the second byte when receiving data (units in kbps). The accuracy is low, but used for the reduction in network line load, and to receive user state information.
Delay	uint32	Network delay (units in milliseconds).
Throughput	uint32	Processing rate (units in kbps). The processing rate of data : Throughput = Total File Size / Response Time (unit per msec)
ServiceQuality	uint32	Shows how well the current transmission speed is guaranteed : (current transmission speed/transmission speed defined in SLA) * 100

4) Step 4: Whenever the top-level CMS sets up a new policy for updating AINP internetworking, the new policy is hierarchically downloaded up to LMSs at end point Thus each LMS that exists hierarchically at each level allocates individually the corresponding local-network policy reflecting new policy of the CMS for the internetworking of guaranteed QoS itself. Finally, each end-point LMS reflects updated policy of LMS at the upper level, and readjusts the user bandwidth again according to a *Red Pin List*[3] where the grade of the user-service class is recorded. As this is a type of pooling method, it is used to prevent competition between users when allocating additional bandwidth. The end-point LMS, for reflecting the next policy of CMS, records user-state information that belongs to low grades of the user-service class after reflecting the policy of CMS internetworking.

5) Step 5: Once the user terminates the service, it is excluded from the AINP's managing target.

---

[3] For computing the service quality of the user, the AINP system uses not only the speed determined by SLA but also the grade of the user-service class. If a resultant service quality is lower than that of SLA standard, the corresponding end-point LMS is called Red Pin itself.

2.2 Agent Based InterNetworking Protocol

In order to guarantee QoS for AINP internetworking, the static/dynamic state information about the user and network environment is required through user agents and LMS agents on AINP network. Static data is the fixed information given in the first communication between the user agent and LMS at the lowest level. Dynamic data is the frequently-updating information which has the state of a user who is receiving the service. The static state information is transferred through the LMS agent after the LMS encapsulates the static factors into a form of XML in the HTTP header according to the request of the user agent connection [8]. The attributes of user static factor are shown in Table 3. In this table, Session ID, Area Code, Available Port and Update Interval are due to the updated policy of the user profile. Also, Business Category, Service Category, Preferred Service Location and Service Profile are the static attributes for monitoring the user network state in real-time. Since each static attribute can have its own weight, the LMS can classify and distribute the network-service quality of each user according to a weighted-sum value per user. Specifically, the service profile is the attribute related to user on demand.

3   Implementation of Agent-Based Internetworking Platform

3.1   AINP System Structure and Implementation

Collecting information through an agent is simple but to process it is a little more detailed and hierarchical, as shown in Fig. 2. In Step 1~ Step 4 explained in Section 2, for communication of the state information between agents, the Common Information Model (CIM) operation [14]~[16], called a CIM client, encodes the attributes in the HTTP header after encapsulating the user state data as static/dynamic attributes with XML [13], as described in Tables 3 and 4. The proposed scheme takes a method of allocating the network resources by each area and entrusting the management of network resources to LMS. Hence, even though there is a little time margin in collecting information, no delay occurs in management because the established policy is reflected immediately after processing the collected data, in view of management.

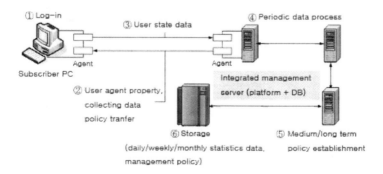

Fig. 2. Process of AINP management system using LMS and CMS

3.2 Common Information Model Implementation

To reduce the server load, the AINP system is distributing data to LMSs that exist hierarchically on the AINP network. For balancing the data load in the proposed AINP system, the user data is processed according to the user's physical location and the user-service (connection) class as follows:

1) The user information is classified due to the user physical location which is distinguished with an area code. The end-point LMS collects and processes the data within the local area.
2) After connecting to the user-access network due to the user's physical location, end-point LMS determines the priority of service quality by the grade of the user-service class. Here the local network bandwidth of each area is allocated according to the network policy of the CMS.

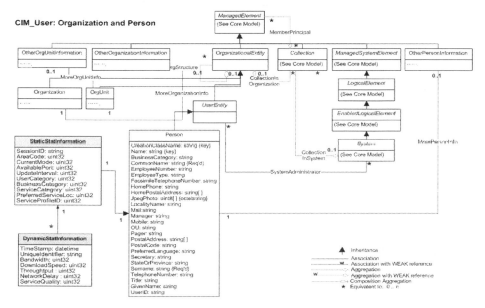

Fig. 3. Expanded model (that which is proposed in this paper is expressed in gray where the field name of the classes not directly related are excluded) of CIM Schema V2.7 [14]

The user data, which is the static/dynamic state information processed in the statistics by time, is passed over from the LMS at the lowest level to the LMS at a higher level. All of LMSs and CMS in the AINP system periodically update the user data, and CMS sets up a policy about the entire network. Finally, CMS downloads this policy to all LMSs for the networking QoS, and each LMS takes charge of its area's own networking control according to the policy. Fig. 3 shows the extended model of the proposed AINP system which is implemented using CIM schema. The CIM is being created and distributed from Distributed Management Task Force [16]. In this figure, it is proposed that the Person object is inherited out of the USER schema and expanded. In addition, the user state information is represented by an attribute adding the user profile information which is dependent on the Person object. This attribute is defined as a static behavior and a dynamic behavior.

4 Conclusion / Expected Effect

In this paper, we proposed an AINP which has a hierarchical infrastructure for internetworking with a guaranteed QoS. In this approach, AINP network is intelligently controlled and managed through network-oriented agents that exist in user terminals and layered-management servers, individually. To implement AINP, we used CIM for object-oriented modeling [14-16]. For encapsulation and transference of network-state information, XML and HTTP were used, respectively. The main advantage of the proposed system was the flexibility and scalability able to expand and reduce depending on the density of subscribers.

Acknowledgments. This work is financially supported by the Ministry of Education , Science and Technology(MEST), the Ministry of Knowledge Economy(MKE) through the fostering project of HUNIC.

References

1. Hia, H.E.: Secure SNMP-Based Network Management in Low Bandwidth Network. Master thesis, Virginia Polytechnic Institute and State University (2001)
2. Whitepaper - QoS protocols & architectures, http://qosforum.com
3. Ferguson, P., Hustom, G.: Quality of Service Delivering QoS on the Internet and in Corporate Networks. Wiley Computer Publishing, Chichester (1998)
4. Xiao, X., Ni, L.M.: Internet QoS: The Big Picture. IEEE Network 13, 8–18 (1999)
5. Marly, N., Chantrain, D., Focant, S., Handekyn, K., Daenen, K., Batsleer, C.: Service Selection in the Access Network. In: IEEE International Conference on Communications 2001, vol. 5, p. 1622 (2001)
6. Korea Internet Dictionary, http://www.nca.or.kr
7. Cheong, F.: Internet Agents: Spiders, Wanderers, Brokers, and Bots. New Riders, Indianapolis (1996)
8. Franklin, S., Graessrt, A.: Is it an Agent, or just a Program?: A Taxonomy for Autonomous Agents. In: Proc. of Third International Workshop on Agent Theories, Architecrures, and Languages, pp. 21–35 (1996)
9. Bellifemine, F., Rimassa, G., Poggi, A.: JADE - A FIPA-Compliant Agent Autonomous Framework. In: Proceedings of the 4th International Conference and Exhibition on the Practical Application of Intelligent Agents and Multi-Agents, UK, vol. 99, pp. 97–108 (1999)
10. Bellifemine, P.A., Rimassa, G.: Developing multi agent systems with a FIPA-compliant agent framework. In: Software - Practice Experience, vol. 31, pp. 103–128. John Wiley Sons, Ltd, Chichester (2001)
11. Bivens, A., Fry, P.H., Gao, L., Hulber, M.F., Szymanski, B.K.: Agent based High Performance Computing. In: Conference on Agent-Based Network Monitoring, Workshop at Agents 1999, pp. 41–53 (1999)
12. Poggi, A., Rimassa, G., Tomaiuolo, M.: Multi-User and Security Support for Multi-Agent Systems. In: Proceedings of WOA 2001 Workshop, Modena, pp. 13–18 (2001)
13. Extensible Markup Language (XML). Version 1.0, W3C Recommendation (2004)
14. Specification for the Representation of CIM in XML. Distributed Management Task Force, Version 2.1, DSP 201 (2003)
15. Specification for CIM Operations over HTTP. Distributed Management Task Force, Version 1.1, DSP 200 (2003)
16. Common Information Model(CIM) Schema Version 2.7. Distributed Management Task Force (2003)

Study on Emergency Services for VoIP Based on IP Network vs. Domestic Emergency Network

IlJin Lee, OkJo Jeong, and ShinGak Kang

Electronics and Communications Research Institute, 138 Gajeongno, Yuseong-gu,
Daejeon, Republic of Korea

Abstract. Enhanced 9-1-1 (E9-1-1) Service is a public safety feature that allows customers to report an emergency or request emergency assistance by dialing the 3-digit telephone number "9-1-1." Emergency calls originated in this manner are then routed to the appropriate Public Safety Answering Point (PSAP) based on the location from which the call was originated. In this paper, we are intended to analyze the structure of system to efficiently provide emergency service using SIP protocol-based internet telephone under IP network environment, and the assessment of its performance.

Keywords: VoIP, Emergency services, IP network.

1   Introduction

Voice of IP (VoIP) technology provides voice service as well as data service via Internet. It has been a promising technology as Internet grows fast and the requirements are increasing. Recently, several protocols have been created to allow telephone calls to be made over IP networks, notably, SIP and H.323. Due to introducing SIP and H.323, There are many changes at internet telephony service. Internet telephony enables a wealth of new service possibility.

Enhanced 9-1-1 (E9-1-1) Service is a public safety feature that allows customers to report an emergency or request emergency assistance by dialing the 3-digit telephone number "9-1-1." Emergency calls originated in this manner are then routed to the appropriate Public Safety Answering Point (PSAP) based on the location from which the call was originated. Depending on the municipal requirements and procedures, a PSAP may transfer the call to the proper public safety agency (e.g., police, fire, emergency medical service), collect and relay emergency information to the proper agency, or dispatch Emergency Service personnel directly for one or more participating agencies.

Because VoIP customers may be fixed, nomadic, or fully mobile, there are a number of challenges associated with providing E9-1-1 Service to VoIP customers. As for wireless callers, identification of a nomadic or mobile VoIP caller's location is critical to routing 9-1-1 calls to the appropriate PSAP, as well as facilitating the dispatch of emergency personnel to the caller's location without having to rely on the caller to provide the location information.

G. Lee, D. Howard, and D. Ślęzak (Eds.): ICHIT 2011, CCIS 206, pp. 31–38, 2011.
© Springer-Verlag Berlin Heidelberg 2011

As the public adopts VoIP, E9-1-1 calls will increasingly originate from VoIP users. Some VoIP telecommunications service provider networks, however, are not natively compatible with the existing E9-1-1 infrastructure.

This paper is intended to analyze the structure of system to efficiently provide emergency service using SIP protocol-based internet telephone under IP network environment, and the assessment of VoIP emergency call vs PSTN emergency call.

2   Requirements of Emergency System for IP-Base Network

Figure 1 illustrates the functional elements and signaling interfaces used to support the i3 solution. The acronyms used to label elements in the diagram are defined in the NENA Master Glossary of this standard.

Several new functions are introduced in the i3 solution that assists in:

- Determination and validation of location information,
- Routing emergency calls to the appropriate interconnection point with the existing infrastructure,
- Providing the interconnection for the IP domain with the existing network elements and databases needed to support delivery of location information to the PSAP.

The IP domain "cloud" in the figure represents the collective set of IP domains, including multiple private and public service provider domains, from which emergency calls might originate, and through which emergency calls are interconnected with the existing emergency services infrastructure that is shown on the right-hand side of the diagram.

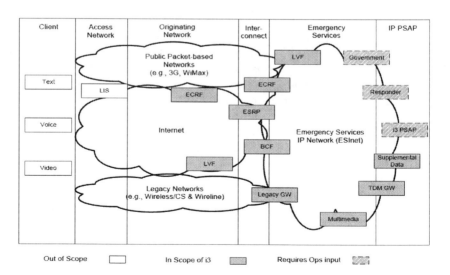

Fig. 1. VoIP Architecture for E9-1-1 service (NENA i3)

2.1   Requirements for Location Information to Support IP-Based Emergency Services

- **Location Determination and Acquisition**

  DA1– The access network shall provide a mechanism for determination and acquisition of location information, and support queries for location.

  DA2 – The location estimate used shall be that associated with the physically (wire, fiber, air) connected network.

  DA3 – Location may be requested at any time. Location information must be associated with the device at the time the location is requested.

- **Location Representation**

  Rep1– Location information may be provided by-value or by-reference; the form is subject to the nature of the request.

  Rep2 – Location determination and acquisition mechanisms must support all location information fields defined within a PIDF-LO. This location information will include a civic location and/or geographic co-ordinates/geodetic shapes.

  Rep3 – Location acquisition mechanisms must allow for easy backwards compatibility as the representation of location information evolves.

  Rep4 – All representations of location shall include the ability to carry altitude and/or floor designation. This requirement does not imply altitude and/or floor designation is always used or supplied.

3   Consideration on VoIP Emergency Call in Korea

In Korea, since 2007 a standard was developed in order to support the emergency call service.
    The main contents of the developed standard for fixed VoIP service is as follows.

  - Emergency Number dialing

  - Location information acquirement

  - Location information Management

  - The caller location information acquisition at PSAP

  - Routing scheme of the emergency call.

In Korea, figure 2 shows the E-911 service architecture for VoIP.

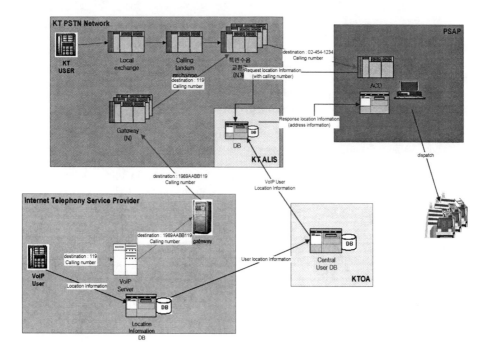

Fig. 2. VoIP Architecture for E9-1-1 service in Korea

Emergency call procedure is as follows.

(1) VoIP User dialing 119

(2) The internet telephone provider inserts 1989AABB into 119 number front.

(3) 1989AABB is routing code for emergency service

(4) The location information of user is delivered by the corresponding gateway which is close in the basis.

(5) The location information of user on the internet telephone provider gateway forwards PSTN exchange.

(6) In the PSTN gate way switch analysis routing code for the emergency call and forwards special exchange for emergency call.

(7) Special exchange forwards emergency call to PSAP

4   Design of Emergency System for IP-Base Network

To cope with the tendency of evolving into an IP network environment in the future, this paper is intended to design the module using constituent elements for the proposed telephone emergency service as well as the detailed module for emergency service under an IP-based network environment.

Figure 3 shows architecture for emergency system based on IP network and figure 4 shows components of emergency system.

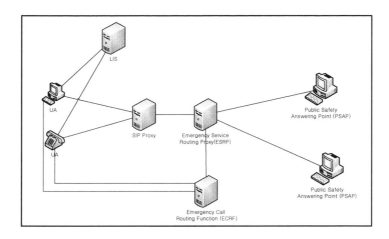

Fig. 3. Architecture for Emergency Service based on IP Network

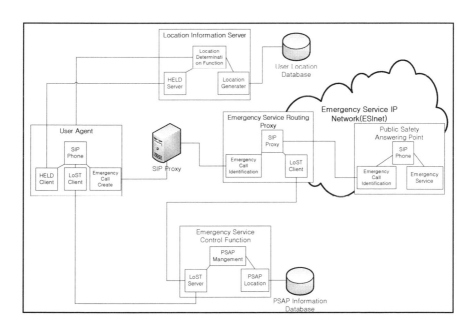

Fig. 4. Component of Emergency Service System based on IP Network

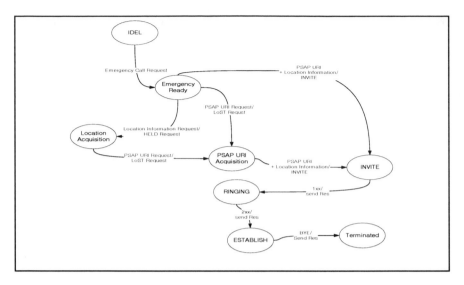

Fig. 5. State Machine of UA

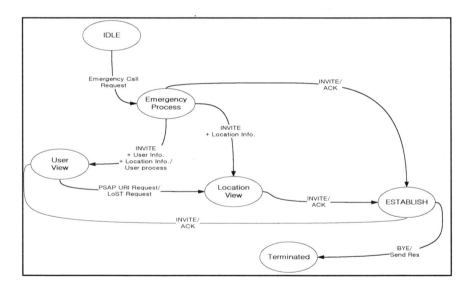

Fig. 6. State Machine of PSAP

To achieve this, at first, the internet telephone that supports emergency service was embodied to enable internet telephone emergency service. The most important point in such designed terminal is that the location information of the terminal, that is, the address information of the caller is directly transmitted by the terminal to a PSAP(Public Safety Answering Point). For this reason, the PSAP does not need to request address information to DB, based on the telephone number of the existing caller where the service is improved from the existing sound-oriented mode to a variety of contents such as sound, character, image, and image conference. As a result, a more prompt and efficient emergency service is being realized. Furthermore, HELD client module was embodied to request the location information of the terminal to the location information server using HELD protocol. The terminal is designed and embodied so that it can store the location information of a terminal using HELD client module upon booting. The test result has proven that the location information of the terminal has been correctly received from the location information server.

Secondly, the location information server that stores and manages the location information of the terminal was designed and embodied. In the location information server, the HELD server module has been embodied such that it provides a response once the terminal requests the location information of a terminal using HELD protocol. In addition, as a method of automatically securing the location according to the mobility of a terminal, it was designed and embodied such that the location of a terminal could be secured when the terminal is transferred in an Ethernet environment using DHCP and SNMP protocols. All the modules embodied have been proven to be satisfactory as shown by the test results.

Thirdly, ECRF(Emergency Call Routing Function) server has been designed and embodied so as to enable the terminal to be connected with the nearest PSAP(Public Safety Answering Point). On this ECRF(Emergency Call Routing Function) server, LoST server module is mounted, where the terminal provides the information of emergency service available in the relevant territory and the URL information of Public Safety Answering Point (PSAP) that handles such emergency service.

Lastly, ESRP(Emergency Service Routing Proxy) server that treats emergency call and PSAP(Public Safety Answering Point) that receives emergency call were designed and embodied. This PSAP provides the function that indicates on the map screen the location information delivered from the terminal. By conducting the relevant test, it was proven that the emergency call generated from the caller has been connected to PSAP.

5   Comparison of Emergency Call Connection Time

Comparison and analysis on the time to take for call delivery was attempted between two methods, which are, inquiring the location information of the caller from PSAP provided in existing PSTN method and directly delivering the location information of the caller from an IP network environment. The result shows that it took 29.41ms for the call of the existing system to be connected, while it only took 1.04ms for the call to be connected when the terminal directly delivers the location information. That is, in the case of the latter, it was found out to have shortened the time by 28.37ms when compared with the existing system.

The proposed system discussed up to now is IP-base network that is to be initiated by the government in the future. In coping with the environment being evolved into a network, it is intended to suggest the structure of the system that enables to provide a more efficient emergency call service under an IP network environment. Such system could be utilized as an excellent candidate technology for the domestic National

Emergency Management Agency and other Public Safety Answering Point (PSAP) to provide emergency services.

Acknowledgments. This research was supported by the ICT Standardization program of MKE (The Ministry of Knowledge Economy)/KCC (Korea Communications Commission).

References

1. NENA, Interim VoIP Architecture for Enchanced 9-1-1 services (i2)
2. NENA Functional and Interface Standards for Next Generation 9-1-1 Version 1.0 (i3), NENA (December 2007)
3. Schulzrinne, H.: A Uniform Resource Name (URN) for Emergency and Other Well-Known Services. Internet Engineering Task Force, RFC 5031 (January 2008)
4. Hardie, T., Newton, A., Schulzrinne, H., Tschofenig, H.: LoST: A Location-to-Service Translation Protocol. Internet Engineering Task Force, RFC 5222 (August 2008)
5. Cueller, J., et al.: Geopriv Requirements. Internet Engineering Task Force, RFC 3693 (February 2004)
6. Peterson, J.: A Presence-based GEOPRIV Location Object Format. Internet Engineering Task Force, RFC 4119 (December 2005)
7. Winterbotton, J., Thomson, M., Tschofenig, H.: GEOPRIV PIDF-LO Usage Clarification, Considerations and Recommendations. Internet Engineering Task Force, draft-ietf-geopriv-pdif-lo-profile-13 (September 2008)
8. Schulzrinne, H.: Dynamic Host Configuration Protocol (DHCPv4 and DHCPv6) Option for Civic Addresses Configuration Information. Internet Engineering Task Force, RFC 4776 (November 2006)
9. Thomson, M., Winterbottom, J., Andrew: Revised Civic Location Format for Presence Information Data Format Location Object (PIDF-LO). Internet Engineering Task Force, RFC 5139 (February 2008)
10. Requirements for a Location-by-Reference Mechanism. In: Marshall, R. (ed.) Internet Engineering Task Force, draft-ietf-geopriv-lbyr-requirements-03 (June 2008)
11. Winterbottom, J., Tschofenig, H., Schulzrinne, H., Thomson, M., Dawson, M.: An HTTPS Location Dereferencing Protocol Using HELD. Internet Engineering Task Force, draft-winterbottom-geopriv-deref-protocol-02 (July 2008)

EOI: Entity of Interest Based Network Fusion for Future Services

Gaurav Tripathi[1], Dhananjay Singh[2], and K.K. Loo[3]

[1] Bharat Electronics Limited, Ghaziabad, India
[2] Natioanl Institute for Mathematical Sciences (NIMS), Daejoen, South Korea
[3] School of Engineering and Information Sciences, Middlesex University, London, UK
gauravtripathy@gmail.com, dan.usn@ieee.org, J.Loo@mdx.ac.uk

Abstract. The Internet has been getting popular to access by today's life. For that, we need to construct a novel model of Internet architecture in a simpler and better way. The present paper talks about a new approach towards future services. This paper tries to build hybrid architecture by combining Joint Directors of Laboratory (JDL) model and SENSEI architecture. This hybrid model recommends data fusion process for Entity of Interest (EOI) based architecture. The EOI based fusion model has supported internet connectivity over things. The EOI has consisted user and machine (objects) in the domain of application. Thus, the proposed novel model will be applicable to all present and future applications domain ranging from energy sectors to the food sectors.

Keywords: Future Internet Architecture, JDL, EOI, Sensor Fusion.

1 Introduction

The Machine-to-Machine communication is likely to be nondeterministic, autonomous but interoperable and event driven depending upon the context and environment. The communications have to be evolutionary in nature for the scope of the communication will always be ever increasing. New participating cntitics will be introduced and thus interfaces will have to evolve. Future environment and contexts and their interfaces systems will also be dynamic in nature. Machine-to-Machine communication will facilitate information flow. The information generated will be huge and it will have to be processed for knowledge generation. The FI model is expected to be an intelligent abstraction of the information scattered all around the world. Presently information is available through the traditional web pages of the servers. The future network will make an attempt to make each EOI to be a part of the system for the Future Internet. The EOI will consist of all human beings as well as objects of interest in the domain. Presently, there are so many machine systems as well as human knowledge systems working in isolation, as well as controlled by each other. They communicate and negotiate with each other. Thus, each system will give the best solutions in terms of its benefits and efficiency. In modern world the importance of decision support systems cannot be ruled out. Future Internet Decision System (FIDS) will play a critical role for all types of decision support systems

G. Lee, D. Howard, and D. Ślęzak (Eds.): ICHIT 2011, CCIS 206, pp. 39–45, 2011.

actively triggering vital steps for decision making in the respective domain. The current systems are highly motivated from the functional Joint Directors of Laboratory (JDL) and Data fusion process Model [3].

The information hungry entities are attempting to grab these resources to make a compatible Future Internet Decision System (FIDS). Thus, the Future services are processing the object of interest to confirm object identification, situation and threat assessments concerning the respective Future Information resources. The machine to machine communication schemes are driven by a dynamic environment of the real sensors world which provides a meaningful Information retrieval system.

In the following sections we described related work, the design of novel services based internet architecture and its methodological schemes. We finally conclude the paper with more information processing, resources.

2 Related Work

The future internet architecture is expected to be an intelligent abstraction of information which is scattered all around the world. The M-to-M communication system is the combination of multiple sensors interactions and thus the emerging technology of data fusion has to be incorporated in the Decision Support System (DSS). Multi sensor data fusion is an emerging technology which is currently applied in range of systems throughout the world. Applications such as medicine treatment, surgery, battlefield applications and many more are being assisted by the Data Fusion Process for DSS. The DSS is, thus, a critical impending factor to decide the actions.

2.1 JDL Model: A Revised View

The JDL model is a highly acclaimed model used in various levels for object identification, assessment of the situation, threat assessment and overall process refinement. The fusion of information in Future Internet System Architecture and JDL model will make system self capable for identification and assessing the overall situation awareness concerning the entities and resources. The process refinement action of data traverses from level 0 to level 4. In starting at level 0, the data is in raw format, which needs to be pre-processed to make it's utilization of further higher levels of data fusion. Level one is associated with object identification of the concerned system. The level two represents aggregation and situation assessment. Moreover, the aggregation is the concept of depicting similar sets of entities under some domain attributes constraints. The data fusion engine operates on positional and identity based two approaches at level 1. The identity fusion is the most important step, which is an input to the further higher levels of JDL model. It is an identity fusion which drives the situational awareness program of the level 2 and further impact assessment of level 3[2].

2.2 SENSEI Model: A Revised View

The SENSEI [3] has developed architecture which is corresponding technological building blocks of Future Internet services for smart environments. The Future Internet networks are trying to bridge the gaps between the real-world resources and digital resources. The Digital Resources (DR) is the representations of the sensors, actuators, and processing elements. In the real world resources consists of Objects of Interest (OoI) which are representations of people, places and things [3].

The resources are providing associate information with the respective entities and attributes. Thus, the SENSEI system can decipher the required resources for the entity level requests. The SENSEI model clearly states that real world entities will be different from the system resources and the software that is implementing it.

The Future Internet model will assist the mapping of data from real world user to the real world resources. These resources will be handled by the appropriate resource hosts which will communicate with the resource end point (REP) manager. A REP is a software process, which represents an interaction of end point for a physical resource [4].

There are multiple scenarios for the interaction of real world entities to the actual resource requested. With the help of web resources user can associate with real-time world entities for sensors or actuators information. The association request is simple requests where literal translations can be done to achieve the valid information for the query or a semantic dictionary approach. The abstractions query has matched through various available resources. Future Internet (FI) system schemes are trying to validate the cause of the request by the requesting real world entity and classified into normal, real and threatening requests.

3   Fusion Architecture of Internet Services

The identification of the real world entities of people, place or things must be done to create a repository of real world entities. The goal of this paper is to present a novel fusion architectural model of Future Internet for M-to M communication system. The fusion architectural model has inspired by taking advantage of the JDL model view of the data in the FI architectures. Most of the assessment techniques in the level 2 of the JDL domain are based on inference engine. The inference engine can be data driven as well as goal driven. The JDL model has implemented at the access layers of the resource hosts to identification of real world entities [9]. The valid relationships between resource seekers entities and resource hosts have analyzed to find the aggregation of similar resource hosts.

The JDL model focuses on the identification of the capability, opportunity and intent of the system objects [4]. It's allow the identification of the object, allow finding the similar sets of activities (aggregation), assess the current situations that can be normal situations or any threatening situations as well as suggest any overall refinements approaches. However, the SENSEI approach is used in real-world objects and real-world resources. It creates resource hosts and resource access points between Objects of Interests and resource hosts. Here, the resource hosts can be sensors and actuators for the information flow.

The JDL model has layered approach towards the modeling of resources. The incorporation of JDL model in Future Internet (FI) architecture is important for identification and awareness stability of FI resource hosts. The FI resources are classified as normal and critical resources. The real-world entities of people, places or things have identified. The real world entities has classified with respect to their attributes. The common attributes categorizations are help to identification of resource seekers entities. The JDL model need to implement at the access layers of the resource hosts known as Resource Access Points which help to identify the real world entities and the demanded digital resources. The entities aggregation, similar resource hosts and value assessments regarding valid relationships between real world entities and resource hosts.

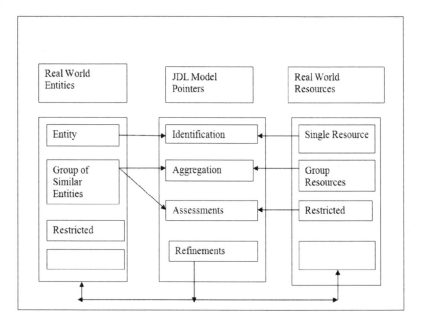

Fig. 1. Fusion Architecture of Future Services

Let us classify the real world objects in to a potential and viable threat matrix with respect to the resource hosts as per Level 3 concepts. Threats to the resource hosts may include destruction of sensors/actuators or any intention of destroying data of other sensors/actuators using some malicious code. The Real World Objects (RWO) will be computed on the three parameters. They are Capability, Opportunity and Intent values. The capability of the RWO has evaluated on the basis of pre-historic databases regarding the RWO and its context. This opportunity has evaluated on the basis of loop holes that can exist at Resource Hosts and Intent values. It is derived from the place of request where RWOP has been originated.

Moreover, in future there are trillions of sensor nodes distributed all over the world will also act as a resource based schemes. A resource host is available in a definite geographical area to maintain a repository of the active sensor nodes in its area of responsibility. Anytime sensor nodes can be added to the respective sensor resource hosts. The real world entities have mapped to the real world sensors through their sensor hosts. These sensor hosts are communicating through an interface of FI to maintain the passage of data and commands from FI users to the FI resource hosts. The real-world entities have dynamically associated with the type of sensors/actuators. A classification of real world objects have maintained beforehand to the type of sensors/actuators available in the FI architecture. Any new resources or entity formed in the FI view and atomically clustered in to mapping of these entities to the one or more resource hosts and then to the sensors. The coupled with object of interest identification, situation and threat assessments of M-to-M communication is driven by a dynamic environment of the real-world sensors to provide a meaningful Information retrieval. However, the novel M-to-M communication model are ready to support all thinkable application domains ranging from food sectors to energy, transport, mobility and logistics, health care, media content, utilities and environment. These services are intelligently gathering information to the Future Internet Users.

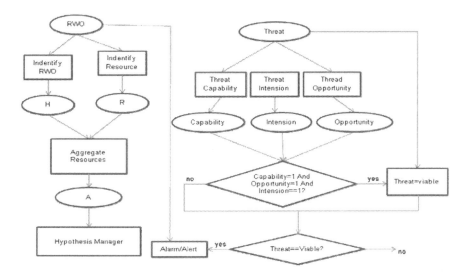

Fig. 2. Entity of Interest Based Fusion Model

The Fig. 2 has explained RWO properties of a threat to the resource hosts. The RWO which possesses all three properties transforms itself from potential threat to viable threat. This viable threat cannot be ignored and has to be neutralised [7]. The current level 3 shall take the notice of our COA (Course of Action) [7] in case of detection of RWO and their activities by means of decision sets. In the figure 2 depicts three levels of the future prediction. *Level one* is associated with RWO identification and Resource Host identification, *Level two* is associated with aggregation of resource hosts and *Level three* consists of safety and threat evaluation of RWO's. In addition to the above points *Hypothesis Manager* will predict the further movement of the RWO. In case of *threats* to resource hosts it will generate *alarm/alerts* which will help the system to negotiate the threatening RWO. Thus, The RWO side is directly proportional to vulnerability of resource hosts.

4   Discussion and Future Works

The idea behind in this paper is to demonstrate the importance of JDL model and SENSEI approach in to the Future Internet architecture. JDL based solution required identification of key attributes from sensors/RWO. It's maintained a unique mapping of resources hosts of RWO to make the FI threat free. The Future Internet architecture is a complex scenario of actual embedded sensors/actuators in to resource hosts. With powerful resource hosts Real World Objects (RWO) has strong accessing techniques. Thus, the intentions of RWO must be deciphered before its access into the Resource Hosts. For that, we have designed a strong Algorithm-1 for FI architecture and the following characteristics to get the hint of scenario of the upcoming events related to RWO.

Algorithm 1

```
Threat_Assesment (RWO)
begin
// Level one
        R <- Indetify_RWO(RWO)
H <- Indentify_Resource_Host(RWO)
//Level Two
A<- Aggregate_Resource_Host(R,H)
Threat<- Hypothesis_Manager(A)
//Level Three
Capability <- Threat_capability(Threat)
Intension  <- Threat_intension(Threat)
Oportunity <- Threat_opportunity(Threat)
if (Capability==1 and Intension==1 and Opportunity ==1)
then Threat = viable
else Threat = potential
if (Threat == viable)
then Alarm/Alert(RWO)
end
```

Capability: The capability of any RWO is identified by all the possible actions that it can take at any given point of time. While interacting with access points to the Resource hosts identification of the RWO and its originating location is important. The past behaviour, if any, must be matched so as to ascertain the credibility of accessing the resources.

Intention: Normally this attribute is associated with humans. Thus, RWO does not have intentions but the humans which are associated with those RWO have the intentions. Intentions can be friendly as well as hostile [7]. Some decision sets regarding the intention of the RWO must be matched with previously history of RWO accessing the Resource Hosts. This shall be resolved by an Inference Engine at the end of Digital Resource Hosts.

Opportunity: An opportunity is anything which can incite the resource to claim its objective [7]. The opportunity for any RWO exists as long as the vulnerability in the system (Resource Hosts to be attacked) exists in any form.

The future work is a prototype implementation in the form of Traffic Monitoring System. The traffic entities are the resource seekers and Traffic Monitoring server will be the resource hosts. This particular model has implemented at the interface end of the resource seekers and resource hosts. It does prove the system in actual scenario.

Capability, Opportunity and Intention of traffic entities has evaluated at the interface end. Thus, to identify the friendly, hostile intention of the resource seekers as well as can be extended to other Machine-Machine communication channels.

5   Conclusions

The idea behind in this paper is to demonstrate the importance of fusion approach to the Future Internet services. We have described JDL based solution to identification of key attributes from sensors/RWO to maintain a unique mapping of resources hosts to RWO to make the FI threat free. Future Internet is a complex scenario of actual sensors/ actuators embedded in to resource hosts. With powerful resource hosts Real World Objects (RWO) have strong accessing techniques. Thus, the intentions of RWO must be deciphered before its access in to the Resource Hosts so as to make the FI architecture strong enough to get the hint of scenario of the upcoming events related to RWO. The application of JDL model at the resource access points (RAP) where RWO tries to access the Resource hosts is not only identify the RWO capability but also their intentions to access the resource hosts. The malicious intentions have classified as a threat to the resources hosts as well as sensors and actuators connected with them to neutralize.

Acknowledgment. This work was supported by NAP of Korea Research Council of Fundamental Science & Technology.

References

1. Hall, D.L., Llinas, J.: Handbook of Multisensor Data Fusion. CRC Press, Boca Raton (2001)
2. Blasch, E., Plano, S.: JDL Level 5 fusion model: user refinement issues and applications in group tracking. In: SPIE, Aerosense, vol. 4729, pp. 270–279 (2002)
3. The SENSEI Real World Internet Architecture: Sensei Consortium – FP7 Project Number 215923 (2009)
4. Bauge, T.: Components for End-to-End Networking, Management and Security and Accounting, SENSEI, Public Deliverable D.3.3 (2009), http://www.sensei-project.eu/
5. Waltz, E., Llinas, J.: Multi-Sensor Data Fusion. Artech House,Inc., Norwood (1990)
6. Hall, D.L., McMullen, S.A.H.: Mathematical Techniques in Multisensor Data fusion, 2nd edn. Artech House, Inc., Boston (2004)
7. Steinberg, A.N.: An Approach To Threat Assessment. CUBRIC, Inc., USA (2005)
8. Steinberg, A.: Threat Assessment Technology Development. In: Proceedings of the Fifth International and Interdisciplinary Conference on Modelling and Using Context (CONTEXT 2005), Paris (2005)
9. Zadeh, L.A.: Fuzzy Algorithms. Information and Control 12, 94–102 (1968)
10. FM 34-130, HeadQuarters, Department of Army, Washington, DC, (July 8, 1994)
11. Shahbazian, E., Rogova, G., de Weert, M.J.: Harbour Protection through Data Fusion Technologies. Springer, Heidelberg (2005)
12. Web Services Description Language, http://www.w3.org/TR/wsdl

Potential Field Based Routing for IPv6 over Low Power WPAN

Sangsu Jung and Dhananjay Singh

Division of Fusion and Convergence of Mathematical Sciences
National Institute for Mathematical Sciences
Daejeon, Republic of Korea
{ssjung,singh}@nims.re.kr

Abstract. Requirements of IPv6 over Low Power WPAN (6LoWPAN) are energy conservation and low protocol complexity. Incorporating the desirable features, potential-field-based routing known as ALFA (autonomous load-balancing field-based any cast routing) achieves autonomous load-balancing and efficient path-length provisioning with little control overhead in a wireless mesh network (WMN), which is a hub-and-spoke type network similarly to 6LoWPAN. In this paper, we provide the protocol architecture of ALFA for 6LoWPAN to consider energy consumption. Distinct from the case of ALFA in a WMN, when the energy level of a sensor node declines to a threshold, the exchange of its potential information holds and its energy shortage is reported to a sink node.

Keywords: IPv6 over Low Power WPAN (6LoWPAN), potential-field-based routing, energy conservation.

1 Introduction

We expect that trillions of devices are connected with each other and form large-scale networks in the future. With the name of "Internet of Things", industry and research communities have developed such kinds of networks. The Internet of Things is nondeterministic and autonomous, but it is interoperable and event driven depending on contexts and environments. This kind of system includes evolutional characteristics for dynamic new environments of various objects and contexts. To support a lot of devices, IPv6 over Low Power WPAN (6LoWPAN) [1] has been proposed. In the Internet of Things based on 6LoWPAN, all things have IP addresses and those require no high level of energy. To deliver messages in the Internet of Things, a routing protocol should select energy-efficient paths with low complexity.

For wireless mesh networks (WMNs) constructed by mesh nodes and mesh gateways, ALFA (autonomous load-balancing field-based anycast routing) [2] has been proposed to achieve autonomous load-balancing and efficient path-length provisioning utilizing only local information. Here, we can consider an application of ALFA in 6LoWPAN. Because 6LoWPAN is a hub-and-spoke type network

G. Lee, D. Howard, and D. Ślęzak (Eds.): ICHIT 2011, CCIS 206, pp. 46–53, 2011.

consisting of sensor nodes and sink nodes, ALFA is suitably adaptable with a little modification. In this paper, we describe how to implement ALFA in 6LoWPAN.

The rest of the paper is as follows. In section 2, we review current routing system in a Linux kernel because we propose to implement ALFA on the top of embedded Linux. Additionally, we explain the basic of ALFA. Section 3 describes the system architecture of ALFA for 6LoWPAN. Finally, section 4 concludes the paper.

2 System Background

In this section, we state Linux kernel architecture, which is used for 6LoWPAN. Furthermore, we provide the key idea of ALFA and consideration of it for 6LoWPAN.

2.1 Routing Operation in a Linux Kernel

The modern networking architecture is classified into two routing functionalities: packet forwarding and packet routing [3]. Packet forwarding is the procedure of receiving a packet, referring a forwarding table, and transmitting the packet to its destination stated in the table. Meanwhile, packet routing is the procedure of constructing the forwarding table. Each node locally processes forwarding and exchanges information to make a routing decision with the help of a routing protocol. The forwarding table indicates the direction of a one-hop packet transmission and the routing table contains next hop information to reach designated destinations.

In current operation systems (OSs), packet forwarding is manipulated in Kernel space and routing is conducted in User space by a daemon program as shown in Fig.1. The forwarding table is placed in the kernel and referred by the routing table. After the kernel receives a packet, it refers to the routing table, and it transmits the packet to its next-hop node through a corresponding network interface. Here, the routing table is constituted by an implemented routing protocol on the top of a routing daemon.

The separation of forwarding and routing with the placement of forwarding inside the kernel and routing in User space [3] make routing processing more efficient. Inside the kernel, forwarding helps a packet rapidly traverses a node.

For reference, the routing process requires complex CPU/memory intensive tasks. To minimize the loads in the kernel, the routing is implemented outside the kernel. This system architecture enables to develop routing functions with no modification of the OS kernel.

2.2 Review of ALFA

By inspiration of nature, ALFA is developed to combine geographic routing and back-pressure routing. ALFA is based on Poisson's equation. Similarly to the motion of a charge in an electrostatic potential field, ALFA forwards a packet to the lowest-energy path through mesh nodes towards a mesh gateway on the direction of the steepest gradient field. The formulation of ALFA is as follows:

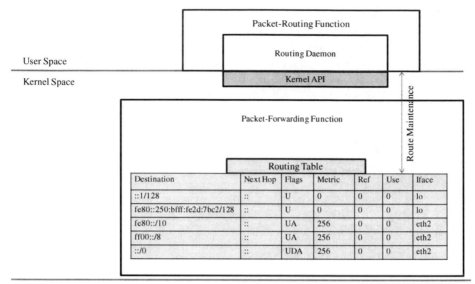

Fig. 1. An example of IPv6 routing architecture in OS kernel

$$\phi(v) = \frac{\sum_{k=0}^{m-1}(\phi_{k+1,v}\vec{r}_{k,v} - \phi_{k,v}\vec{r}_{k+1,v}) \cdot (\vec{r}_{k,v} - \vec{r}_{k+1,v}) + \alpha q(v)}{\sum_{k=0}^{m-1}\| \vec{r}_{k,v} - \vec{r}_{k+1,v} \|^2}, \tag{1}$$

where the potential $\phi(v)$ of node v are affected by potentials $\phi_{k,v}$ of its one-hop neighbors, real-space vector $\vec{r}_{k,v}$ from node v to node k, and its queue length $q(v)$. Here, node v has m one-hop neighbors. By a few iteration of (1) in each node, every node has a converged potential value. Then, a routing decision for uplink is to select a one-hop neighbor with the lowest potential towards a mesh gateway:

$$\arg\max_{k \in K_v} \frac{\phi(k) - \phi(v)}{\| \vec{r}_k - \vec{r}_v \|}. \tag{2}$$

According to the tuning parameter α, the behavior of ALFA is determined. In a low value of α, ALFA acts as geographic routing. Whereas in a high value of α, ALFA operates as back-pressure routing. Therefore, we can manipulate a routing behavior considering a network characteristic. With the help of a finite element method (FEM) and a local equilibrium method (LEM), ALFA utilizes only one-hop local information so that the routing control overhead is $O(1)$.

3 System Architecture

In this section, we design our system as shown in Fig. 2. We refer our system as ALFA-S (sensor networks) system. Similar to conventional routing architectures, we

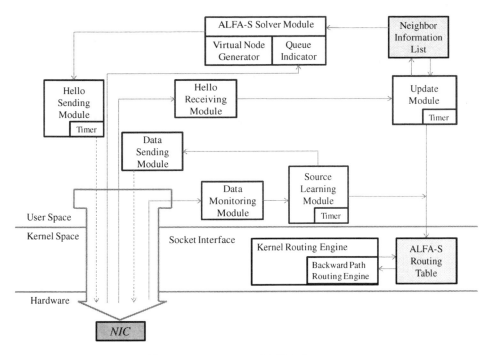

Fig. 2. The overview of AFLA-S system

implement ALFA-S system in two space levels: user space and kernel space. User space consists of Hello Sending Module, ALFA-S Solver Module, Neighbor Information Module, Hello Receiving Module, Update Module, Data Monitoring Module, Source Learning Module, and Data Sending Module. In addition, kernel space contains Kernel Routing Engine with Backward Path Routing Engine and ALFA-S Routing Table. We classify these modules of ALFA-S system into data entity and functional entity. (Fig. 3) To enable ALFA-S system, we modify a protocol stack of 6LoWPAN represented in Fig. 4. Here, a 'hello' message contains location, potential, and energy state information. The energy state information is represented by 0 or 1. The value of 0 implies energy shortage, whereas that of 1 is normal energy condition. When a one-hop neighbor has the value of 0, a node does not consider the potential value of the energy-insufficient node. Furthermore, this information is forwarded to a sink and thus a manager takes an action.

ALFA-S system focuses on anycast uplink routing for data gathering from sensor nodes to a sink node. To support downlink and unicast routing, ALFA-S system operates as follows: for downlink traffic from a sink node to a sensor node, ALFA-S system adopts source-learning based forwarding. In other words, the utilized paths of packets for uplink are recorded in sensor nodes and sink nodes; and packets for downlink exploit reverse-paths of recent uplink routing. In ALFA-S system, Source Learning Module is responsible for manipulating downlink routing. Source Learning Module maps the source and previous-hop addresses of a packet in an uplink-path into the destination and the next-hop addresses for a downlink-path. This downlink routing scheme can properly adapt to network traffic conditions referring to the recent

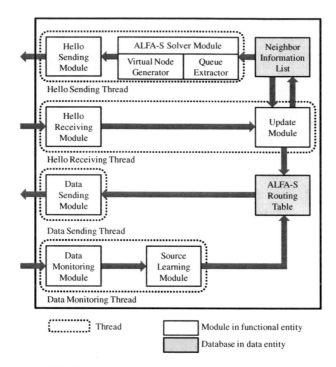

Fig. 3. The operation modules of AFLA-S system

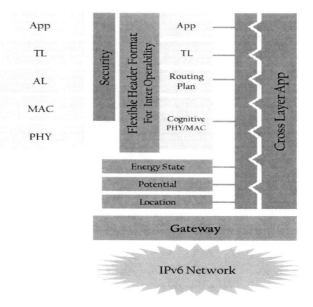

Fig. 4. The protocol stack for ALFA-S system

uplink routing path information. Furthermore, it requires little control overheads similarly to source routing. This scheme is also relevant to a previously suggested approach [4], which underscores that source routing is a well-compromised method for downlink routing overcoming on-demand routing and proactive routing. On the other hand, insufficient uplink traffic can prevent rapid adaptive load balancing, e.g. in cases of IPTV and other one-way traffic. To solve this generic problem, Source Learning Module in each sensor node periodically generates an uplink-refresh message towards a sink node for an updated-downlink path.

For unicast routing, we propose the incorporation of uplink and downlink routing schemes. Uplink routing refers to many-to-one or many-to-few communications from sensor nodes to sink nodes, and downlink routing means vice versa. Unicast routing is for one-to-one communication. If a sensor node receives a packet, which has the external network prefix of a destination, the node forwards the packet with the same policy of uplink routing. In a while, if a sensor node receives a packet, which has the internal network prefix of a destination with the same network prefix of the node, the node refers to its routing table managed by Source Learning Module, and delivers it based on the downlink routing policy of ALFA-S system. When there is no matching information for downlink of a unicast packet, a node forwards the packet to its default next hop node (gateway). The iteration of this process helps eventually find an appropriate path because every sensor node sends a packet to a sink node at least once; there always exists a path of downlink for each sensor node.

3.1 Data Entity

Data entity manages data storage for ALFA-S system. ALFA-S system includes Neighbor Information List and ALFA-S Routing Table to perform routing as other systems.

Neighbor Information List contains IP addresses, location information, and potential values, energy-state of one-hop neighbors of a node. Neighbor Information List is referred by ALFA-S Solver Module to calculate a potential of the node. Update Module deals with information for data entity. After a node receives a 'hello' message from its one-hop neighbors, Update Module renews current Neighbor Information List. In the case of a 'hello' message receiving from a new node, Update Module reflects the new node in its one-hop neighbor list. If a node receives no 'hello' message from a node, which is included in a current one-hop neighbor list, for pre-defined period, t_u, it removes the non-existent node in its one-hop neighbor list.

On the other hand, ALFA-S Routing Table includes the list of next forwarding candidates with IP addresses, for uplink and downlink routing. Here, we state the next forwarding nodes as default gateways because the subnet of a final destination is normally not matched with that of a source node in 6LoWPAN due to the aim of Internet access. Update Module refreshes default gateways for an uplink packet as reported from ALFA-S Routing Table of a node; so does Source Learning Module for a downlink packet. Source Learning Module continuously updates and records the source and the previous-hop-node addresses of an uplink packet. ALFA-S Routing Table is referenced by Data Sending Module for packet forwarding of a node.

3.2 Functional Entity

Functional entity is composed of Hello Sending Thread, Hello Receiving Thread, Data Sending Thread, and Data Monitoring Thread.

First, Hello Sending Thread consists of Hello Sending Module and ALFA-S Solver Module, and it is iterated every t second. Hello Sending Module of a node acquires the potential value of a node from ALFA-S Solver Module, and broadcasts 'hello' message to one-hop neighbors every t second.

The 'hello' message contains the IP address, location information, potential value, and energy state of the node, which are used for potential calculation in other nodes to decide routing paths. ALFA-S Solver Module obtains the potential of a node based on (1). For this purpose, ALFA-S Solver Module refers to Neighbor Information List to extract the location, potential, and energy state information of the one-hop neighbors of the node. During this process, it examines the possibility of potential calculation. In order to operate (1), there should be sufficient neighbors to form at least one triangle. For the case of insufficient neighbors, Virtual Node Generator helps provide a solution for calculation with virtual nodes as mentioned in [5]. To get the queue length of a node, ALFA-S Solver Module uses Queue Extractor. Queue Extractor operates its own system algorithm to obtain queue information because normal commercial sensor devices provide no device driver which can directly extract queue length information. For reference, a sink node in ALFA-S system does not run ALFA-S Solver Module because it should have its pre-defined fixed potential.

Second, Hello Receiving Thread is divided into Hello Receiving Module and Update Module. The thread runs every $t_h / | K_n |$ where t_h represents the operating cycle of Hello Sending Thread and $| K_n |$ describes the number of one-hop neighbors. Hello Receiving Module delivers the information of a one-hop neighbor to Update Module after receiving a 'hello' message. Inside the thread, Update Module refreshes Neighbor Information List based on the information from Hello Receiving Module. If the appearance or disappearance of a node occurs, Update Module informs Neighbor Information List to update it. Further, it decides the next forwarding node by (2) and reports it to ALFA-S Routing Table.

Third, Data Sending Thread takes in charge of sending a data packet referring to information from ALFA-S Routing Table. Data Sending Thread is executed for both uplink and downlink data packets.

Finally, Data Monitoring Thread includes Data Monitoring Module and Source Learning Module. When Data Monitoring Module deals with data packets, Source Learning Module drives the source and previous-hop addresses of a packet for source-learning based forwarding of downlink and unicast routing. Subsequently, the source and previous-hop addresses are mapped to a destination and next-hop addresses for downlink and unicast routing.

In a conventional IP header, there is no previous hop IP address. To enable source-learning based forwarding, ALFA-S system utilizes Reverse Address Resolution Protocol (RARP) referencing a previous hop MAC address. By this process, Source Learning Module maintains a forwarding information base (FIB) at every sensor node and sink node. Source Learning Module also conveys the FIB to ALFA-S Routing Table. To provide an appropriate path for downlink routing, Source Learning Module continuously monitors the update state of the FIB. If there is an idle period of uplink traffic for a pre-defined period, t_i, Source Learning Module calls Data Sending Module to send a uplink-refresh message to a sink.

4 Conclusions

In this paper, we introduce the system architecture of ALFA for 6LoWPAN, which finds the shortest-path consuming low-energy in each sensor node. We describe practical implementation detail of ALFA for 6LoWPAN. As an additional function, ALFA in 6LoWPAN reacts to the energy-level of a sensor node.

In future work, we will consider the mobile-capability of our scheme and implement it in a practical test-bed. Furthermore, we will provide specific design guidelines for real applications such as a ubiquitous health care system, an emergence-notification system, and other sensor based systems.

Acknowledgments. The work was supported by NAP of Korea Research Council of Fundamental Science & Technology.

References

1. Kushalnagaret, N., et al.: IPv6 over Low-Power Wireless Personal Area Networks (6LoWPANs): Overview, Assumptions, Problem Statement, and Goals. IETF RFC 4919 (2007)
2. Jung, S., et al.: Distributed Potential Field Based Routing and Autonomous Load Balancing for Wireless Mesh Networks. IEEE Comm. Lett. 13, 429–431 (2009)
3. Peterson, L., Davie, B.: Computer Networks, 2nd edn. Morgan Kaufmann Publishers, San Francisco (2000)
4. Baumann, R., et al.: Routing Packet into Wireless Mesh Networks. In: Proc. of IEEE WiMob (2007)
5. Jung, S., et al.: Greedy Local Routing Strategy for Autonomous Global Load Balancing Based on Three-Dimensional Potential Field. IEEE Comm. Lett. 14, 839–841 (2010)

Throughput Analysis of Wireless Mesh Network Test-Bed

Madhusudan Singh, Song-Gon Lee, Whye Kit Tan, and Jun Huy Lam

Dept. of Ubiquitous IT, Division of Computer & Information Engineering,
Dongseo University, Busan- 617-716, Korea
sonu.dsu@gmail.com, nok60@dongseo.ac.kr, timljh@gmail.com,
whyekit.tan@gmail.com

Abstract. Wireless mesh networks (WMNs) is bringing wide popularity as a flexible and cost-effective access in wireless network technology. Many researchers had done their wireless mesh network experiments through simulations. The simulation results and real world implementation results have big differences. In this paper, we have set up the testbed in our lab. Our experiment is based on IEEE 802.11s draft 4.0. In our experiment, we select the best path with the help of modified air time metric for both reactive and proactive path discovery. In this paper, we have evaluated the throughput results of different data packet size for UDP transmission protocols.

Keywords: IEEE 802.11s, HWMP, Airtime link metric, UDP, Packet size.

1 Introduction

WMNs are emerging as a low cost and flexible wireless network technology with self-organizing infrastructure [1]. Wireless mesh network have been applied feasibly to the areas such as, in military, rural area, building automation etc.

Most of the proposals results are evaluated through simulation or analytically. However, simulation and analytical results are not as good as real implementation results. WMNs real implementation results are very hard to be found because 802.11s do not have final draft [2]. This had motivated us to implement WMNs according to the current status of IEEE 802.11s draft 4.0. In this paper, we mentioned the overview of IEEE 802.11s draft in section 2. We defined about test-bed implementation details in section 3. In the section 4, we have discussed the performance analysis and in the section 5 conclude the article.

2 Overview of IEEE 802.11s (WMNs)

In this section, we discuss about the overview of the IEEE 802.11s WMNs features. Our implementation is based on the unofficial 802.11s draft 4.0 [3], which is available by the time of our implementation.

G. Lee, D. Howard, and D. Ślęzak (Eds.): ICHIT 2011, CCIS 206, pp. 54–61, 2011.
© Springer-Verlag Berlin Heidelberg 2011

In wireless mesh network, mesh point (MP) is the base of the network. The mesh point plays many roles in the network such as becoming the portal, access point, and simple mesh point [4]. A MP that acts as the portal is connected to the wired network and it provides internet connectivity to its network members. Each network has one or more mesh portals and each MP can select its portal according to its connectivity to the portals [5]. Mobile stations can also connect to portal through the WMN. The stations are not aware of the mesh features and the MP shall works as the proxy for the stations' connections [6].

The hybrid wireless mesh protocol (HWMP) is the default path selection algorithm for wireless mesh network [7].The HWMP is combination of both reactive and proactive routing protocol. Therefore, the HWMP holds flexibility of both on-demand route discoveries for connection between network members and proactive tree formation for the connections to the portal [8]. Mesh portal is connected with wired Ethernet and works as the root [9]. All other MPs connected with each other through the root. The HWMP tree connects all MPs with the help of root (Mesh Portal) [10].

The IEEE 802.11s draft defines default airtime link metric for the best path selection. Other metrics can also be used for path selection. The formula for Airtime Link Metric is shown in equation (1).

$$C_a = \left[O + \frac{B_t}{r} \right] \frac{1}{1 - e_{pt}} \tag{1}$$

Where O is a constant value, B_t is the test frame length r is the transmission rate (in Mbps) and e_{pt} is the data loss ratio [5].

Some researcher has proposed some modified metric for path calculation. In our work, we had used another metric and detailed information about it is discussed in the next section.

3   Test-Bed Implementation

In this paper, we have implemented 802.11s prototype with the help of existing hardware (desktop PCs and laptops with Atheros chipset based wireless NICs). The implementation specifications for the stations are described below. Ubuntu Linux operating system with kernel 2.6 had been used for the MPs. We have used Multiband Atheros Driver for Wi-Fi (Madwifi) driver [10] to create several virtual access point (VAP). We have used jperf [11] java based software for the throughput result. The Jperf is a graphical user interface using Iperf, which allows us to calculate the throughput results.

3.1   Network Topology

All the stations had been placed in the same room without any physical separation. Only the mesh portal is connected to the wired LAN. The position of each station is shown in Fig. 1.

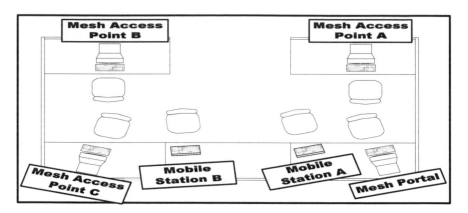

Fig. 1. Real-time network process

We had made a simple test with 4 PCs having the mesh features and two laptops act as the legacy stations without mesh features. One of the PCs was set as the mesh portal point (MPP), three other PCs as mesh access point (MAP) and the laptops as Mobile stations.

3.2 HWMP Specification

The hybrid wireless mesh protocol (HWMP) is default routing protocol used in WMNs. The HWMP is a combination of proactive and reactive routing protocols.

Once the mesh software is started, mesh portal will start to form the tree for the mesh network. The tree is formed based on the metric value or the radio quality of each link. Since the mobile station is not part of the mesh network, it was identified as a proxy station connected to the mesh access point A. The mesh portal gains information about the path metric by broadcasting the path request frames to all mesh access points within the mesh network. All the MAPs will then reply to the request and paths between the stations and the portal are formed [9].

The reactive routing is triggered by the mesh software after the proactive tree is formed, so that the path formation can be observed. The formation of reactive paths for this test is very similar to the formation of proactive paths. Path request frames will be unicast to all MAPs that it can detect and the Maps will reply to the request.

3.2.1 Path Request (PREQ) and Path Reply (PREP) Operation
The main function of this frame is to let all the stations that it passed by to learn about the next hop MAC address when the destination MAC address is given. The operation of PREQ is shown in Fig. 2(A) by assuming that a mesh STA A initiated the path formation. The PREQ is assumed to past through two intermediate stations before reaching the destination.

The operation is similar to the PREQ where the destination station will reply to the originator station. The operation is shown in Fig. 2(B). When the PREP frame reaches the originator station, a bidirectional path is formed.

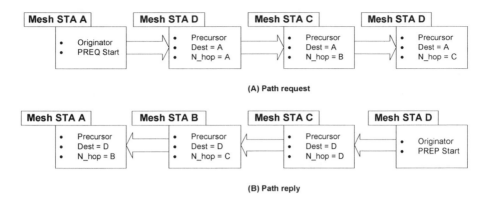

(A) Path request

(B) Path reply

Fig. 2. Path request and path reply for path selection

3.3 Metric Methodology

Research has been done by Rosario G. Garroppo et al [1] stating that the default airtime link metric causes the metric value to change too often and they had proposed their own metric formula. The test done in our lab is using the exact metric formula.

The formula is shown in equation 2, where S is the signal strength and ε_r is the error rate.

$$metric_{Link} = [\frac{2}{1-\varepsilon_r}][\frac{125}{S}+1.6]$$ (2)

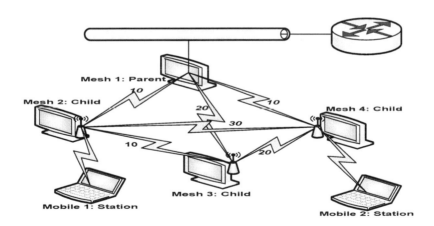

Fig. 3. Path metric between nodes

In the case of multiple paths, the best path shall be chosen. This is done by comparing the metric values of the path. In Fig. 3 we can easily see that all path metric values from Mobile1 to Mobile2.

Different metric values are shown in below.

1. Mobile 1 – Mesh 2 – Mesh 4- Mobile 2 = 30
2. Mobile 1-Mesh 2-Mesh 1- Mesh 4 – Mobile 2 = 20
3. Mobile 1 – Mesh 2 – Mesh 3 – Mesh 4 – Mobile 2 = 30

The lowest path metric value Between Mobile 1 to mobile 2 is Mobile 1-Mesh 2-
Mesh 1- Mesh 4 – Mobile 2 = 20. The final best path selection is shown in Fig. 4.

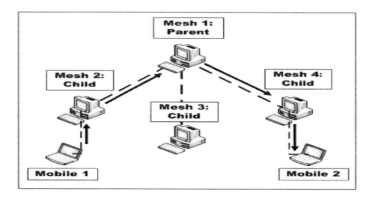

Fig. 4. Best path selection

4   Performance Analyses

In this section, we have defined an example of our test based. In our testbed, we use 4
PCs and 2 laptops as stations. Each system has been assigned a unique name, and IP
address. Every system has its own MAC address. We can see the details in table 1.

Table 1. Networks computer information (ID, IP address, MAC Address)

Name	IP Address	MAC Address
Mesh 1 (MPP)	192.168.1.1	00:24:73:1E:D7:BD
Mesh 2 (MAP)	192.168.1.16	00:24:73:1E:EB:3F
Mesh 3 (MAP)	192.168.1.5	00:24:73:1E:D6:6D
Mesh 4 (MAP)	192.168.1.10	00:24:73:1E:D7:4D
Mobile 1 (Station)	192.168.1.30	74:F0:6D:32:1B:1F
Mobile 2 (Station)	192.168.1.25	74:F0:6D:31:23:A6

During implementation, a mesh portal point (Mesh1), three mesh points (Mesh2,
Mesh3 & Mesh4), and two mobile stations (Mobile1 and Mobile2) had been used.
The tree was created after the initialization (Mesh1- Mesh2- Mesh4 and Mesh1-
Mesh3). The application required to get the MAC address of mobile1 and Mobile2,
whose IP address is known. The mobile1 has broadcast path request (PREQ) message,
which received by Mesh4. This node unicast PREQ to final destination the mesh

portal point (MPP). The message frame was thus sent to Mesh2 which forwards it to Mesh1. Mesh1 has information about all mobile stations of the network and replies to mobile2 MAC address. After Path creation between Mobile1 to Mobile2 starts the communication.

The experiment had been done according to the experimental scenario defined below. We had made settings and parameters for UDP test were defined. The experiment had been initialized with UDP test. We had used some common settings for different packet sizes. In table 2 has shown the common parameters of experiment.

Table 2. Common settings for different data packets

Transport Protocol	UDP
UDP Bandwidth	200 kbps
Buffer size	41 kbytes
Number of data transmission for each trail	10
Number of trails	3

The UDP bandwidth had been set to 200 Kbits/sec at the client side thus the client will constantly send out data frames at the stated speed. The bandwidth of the path is determined by the datagram ratio where higher datagram ratio means a lower path bandwidth with a maximum bandwidth of 200 Kbits/sec if no data is loss. For experiment we had used several datagram sizes (100 bytes, 200 bytes, 300 bytes, 400 bytes, 500 bytes, 1000 bytes, 1500 bytes, 2000 bytes, 2500 bytes and 3000 bytes). We have defined the common buffer size for experiment is 41 kbytes. The buffer size is the just buffer which packets are received in, and limits the maximum receivable data packets size. We have fixed 10 common transmission times for every trails. We did every packet size test to 3 times trails. In table 3 is shown average throughput result of different packet size after experiment. We can easily find 200 bytes data packets throughput result is best result in all packet size. During 200 bytes packets size data loss is very less and data transmission is very high. Within 100 to 500 bytes packet loss ratio and time variation are very less.

Table 3. Average throughput results of different UDP data packets

Datagram size (bytes)	Bandwidth (Kbits/sec)	Jitters (ms)	Data Loss Rate (Loss/Total) (percentage)	Number of Frames
100	196	0.805	492/25001 (2%)	25001
200	198	1.158	125/12501 (1%)	12501
300	196	0.338	168/8335 (2%)	8335
400	193	0.906	224/6251 (3.6%)	6251
500	187	15.659	273/5001 (5.5%)	5001
1000	172	25.525	327/2501 (13%)	2501
1500	185	9.365	122/1668 (7.3%)	1668
2000	163	37.998	228/ 1251 (18%)	1251
2500	183	21.084	83/1001 (8.3%)	1001
3000	192	7.702	34/835 (4.1%)	835

The result shows that jitters and datagram loss ratio increases with increasing datagram size. The jitters or packet delay variation increased sharply when the datagram size exceeds the maximum size allowable. This causes the high latency as shown by the jitters. Surprisingly, the performance of the network depends on the frame size and not the number of frames.

In Fig. 5 shown average throughput results of different UDP data grams and Fig. 6 is shown the average jitter (packet delay variation time) value of each datagram size. Packet jitter is expressed as an average of the deviation from the network. 100 to 200 datagram packets packet delay time is very less compare to 500 to 1000 datagram packets.

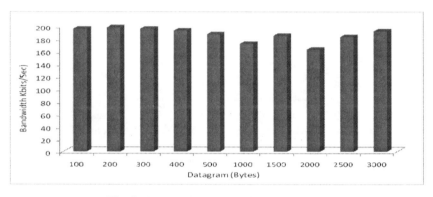

Fig. 5. Average throughput of UDP datagram

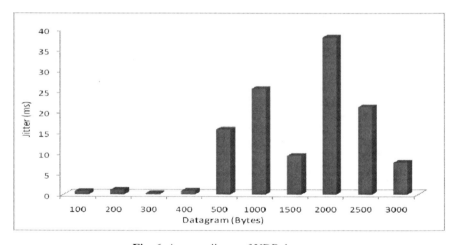

Fig. 6. Average jitters of UDP datagram

5   Conclusion

This paper presented an experimental throughput analysis of different data packet size for UDP transmission. Most of researcher did the wireless mesh network throughput analysis through the simulation. The simulation analysis is easier to be used but the result is often less reliable compared to real implementation. Therefore, a real implementation test is assumed to be able to provide a more reliable result.

Our testbed is constructed based on the IEEE 802.11s draft 4.0. The network formed consisted of four PCs and two laptops. In our experiment, the best path selection is done by using the modified air time metric which is claimed to be able to reduce the PREQ and RREQ frames flooding. In our experiment, the non-mesh stations information are stored through the proxy mechanisms in HWMP tree table whenever HWMP tree is available in the network.

In our experiment, we had shown the throughput results and packet delay variation on different datagram sizes. We had found the performance of the network depends on the frame size and not the number of frames in UDP transmission.

Acknowledgments. This work was supported by the 2011 National Research Foundation of Korea.

References

1. Garroppo, R.G., Giordano, S., Tavanti, L.: Implementation frameworks for IEEE 802. In: Proceedings of Computer Communications, vol. 33, pp. 336–349 (2010)
2. IEEE P802.11s Task Group s, IEEE Unapproved draft standard P802.11s/D4.0 (2010)
3. Wang, X., Lim, A.O.: IEEE 802.11s wireless mesh networks: Framework and challenges. Journal Ad Hoc Networks 6(6), 970–984 (2008)
4. Garroppo, R.G., Giordano, S., Lacono, D., Tavanti, L.: On the development of a IEE 802.11s Mesh Point prototype. In: ACM Tridentcom 2008, Innsbruck, Austria, pp. 969–978 (2008)
5. Xu, X., Liu, Z.: Design and Implementation of Wireless Mesh Network Testbed Based on Layer 2 Routing. In: The 4th International Conference on Wireless communications. Networking and Mobile Computing (WiCOM 2008),, Dalian, China, pp. 1–4 (2008)
6. Tsai, T.J., Chen, J.W.: IEEE802.11 MAC Protocol over Wireless Mesh Networks: Problems and Perspectives. In: Proceedings of the 19th International Conference on Advanced Information Networking and Applications (IEEE AINA 2005), Washington DC, USA, vol. 2, pp. 60–63 (2005)
7. Iannone, L., Kabassanov, K., Fdida, S.: The MeshDVNet Wireless Mesh Network Testbed. In: WiNTECH 2006, ACM International Workshop on Wireless Network Testbeds, Experimental, Los Angeles, California, USA, pp. 107–108 (2006)
8. Li, V.O.K., Cui, L., Liu, Q., Yang, G., Zhao, Z., Leung, K.: A Heterogenous Peer-to-Peer Network Testbed. In: ICFUN 2009, The First International Conference on Ubiquitous and Future Networks 2009, Hong kong, China (2009)
9. Singh, M., Lee, S.G., Tan, W.K., Lam, J.H.: Cluster-based routing scheme for Wireless Mesh Networks. In: 13th International Conference on Advanced Communication Technology (ICACT), Phonix Park (Korea), Februry 13-16, pp. 335–338 (2011)
10. Madwifi:multiband atheros driver for wifi, http://madwifi-project.org
11. http://code.google.com/p/xjperf/

Cooperative Heterogeneous Network Interworking between WiMAX and WiFi

Hae Jung Kim[1], Chan Jung Park[1], and Ronny Yongho Kim[2],[*]

[1] Gangneung-Wonju National University
Department of Computer Science and Engineering,
Wonju, Gangwon, 220-711 Korea
kimsam72@hanmail.net, cjpark@gwnu.ac.kr
[2] Kyungil University
School of Computer Engineering,
Gyeongsan, Gyeongbuk, 712-701 Korea
ronnykim@kiu.ac.kr

Abstract. As more wireless devices are being connected to the Internet, data traffic is growing and expected to increase beyond wireless network's capacity in near future. Considering radio spectrum is limited, the efficient way to support increasing data traffic in cellular networks such as mobile WiMAX, operating in license band, is using WiFi network operating in license exempt band. By employing WiFi, mobile WiMAX network can offload some traffic to WiFi networks. In terms of admission control and Quality of Service (QoS) control, WiFi employs simple strategy as compared with cellular network. However, when WiFi is used for traffic offloading of cellular networks, cellular users' QoS should be continuously guaranteed in the WiFi networks. Therefore, clear interworking scenarios and control procedures between WiFi and WiMAX should be defined. In this paper, we present two most prospective collaborative heterogeneous interworking scenarios between WiFi and WiMAX, and collaborative heterogeneous architecture and protocols are proposed. Efficient and easy handover scheme in the overlay collaborative heterogeneous network interworking is also proposed and its benefits are verified with simulations.

Keywords: 4G, IMT-Advanced, IEEE 802.11, IEEE 802.16, WiFi, WiMAX, Heterogeneous Network.

1   Introduction

Several recent studies have pointed out there will be explosive growth in mobile data demand driven by compelling communication capable devices such as smart phones, notebook computers, netbooks and tablets. Globally, mobile data traffic will double every year through 2013, increasing 66 times between 2008 and 2013. Mobile data traffic will grow at a Compound Annual Growth Rate (CAGR) of

---

[*] Corresponding author.

G. Lee, D. Howard, and D. Ślęzak (Eds.): ICHIT 2011, CCIS 206, pp. 62–69, 2011.
© Springer-Verlag Berlin Heidelberg 2011

131 % between 2008 and 2013, reaching over 2 exabytes per month by 2013 [1]. Therefore, a critical challenge for future broadband wireless networks, such as mobile WiMAX, is to provide significantly enhanced capacity to meet this exponential growth in demand. Also, future networks must drastically reduce cost/bit, while adding new services [2]. Heterogeneous networks interworking, so called multiple radio access technology (Multi-RAT) operation, represents a disruptive approach towards capacity enhancements and cost/bit decrease, which efficiently utilizes all spectral resources in the system [3]. In addition, other metrics such as client quality of service(QoS), coverage etc., can also be enhanced with the heterogeneous networks interworking.

Mobile WiMAX 2.0 based on IEEE 802.16m [4] is selected as one of IMT-Advanced technologies, also known as 4G, by ITU-R in October 2010. Mobile WiMAX is expected to provide enhanced broadband packet data services as a future broadband wireless network. WiFi, based on IEEE 802.11 technology [5], is getting more popular than ever, thanks to its simple and easy deployment and excellent data packet processing performance. As other future broadband wireless networks, mobile WiMAX needs to be further enhanced to provide significantly enhanced capacity and reduce cost/bit drastically. In order to provide such capabilities, collaborative heterogeneous network operation, that is Multi-RAT operation with other wireless systems working in license exempt band, such as WiFi, is required.

Primary goals of collaborative Multi-RAT operation are: 1. network capacity increase, 2. cost/bit decrease. Because WiFi based on IEEE 802.11 wireless access technology employs license exempt spectrum, WiFi is one of the best candidates for collaborative heterogeneous network interworking with the broadband wireless networks like mobile WiMAX based on IEEE 802.16 wireless access technology. When WiFi is used collaboratively with the broadband wireless networks within a cell of the broadband wireless network in a overlay network manner, there exists two benefits: 1. the cellular network can offload some data traffic to WiFi and 2. heterogeneous mobile station (MS) can utilize more bandwidth yielding higher throughput. Therefore, mobile WiMAX's collaborative Multi-RAT operation with WiFi can achieve the both primary goals.

In this paper, we present prospective collaborative Multi-RAT operation scenarios between WiFi and WiMAX and collaborative Multi-RAT architecture and protocols are proposed. The remaining part of the paper is organized as follows. In Section 2, two prospective collaborative Multi-RAT operation scenarios are discuss. We propose Multi-RAT protocol architecture of Multi-RAT network components and functions of the Virtualization layer which plays key role for collaborative Multi-RAT operation. In Section 3, Multi-RAT protocols are proposed and discussed including proposed *timer based Multi-RAT handvoer* scheme. In Section 4, performance of the proposed HetNet handover scheme is evaluated by simulation and its advantages are discussed. Finally, we conclude the paper in Section 5.

2   Multi-RAT Scenarios and Protocol Architecture

2.1   Multi-RAT Scenarios

Fig. 1 (a) illustrates two prospective scenarios of cooperative Multi-RAT operation between WiFi and WiMAX. From the perspective of the WiMAX network, because the MS is redirected to use WiFi link to offload WiMAX BS's traffic or to achieve better throughput, the WiMAX network is responsible to guarantee QoS of the WiMAX MSs which use WiFi link for the sake of WiMAX BS. Therefore, WiMAX BS establishes and maintains control connections with Het-Net APs, which are new type of APs providing cooperative Multi-RAT services, to meet the required QoS levels of WiMAX networks. In this paper, two cooperative scenarios for WiMAX and WiFi multi radio operation (HetNet Type I and HetNet Type II) are defined and protocol architecture for HetNet devices and detailed HetNet protocols are proposed and evaluated.

The scenarios of HetNet Type I and HetNet Type II are:

▷ HetNet Type I: In this scenario, HetNet AP, which is a multi mode device with Ethernet, WiFi and WiMAX and it offloads traffic of WiMAX networks by providing WiFi link to MSs. MSs are able to increase its throughput by utilizing HetNet AP with good link quality. HetNet AP's call admission control and QoS control should be controlled by the WiMAX BS through WiMAX link. In this scenario, HetNet AP performs like a WiMAX femtocell BS with over the air control link with the BS.

▷ HetNet Type II: HetNet AP performs like a relay node which receives the data from the WiMAX BS and relays the data to the MS through WiFi link. In this case, cooperating HetNet AP is selected based on the link quality between the BS and the HetNet AP and between HetNet AP and the MS. The MS can receive data from the BS in multiple paths: direct link from the BS and indirect link via WiFi.

2.2   Multi-RAT Protocol Architecture

In order for the NetNet devices (HetNet AP Type I, HetNet AP Type II and HetNet MS) to cooperatively operate with WiMAX and WiFi, devices have to support multiple access network interfaces, such as IEEE 802.3, IEEE 802.11 and IEEE 802.16. For seamless services across different access network interfaces, there should be Virtualization layer, which is also called 2.5 layer because it can be located between layer 2 (data link layer) and layer 3 (network layer). Protocol architecture for HetNet AP Type I, Type II and HetNet MS are shown in Fig. 1 (b). Virtualization layer is located above Medium Access Control (MAC) layer and manages different interfaces. The Virtualization layer is similar to the Media Independent Handover (MIH) function defined in IEEE 802.21 in terms of its location and some handover related functions. However, the Virtualization layer in this paper has more cooperative Multi-RAT interworking functions than MIH. Some of such functions are: 1. Data encapsulation and decapsulation between access network technologies. 2. QoS and admission control procedure across access network technologies.

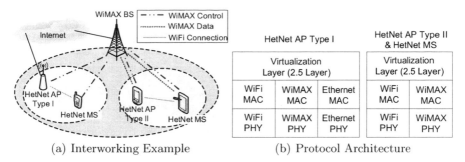

| | (a) Interworking Example | (b) Protocol Architecture |

Fig. 1. Interworking example and protocol architecture of heterogeneous network interworking

3   Multi-RAT Protocols

Multi-RAT protocols for cooperative heterogeneous network interworking between IEEE 802.11 and IEEE 802.16 networks are provided in this section. In the design of Multi-RAT protocols, in order to minimize the impact on the existing standard, existing control messages and procedures are reused.

3.1   HetNet AP Type I

Fig. 2 (a) shows control procedures to establish a session, deliver data and terminate the session when HetNet AP Type I is involved. When traffic to the HetNet MS arrives at the HetNet AP Type I through IEEE 802.3, the Virtualization layer notifies the IEEE 802.16 BS of the packet arrival by initiating service flow creation procedure with the IEEE 802.16 BS. When the IEEE 802.16 BS receives the service flow creation request, it initiates a service flow creation procedure with the HetNet MS in order to create the same service flow as the one with the HetNet AP Type I. After finishing the service flow creation between the HetNet AP Type I and the IEEE 802.16 BS and between the IEEE 802.16 BS and the HetNet MS, IEEE 802.11 QoS parameters are mapped to the established service flow. In this way, QoS in the HetNet AP Type I is controlled by the IEEE 802.16 BS. The received data packet is first formatted in IEEE 802.16 packet format following the QoS parameters established through the service flow creation. Then, with the involvement of Virtualization of HetNet AP Type I, IEEE 802.11 MAC encapsulates the IEEE 802.16 data into IEEE 802.11 packet. The encapsulated IEEE 802.11 packet is delivered to the HetNet MS over IEEE 802.11 link. After delivering the data over IEEE 802.11 link, the service flow can be terminated. The service flow can be normally terminated by the HetNet AP Type I because the HetNet AP Type I knows when data transmission ends. As we can see from this example, IEEE 802.16 BS can control QoS of data transmitted from HetNet AP Type I to HetNet MS by establishing same QoS parameters with HetNet AP Type I as the one the IEEE 802.16 BS would establish with the HetNet MS through the IEEE 802.16 service flow creation procedure. Uplink procedure is similar to the downlink procedure except that the HetNet MS initiates service

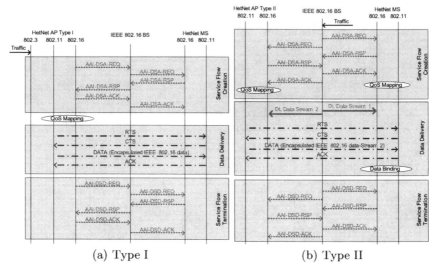

Fig. 2. Cooperative Heterogeneous Network Interworking between IEEE 802.11 and IEEE 802.16: Downlink

flow creation and termination procedures because it is aware when traffic starts and ends.

3.2   HetNet AP Type II

Fig. 2 (b) shows control procedures to establish a session, deliver data and terminate the session when HetNet AP Type II is involved. In downlink of this case, traffic to the HetNet MS arrives at the IEEE 802.16 BS because the HetNet AP Type II performs like a relay node using WiFi link. In order to notify the HetNet MS of new packet arrival and also notify HetNet AP Type II of packet relay, the IEEE 802.16 BS establishes service flows with the HetNet MS and the HetNet AP Type II with a same service flow identification (SFID) which makes the HetNet MS possible to combine received packets from the IEEE 802.16 BS and HetNet AP Type II over IEEE 802.11 link. After the service flow establishment, the HetNet AP Type II and the HetNet MS can map QoS parameters of IEEE 802.11 link to the ones established with the IEEE 802.16 BS during the service flow creation. The IEEE 802.16 BS now can transmit downlink packet data using direct link and ralay link via the HetNet AP Type II to the HetNet MS. The HetNet AP Type II relays the received packet from the IEEE 802.16 BS using the IEEE 802.11 link to the HetNet MS. The HetNet MS is able to combine received packets directly from the IEEE 802.16 BS and indirectly via the HetNet AP Type II. After transmission of downlink data, the IEEE 802.16 BS initiates the service flow termination procedure with both the HetNet MS and the HetNet AP Type II. Uplink procedure is similar to the downlink procedure except that the HetNet MS initiates service flow creation and termination procedures because it is aware when traffic starts and ends.

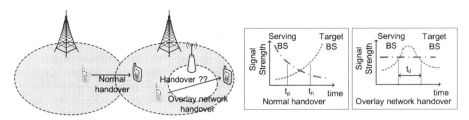

Fig. 3. Handover schemes: normal handover, overlay heterogeneous handover. Timer based handover scheme can prevent MSs' frequent handover between WiFi and WiMAX.

3.3   Timer Based HetNet Handover

In Multi-RAT operation, especially under overlay configuration, mobility management should be designed differently from typical handover protocol [6]. In case of normal handover between homogeneous cells, if the received signal strength indication (RSSI) from the serving BS is below a certain value, handover procedure is triggered [7]. If the received signal from the handover target BS is above a certain value, the MS performs handover to the target BS as shown in Fig. 3. In the Fig. 3, at time t_p, RSSI of the serving BS gets worse than the handover preparation threshold and at time t_h, RSSI of the target BS gets better than the handover execution threshold and the MS performs handover to the target BS. However, in case of Multi-RAT handover, new handover trigger conditions are required. Even though HetNet MSs enter into coverage area of HetNet AP which can provide better services, the received signal from the serving BS, usually macro cell BS, might be still good enough not to trigger handover procedure. Therefore, RSSI from the HetNet AP should trigger handover procedure. There is another parameter we should consider. Because frequent handover between HetNet AP and macro BS might be caused due to HetNet MSs' mobility, MSs' mobility should be considered. Therefore, we propose *HetNet_handover_timer* based handover scheme which can prevent HetNet MSs from performing frequent handovers between macro BS and overlaid HetNet AP. When the signal quality of the HetNet target AP gets better than the HetNet handover preparation threshold, the HetNet MS starts *HetNet_handover_timer*. If the HetNet MS still resides in the target HetNet AP coverage area, after expiration of the *HetNet_handover_timer*, it performs handover. In this way, HetNet MSs' frequent handover between HetNet AP and macro BS can be prevented. Since *HetNet_handover_timer* depends on moving speed, it can be optimally selected considering HetNet MS's moving speed.

4   Performance Evaluation of the Proposed Timer Based HetNet Handover Scheme

In this section, performance of the proposed *HetNet_handover_timer* based handover scheme is evaluated with simlations and its benefits are discussed.

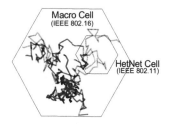

Fig. 4. Network topology for performance evaluation

4.1   Evaluation Network Topology

Network topology to be used to show the benefits of using the proposed *Het-Net_handover_timer* based handover scheme is shown in Fig. 4. Within a macro cell (IEEE 802.16), HetNet cells are located. For simplicity, one HetNet cell is shown in Fig. 4. It is assumed that there are enough number of HetNet cells which can cover a coverage area of macro cell. However, in order to see the effects of HetNet MSs' mobility in relation to handover, i.e., sojourn time in a small HetNet cell, when a HetNet MS moves out of one HetNet cell, it performs handover to a large macro cell. MS's movement is measured in every second and marked with circles. Fig. 4 shows the trace of 5 HetNet MSs moving at random speed for 1000 seconds.

4.2   Mobility Model and MSs' Mobility

We consider the motion of one MS around the service area of a cell randomly. Its initial speed (in km/h) and direction (in degrees) are generated with a uniform distribution of U[10,80] and U[0,360], respectively. The MS will change its speed and direction after a certain amount of time with an exponential distribution, with a mean value of 10 seconds. The new speed is uniformly generated with U[10,80] if the current speed is below 10 km/h; otherwise, it is obtained using U[v-10,v+10], where v is the current speed. The new direction is obtained from a Gaussian distribution with the mean as the current direction, and a standard deviation of 40 degrees. HetNet MSs are grouped into three classes: low mobile class (0km/h 30km/h), medium speed class (30km/h 80km/h) and high speed class (80km/h 150km/h).

4.3   Sojourn Time of Different Speed Class

In order to see the distribution of HetNet MS while moving within its coverage area of WiFi small cell and WiMAX large cell, 500 HetNet MSs per mobility class is simulated. 500 HetNet MSs initiate movement within a certain WiFi small cell at random at the same time and move around per the employed mobility model. As we can see from the results shown in Fig. 5, as speed increases, HetNet MS population decreases faster. Many users in the WiFi small cell hand over to the WiMAX large cell and finally moving out to other WiMAX large cells. The sojourn time of HetNet MSs in a WiFi cell gets smaller as mobile's moving

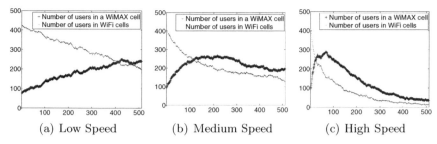

(a) Low Speed   (b) Medium Speed   (c) High Speed

Fig. 5. Number of users in WiFi cells and a WiMAX large cell. 500 users initiate mobility within a certain WiFi cell at time 0.

speed increases. Therefore, we can find some optimal value which can reduce unnecessary handover to a WiFi cell so that HetNet MSs can receive services for a long time without experiencing handover interruption.

5   Conclusion

In this paper, two most prospective collaborative WiFi and WiMAX heterogeneous interworking scenarios are presented and Multi-RAT interworking architecture and protocols are proposed. The proposed architecture and protocols are designed in order to minimize the impact on the existing architecture and protocols by reusing the existing control messages and procedures. We also proposed the simple and efficient heterogeneous network handover scheme in overlay network considering MSs' mobility classes. We showed the advantages of using the proposed scheme with the simulation. In the simulation, we defined mobility model and users are classified in three classes depending on their moving speed.

References

1. Morgan Stanley Research, The Mobile Internet Report: Ramping Faster than Desktop Internet, the Mobile Internet Will Be Bigger than Most Think (December 15, 2009)
2. IEEE 802.16ppc-10/0016r1, Future 802.16 network: Challenges and Possibilities (March 2010)
3. IEEE 802.16ppc-10/0002r3, Hierarchical Network Ad Hoc Study Report (March 04, 2011)
4. IEEE P802.16m/D12, IEEE 802.16m DRAFT Amendments for Local and Metropolitan Area Networks, (February 17, 2011)
5. IEEE Standard for Local and Metropolitan Area Networks-Part 11:Wireless LAN Medium Access Control (MAC) and Physical Layer (PHY) specifications, IEEE Std 802.11-2007 (June 18, 2007)
6. Kim, R.Y., Jung, I., Kim, Y.Y.: An Improved Cross-Layering Design for IPv6 Fast Handover with IEEE 802.16m Entry Before Break Handover. IEICE Transactions on Fundamentals of Electronics Communications and Computer Sciences (August 2010)
7. Kim, R.Y., Jung, I., Yang, X., Chou, C.-C.: Advanced Handover Schemes in IMT-Advanced Systems. IEEE Communications Magazine (August 2010)

Texture-Based Pencil Drawings from Pictures

Yunmi Kwon[1] and Kyungha Min[2]

[1] Dept. of Computer Science, Graduate School, Sangmyung Univ., Seoul, Korea
flyaway20@naver.com
[2] Division of Digital Media, Sangmyung Univ., Seoul, Korea
minkh@smu.ac.kr

Abstract. We re-render a black-and-white picture as a pencil drawing by overlapping pencil textures along the flow of the picture. The pencil textures are generated by overlapping pencil stroke textures captured from real strokes. An input picture is segmented into iso-tonal regions, each of which is textured in a direction determined by an algorithm for detecting smooth and feature-sensitive flow on the picture. Finally, we add contours and feature lines to emphasize salient shapes in the scene. We present several images that show the degree of realism which can be achieved.

Keywords: non-photorealistic rendering, pencil drawing.

1 Introduction

In NPR, pencil drawing is recognized as one of the important techniques for depicting objects. The pencil drawing in fine arts has a very wide range of applications, from rough sketches to detailed illustrations. There are three important aspects of a pencil drawing: *shading*, *line* and *hatching* [1]. The shading, which we will call *tone*, represents the intensity of light coming from part of a scene. Lines, in the form of contours and feature lines, emphasize salient shape information. Hatching consists of series of parallel or orthogonal pencil strokes drawn close together and its pattern produces the characteristic appearance of a pencil drawing. Our aim is to process a black-and-white picture in order to produce a pencil drawing that possesses these three aspects.

There have been many attempts to develop non-photorealistic rendering techniques which mimic pencil drawings. Existing methods are largely based on one of four approaches: modeling individual strokes, using line integral convolution filters, tonal mapping, and physical modeling. In stroke-based schemes, brush-strokes are created along a direction field derived from an image. Originally designed to mimic painting techniques, this approach can be extended to similar media. For instance, Salisbury et al. [3] modeled pen-and-ink illustration, Matsui et al. [12] produced crayon effects by overlapping textures, and Murakami et al. [13] explored the effects produced by a range of media, including pastels, charcoal and crayons. But stroke-based schemes are not a good match with the hatching patterns seen in monochrome pencil drawings. Line integral convolution (LIC) [2] is a technique that can be used to visualize vector

G. Lee, D. Howard, and D. Ślęzak (Eds.): ICHIT 2011, CCIS 206, pp. 70–77, 2011.

fields as textures, and thus offers a way of producing hatching. Mao et al. [8] pioneered this application of LIC. They segment an image into a number of regions, each of which is then rendered with strokes in a uniform direction; but this uniformity is not very realistic. Li and Huang [9] got around this problem by using gradient vectors and edges to determine stroke directions. This gives more variety, but is unable to express tonal details. Yamamoto et al. [10] added contours and paper surface effects, and also showed how to overlap textures to match the intensities of different regions of an image. But this scheme still suffers from the lack of realistic pencil strokes, detailed shading, and contours. Schemes based on tonal maps build textures with different tones to represent variations in intensity. They can achieve accurate shading and attractive hatching, and also reproduce feature lines such as contours. However, the input to most of these schemes must be in 3D; and many of them are also designed to run in real time, which limits their sophistication. In the present context, the most relevant work is probably that of Lake et al. [6] and Lee et al. [14], who rendered triangular meshes with various tones produced by overlapping a simple pencil texture; but this initial texture is fixed. At the other extreme, in terms of computation time, are techniques that involve modeling the physical processes of pencil drawing. This gives a realistic result, but is difficult for a user to control, as well as being slow. A typical contribution is that of Sousa and Buchanan [4], who constructed a physically-based model of a graphite pencil drawing on paper, and used it to render scenes represented in 3D. They went on [5] to model the action of blenders and erasers.

The availability of 3D data naturally assists the rendering process, but the ability to re-render a photograph is arguably more useful. We therefore focus on the creation of pencil drawings from photographs. Our scheme, outlined in Fig. 1, consists of the following steps: First, we create a set of pencil textures from overlapping strokes. An input image is segmented into several isotonal regions and the flow in the input image is then analyzed. A pencil texture is applied to the region multiple times. The region of the brightest tone is rendered for one time, while the region of the darkest tone is rendered for n times, where n denotes the levels of tones. The direction of flow is exploited in applying textures to the regions. In a subsequent process, the edges are found in the image, and then a stroke-based method is used to draw contours and feature lines.

The distinctive features of our scheme are as follows:

1. Applying a pencil texture according to the tone for multiple times re-renders an input image as a pencil drawing of smooth tone and convincing hatching patterns.
2. Deriving the drawing direction from the smooth tangent flow in the image improves the hatching effect.
3. Adding contours and feature lines conveys the most significant shape information with greater clarity.

This paper is organized as follows. We suggest the algorithm of generating pencil textures in Section 2. In Section 3, we propose how the drawing directions are estimated. We explain the pencil drawing scheme in Section 4. We show results and give discussions in Section 5. Finally, we make a conclusion and propose future works in Section 6.

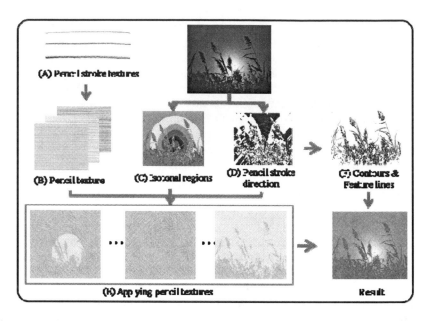

Fig. 1. An overview of our pencil drawing algorithm. Block A and B are explained in Sec. 2, C and E in Sec. 4, D in Sec. 3.

2 Generating Pencil Textures

The first step in generating pencil textures is to capture real pencil strokes as shown in the top row of Fig. 2. The pencil strokes are generated by varying types of pencils, thickness of the tips, and pressures. We then overlap individual pencil strokes to build pencil textures with different tones and hatching patterns as shown in the middle row of Fig. 2. The bottom row of Fig. 2 shows the details of the generated pencil textures.

Lake et al. [6] created a texture map expressing a range of tones by combining several pencil textures in both parallel and orthogonal directions. Tone is determined only by the density of the strokes, and darker tones are created by strokes which are orthogonal. Praun et al. [7] vary the tone by increasing the density of strokes without overlapping textures. Lee et al. [14] generated different tones by overlapping a stroke texture using a decrement model:

$$c_t' = c_t - \mu_b\, c_a, \text{where } c_a = c_t(1.0 - c_s).$$ (1)

The variables c_t and c_t' respectively are the current and updated texture tones. c_s' is the tone of the capture stroke used in generating a pencil texture, and μ_b is used to control the rate of darkening as textures are overlapped.

An important distinctive point of our approach is that we do not present a spectrum of pencil textures of different tones. We create pencil drawings of various tones by overlapping a pencil texture multiple times. This strategy allows us to re-render an

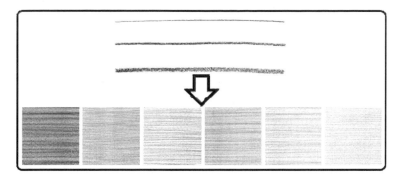

Fig. 2. Generating pencil textures: (Top row) pencil stroke textures, (Bottom row) generated pencil textures

image into a pencil drawing whose tone is smoothly changing. By using different pencil textures, we can create pencil drawings of different appearances from an input image.

We modified Eq. (1) in [14] to build different pencil textures from a pencil stroke texture, as follows:

$$c_k = (1.0 - \mu_b(1.0 - c_0))^k, \tag{2}$$

where c_k is the tone of a texture generated by k overlapping applications of a pencil stroke texture whose tone is c_0. The parameter μ_b also plays a role in controlling the tone of the resulting textures, and μ_b can be determined as follows:

$$\mu_b = \frac{1.0 - c_n^{1/k}}{1.0 - c_0}. \tag{3}$$

3 Estimating Drawing Directions

The direction of stroke textures must (i) be defined for all the pixels in an image, (ii) be smooth, and (iii) preserve the features of the objects in the scene. In previous schemes for converting images to pencil drawings, drawing directions were determined by users [3, 8, 10, 11, 16], from gradient fields [9], or from the contours of objects [12]. We surveyed filter-based edge extraction schemes across several areas of NPR, and decided to use edge tangent flow (ETF) [15]. ETF extracts edges using a DoG (difference of Gaussians) filters and creates a feature-preserving tangent vector fields which fulfills our requirements. However, the basic ETF scheme only derives tangent on pixels close to feature edges. We resolved this limitation by propagating the tangent vectors across the image using the following procedure, which is summarized in Fig. 3.

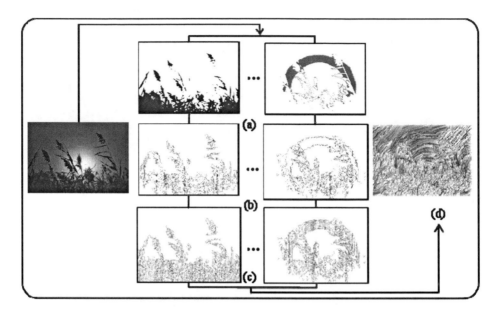

Fig. 3. Estimating drawing directions: (a) an iso-tonal region segmented from an image; (b) its tangent vectors estimated using ETF (blue vectors); (c) drawing directions (red vectors); (d) all the drawing directions in an image

Step 1. Sort the pixels in the image into two groups: $P = \{ p_0, ..., p_n \}$ consists of those pixels whose tangent $t(p)$ are determined by ETF, and $Q = \{ q_0, ..., q_m \}$ are those pixels with undefined tangent. Note that $t(p_i)$, $p_i \in P$, is non-zero, and $t(q_j)$, q_j Q, is zero.

Step 2. Approximate the boundary of Q as a group of pixels q_k which have $p_i \in P$ as one of their eight neighborhoods.

Step 3. Compute a tangent vector at each q_k as the sum of the tangent vectors at the neighbor pixels of q_k.

Step 4. If a non-zero tangent vector is assigned at q_k, q_k is moved from Q to P.

Step 5. Repeat steps 2 to 4 until tangent vectors are defined at all the pixels of the image.

4   Applying Pencil Textures

We apply a pencil texture to an image along the direction of flow estimated in Section 3. Suppose we segment an image into n isotonal regions, where 1st region is the darkest and n-th region is the brightest. We denote an *accumulated isotonal region* as a set of isotonal regions whose tone is brighter than a specific isotonal region. The i-th isotonal region, for example, denotes a set of isotonal regions between i-th and n-th region. A pencil texture is applied for n times. i-th accumulated isotonal region is drawn at i-th application. Therefore, a region of k-th tone is drawn for k times. At each drawing, the directions for each pixel are different, since the pixel is covered with a pencil texture which is drawn along the flow at a sampled pixel. A pencil texture is drawn using the following procedures:

Step 1. For i-th drawing, draw a pixel from those belong to the regions between i-th and n-th tone.

Step 2. Sample a pixel in the target region and apply our pencil texture at the pixel along the direction of flow on the pixel. Therefore, any pixels that lie inside the range of the texture are drawn by the texture.

Step 3. Repeat steps 1 and 2 until all the pixels in the regions between i-th and n-th tone are covered with pencil textures.

Step 4. Apply the decrement model in Eq. (1) in [14] to any pixel that is covered by more than twice. Fig. 4 shows the results of each step.

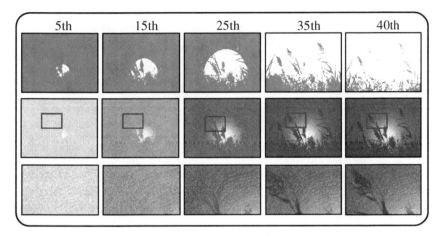

Fig. 4. Procedures of applying pencil textures on an image segmented into 40 isotonal regions: (1st row) an accumulated iso-tonal region; (2nd row) the pencil drawing images at each step (scaled in 10%) (3rd row) the regions inside the red box in column (b) are shown (scaled in 50%)

5   Implementation and Results

We implemented our algorithm on a PC with an Intel Pentium QuadCore^{TM} Q6600 CPU and 4GByte of memory. The programming environment was Visual Studio 2008 with the OpenGL libraries. The grayscale pictures to test our algorithm are in Fig. 5, and the resolutions of the input pictures and the computation times are given in Fig. 5. These times include estimating drawing directions, applying pencil textures and drawing contours and feature lines; but they exclude the time for generating pencil textures, which takes several seconds. The results shown in Fig. 5 were generated using several pencil textures. The result images in original size are presented in the accompanying image files.

Our scheme generates reasonably convincing pencil drawings in a straightforward way. However, to apply pencil textures is a matter of concern. Good results are obtained on the images containing areas of similar textures, which naturally favor the

Fig. 5. Pencil re-rendered images

use of a tonal map. But our algorithm has no mechanism to aggregate small-sized features, and thus the pencil effect is so fine as to become almost invisible in a very detailed image.

6   Conclusion and Future Work

We have presented a scheme for re-rendering pictures as pencil drawings. Our scheme overlaps a pencil texture for multiple times along the direction of flow in the image.

An image is segmented into several isotonal regions. In the image, the direction in which the texture is to be drawn is determined using a feature-sensitive analysis of flow. Our scheme also draws contours and feature lines which convey the salient features of objects more effectively.

Urgent future work is to use GPU programming to increase speed. Another challenge is to improve our scheme to render detailed and unorganized textures in a bolder manner. We also plan to extend our scheme to mimic drawings produced with colored pencils and to video.

Acknowledgments. This research is also supported by Basic Science Research Program through the Korea Research Foundation (KRF) funded by the Ministry of Education, Science and Technology at 2011.

References

[1] Franks, G.: Pencil Drawing: Learn to Draw 12 Classic Subjects. Springer, Heidelberg (2006)

[2] Cabral, B., Leedom, C.: Imaging vector field using line integral convolution. In: Siggraph 1993, pp. 263–270 (1993)

[3] Salisbury, M., Wong, M., Hughes, J., Salesin, D.: Orientable textures for image-based pen-and-ink illustration. In: Siggraph 1997, pp. 401–406 (1997)

[4] Sousa, M.C., Buchanan, J.: Computer-generated graphite pencil rendering of 3D polygonal models. In: Eurographics 1999, pp. 195–207 (1999)

[5] Sousa, M.C., Buchanan, J.: Observational model of blenders and erasers in computer-generated pencil rendering. In: Graphics Interface 1999, pp. 157–166 (1999)

[6] Lake, A., Marshall, C., Harris, M., Blackstein, M.: Stylized rendering techniques for scalable real-time 3D animation. In: NPAR 2000, pp. 13–20 (2000)

[7] Praun, E., Hoppe, H., Webb, M., Finkelstein, A.: Real-time hatching. In: Siggraph 2001, pp. 579–584 (2001)

[8] Mao, X., Nagasaka, Y., Imamiya, A.: Automatic generation of pencil drawing using LIC. In: ACM Siggraph 2002 Abstractions and Applications, p. 149 (2002)

[9] Li, N., Huang, Z.: A feature-based pencil drawing method. In: The 1st International Conference on Computer Graphics and Interactive Techniques in Australasia and South East Asia 2003, pp. 135–140 (2003)

[10] Yamamoto, S., Mao, X., Imamiya, A.: Enhanced LIC pencil filter. In: International Conference on Computer Graphics Imaging and Visualization 2004, pp. 251–256 (2004)

[11] Yamamoto, S., Mao, X., Imamiya, A.: Colored pencil filter with custom colors. In: Pacific Graphics 2004, pp. 329–338 (2004)

[12] Matsui, H., Johan, H., Nishita, T.: Creating colored pencil images by drawing strokes based on boundaries of regions. In: International on Computer Graphics 2005, pp. 148–155 (2005)

[13] Murakami, K., Tsuruno, R., Genda, E.: Multiple illuminated paper textures for drawing strokes. In: International on Computer Graphics 2005, pp. 156–161 (2005)

[14] Lee, H., Kwon, S., Lee, S.: Real-time Pencil Rendering. In: NPAR 2006, pp. 37–45 (2006)

[15] Kang, H., Lee, S., Chui, C.: Coherent Line Drawing. In: NPAR 2007, pp. 43–50 (2007)

[16] Xie, D., Zhao, Y., Xu, D., Yang, X.: Convolution Filter Based Pencil Drawing and Its Implementation on GPU. In: Xu, M., Zhan, Y.-W., Cao, J., Liu, Y. (eds.) APPT 2007. LNCS, vol. 4847, pp. 723–732. Springer, Heidelberg (2007)

Camera Colour Fidelity
through Adapting ICC Profiles

Guy K. Kloss

School of Computing + Mathematical Sciences,
Auckland University of Technology, Auckland, New Zealand
guy.kloss@aut.ac.nz

Abstract. Observing colour robustly in images captured with off-the-shelf cameras can be a challenge. Particularly if the observation is conducted over a long time, while encountering light changes. Factors playing into using this colour information directly for quantitative reasoning: unknown processing in the camera's firm ware, unpredictable variations in illumination conditions, and colour descriptors possibly not comparable with those of other devices. The research presented outlines an approach based on standard compliant colour management as specified by the International Colour Consortium (ICC). Initially, a device is characterised, resulting in an ICC profile for the starting conditions. Software tracks the changes of a number of coloured objects in the scene. With these changes a colour corrective transformation is fused into the ICC profile, compensating for the colour shift induced over time. An ICC profile transformation yields device independent CIE LAB colour space information. This paper describes background and process of the colour correction – including profile adaptation without process interruption – along with validation experiments conducted.

Keywords: Colour Management, ICC Profiles, Dynamic Adaptation, Colour Correction, Colour Fidelity, Camera.

1 Introduction

Many technical systems involving camera based image capturing operating within not fully controlled light situations. These may have to identify colour coded markers, perform industrial identification or classification (e. g. quality control), etc. A commonality in these scenarios: They demand online or near real-time video image processing, are used in indoor and/or outdoor environments with usually slowly changing conditions, and are built using relatively cheap commodity colour cameras (web camera type devices).

A common approach to solve the problem of normalising the camera's colour input is *colour constancy* [1]. Colour constancy algorithms usually infer light properties (such as white and possibly black point), and correct for them. Additionally, colour channels are mostly scaled to normalise the device's gamut size. These operations are applied directly to the device specific colour spaces (e. g. RGB), disregarding individual device capturing characteristics.

G. Lee, D. Howard, and D. Ślęzak (Eds.): ICHIT 2011, CCIS 206, pp. 78–87, 2011.
© Springer-Verlag Berlin Heidelberg 2011

Contrary to this, the industry has adopted *colour management* [2], which aims at controlled conversions between colour representations of devices, while retaining colour fidelity as far as possible. This is conducted by obtaining colour chatracteristics for in- and output devices (ICC profiles [3]). These profiles are used to convert between a device dependent input or output space and an intermediate profile connection space (PCS).

This paper presents research aiming to make colour based reasoning in long time observations more robust, while tolerating changes in the scene's illumination. Such changes are induced by nature (time of day, overcast or clear, blue sky) and artificial sources (incandescent or fluorescent lamps). Applying the techniques outlined can yield (at no additional effort) colour information in a colour space much more suitable for colour based reasoning than the commonly available camera encodings.

The following Sect. 2 introduces colour space encodings and colour adaptation mechanisms. Sect. 3 then describes the core of this research: how colour management techniques can be applied to correct for colour shifts when the illumination changes with time. Lastly, validation results from applying these new techniques are summarised in Sect. 4.

2   Colour Adaptation Approach

Colour adaptation is a (normalising) colour transformation, applied to an individual colour represented in a specific colour space, resulting in a colour of the same colour space. Firstly, this section introduces different concepts of colour representations (Sect. 2.1). This is important in understanding the concepts of colour management (Sect. 2.2), and extending them for dynamically changing environments (Sect. 2.3).

2.1   Colour Spaces

Besides actual colour space encoding (e. g. bit depth, number formatting, channel order), colour spaces can be represented in different ways. Some of these representations may include the number of channels (e. g. RGB vs. CMYK), additive/subtractive encoding (e. g. RGB vs. CMYK)), coordinate system (e. g. Cartesian RGB vs. cylindrical HSV), etc. All colour spaces discussed here can be represented as an n-tuple for n colour encoding components.

Important concepts for this paper are *device dependent* and *device independent* colour spaces, as well as *perceptually linearised* colour spaces:

Device dependent colour spaces constitute the native colour representation of a physical device. These contain the direct reading obtained from a device, or a value encoded for screen output. Device dependent spaces are largely influenced by physical factors, such as pigments used in the sensor, paper or pigments (used in ink) of a printer, colouring of a display's light emitting cells, etc. Due to physical difference in devices (even production differences among identical models), colour tuples of such colour spaces do vary. Therefore, dependent colour descriptions are *only* meaningful for *one* specific device under *one* specific condition (illumination of scene, paper in printer). Commonly encountered device dependent spaces are for example RGB, CMYK and YUV.

Device independent colour spaces are based on independently defined standards, usually in terms of spectral descriptions. Therefore, independent colour representation is meaningful regardless of physical device involved. They are intended as a vehicle to communicate or compare colour robustly. By principle, no colour representation for a normal capturing device can be obtained in a device independent colour space. Such family of standards is established by the International Commission on Illumination (CIE). All CIE colour spaces are derived from CIE XYZ, which is defined on the basis of standardised human eye responses. The most notable of these is CIE $L^*a^*b^*$ (or CIE LAB) [4]. Another prominent device independent colour space is sRGB, which is however encumbered by problems outside the scope of this paper.

Perceptually linearised colour spaces take a complimentary approach in their representation. The values in device colour are commonly defined in terms of values representing a level in a physical system. These values tend to be far from proportional to human perception. To improve usability, colour spaces reflecting perceptual linear scaling of colour representation have been derived. In these representations similar distances (ΔE) within the colour space reflect similar visual differences to the human eye. Notable perceptually linearised colour spaces are CIE's $L^*a^*b^*$ and $L^*u^*v^*$.

The before mentioned PCS is either in CIE LAB or CIE XYZ format. As CIE LAB is derived from CIE XYZ, all profiles are interoperable with each other through the PCS. $L^*a^*b^*$ is particularly useful for colour correction, as colour transformations on a linear space become easier. For colour based reasoning it is very suitable as well: One the one hand due to the linear nature (differential operations are independent of the locations within the colour space), and on the other hand as absolute values can be used independently of the device type used (camera, display, printer, etc.). Lastly, $L^*a^*b^*$ is an *opponent colour space*, in which chromaticity is split into two orthogonal components (red–green vs. blue–yellow axis), while the third is the lightness component. This pair-wise orthogonal system is very convenient for quantitative reasoning on colour.

2.2   Colour Management

The primary goal of colour management is to obtain a good match of colour appearance across colour devices; for example, a video should appear the same colour on a computer monitor, a TV screen and on a printed frame of video. Colour management tries to achieve the same appearance on all of these devices, provided the devices are capable of delivering the needed colour intensities. Handling of colour on computers is complicated, given the wide variety and near infinite combinations of video cards, displays, printers, ink and cameras. Join each input and output device individually leads to unmanageable numbers of conversion pairings. Colour management implemented by means of ICC profiles can bring some order. These profiles describe a conversion between PCS and the device dependent colour space. Therefore, per device and conditions only one profile is needed, and conversion pipelines are passing through the independent, canonical PCS.

ICC input profiles (e. g. for cameras, scanners) are derived through a *characterisation* process. A *target* consisting of many coloured patches is captured (see example in Fig. 4). The device independent colours of the patches are precisely known (e. g. through individual measurement with a spectrophotometer), and device colour is obtained from the captured image. The *profiling* process determines transformation tables used by a *colour management module* for colour conversion between device space and PCS. Output profiles are derived similarly, only that known colour tuples are sent to the device, and the actual output (e. g. on screen or paper) are photometrically measured. The derived transformation tables are encoded within the standardised ICC profile format (see Sect. 3.3).

2.3   Static vs. Dynamic Chromatic Adaptation

Colour management workflows are commonly used in "static" environments. These do not encounter changes in conditions invalidating the ICC profile for the case. Within such environments the input profile for the camera is determined initially, and then applied to each frame captured during operation (see top of Fig. 1).

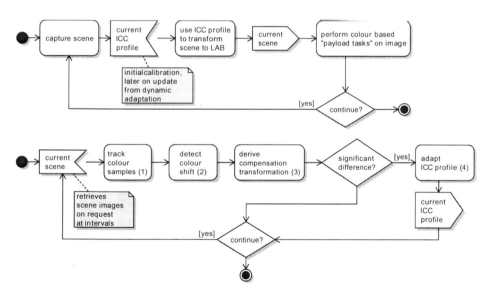

Fig. 1. Colour managed image processing (top) and ICC profile adaptation process (bottom)

Unfortunately not all environments are that stable, and classic colour management approaches are not applicable. Such situations either demand to resort to alternative approaches (e. g. colour constancy) or to modify the process as described in the following section.

3   Dynamic Colour Correction

A camera ICC profile becomes unsuitable upon changes in the illumination conditions of a scene. Ideally, the ICC profile in use can be replaced by an alternative suitable for the current conditions, and processing continues without disruptive re-characterisation. Common environments often expose slow changes compared to the frequency of image acquisition. Therefore, the ICC profile needs to be updated seldom, and a (current) ICC profile can be seen as an initial approximation of the $L^*a^*b^*$ PCS. Upon this approximation, a decoupled process can compute a corrective colour transformation without the need of real-time capability. This transformation can be used to update an ICC profile, which then can be "injected" back into the real-time image processing loop, ready for application to subsequent frames (see bottom of Fig. 1).

3.1   Gathering Colour Information

A full device characterisation (ICC profiling) is very disruptive (capturing of complete test chart with extensive table computation). We were seeking alternative approaches to gather information useful for a limited process to just update a profile. A sufficient number of data points in colour space needs to be gathered, that can be used to relate a previously known to a current colour value. Increasing number and accuracy of these data points increases the potential for computing a suitable colour correction.

In many applications it is possible to gather a number of colour samples: For example black and white points using approaches from colour constancy, plus a few stable colour samples. These might result from existing "payload" tasks in the processing loop (e. g. tracking of coloured objects), or using other techniques. Collection of such colour information is a topic of very application specific further research. A number of about six colour samples (of reasonable accuracy) can already be sufficient for the approaches discussed here.

3.2   Compensating for Colour Shifts

To compensate for distortions or shifts in colour space, a "distortion model" for the deformation has to be established, which then can be parameterised for a corrective transformation. A series of experiments under controlled laboratory conditions (14 illuminants, incl. incandescent, fluorescent and LED illuminants, under different brightnesses, and a variety of colour filters) was set up to analyse these distortions. As scene content a colour test chart suitable for photography was used (Christophe Métairie Photographie DigitaL TargeT 003) to have precise knowledge of all colours. All scenes were captured with fixed settings (automatic exposure/adaptation modes off). The images were transformed to $L^*a^*b^*$ PCS colours using an ICC profile obtained from one condition declared as reference (a fluorescent illuminant). Colour perception errors encountered in the charts patches (due to use of an unsuitable ICC profile) are qualitatively visualised in $L^*a^*b^*$ space of Fig. 2 (left). See the Ph. D. thesis of Kloss [5] for further information of the test series.

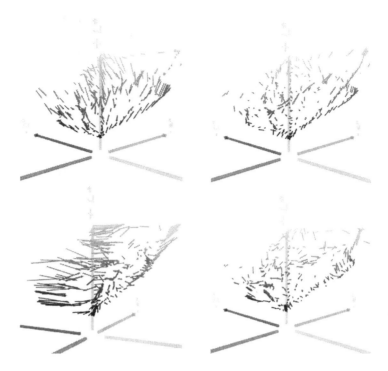

Fig. 2. Shift vector field of the 285 colour patches relative to ICC profiled reference conditions in $L^*a^*b^*$ colour space. Uncorrected colour appearance (left) and corrected appearance (right). For a less bright fluorescent illuminant (top) and a similarly bright incandescent illuminant (bottom).

It proved to be difficult to determine general deformation models from this. As a solution, the common engineering approach was used: Firstly apply a Taylor series approximation up to the linear term (affine transformation). A verification for the suitability of corrective linear transformations on a $L^*a^*b^*$ PCS has been conducted by porting colour constancy algorithms to $L^*a^*b^*$ space [5].

A 3-D linear vector field regression [6] was used to derive the affine transformation from over-defined data (see Sect. 3.1), as an affine transformation in $\mathbb{R}^3$ only requires four pairs of data points. Additional data points are useful to reduce the overall error induced by noisy measurements. The affine transformations were computed from two sets of data points: A highly over-defined case (using all 285 available colour patches), and a slightly over-defined case (using only six: white, black, a medium saturated red, green, blue and yellow).

3.3 Integrated in Colour Management

Look-up tables for the processing elements are the "active ingredients" of an ICC profile. These can be per-channel look-up tables (1-D LUTss: "A" and "B" curves), as well as multi-dimensional colour look-up-tables (CLUT: commonly used for camera input profiles in 3-D to 3-D). Other elements ("M" curves and matrix) are only used for simpler (and less capable) matrix profiles. Such unused elements are disabled (set to a default identity transform). Fig. 3 shows the typical transformation chain (unused elements faded to grey) [3]. The CLUT at the "heart" of the chain contains the main transformation data for mapping complex device-to-PCS relationships. For good quality profiles it often consists of 17^3 3-tuples.

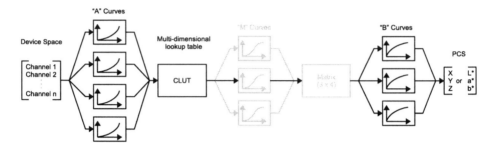

Fig. 3. Chain of processing elements for a transformation from device colour space to PCS (for three or more colour channels)

The affine transformation needs to be applied to the whole colour space, so all (equally spaced) CLUT tuples are updated to fit the corrected transformation. So, every tuple element from the CLUT is extracted, "B" curve transformation are applied, and the corrective affine transformation is applied afterwards. To update the CLUT, an inverse "B" curve transformation is applied to the adapted colour tuple, and the result replaces the original colour tuple in the CLUT. This new ICC profile provides a convenient one-step colour correction process, as the affine transformation has been "smeltered" into it.

3.4   Capturing Workflow

In a practical workflow, determining a corrective affine transformation (Sect. 3.2) and updating the ICC profile can be performed in a detached adaptation process (see Fig. 1, bottom). This way the commonly static chromatic adaptation using ICC based colour management has been upgraded to be capable for dynamic chromatic adaptation (see Sect. 2.3). The process needs "boot strapping" through an initial full device characterisation for a good quality starting ICC profile. This of course depends on a working approach to track colour samples for the compensation transformation (Sect. 3.1).

4   Results

Resulting colour samples after correction through an affine transformation (Sect. 3.2) are plotted in $L^*a^*b^*$ space for all 285 colour samples of the test chart. The relative colour error for each colour patch before (left) and after the

Fig. 4. Appearance of the 285 colour patches with the reference profile applied (top row), and corrected using the corrected profile (centre row) for four different illuminants. For comparison the image taken under the reference illuminant (bottom). The first and third illuminant are less bright, the first and second under slightly "colder" light, the third with a similar colour temperature, and the last slightly "warmer" (also the only incandescent, all others fluorescent).

application (right) of the corrective transformation is qualitatively apparent from Fig. 2. Resulting images are shown directly in Fig. 4.

Furthermore, a more quantitative evaluation has been performed. It was conducted by computing the Euclidean colour errors ΔE_{ab}^* in $L^*a^*b^*$ space for all 285 patches of the used test target. $L^*a^*b^*$ is scaled in a way that a Euclidean distance $\Delta E_{ab}^* = 1$ is just identifiable by a human observer. To establish a lower bound, the same deviations have been computed after applying a white point correction (using a Bradford transformation) only. Such a white point correction can be seen as representative for common colour constancy approaches as well. Table 1 charts these results, averaged over all 13 illumination cases (excluding the reference conditions). It is evident from the figures, visible improvements of a magnitude of 3–4 ΔE_{ab}^* have been achieved over white point correction only (colour constancy).

Finally, practical usability of this colour correction approach was evaluated through two test cases. The first one employs *a priori* knowledge of specific colours re-occurring in a scene (problem domain of robot soccer [7]). Card stock used for markup of the robots (five colours) and the colour of the field and field markings (black and white) were easily identifiable for tracking. In this case (the more likely) illumination intensity changes were analysed rather than changes in illuminant. A light level variation also influences colour saturation, and potentially also alters slightly the camera's capturing characteristics.

Table 1. Averages and standard deviations of colour patch deviation after corrective affine transformation

Transformation	avg. ΔE_{ab}^*
White point correction only	11.4 (5.6)
Affine transformation in $L^*a^*b^*$ (285 patches)	7.0 (3.7)
Affine transformation in $L^*a^*b^*$ (6 patches)	8.0 (4.3)

We found the usable range of light intensities to be more than doubled, while maintaining a working system. Also classifier training using a genetic algorithm was improved when using colour corrected imagery. Lastly, switching to a^*b^*-chromaticity (chromaticity channels of $L^*a^*b^*$) from previously used rg-chromaticity showed some advantages in training and classification.

Another practical test case was conducted using long time (approx. 9 h) observations of a scenes using a static perspective, while applying continuous correction of the ICC profile [5]. The scene was subject to significant changes during the day's light cycle. In this case no *a priori* knowledge was used to track colour changes (see Sect. 3.1). As an alternative, each frame was segmented into 192 square segments. Average colours for each of these segments were tracked, while those exhibiting an "excessive" relative change were discarded (usually indicating local scene changes). The colour samples were correlated with their counterparts from earlier frames to compute the affine transformation through regression. Results from applying this largely over-defined regression problem showed that some improvement could be recorded, but correction results were not good under all conditions. Such simple approaches for scene segmentation are not sufficient, and further research needs to be conducted.

5   Conclusion

We have built a system that perceives colour – and does that as precisely as possible – while adapting to light changes, when encountering common natural and artificial light sources. It obtains device independent and visually linear colour descriptors (in CIE LAB colour space). The motivation for using this colour space is driven by its beneficial properties for analysing and comparing colours. Therefore, we have also based the time adaptive colour correction process to operate on this colour space.

By applying this technique, practical colour correction is departing from "tinkering" on unsuitable, device dependent, highly non-linear device colour spaces towards generally accepted operations, founded on colour science. Using ICC profiles furthermore enables applications to rely on standard compliant procedure, using a long time industry accepted approach. Processing can take advantage of one of many efficient implementations of colour management engines available (commercially as well as in open source). Colour corrected processing therefore can become easy to implement without major modification of current approach.

The approach introduced has been successfully verified by means of various qualitative and quantitative analyses. Adaptive colour correction discussed consider the additional factor "time," which can be used to extract further input data as opposed to colour constancy, which only bases corrections on one image at a time. As a hypothesis for this technique, changes in light conditions occur comparatively slow relative to the rate at which images captured can be used to track physical changes in the scene. This difference in rates can be used to offload the slower ICC profile adaptation to a separate process, while applying ICC profile correction in real-time.

Future work includes the application of the discussed colour correction approach in further real world systems. Along with this, application specific data collection techniques need to be developed, employing domain specific knowledge to identify suitable objects for colour tracking.

References

1. Ebner, M.: Color Constancy. Imaging Science and Technology. Wiley-IS&T, West Sussex (2007)
2. Has, M., Newman, T.: Color Management: Current Practice and The Adoption of a New Standard. Whitepaper 1, International Color Consortium (1995), http://www.color.org/wpaper1.html
3. International Color Consortium: Specification ICC.1:2004-10 (Profile version 4.2.0.0) [ISO 15076-1:2005] (2006), http://color.org/icc_specs2.xalter
4. Hoffmann, G.: CIE Color Space. Technical report, Fachhochschule Emden (2000), http://www.fho-emden.de/~hoffmann/ciexyz29082000.pdf
5. Kloss, G.K.: Adaptation of Colour Perception through Dynamic ICC Profile Modification. Doctor of Philosophy (Ph. D.), Massey University, Auckland, New Zealand (2010), http://hdl.handle.net/10179/1683
6. Kloss, G.K., Kloss, T.F.: n–Dimensional Linear Vector Field Regression with NumPy. The Python Papers Source Codes 2 (2010), http://ojs.pythonpapers.org/index.php/tppsc/issue/view/20
7. Kloss, G.K., Shin, H., Reyes, N.H.: Dynamic Colour Adaptation for Colour Object Tracking. In: Proceedings of the 24th International Image and Vision Computing New Zealand (IVCNZ), Wellington, New Zealand, pp. 340–345 (2009)

Breast Mass Segmentation in Digital Mammography Using Graph Cuts

S. Don[1], Eumin Choi[2], and Dugki Min[1,*]

[1] School of Computer Science and Engineering, Konkuk University, Seoul,
133-701, Korea
{donsasi,dkmin}@konkuk.ac.kr
[2] School of Business IT, Kookmin University, Jeongneung-dong, Seongbuk-gu,
136-792, Korea
emchoi@kookmin.ac.kr

Abstract. This paper presents a novel method for the segmentation of breast masses on a mammography. Accurate segmentation is an important task for the correct detection of lesions and its characterization in computer-aided diagnosis systems. Many popular methods exist, of which most of them rely on statistical analysis. Similar to other methods, we propose a graph theoretic image segmentation technique to segment the breast masses automatically. This method consists of two main steps. First we introduce a thresholding method to obtain the rough region of the masses by eliminating all other artifacts. Then, on the basis of this rough region, the graph cuts method was applied to extract the masses from the mammography. The results were evaluated by an expert radiologist and we compared our proposed method with the level set algorithm, which shows the highest success rate. In contrast, we experiment our method on two different databases: DDSM and MiniMIAS. Experimental results show that the proposed method has the potential to detect the masses correctly and is useful for CAD systems.

Keywords: Segmentation, Mammogram, Graph Cuts.

1 Introduction

Breast cancer is the second most major health problem in developed countries. According to the latest study from NCI (U.S.National Cancer Institute) more than 207,090 new cases were reported and about 39,840 were the death statistics in 2010. Nearly 1.4 million US women have a family history of breast cancer [1, 2]. As there is no effective method for its prevention, the diagnosis of breast cancer in its early stage of development has become very crucial for the prevention of cancer. Computer-aided diagnosis (CAD) systems play an important role in earlier diagnosis of breast cancer. Another method such as a biopsy is used for detection of breast cancer, where the patient undergoes a surgical procedure. Many CAD based systems were used by the radiologist for the diagnosis of breast masses. The limitation of these systems is

---

* Corresponding author.

G. Lee, D. Howard, and D. Ślęzak (Eds.): ICHIT 2011, CCIS 206, pp. 88–96, 2011.

mainly at breast mass detection [3]. Therefore, accurate segmentation of a breast mass is an important step for the diagnosis of breast cancer in mammography. Breast mass is a localized swelling, which is described by its characteristics. An accurate contour extraction of the actual mass which enables to distinguish between lesion and the normal tissues. Usually these boundaries are of varying size and shape. Because of this, breast mass segmentation is a challenging task. Several methods have been proposed for the segmentation of the masses from mammograms. The most common approaches are region based method and edge based method. The purpose of the region based method is to identify the region based on some predefined conditions. The conditions can vary from pixel intensities to texture features. Kostas Haris et.al [4] proposed a hybrid multidimensional segmentation technique which combines edge and region based through the morphological algorithms of watersheds to segment the mass region. Enmin.et.al [5] proposed an automated segmentation method which combines edge gradient, pixel intensities as well as shape. Enmin Song et.al [6] used a hybrid approach to segment the mass based on template matching and dynamic programming. Arnau Oliver et.al [7] develop a method which automatically estimates the density of a breast based on a statistical approach. Eltonsy et.al [8] proposed the detection of masses by identifying the presence of concentric layers with morphological characteristics and lower relative incidence in the breast region. The method performs well in identifying malignant masses. A fuzzy based region growing method that considers uncertainty present around the boundary was suggested by Guliato et.al [9]. This method has an accuracy of classifying the masses and tumor as benign or malignant with an accuracy of 0.8 for sensitivity and 0.9 for specificity. Zhang et.al [10] used the whole mammograms instead of manually selecting the regions for growing. The pixels with local maximum gray level are considered as seeds, from which many candidate objects are grow using the modified region growing algorithm, which acquired an accuracy of 90%. The main challenges in mass segmentation are masses may in contact with surrounding breast tissues which can have similar intensity values. Chu et.al [26] proposed a graph based segmentation algorithm in which region growing were represented by a growing tree whose root is selected as the seed. Leaves which have the ability to grow in the connected area. The author concludes that the new graph based segmentation has closer match with the radiologist outlines. Considering an image as graph and partition them to a different group is a tedious task. Shi.et.al [11] proposed a normalized cut which focuses on global impression of the image data rather than local features. This method was tested on real and synthetic images which satisfy the extraction of big pictures of a scene. Edge based mass detection limits the researchers due to the difficulty in extracting the boundaries between the masses. Typically algorithm works on detecting the edges and grouping into contours. This can be done based on filtering and active contour models. Petricks et.al [12] introduced a new approach for segmenting the masses from a digitized mammogram using adaptive density weighted contrast enhancement filter based on Laplacian-Gaussian edge detection. Initially the image is filtered globally and then segments the image with Laplacian Gaussian edge detection. A set of 84 images was considered in this experiment. By avoiding the filtering techniques, Hong [13] introduced ISO-level contour map, in which region forms a dense quasi-concentric pattern of contours. The structure of the image was analyzed using the inclusion tree that is a hierarchical relationship between contours.

This method is applicable for detecting breast region, pectoral muscle and dense mass. An automated method for segmenting the mass was introduced by Jiazheng Shi et.al [14]. The method was based on level set algorithm and introduced two types of image features related to the presence of microcalcification. For the classification purpose they used linear discriminate analysis with a stepwise feature selection, and achieved a view based Az value 0.83. Dynamic programming is an optimization algorithm. Timp et.al [15] introduced an automated segmentation techniques based on dynamic programming. This method achieves a success rate of 99.9% to obtain a close contours. They compared with region growing and the discrete contour model. The mean overlap percentage of dynamic programming was 69% where as in the case of region growing and discrete contour model was 60% and 58 %. Mass segmentation is still an important field. Although many algorithms exist, still we need to improve the accuracy of extraction. The proposed method employed the fact that the boundaries of the mass are detected more precisely. This method relies on contrast enhancement followed by binary threshold and finally the graph theoretic image segmentation techniques to segment the mass. To establish the advantage of this method, we evaluated its performance on the standardized image database using AOM and CM coefficient. We compare our test results with those obtained using level set. The proposed method has been evaluated using two globally available databases: Mini-Mammographic Image Analysis Society database [23] and Digital Database of Screening Mammography [24]. Section 2 describes the proposed segmentation method. Experimental results were validated in Section 3. Finally conclusions are given in Section 4.

2   Methods

The proposed method of breast mass segmentation mainly consists of three groups of operations; noise removal to enhance the quality of the image, thresholding to obtain preliminary mask and segmenting the mass using graph cuts. Figure.1 shows an overview of the proposed method.

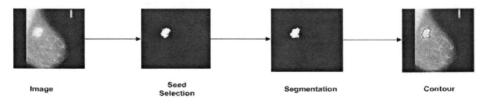

Image   Seed   Segmentation   Contour
               Selection

Fig. 1. The flow pipeline of proposed mass segmentation

2.1   Noise Removal

The extraction of mass can be influenced by noise, causing mass contour to be discontinuous. Because of this reason, it is necessary to suppress the noise. The Gaussian filter algorithm is applied in our case to reduce the image noise, as expressed in Eq.1.

$$G(x, y) = \frac{1}{2\pi\sigma^{2^e}} - \frac{x^2 + y^2}{2\sigma^2} \tag{1}$$

The variable x and y are the distance from the origin in both vertical and horizontal axis; σ is the standard deviation of the Gaussian distribution. Mammogram images usually contain artifacts such as markers and labels. These artifacts can mislead the segmentation process. To avoid this, [16] [17] used connected component labeling algorithm to find the largest region, in the breast region from the filtered image. This approach is excellent for finding non-connected objects in images. Each image is represented by an array of image elements. The algorithm assigns labels to the black image elements in such a way that adjacent black image pixels are grouped by the same label. An example for artifacts removal from a mammogram images is shown in Figure.2. Figs 2a and 2b are the two images, and results of the CCL algorithm as shown in Figs 2c and 2d.

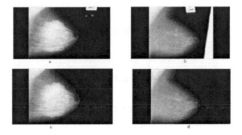

Fig. 2. Artifacts removal

2.2 Obtaining Preliminary Mask

To obtain the rough region of the breast mass, it must satisfy certain conditions. Select only the masses from the region of interest, i.e. The breast lesions. For images like mammogram there will be always abrupt changes in the intensity peak values. Threshold values that are too low or too high may lead to miss classifying of the mass area, figure 3. This happens because of the different types of tissue densities. Mudigonda et.al[18] used multilevel thresholding by varying levels of intensities to generate the contour. We found that good multilevel thresholding values lies between at $0.71 \leq T \leq 0.81$. The above parameter value chosen based on the observation of the histogram, as explained later on seed selection.

Fig. 3. (a) Original image b) low threshold c) medium threshold d) high threshold

2.3 Segmentation with Graph Cuts

A graph cuts framework for object segmentation was introduced by Boykov et.al [19]. Graph cuts have the ability to solve a wide range of problems in computer vision, which includes image segmentation by maintaining a global minimum energy function. This method considers the image as a weighted undirected graph G= {V, E, W}, where V and E are defined as a set of vertices and edges, and W is the edge weight. There are two special designated nodes S and T which represents the foreground and the background. The algorithm needs a set of sample for object S and sample of the background object as T. Neighboring pixels are connected by edges. Two kinds of links are defined t-links and n-links. Edges between pixels are called n-links and t-links are used to connect pixels to designated nodes. The goal is to choose the appropriate seed region and an edge weight, so that we can have the best cut that would give an optimal segmentation with minimum energy value. We introduced a semi automatic threshold method in order to choose the seed regions.

Seed selection

In order to choose the seed region automatically, we introduced thresholding method. The output of thresholding method is a binary segmentation (S,T),where the foreground region corresponds to S which contains all breast rough mass and T contains the background region. Since the background is equal to the non mass region, we consider background seed equal to the background mask of the image. Estimating the foreground seed points are more challenging due to the presence of the pectoral muscles and other irregular microcalcifications which are not the part of the lesion. In our implementation after several trials we found the threshold range which can preserve the breast mass and remove pectoral muscles and all other small microcalcifications. We applied morphological operator for discarding all other irrelevant regions from the image based on the area of size. The resultant masses were considered as a foreground seed point for the segmentation purpose. The implementation of the graph cuts algorithm as introduced by Boykov et.al [19], the cost function introduced for the segmentation is in Eq.2.

$$E_t = E_{region} + E_{boundary} \tag{2}$$

Where E_{region} assumes the penalties for assigning the pixels to "object" or "background". And $E_{boundary}$ is the smoothing term evaluates the penalties for assigning two neighboring pixels to different regions. These two terms repeated with λ which specifies the importance of region properties and λ must be ≥ 0. Thus

$$E_t = \lambda.\sum_{p\in I} R_p(A_p) + \sum_{(p,q)\in N} B_{(p,q)} \tag{3}$$

Once the graph was constructed based on the corrected weight function. A min-cut/max flow optimization algorithm is applied to the graph. We used the graph cuts algorithm from Boykov and Kolmogorov [20]. Since mammogram masses have high intensity values, directed edge weight would be the excellent approach to cuts from brighter region of darker tissue in the background. Defined as

$$w_{(p,q)} = \begin{cases} 1 & \text{If } I_p < I_q \\ \exp\left(-\dfrac{(I_p - I_q)^2}{2\sigma^2}\right) & \text{If } I_p > I_q \end{cases} \qquad (4)$$

The edge weight considered for the segmentation of the region term as(Eq.5) ,where I is the original image, μ is the mean and σ >0.

$$R(S) = \exp\left(-(I_p - \mu)/2\sigma^2\right) \qquad (5)$$

3 Experimental Results

The proposed method was tested on a set of 90 images randomly chosen for the experiment from MiniMIAS and DDSM databases. Our algorithm has successfully detected the masses. The result produced by our proposed method was verified under the supervision of radiologist. Fig.4 shows some examples of detected masses from the two databases. In order to provide a comparison we compare our test result with a level set algorithm, be considered as a popular algorithm for image segmentation. The results show that our approach has very close matching with ground truth image, where as in the case of the level set algorithm was over segmented. We evaluated the performance of the segmentation by considering two quantitative measures. First quantitative measures are area overlap measures (AOM) [21] and the second one is combined measures (CM) which calculates the level of agreement. For each of the segmented masses we calculated the area overlap measures and combined measures.

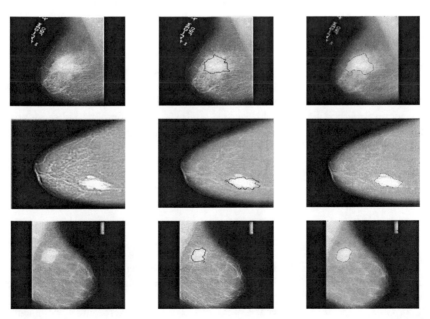

Fig. 4. Performance comparison of different segmentation algorithm column (1) original, (2) Graph Cuts and (3) Level set

$$AOM = \frac{A_s \cap A_g}{A_s \cup A_g}, \tag{6}$$

Where A_s denotes the area of segmented mass by the algorithm and A_g is the ground truth image provided by the radiologist. The combined measure is the combination of under segmentation, over segmentation and the area overlap measure.

$$CM = \left[AOM + \left(1 - \frac{A_g - A_s}{A_g} \right) + \left(1 - \frac{A_s - A_g}{A_s} \right) \right] / 3 \tag{7}$$

Which shows the performance level of the graph cuts approach is higher than the level set method. The max average value of AOM and CM obtained for graph cuts is 0.962 and 0.985. The results are shown in Table.1. The limitation of the graph cuts approach is that they are very sensitive to the seed selection. Another graph theoretic segmentation approach was the isoperimetric graph partition approach proposed by Grady and Schwartz [25]. The advantages of the graph cut approach over isoperimetric graph partition are that graph cuts are applicable to ND where as in the case of ISO which limits to 2D images.

Table 1. Performance Measures AOM, CM

Method	AOM (Max)	CM (Max)
Proposed	0.962	0.985
Level set	0.945	0.961

4   Conclusion

This paper presents a method to detect and segment the masses from the mammography. A hybrid method based on graph cuts has been developed for the segmentation. This method consistently performs well in most of the cases. An initial seed selection method based on histogram analysis was inherited. Subsequently, the graph cuts approach is used to segment the mass region of the mammogram. The proposed method demonstrates that graph cuts based mass segmentation provides an informative tool for the analysis of mammogram images. Finally, as a future work we will consider more samples and consider new algorithms for the extraction of the masses from mammography images.

References

1. Smith, R.A., Cokkinides, V., Brawley, O.W.: Cancer Screening in United State, 2010: A review of current American Cancer Society Guidelines and Issues in Cancer Screening. CA Cnacer J. Clin. 60, 99–119 (2010)
2. Lee, C.-S., Kanmaz, T.J.: Recent Controversies in Cancer Screening Recommendations US Pharm., vol. 35(11), pp. 3–8 (2010) (oncology Supply)

3. Philpotts, L.E.: Can Computer-aided Detection be Detrimental to mammographic Interpretation. Radiology, 17–22 (2009)
4. Haris, K., Efstratiadis, S.N., Maglaveras, N., Katsaggelos, A.K.: Hybrid Image Segmentation Using watersheds and Fast Region Merging. IEEE Trans. Image Proc. 7(12), 1684–1699 (1998)
5. Song, E., Jiang, L., Jin, R., Zhang, L., Yuan, Y., Li, Q.: Breast mass Segmentation in Mammography Using Plane Fitting and Dynamic Programming. Academic Radiology 17(7), 826–835 (2009)
6. Song, E., Xu, S., Xu, X., Zeng, J., Yuan, Y., Hung, C.-C.: Hybrid Segmentation of mass in mammograms Using Template matching and Dynamic Programming. Academic Radiology 17(11), 1224–1414 (2010)
7. Oliver, A., Freixenet, J., Bosch, A., Raba, D., Zwiggelaar, R.: Automatic classification of breast tissue. In: Marques, J.S., Pérez de la Blanca, N., Pina, P. (eds.) IbPRIA 2005. LNCS, vol. 3523, pp. 171–175. Springer, Heidelberg (2005)
8. Eltonsy, N.H., Tourassi, G.D., Elmaghraby, A.S.: A Concentric Morphology Model for the Detection of masses in mammography. IEEE Trans. on MedicImaging 26(6), 880–889 (2007)
9. Guliato, D., Rangayyan, R.M., Carnielli, W.A., Zuffo, J.A., Desautels, J.L.: Segmentation of Breast Tumors in mammograms using fuzzy sets. Electron. Imaging 12, 369 (2003)
10. Han, Z., Foo, S.W., Krishnan, S.M., Thug, C.H.: Automated Breast masses Segmentation in Digitized mammograms. In: IEEE International Workshop in Biomedical Circuits and Systems, pp. S2, 2–4 (2004)
11. Shi, J., Malik, J.: Normalized Cuts and Image Segmentation. IEEE Trans. on Pattern Analysis and Machine Intelligence 22(8), 888–905 (2000)
12. Petrick, N., Chan, H.P., Sahiner, B.: Automated Detection of Breast masses on Digital mammograms Using Adaptive density- weighted contrast enhancement Filtering. In: Proc. SPIE, pp. 590–597 (1995)
13. Hong, B., Brady, M.: A topographic representation for mammogram segmentation. In: Ellis, R.E., Peters, T.M. (eds.) MICCAI 2003. LNCS, vol. 2879, pp. 730–737. Springer, Heidelberg (2003)
14. Shi, J., Sahiner, B., Chan, H.-P., Ge, J., Hadjiiski, L., Helvie, M.A., Nees, A., Wu, Y.-T., Wei, J., Zhou, C., Zhang, Y., Cui, J.: Characterization of Mammographic masses based on level set segmentation with new Image features and patient Information. Med. Phys. 35(1), 280–290 (2008)
15. Timp, S., Karssemeijer, N.: A new 2D segmentation method based on dynamic programming applied to computer aided detection in mammography. Med. Phys., 958–971 (2004)
16. Davies, E.R.: Machine Vision: Theory, Algorithm, Practicalities. Elsevier, Amsterdam (2005)
17. Bhattacharya, P.: Connected Component Labeling for Binary Images on a reconfigurable mesh architecture. Journal of Systems Architecture, 309–313 (1996)
18. Mudigonda, N.R., Rangayyan, R.M., Leo Desautels, J.E.: Detection of breast masses in mammograms by density slicing and texture flow field analysis. IEEE Trans. on Medic. Imag. 20(12), 1215–1227 (2001)
19. Boykov, Y.Y., Marie-Pierre Jolly, M.P.: Interactive graph cuts for optimal boundary and region segmentation of objects in N-D images. In: Proc. ICCV, pp. 105–112 (2001)
20. Boykov, Y.Y., Kolmogorov, V.: An experimental comparison of Min-Cut/Max-Flow algorithm for energy minimization in vision. IEEE Trans. pattern Anal. Mach. Intell. 26(9), 1124–1137 (2004)

21. Xu, S., Liu, H., Song, E.: Marker Controlled watershed for lesion segmentation in mammograms. J. Digital Imaging (2011) (published online)
22. Jeffreys, M., Warren, R., Davey Smith, G., Gunnell, D.: Breast Density: agreement of measures from film and digital image. The British Journal of Radiology 76, 561–563 (2003)
23. Suckling, J., Boggis, C.R.M., Stamatakis, E., Taylor, P.: The Mammographic Image Analysis Society Digital Mammogram Database. International Congress Series 1069, 375–378 (1994)
24. Heath, M., Bowyer, K., Kopans, D., Moore, R., Kegelmeyer, P.J.: The Digital database for screening mammography. In: Proc. Int. Workshop Dig. Mammography, pp. 212–218 (2000)
25. Grady, L., Schwartz, E.L.: Isoperimetric Graph Partitioning for Image Segmentation. IEEE Trans. Patten Anal. Mach. Intell. 28(3), 469–475 (2006)
26. Chu, Y., Li, L.: Graph-based region growing for mass segmentation in digital mammography. Medical Imaging 4684, 1690–1697 (2002)

ScaleLoc: A Scalable Real-Time Locating System for Moving Targets

Si-Young Ahn[1], Jun-Hyung Kim[2], Jun-Seok Park[1],
Ha-Ryoung Oh[1], and Yeong Rak Seong[1]

[1] Department of Electrical Engineering, Kookmin University,
861-1 Jeongneung-Dong Seongkbuk-Gu Seoul, Korea
[2] The graduate school of education, Kyunghee University,
1 Hweki-Dong Dongdaemun Seoul, Korea
onsaiahn@gmail.com, jhkim@khcu.ac.kr,
{jspark,hroh,yeong}@kookmin.ac.kr

Abstract. In this paper an inexpensive, low-power and scalable location system for mobile objects under indoor environments, ScaleLoc, is proposed and its performance is evaluated by simulation. IEEE 802.15.4a (Chirp spread spectrum) protocol is adopted and the location of a target is determined with trilateration. As the number of mobile objects increases, the performance may deteriorate due to communication collisions. For graceful degradation and system scalability, an arbitration scheme is adopted. Sufficient information for localization such as the node-ids and locations of reference nodes is supplied by arbitration nodes.

Keywords: RTLS, location estimation, simulation, discrete event system.

1 Introduction

Location-Based Service (LBS) has been used in a variety of contexts, such as object tracking, entertainment, work, etc. RTLS's (Real-Time Locating System) may use diverse technologies according to the requirements and environment [1], [2], [3], [4], [5]. Several methods such as Time of Arrival (ToA), Time Difference of Arrival (TDoA), Angle of Arrival (AoA), Two Way Ranging (TWR), Symmetrical Double Sided Tow Way Ranging (SDS-TWR), etc. are used for ranging [6]. Location of target object can be calculated by using triangulation, trilateration based on the measured ranging information. Each RTLS has the pros and cons. GPS is one of the most frequently used techniques for localization, but it cannot be used for indoor applications. Furthermore, some communication link is needed to monitor the locations of objects in a control center.

In this paper, ScaleLoc is developed for real-time monitoring the locations of mobile targets scattered in an (indoor) service area. The system is required to be inexpensive, low-power and scalable. Sensor network technology is suitable for these requirements. IEEE 802.15.4a protocol defines the protocol and compatible interconnection for data communication devices using low-data-rate, low-power and low-complexity, short-range Radio Frequency (RF) transmissions in a Wireless

G. Lee, D. Howard, and D. Ślęzak (Eds.): ICHIT 2011, CCIS 206, pp. 97–103, 2011.
© Springer-Verlag Berlin Heidelberg 2011

Personal Area Network (WPAN) [7]. So, TWR of IEEE 802.15.4a protocol is used for ranging. Location of a target node is determined with trilateration. Since at least three distances are needed for this scheme and multiple distances cannot be measured at the same time, it is crucial to limit the time interval between consecutive ranging for a mobile object. As the number of objects increases, the probability of communication collision increases. IEEE 802.15.4a protocol provides CSMA-CA scheme. Devices must wait for a random back-off time if a channel isn't free, i.e. when a carrier is sensed. So localization error may be affected by neighboring nodes when target nodes move. For graceful degradation and system scalability, an arbitration scheme is adopted to ScaleLoc. Sufficient information for localization such as the node-ids and locations of reference nodes is supplied by arbitration nodes.

This paper is composed of the following. Chapter 2 explains the architecture, arbitration and operation procedure of ScaleLoc. The method for measuring the performance and simulation result is given in chapter 3. Lastly, Chapter 4 draws the conclusion of this study.

2   ScaleLoc

In this section, a geographical architecture of ScaleLoc for scalability is suggested, considering channel management and interference from neighboring cells. Micro-cell and macro-cell are defined, and then the system may be constructed by arranging macro-cells repetitively as needed to cover the service area. The proposed arbitration procedure and operation procedure is explained.

2.1   Macro-cell and Micro-cell

The micro-cell is the basic module of ScaleLoc. In a micro-cell, four reference nodes (r-node) for localization of mobile target nodes(t-node) are placed at each corners. At the center of the micro-cell, an arbitration node (a-node) is located. Two channels called measurement channel (m-channel) and arbitration channel (a-channel) respectively, are allocated to a micro-cell. With this channel scheme, arbitration and ranging operations can be performed independently without any intervention.

The macro-cell is a two dimensional array of micro-cells as shown in Fig. 1. In a macro-cell, distinct channels are allocated to each micro-cell to reduce interference. This prevents any node of a neighbor cell from using the same communication channel. ScaleLoc may be constructed by arranging macro-cells repetitively as shown in Fig. 1. As a result, every micro-cell or macro-cell may be considered to be independent and this makes ScaleLoc scalable. The size of a macro-cell (the number of micro-cells in a macro-cell) depends on the number of available channels. For example, a macro-channel of size 2 x 2 is possible if more than 8 channels are available and 3 x 3 is possible if more than 18 channels are available. As 14 channels are available in IEEE 802.15.4a, a 2 x 2 macro-cell is chosen in ScaleLoc as shown in Fig. 1.

The geographical size of a micro-cell is dependent on the signal strength of the nodes. Once the size is determined, the signal strength must be tuned so that every t-node in an arbitrary position can stably communicate with any r-node or a-node in the same micro-cell, since RF signal may be affected by EMI/EMC environment. If the signal power is too high compared to the optimal value, the performance may be degraded, since communication in a neighbor micro-cell using the same channel can be interfered.

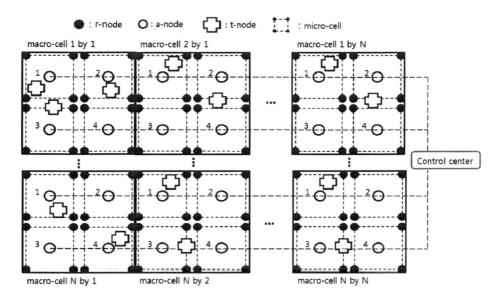

Fig. 1. An example architecture of ScaleLoc with 2 x 2 sized macro-cells

2.2   Arbitration

ScaleLoc determines the location of a target node with trilateration. At least three distances to r-nodes whose positions are known in advance are needed and multiple distances cannot be measured concurrently in IEEE 802.15.4a protocol. Since the targets are assumed to be mobile, it is crucial to limit the time interval between the consecutive ranging for a mobile object. It is desirable to make a localization of a t-node atomic, preventing other t-nodes from interrupting the session. As the number of objects increases, the probability of communication collision or interrupt increases. When this type of interrupt occurs, the t-node must back-off and wait. For graceful degradation and system scalability, an arbitration scheme is adopted to ScaleLoc.

As targets are mobile, an r-node cannot decide whether a particular t-node is in its micro-cell or not. So it is natural that a t-node initiates the operation. Sufficient information for localization such as the node-ids and locations of r-nodes is supplied by arbitration nodes. This procedure is described in detail in subsection 2.3.

2.3   Operation Procedure

The operation procedure of ScaleLoc is divided into three sessions: arbitration session, ranging session, and reporting session. In an arbitration session, t-node transmits a request message to a-node through a-channel and receives a response to acquire grant for distance measurement. The following pseudo-code presents the operation procedure of ScaleLoc (t-node). Information on m-channels, IDs and locations of r-nodes is attached to grant message as discussed in subsection 2.2.

Collisions may occur in an arbitration session in despite of the use of CSMA-CA in IEEE 802.15.4a protocol. But collisions or waiting for a clear channel do not degrade the performance severely, since arbitration of a t-node and ranging of other t-node can proceed concurrently.

```
t-node ()                // size 2*2 assumed
  while (period for localization) wait
  for(i=0; i<4; i++) {
/*the start of arbitration session*/
    if(a-channel(i)!=FREE)
      Random Back-off;
    else {
      Send a Request message through a-channel(i);
      if(!ack message received)
        Random Back-off;
      else {
        Extract m-channel, IDs and positions of r-node;
/*the end of arbitration session*/
/*the start of ranging session*/
        for(j=0; j<4; j++)
          Measure distance to r-node(j);
        Calculate my location;
/*the end of ranging session*/
/*the start of reporting session*/
        if(a-channel(i)!=FREE)
          Random back-off;
        else
          Send location to a-node through a-channel(i)
/*the start of reporting session*/
      }
    }
  }
end.
```

In a ranging session, t-node, which was authorized through the arbitration session, communicates with r-nodes through m-channel to measure distances. T-node measures the distance from 4 r-nodes that are located in vertices. Trilateration is used to calculate its own location based on measured distances. In a reporting session, the calculated location is transmitted to the control center for monitoring through a-channel.

3   Simulation Results

3.1   Performance Metric

Many performance metrics can be used to evaluate RTLS's. In this paper, average localization error is used as performance metrics. Localization error is defined as the difference between the real and estimated location.

The localization error representatively stems from a ranging error. It depends on the applied technology. If we focus on TWR of IEEE 802.15.4a protocol, major factors include clock offset, clock drift, multipath fading, etc. Since an analysis of ranging error is out of the scope of this study, it is assumed that distances are measured ideally, i.e. ranging error is zero in our performance evaluation. RTLS's for mobile target based on TWR of IEEE 802.15.4a protocol may suffer from another type of error even with the ideal ranging. For trilateration, at least three distances to reference points (r-nodes in ScaleLoc) must be determined. This means that at least three measurements are carried sequentially, and there may be some time interval between consecutive measurements while the target is moving. Fig. 2 shows this scenario graphically. Fig. 2(a) shows an ideal case of a target is in stationary position or four measurements are carried out concurrently with ideal ranging. Four circles intersect on a point, and the location may be calculated with trilateration. On the other hand, when a target moves as in Fig. 2(b), the circles may not intersect on a point. In this case some calculation or estimation is needed to determine the location. This type of error is named 'mobility error' in this paper.

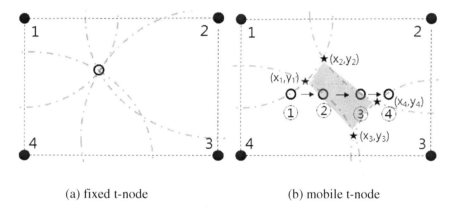

(a) fixed t-node   (b) mobile t-node

Fig. 2. The trilateration of fixed and mobile t-node with ideal ranging

Mobility error is defined as the geographical distance between the estimated position and real position, the center between starting point (▢ in Fig. 2) and ending point (④ in Fig. 2). We can easily imagine that mobility errors become severe as the speed or interval of measurements increase.

3.2   Simulation Results

The performance metrics of the proposed ScaleLoc is analyzed using simulation. For the simulation, Modeling of ScaleLoc is performed using DEVS formalism that describes a discrete event system using hierarchical and modular method [8]. DEVS model can be easily simulated with DEVS formalism in an abstract simulator. In this paper, ScaleLoc is simulated using DEVSim++ [9] which is based on C++ language.

To evaluate the performance of ScaleLoc, those of ScaleLoc and non-arbitration grid locating system (NaGLS) are simulated and compared. In NaGLS, reference nodes are located at every vertex of grids without an arbitration node. Thus, one communication channel is used for measuring distance. The same ranging and location methods are used for both systems.

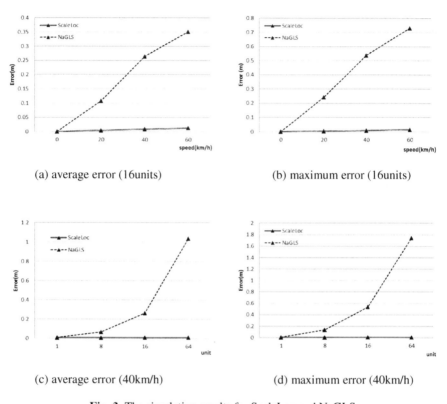

(a) average error (16units)   (b) maximum error (16units)

(c) average error (40km/h)   (d) maximum error (40km/h)

Fig. 3. The simulation results for ScaleLoc and NaGLS

The time interval between reception and transmission of a communication packet is assumed to be zero. The back-off time is based on the value in IEEE 802.15.4a and the time for waiting response is configured as 100msec. We compare the average and maximum value of performance metrics that are simulated in several times. The speed of t-nodes varies respectively such as 0, 20, 40 and 60km/h. The population of t-nodes in a micro-cell is also simulated with 1, 8, 16 and 64 respectively. Simulation results show that performances NaGLS is much more sensitive to the speed and the number of t-nodes, compared to those of ScaleLoc. ScaleLoc can be installed in construction site and the indoor parking lot to monitor workers or vehicles.

4   Conclusion

In this paper, an inexpensive, low-power and scalable RTLS, ScaleLoc is proposed and its performances are investigated by simulation. In order to meet the requirements, ScaleLoc is developed based on IEEE 802.15.4a. ScaleLoc can monitor mobile targets even in indoor environments. The system may be constructed just by arranging macro-cells as needed. And the arbitration scheme is adopted to limit the mobility error. While the average error of NaGLS is 0.263m When the number of t-node is 16 and speed is 40km/h, ScaleLoc's is 0.009m. Simulation results show that performances of ScaleLoc are much more insensitive to the speed and the number of t-nodes. ScaleLoc can be installed in construction site and the indoor parking lot to monitor workers or vehicles.

Acknowledgments. This study was supported by research funds from Guwon Scholarship Foundation.

References

1. Ni, L.M., Liu, Y., Lau, Y.C., Patil, A.P.: LANDMARC: Indoor Locating Sensing Using Active RFID. In: The First IEEE International Conference on Pervasive Computing and Communications, pp. 407–415 (2003)
2. Priyantha, N.B., Chakraborty, A., Balakrishnan, H.: The Cricket Location-Support System. In: 6th ACM International Conference on Mobile Computing and Networking, Boston (2000)
3. Bahl, P., Padmanabhan, V.N.: RADAR: An In-Building RF-based User Location and Tracking System. In: 9th Annual Joint Conference of the IEEE Coumputer and Communication Societies, vol. 2, pp. 775–784 (2000)
4. Fontana, R.J., Gunderson, S.J.: Ultra-wideband precision asset location system. In: Digest of IEEE conference on Ultra Wideband Systems and Technologies, pp. 147–150 (2002)
5. Harter, A., Hopper, A., Steggles, P., Ward, A., Webster, P.: The Anatomy of a Context-Aware Application. Jounal of Wireless Networks 8, 187–197 (2002)
6. Mehmood, H., Tripathi, N.K., Tipdecho, T.: Indoor Positioning System Using Artificial Neural Network. J. Computer. Sci. 6, 1219–1225 (2010)
7. IEEE Computer Society, Part 15.4: Wireless Medium Access Control (MAC) and Physical Layer (PHY) Specifications for Low-Rate Wireless Personal Area Networks (WPANs) Amendment 1: Add Alternate PHYs (2007)
8. Zeigler, B.P.: Multifacetted Modeling and Discrete Event Simulation. Academic Press, Lodon (1984)
9. Kim, T.G.: DEVSim++ User's Manual: C++ Based Simulation with Hierarchical Modular DEVS Model, Computer Engineering Lab. Dept. of Electrical Enginerring, KAIST (1994)

Research on Advanced Performance Evaluation of Video Digital Contents

Seok-Hoon Kim and Byung-Ryul Aan

Department of Electrical and Computer Engineering, SungKyunKwan University,
Seoul, Republic of South Korea
{shkim,byahn}@cpcmail.or.kr

Abstract. The purposes of this study are to present performance evaluation improvements of filtering technology needed for technological protection measures as a means of copyright protection technology of a variety of contents distribution environment and to draw test experiment evaluation results targeting P2P· Web-hard· Torrent service etc. Though the filtering techniques different by contents genre such as a sound source stabilized image· video· text· games· SW etc. have been proposed, this study attempts to propose improved performance evaluation measurement items for performance evaluation of existing video filtering technology, limiting to the field of video. This study attempts to define performance evaluation items of existing filtering such as tenacity and reliability, time taken for feature point extraction and recognition, consistency on providing same results, minimum unit needed for recognition, filtering on random part, accuracy for time information and time precision, media type classification, size of feature information, size of feature DB according to evaluation item by improved evaluation method and to conduct a performance evaluation experimental test for technical measurement against data set needed for evaluation, methods for quantization of evaluation results.

Keywords: DRM, Filtering, Hash, Watermark, Evaluation, Digital Contents, DNA, Tenacity, Reliability.

1 Introduction

As DRM(Digital Rights Management) which was previously the most powerful means protecting digitized various contents[1] becomes non-DRM, contents filtering technology, one of the alternative technologies that can protect contents, is being applied [3]. The filtering technology recognized as an alternative technology of technological protection measures to protect copyrights is applied as the copyright protection technology of online service providers of special types (Web-hard, P2P, Torrent service etc.) making contents sharing as a major objective.

Since the contents filtering service is applied as other types (Keyword, Hash, Feature Point, Contents-DNA etc.) and algorithm (Negative filtering algorithm, Positive filtering algorithm etc.), provision of applicable test experiment environment and objective assessment items should be preceded for meaningful performance evaluation.

G. Lee, D. Howard, and D. Ślęzak (Eds.): ICHIT 2011, CCIS 206, pp. 104–112, 2011.

This study attempts to present improved performance evaluation items and criteria, evaluation methods etc. to systematically evaluate technical performance on contents filtering of technological protection measures through previous literature review of video filtering technology to evaluate performance of the video contents filtering service not dependent on specific technologies [5, 6].

2   Video Filtering Technology

Filtering in copyright protection technology collectively calls several methods to recognize specific contents in online service or restrict access to shared contents of users to prevent distribution of illegally shared works [8, 9]. That an online service provider deletes contents through his own search can be called filtering and preventing users from searching contents by setting search forbidden words in P2P service is also filtering. There are many methods in filtering method but the scope of this study among them is audio and video filtering technology currently widely used [2]. The contents filtering technology is the technology to build meta information DB by extracting unique features from digital contents and to filter specific contents by searching contents and then identifying them in DB as shown below. For video contents filtering, features for identifying contents are needed and these features should satisfy the following properties [10, 11].

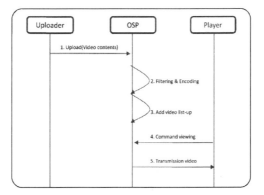

Fig. 1. Video filtering technology flow diagram

Tenacity: Features should represent similarity for perceptually similar contents. That is, similar features should be retained for changes such as compression, re-sampling etc. maintaining perceptual characteristics of original contents.

Mutual independence: Contents with different information should have different features and same but modified contents should have features with very high similarity.

Search efficiency: The features within database should be made by considering structure for efficient search.

To configure contents filtering system based on features satisfying the above characteristics, first, the process creating feature DB is required and after this process, identification of contents is possible. Contents filtering technology means a technology analyzing contents into hash function or content-based feature-point and making DB and then by searching DB by same feature point for random contents and recognizing contents, blocking or allowing transfer according to the result.

Fig. 2. Feature DB creation and search

Contents filtering using hash function have advantages of simple embodiment and fast recognition speed but hash values different according to small transformation of contents make it to recognize. Therefore, a method using content based feature point strong to transformation is mainly used [4].

3   Finger Print Based Filtering Technology

3.1   Watermark Based Finger Print Filtering

Water mark based finger print is identification information insertion method and is the technology proving a contents illegal distributor to be tracked by inserting information of a contents purchaser into contents. It is very similar to digital water marking but inserted information is different [7]. There is a disadvantage that since water mark information should be already inserted in the contents creation process, water mark information inserted in all contents should exist for an experiment.

3.2   Feature-Point Based Finger Print Filtering

This is the method to find out whether two contents are identical or similar by finding feature point in contents of music, video etc. and saving it in DB and then comparing it with feature point extracted from other contents [5]. The feature point used here means frequency or screen conversion Information, location information, color information of a sound source etc. the unique characteristics that music or video files have and this is called sound source DNA, video DNA. Like checking the identity of a person through a fingerprint [10], such contents can be recognized objectively through feature point of contents.

4   Video Filtering Performance Evaluation Method

4.1   Tenacity

Even after digital video contents are transformed variously, feature information similar to the original should be able to extract and the kinds of distortion (transformation) of video contents to evaluate this tenacity are as follows.

- Text & Logo overlay
- Compression, Re-compression, Multi-tier compression
- Resolution change
- Frame rate change
- Capturing on camera
- Color to monochrome conversion
- Brightness change.

4.2   Reliability

It means accuracy of video contents filtering technology and consists of items to evaluate accuracy on recognition results and distinction degree with other contents and performance on algorithm etc.

- Recognition rate (%): As a result of search, rate representing information of question video exactly from feature DB into results

$$Ra = \frac{a}{N} \times 100, \quad R_{a_r} = \frac{\left(\Sigma_{i=1}^{n} \frac{a_i}{N_i} \times 100\right)}{n}$$

Ra: Recognition rate
a: Number of question video detected exactly
N: Number of question video
R_{a_r}: average recognition rate
a_i: Number of question video detected exactly
N_i: number of question video

- False positive rate (%): As a result of search, rate representing information different from question video into results.

$$R_f = \frac{f}{N} \times 100, \quad R_{f_r} = \frac{\left(\Sigma_{i=1}^{n} \frac{f_i}{N_i} \times 100\right)}{n}$$

R_f: False positive rate
f: Number of false recognition detection video
N: Number of question video
R_{f_r}: Average false positive rate
f_i: number of false recognition detection video
N_i: Number of question video

- False negative rate (%): As a result of search, rate representing the not recognizing result though there is information of question video in feature DB.

$$R_m = \frac{m}{N} \times 100, \quad R_{m_r} = \frac{\left(\Sigma_{i=1}^{n} \frac{m_i}{N_i} \times 100\right)}{n}$$

R_m: False negative rate
m: Number of undetected video
N: Number of question video
R_{m_r}: average recognition disallowance rate
m_i: number of undetected video
N_i: number of question video

4.3   Processing Speed

To evaluate processing speed, the following two evaluation methods are used.

• **Feature information extraction time:** Measures feature information extraction time including decoding about video based on 640x480 resolutions.

• **DB search time:** Measures time taking search results for question video in 1000 hours-sized DB.

The processing speed of recognition technology consists of feature point extraction time and search time until returning recognition results. It indicates the average time for a certain size of audio contents set as ms unit. Video DB search time compares based on time taken in comparing question video of three minutes long and 50 hours of database by referring to requirements of video detection algorithm suggested in MPEG Video Group. Evaluation score is calculated as follows. (t: time required (second))

$$100 : t < 1, \qquad 90 - (t-1)\frac{20}{60} : 1 \le t < 60,$$

$$70 - (t-1)\frac{50}{4 \times 60} : 60 \le t < 3, \qquad 0 : 300 \le t$$

4.4   Other Evaluation Criteria

The evaluation methods for consistency, media type, minimum unit, partial matching, time and information amount etc. are as follows. The evaluation methods for consistency, media type, minimum unit, partial matching, time and information amount etc. are as follows.

• **Consistency:** It measures whether feature information extracted from random location of video identifies the video exactly. Since consistency is the condition that same feature value should be extracted in repetitive feature point extraction for random audio, a method to compare random audio by extracting more than 10 times of features is used. Consistency evaluation indicator is calculated as follows.

Consistency evaluation indicator = Number of groups with different characteristics/ 10

If it is a technology that consistency is maintained perfectly, the value is 1/10. Although it is the same question video set targeting database of reference video, n times of detection experiments are conducted changing input order. Same recognition rate, False positive rate, False negative rate are measured.

$$\text{Consistency evalutation score} = \frac{C_a + C_f + C_m}{3nN} \times 100$$

C_a: consistency evaluation score
C_f: Number of question video with same false positive rate
C_f: Number of question video with same False negative rate

• **Media type:** It determines that media using for feature information extraction of video is audio or video (pure image) or the combination of audio and video.

• **Minimum unit:** It measures the size of question video for search as time unit (sec, min). It indicates the minimum unit of audio contents needed for recognition as sec. Targeting many h second long question video, new question video is made (N x r) increasing 1 second unit from the starting point and recognition rate is measured after carrying out extraction rate against the whole reference video. If the time when recognition rate becomes more than 98% for the first time, is after t second from the starting point of question video, the minimum unit of the original work needed for feature extraction is evaluated as t second.

The evaluation score of minimum unit evaluation item can be calculated as follows.

$$100 - \frac{t-1}{h}$$

t: minimum time for recognition (second)
h: full time of question video (second))

• **Partial matching:** To evaluate partial matching, the following evaluation methods are used. It is calculated as matching rate (%) expressing the degree of partial matching for an original song. It is the evaluation conducting recognition at other random points, not the location of contents whose feature points are registered and after cutting random length about a certain size of audio contents set from the start, it is configuration test set again. The length should be between 1 sec and 10 sec.

• **Time accuracy:** Time accuracy expresses tolerance that question video found location information on temporal axis of original video as time unit. As evaluation scale for the location of question contents in audio contents found by recognition result, error values for the location of actual original contents and recognition location are represented by ms. Setting the tolerance of time information as $|\varepsilon|$ and evaluation score of time accuracy is calculated as follows.

$$100 : 0 < |\varepsilon| \le \frac{1}{60},$$
$$90 : \frac{1}{60} < |\varepsilon| \le 0.5,$$
$$70 : 0.5 < |\varepsilon| \le 5,$$
$$50 : 5 < |\varepsilon| \le 5,$$
$$50 - (|\varepsilon| - 30) \times \frac{50}{270} : 30 < |\varepsilon|,$$
$$0 : 300 < |\varepsilon|$$

• **Time precision:** Time precision is expressed as minimum time unit (ms, sec and min) that question video expresses location information on temporal axis of original video. It is an indicator expressing to what extent location of question contents can be represented precisely in audio contents found by recognition results and is expressed as sec or ms unit according to technology. It evaluates the degree of finding exact location of video in reference video (100: frame unit, 90: second unit, 50: minute unit).

• **Feature point information amount:** Feature point information amount uses the following evaluation methods. The size of feature point extracted from original contents (findings showed that 3~5Kb are appropriate) is evaluated by Kbyte and for size of extracted feature information, feature point is extracted for a certain size of audio contents set and the total size is expressed as KB. Feature point DB size uses the following evaluation method. The size of feature point DB extracted from 1,000-hour quantity of original contents is evaluated as M-byte. For size of certain DB for recognition, after extracting feature point for a certain size of audio contents set and configuration DB, the total size is expressed as KB. If there is a memory-residing portion for search, the size is also expressed by KB and added. Based on 1 hour long video, compare the sizes of feature information. Evaluation measures are as follows. (s: file size (Kb))

$$100 : s \leq 50,$$

$$80 : 50 < s \leq 500,$$

$$50 : 500 < s \leq 1000,$$

$$20 : 1000 < s$$

4.5   Performance Evaluation Result Calculation

Evaluation figure by each performance evaluation item is calculated as the follow equation. Evaluation index per sub classification:

$$\{S(i)\} = \{S(1), S(2), S(3), \ldots\ldots, S(n-1), S(n)\}$$

Middle classification evaluation figure is calculated after calculating evaluation index of sub classification evaluation elements and then giving weight by middle classification and evaluation index by top classification is calculated as follows. Evaluation result by each item

$$M_i(tot) = \left\{ \sum_{i=1}^{a} \frac{S_i(a)}{a} + \sum_{i=1}^{b} \frac{S_i(b)}{b} + \ldots + \sum_{i=1}^{x} \frac{S_i(x)}{x} \right\}$$

5   Experimental Test of Video Filtering Technology Performance

Database for video contents filtering performance evaluation is composed of video files of the following specification. In this test, original video contents experimental environment is as follows.

· Video contents for broadcasting: Color/FullHD (1280x720)/1.5Mbps/ over 30fps /MPEG2
· Feature DB is composed of feature information extracted from over 1,000 hours of video files.
· More than 1,000 hours of video composing feature DB consists of at least 5 different genres by genre among the following configuration.

The evaluation test for video filtering technology performance evaluation is limited to limited number of four kinds of scenarios and distorted environment was assumed by shooting original video with camcorder and a variety of transformation attacks for result format were attempted and the number of cases of the experiments is as follows.

- Test_case_1:   Projector camcorder shooting + MPEG2 compression
- Test_case_2:   Test_case_1 + compression format conversion (MPEG1/4)
- Test_case_3:   Test_case_1 + frame rate conversion (30fps → 24fps)
- Test_case_4:   Test_case_1 + size conversion (1280x720/ 800x600/640x480)

	Case_1	Case_2		Case_3	Case_4		
		MPEG-1	MPEG-4		1280*720	800*600	640*480
Sample1	O(79.8%)	O(78.8%)	O(79.1%)	O(75.5%)	O(76.5%)	O(70.2%)	O(54.0%)
Sample2	O(98.0%)	O(60.5%)	O(58.5%)	O(84.3%)	O(84.3%)	O(73.9%)	O(72.5%)
Sample3	O(83.8%)	O(78.3%)	O(49.4%)	O(66.6%)	O(92.7%)	O(84.4%)	O(67.2%)
Sample4	O(99.7%)	O(99.7%)	O(74.9%)	O(98.3%)	O(92.5%)	O(68.5%)	O(49.2%)
Sample5	O(100%)	O(99.0%)	O(96.1%)	O(46.8%)	O(96.8%)	O(72.6%)	O(77.4%)
Sample6	O(98.3%)	O(83.1%)	O(71.9%)	O(69.2%)	O(71.2%)	O(73.2%)	O(70.5%)
Average	**93.3%**	**83.2%**	**71.7%**	**73.5%**	**85.7%**	**73.8%**	**65.1%**

O: All inserted 40bits are detected (detection rate, %),

$$\text{Detection rate}(\%) = \frac{2\text{bit} - \text{Number of detection success}}{2\text{bit} - \text{Number of detection attempts}}$$

6   Conclusion

This study attempts to prepare measures for evaluation measurement of filtering based technology being applied as a means of technological measures for an online service provider to protect contents focusing on video among various contents such as music, movies, broadcasting, publication, games, cartoon, SW etc. If video A is distorted and the result B is created, the accuracy to recognize that B is A can be seen as very significant research access point.

Considering that this is the experiment of evaluation item area limited according to each sample video, when attempting transformation attack or conversion to some degree, detection rate for original video of Case1 was the highest   93.3% and in case of frame size conversion, as lowering the screen resolution, average recognition detection result was measured (85.7% → 65.1%).

Though they are limited experimental evaluation items, if video filtering technology is measured objectively based on measurement result between each evaluation item, it can be used as data to improve the efficiency of recognition and search and prevent illegal leak and detection of similarity with original video.

References

1. Jonker, H.L.: Security of Digital Rights Management Systems. Master's technische Universisteit Eindhoven (August 2004)
2. Schonberg, D.: Fingerprinting and Forensic Analysis of Multimedia, MM 2004, New York, USA (October 2004)
3. Jean Camp, L.: First Principles of Copyright for DRM Design. In: IEEE Internet Computing. Harvard University, Harvard (2003)
4. Cohen, J.: DRM and privacy. Berkeley Technol. Law J. 18, 575 (2003)
5. Fox, B., Brian La Macchia, B.: Encouraging recognition of fair uses in DRM systems. Communication. ACM 46(4), 61–63 (2003)
6. Burk, D., Cohen, J.: Fair use infrastructure for rights management systems. Harvard J. Law Technol. 15(1), 42–83 (Fall 2001)
7. Multimedia Content Screening using a Dual Watermarking and Fingerprinting System, Multimedia 2002, December 1-6, p. 251. Juan-les-Pins, France (2002)
8. Cheng, Y.: Music Database Retrieval Based on Spectral Similarity. In: Haitsma, J., Kalker, T., Oostveen, J. (eds.) International Symposium on Music Information Retrieval (ISMIR) 2001, Robust Audio, Bloomington, USA (October 2001)
9. Hashing for Content Identification, Content Based Multimedia Indexing 2001, Brescia, Italy (September 2001)
10. Haitsma, J., Kalker, T., Oostveen, J.: Robust Audio Hashing for Content Identification, In: Content Based Multimedia Indexing 2001, Brescia, Italy (September 2001)

High Resolution Image Reconstructed by ARPS Motion Estimation and POCS

Wonsun Bong and Yong Cheol Kim

Department of Electrical and Computer Engineering, University of Seoul, Korea
gaam@uos.ac.kr, yckim@uos.ac.kr

Abstract. In POCS (projection onto convex sets)-based reconstruction of HR (high resolution) image, the quality of reconstructed image is gradually improved through iterative motion estimation and image restoration. The amount of computation, however, increases because of the repeated inter-frame motion estimation. In this paper, an HR reconstruction algorithm is proposed where modified ARPS (adaptive rood pattern search) and POCS are simultaneously performed. In the modified ARPS, the motion estimates obtained from phase correlation or from the previous steps in POCS reconstruction are utilized as the initial reference in the motion estimation. Moreover, estimated motion is regularized with reference to the neighboring blocks' motion to enhance the reliability. Computer simulation results show that, when compared to conventional methods which are composed of full search block matching and POCS restoration, the proposed method is about 30 times faster and yet produces HR images of almost equal or better quality.

1 Introduction

A high resolution (HR) image can be reconstructed by interpolating the motion-calibrated low resolution (LR) frames. In normal case, interpolated images are unnatural and deteriorative. A better-quality HR image can be obtained by eliminating blurs and noise in the LR frames and then integrating them. In the reconstruction of a HR image out of several LR frames, the redundant information spread over frames is fused into a single image. From the standpoint of HR reconstruction, each LR image frame can be regarded as a blurred version of a single HR image, with minor difference due to the inter-frame motion. We propose an algorithm that reconstructs a HR image by recovering details in the image, out of a sequence of LR frames.

There have been several works on HR reconstruction from an LR image sequence. Among them are non-uniform interpolation[1], stochastic regularization [2,3,4], reconstruction in frequency domain[5] and POCS [1,6,7]. POCS utilizes some characteristics of a real-world scene as constraints to reconstruct HR image. Stark and Oskoui[8] applied POCS method to the reconstruction of HR image. Tekalp[6,7] extended their work into restoring of an image which is noisy and blurred with motion.

G. Lee, D. Howard, and D. Ślęzak (Eds.): ICHIT 2011, CCIS 206, pp. 113–120, 2011.

In the reconstruction of a HR image, the accuracy of the inter-frame motion in LR frames is a critical factor. Motion estimated over the reconstructed frames with finer details would be better than estimates over the coarse LR frames. With this motivation, Mateos[2] and Hardie[3] proposed a Baysian method of estimating motion vectors and HR image simultaneously from a compressed LR sequence.

POCS-based reconstruction produces a high quality HR image provided that the estimated motion is accurate. Feature matching, optical flow and phase correlation are frequently used motion estimation techniques. Block matching algorithm (BMA) has been widely used in extracting the motion vectors for video compression, due to the acceptable performance and reasonable complexity [9]. BMA, however, are heavily time-consuming and prone to error. In this paper, we propose a way of integrating POCS-based reconstruction and motion estimation by phase correlation and adaptive rood pattern search(ARPS) [10].

Though ARPS is fast in motion estimation, its accuracy is highly dependent on the initial reference motion vector. In our work, we perform phase correlation to get a rough estimate of global inter-frame motion. Then, the motion estimates from phase correlation is used as the initial reference motion vector. In order to improve the reliability, the error-prone motion vectors are regularized with the constraints of the motion vectors of the neighboring pixels. Experimental results show that ARPS with phase correlation reconstructs a HR image with better quality than ARPS alone.

The organization of this paper is as follows. In Sec. 2, the model between HR and LR is described with POCS method in Sec. 2. The proposed HR reconstruction method is described in Sec. 3. Experimental results are presented in Sec. 4. Finally, we draw a conclusion in Sec. 5.

2   HR POCS from LR

Let's assume that the LR images we get from a LR sensor are the blurred version of the natural scenes, which are originally HR frames. A HR image sequence is composed of K consecutive frames each of which is of $(L_1M \times L_2N)$ size.

$f_k(u, v)$ denotes the k-th frame of the HR sequence. The k-th LR frame $g_k(m, n)$ is obtained by first blurring $f_k(u, v)$ and then downsampling it with a factor of $L_1 \times L_2$. The blurring can be modeled as a low pass filter. Eq. (1) denotes the vector expressions of the $f_k(u, v)$ and $g_k(m, n)$.

$$\mathbf{f}_k = [f_{1:k}, f_{2:k}, \ldots, f_{L_1ML_2N:k}]^t$$
$$\mathbf{g}_k = [g_{1:k}, g_{2:k}, \ldots, g_{MN:k}]^t, k = 1, 2, \ldots, K \tag{1}$$

Let the r-th HR frame be the reference frame, which is to be reconstructed. Let $d_{r,k}(i)$ denote the motion vector for the i-th pixel in the k-th HR frame to the corresponding pixel in the r-th HR frame.

The formation of $\mathbf{g}_k$ from $\mathbf{f}_r$ is shown in Eq. (2). $\mathbf{M}(\mathbf{d}_{r,k})$ is the motion compensation operator from $\mathbf{f}_r$ to $\mathbf{f}_k$. $\mathbf{B}$ is the $(L_1ML_2N \times L_1ML_2N)$ operator for sensor blurring. $\mathbf{D}$ denotes the $(MN \times L_1 M L_2 N)$ operator for down sampling. $\mathbf{n}_k$ is a Gaussian noise of zero-mean and a variance of σ_n^2.

$$\mathbf{g}_k = \mathbf{D} \cdot \mathbf{B} \cdot \mathbf{M}(\mathbf{d}_{r,k})\mathbf{f}_r + \mathbf{n}_k \tag{2}$$

POCS is a widely used technique to restore unknown signal by particular properties of desired signal[8]. The process of POCS is summarized in Eq. (3).

P_j denotes the projection operator onto the j-th convex set and λ_j is relaxation parameter which adjusts the convergence rate. $\mathbf{I}$ represents the identity operator. The initial signal, $\mathbf{f}^{(0)}$, is iteratively projected and $\mathbf{f}^{(n)}$ is the n-times projected version of $\mathbf{f}^{(0)}$.

$$\begin{aligned}
\mathbf{f}^{(n+1)} &= T_m T_{m-1} \ldots T_1 \mathbf{f}^{(n)}, \\
T_j &\equiv \mathbf{I} + \lambda_j(P_j - \mathbf{I}), \ 0 < \lambda_j < 2, \ 1 \le j \le m
\end{aligned} \tag{3}$$

The projection operator $r_k(m,n)$ represents the deviation between g_k and the synthetic LR from the degradation process. The projection operator $P_{m,n;k}$ $[f_r(u,v)]$ gradually drives $|r_k(m,n)|$ to stay within δ_0[6,7]. For details, please refer to [9].

$$C_{m,n;k} = \{f_r(u,v) : |r_k(m,n)| \le \delta_0\} \tag{4}$$

$$r_k(m,n) = g_k(m,n) - \sum_u \sum_v h_k(m,n;u,v)f_r(u,v) \tag{5}$$

$$P_{m,n;k}[f_r(u,v)] = \begin{cases} f_r(u,v) + \frac{r_k(m,n)-\delta_0}{\sum_o \sum_p h_k^2(m,n;o,p)} h_k(m,n;u,v) & \text{if } r_k(m,n) > \delta_0 \\ f_r(u,v) & \text{if } |r_k(m,n)| < \delta_0 \\ f_r(u,v) + \frac{r_k(m,n)+\delta_0}{\sum_o \sum_p h_k^2(m,n;o,p)} h_k(m,n;u,v) & \text{if } r_k(m,n) < -\delta_0 \end{cases} \tag{6}$$

3   Motion Estimation by Regularization

3.1   Phase Correlation and ARPS

In phase correlation, motion estimation is obtained from the phase difference since a displacement in spatial domain produces a phase offset in frequency-domain. Phase correlation provides reliable results when the dominant motion is global translation. When local motion is non-negligible, phase correlation does not produce reliable motion vectors. This deficiency of phase correlation can mitigated by integrating ARPS and registration of aliased image [11]. When motion from phase correlation is used as the initial reference motion vector in ARPS, the accuracy of ARPS is highly improved.

ARPS is fast and the accuracy of motion estimation is comparable to that of BMA. ARPS consists of two phases. The first phase searches for reference motion by using large diamond search pattern (LDSP). In the second phase,

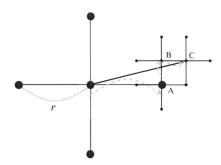

Fig. 1. Gradual motion estimation in ARPS

minute motion is estimated by applying small diamond search pattern (SDSP), as shown in Fig. 1. In LDSP, the values of sum of squared difference (SSD) for five candidate points are compared. The point with the smallest SSD (point A) is chosen as the estimated motion. Then, detailed motion B is searched for by applying SDSP to point A. Again, detailed motion C is searched for by applying SDSP to point B. This process is repeated until the newly found point does not produce smaller SSD than the previous step.

In this paper, we do not perform LDSP. Instead, the reference motion provided to SDSP is the motion vector from phase correlation or the motion vector reconstructed in the very previous step of POCS.

3.2   Motion Regularization

The error in the estimated motion vector can be regularized with smoothness assumption. In Eq. (7) and Eq. (8), l represents the block number and $\bar{\mathbf{d}}_{l;r,k}$ is the mean motion vectors of the neighbors of the l-th block.

$$\hat{\mathbf{d}}_{l;r,k} = \arg \min_{\mathbf{d}}\{L\} \tag{7}$$

$$L = \|\mathbf{DBM}(\mathbf{d})\mathbf{f}_r(l) - \mathbf{g}_k(l)\|^2 + \gamma\|\mathbf{d} - \bar{\mathbf{d}}_{l;r,k}\|^2 \tag{8}$$

The first term in Eq. (8) is the error between the motion compensated block of the degraded image and its corresponding block in g_k. The second term is the deviation between the block motion vector and the mean motion vector of the neighboring blocks. Fig. 2 shows the proposed HR reconstruction algorithm. Large blocks and small blocks represent HR images and LR images, respectively. Summary is as follows:

Step 1. Get the initial HR, $\mathbf{f}_k^{(0)}$, by bilinear interpolation of LR $\mathbf{g}_k$.
Step 2. Estimate the motion vector by ARPS between frames $\mathbf{f}_k$ and $\mathbf{f}_{k+1}$.
Step 3. Compute the residual between $\mathbf{f}_k^n$ and LR frames by Eq. (5) and project $\mathbf{f}_k^n$ onto each block set by Eq. (3).
Step 4. Reconstruct a HR image, $\mathbf{f}_k^{n+1}$, by applying POCS in Eq. (6).
Step 5. Terminate if Eq. (4) is satisfied, or go back to **Step 2**.

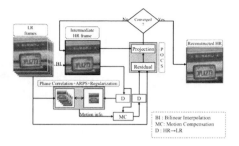

Fig. 2. The process of LR frame formation from of a natural HR image sequence

4   Experimental Results

We tested our HR reconstruction algorithm on two types of LR image sequences. One LR sequence is generated by shifting a HR image with known incremental motion and then downsampling. The other LR sequence is obtained from downsampling a sequence of HR images taken by a digital camera. The performance measure is visual quality inspection and PSNR with reference to the original HR image.

Performance of the proposed method is compared with three other methods. The first method is a simple bilinear interpolation of the LR image (labeled as "BI"). In the second method, motion estimation is by full search of BMA of the initial bilinear interpolated frames and then POCS reconstruction is performed iteratively (labeled as "FS").

In the experiments, the relaxation parameter λ_j in Eq. (3) is set to be 0.1. The threshold δ_0 is 0.01 and the number of POCS iteration is set to 10.

4.1   Text-A

The motion in *Text-A* sequence is pure translation. LR image frames are obtained from blurring and downsampling of HR sequence taken by a digital camera. The downsampling factor is 4. As shown in Fig. 3, the proposed method produces a HR image with FS quality.

4.2   Mobile

Mobile sequence is CIF-size MPEG test sequence. The motion is pure translation. LR image frames are obtained from blurring by 4 x 4 LPF and then downsampling by a factor of 2. As shown in Fig. 4, the proposed method produces a HR image with FS quality.

4.3   Pentagon

In this experiment, a 384 x 384 image is used as the original HR image, which is obtained from taking some region in the original 1024 x 1024 *Pentagon* image.

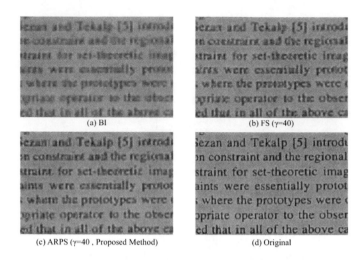

Fig. 3. The reconstructed *Text-A* HR images after 10 iterations

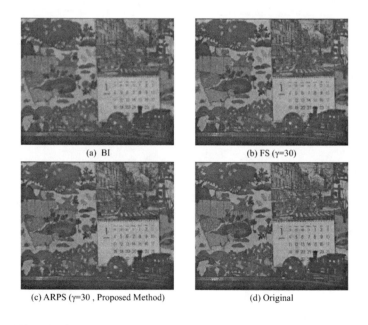

Fig. 4. The reconstructed *Mobile* HR images after 10 iterations

Five frames of LR images are as follows: We blurred the original image with 3×3 uniform LPF, added Gaussian noise with 30 dB SNR and then downsampled with a factor of 3. The 128×128 LR frames have displacements of $(0,0)$, $(2/3,2/3)$, $(1,4/3)$, $(4/3,4/3)$ and $(4/3,2)$ in LR resolution with respect to the reference image.

(a) BI   (b) FS (γ=80)

(c) ARPS (γ=80, Proposed Method)   (d) Original

Fig. 5. The reconstructed *Pentagon* HR images after 10 iterations

Fig. 5 shows a comparison of the reconstructed HR images, after 10 iterations. The proposed method produces better visual quality than "BI" and produces a HR image with FS quality, especially at the region boundaries.

5 Conclusions

Accuracy of the estimation of inter-frame motion is a crucial factor in HR image reconstruction. With a motivation that motion over the reconstructed frames would have finer details, we developed a POCS-based HR reconstruction algorithm, where the motion vectors are estimated from the HR frames which are being reconstructed. In this paper, we used the motion estimate from phase correlation as the initial reference motion in ARPS. As a result, the computation time for motion estimation is reduced to 1/30 of the full search BMA.

We tested our method on LR frames with known motion and with unknown motion. Experimental results show the proposed method has a better performance both in visual quality and PSNR of the reconstructed HR frames. Hence, we conclude that the proposed ARPS with motion regularization is very effective for HR reconstruction.

References

1. Borman, S., Stevenson, R.: Spatial resolution enhancement of low-resolution image sequences: A comprehensive review with directions for future research. Lab. Image and Signal Analysis, University of Notre Dame, Tech. Rep. (July 1998)
2. Mateos, J., Katsaggelos, A., Molina, R.: Simultaneous motion estimation and resolution enhancement of compressed low resolution video. In: Proceedings of International Conference IEEE Image Processing, 2000, vol. 2, pp. 653–656 (September 2000)

3. Hardie, R., Barnard, K., Armstrong, E.: Joint map registration and high-resolution image estimation using a sequence of undersampled images. IEEE Trans. on Image Processing 6, 1621–1633 (1997)
4. Schultz, R., Stevenson, R.: Extraction of high-resolution frames from video sequences. IEEE Trans. on Image Processing 5, 996–1011 (1996)
5. Kim, S., Bose, N., Valenzuela, H.: Recursive reconstruction of high resolution image from noisy undersampled multiframes. IEEE Trans. on Acoustics, Speech and Signal Proc. 38, 1013–1027 (1990)
6. Patti, A., Sezan, M., Tekalp, M.: Super resolution video reconstruction with arbitrary sampling lattices and non-zero aperture time. IEEE Trans. on Image Processing 6, 1064–1078 (1997)
7. Tekalp, A., Ozkan, M., Sezan, M.: High-resolution image reconstruction from lower-resolution image sequences and space-varying image restoration. In: IEEE ICASSP 1992, vol. 3, pp. 169–172 (1992)
8. Stark, H., Oskoui, P.: High-resolution image recovery from image-plane arrays, using convex projections. J. Opt. Soc. Amer. A 6, 1715–1726 (1989)
9. Kim, B., Song, H., Kim, T., Kim, Y.: Reconstruction of an hr image by simultaneous pocs and regularized block matching. In: SPIE, vol. 6794 (2007)
10. Nie, Y., Ma, K.: Adaptive rood pattern search for fast block-matching motion estimation. IEEE Trans. on Image Processing 11, 1442–1448 (2002)
11. Vandewalle, P., Sbaiz, L., Susstrunk, S., Vetterli, M.: Registration of aliased images for super-resolution imaging. In: SPIE, vol. 6077, pp. 13–23 (2006)

Implementation of a Real-Time Image Object Tracking System for PTZ Cameras

Sang-Gu Lee[1] and R. Batkhishig[2]

[1,2] Department of Computer Engineering,
Hannam University, Daejon, 306-791, Korea
sglee@hnu.kr

Abstract. In this paper, we implement a real-time surveillance monitoring and image object tracking system using PTZ (Pan-Tilt-Zoom) camera. For image object tracking, we use the mean shift tracking algorithm based on the color image distribution of detected object. Mean shift algorithm is efficient for real-time tracking because of its fast and stable performance. In this system, MatLab language is used for clustering moving object and accessing the PTZ protocol and RS-485 communication for controlling the position of PTZ cam-era in order to arrange the moving objects in the middle part of the monitor screen. This system can be applied to an effective and fast image surveillance system for continuous object tracking in a wider area.

Index Terms: Image object tracking, PTZ camera, Mean shift algorithm.

1 Introduction

PTZ cameras can control the pan, tilt and zoom operations of the camera lens through a surveillance DVR or computer system. PTZ cameras have the ability to moving right, left, up, down and even zoom. Generally, PTZ cameras cover very large areas compared with fixed cameras which would not be able to do. Therefore, recently PTZ cameras have been used in many applications for security operations of moving object tracking through manually operating or automatic control. In this paper, we implement an image object tracking system for PTZ camera.

The proposed system uses the mean shift tracking algorithm based on the color image distribution of moving object. Mean shift algorithm is very efficient for real-time image tracking because of its fast and stable performance. In this system, we use MatLab language for detecting moving object and accessing the PTZ protocol and RS-485 communication for controlling the position of PTZ camera in order to arrange the position of the moving object in the middle part of the monitor screen. The proposed system can be applied to an effective and fast image surveillance system for continuous object tracking in a wider area.

G. Lee, D. Howard, and D. Ślęzak (Eds.): ICHIT 2011, CCIS 206, pp. 121–128, 2011.
© Springer-Verlag Berlin Heidelberg 2011

The organization of this paper is as followings. Chapter 2 tackled mean shift algorithm. In Chapter 3, the proposed system is discussed. Experimentation and results are discussed in Chapter 4. And finally, Chapter 5 draws a conclusion.

2   Mean Shift Algorithm

Mean shift algorithm is a general non-parametric mode clustering procedure. Mean shift image segmentation has 2 main steps as discontinuity preserving filtering and mean shift clustering [1]. Using Mean shift algorithm for real-time object tracking is reported in [2,3].

Mean shift algorithm is as follows.

Given n data x_i in d dimensional space, the multivariate kernel density estimator $\tilde{f}_{h,k}(x)$ computed at point x is

$$\tilde{f}_{h,k}(x) = \frac{1}{nh^d} \sum_{i=1}^{n} k(\frac{x - x_i}{h}) \tag{1}$$

where h represents the bandwidth.

The estimating the density gradient is represented as

$$\nabla \tilde{f}_{h,k}(x) = \frac{1}{nh^d} \sum_{i=1}^{n} \nabla k(\frac{x - x_i}{h}) \tag{2}$$

The successive locations $\{y_i\}$ of the Kernel G are then

$$y_{j+1} = \frac{\sum x_i g(\left\|\frac{y_i - x_i}{h}\right\|^2)}{\sum g(\left\|\frac{y_i - x_i}{h}\right\|^2)} \tag{3}$$

The procedure of mean shift algorithm for a given point x_i is as follows.

1. Computing the mean shift vector $mv(x_i^t)$
2. Translating density estimation window

$$x_i^{t+1} = x_i^t + mv(x_i^t)$$

3. Repeating step 1 and 2 until convergence

$$\nabla f(x_i) = 0$$

Refer to [1,2, and 3] about the details of the mean shift algorithm.

3   The Proposed System

In the proposed system, images are captured from PTZ camera. These images are transferred to PC, and MatLab program extracts moving object and eliminates the back-ground images. Then, the preprocessing procedures such as filtering and morphological computations (erosion, dilation, open and close operations) are performed. Mat-Lab program send the packets such as P, T and Z data to the PTZ camera. In the monitor, the moving objects are displayed in the middle part of the screen. Fig. 1 shows the entire hardware structure in this system. We use SPD-1000 PTZ Dome camera [4]. A frame size of 320 x 240 image data is used.

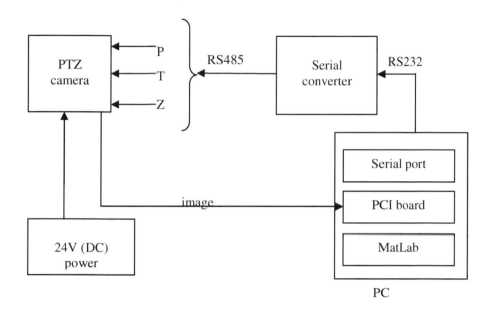

Fig. 1. Entire hardware structure

Table 1 shows the transferring packet format and protocol for controlling the motors of PTZ camera. For example, if we send the packet in the Table 2, pan of PTZ camera is moving to right direction in speed 34. Table 3 shows the commands for PTZ control.

Table 1. Command packet format

Byte 1	Byte 2	Byte 3	Byte 4	Byte 5	Byte 6
STX	Cam ID	Host ID	Cmd-1	Cmd-2	Data 1

Byte 7	Byte 8	Byte 9	Byte 10	Byte 11	
Data2	Data 3	Data 4	ETX	Check-sum	

Table 2. A protocol example

Command	Protocol	Comments
Pan right	A0 01 00 00 02 22 00 00 00 AF DA	Speed 34

Table 3. Commands for PTZ control

Command	Protocol								Comments
OSD ON	A0	01	00	00	B1	00	00	00	Menu On
	00	AF	4D						
One shot AF	A0	01	00	E0	15	00	00	00	
	00	AF	09						
Call Zoom Pos.	A0	01	00	E0	02	00	00	00	
	00	AF	1C						
Call Zoom Pos.	A0	01	00	E0	01	00	00	00	
	00	AF	1D						
Set Preset 3	A0	01	00	00	03	02	00	00	
	00	AF	F9						
Call Preset 3	A0	01	00	00	07	02	00	00	
	00	AF	F5						
Zoom Tele	A0	01	00	00	20	00	00	07	
	00	AF	D7						
Zoom Stop	A0	01	00	00	00	00	00	00	
	00	AF	FE						
Pan Right	A0	01	00	00	02	22	00	00	Speed 34
	00	AF	DA						

4   Experimentation and Results

In this system, the entire flowchart for the PTZ tracking system is shown in Fig.2.

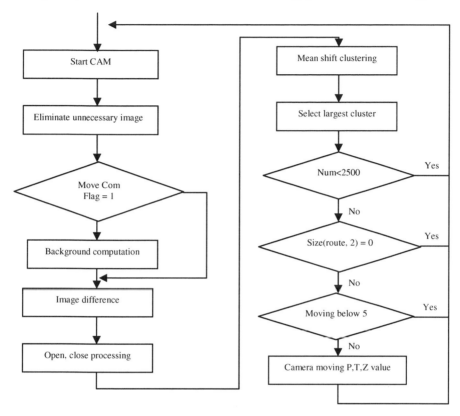

Fig. 2. Flowchart

In this system, PC development environment is in Table 4.

Table 4. Development environment in PC

	Development environment
Hardware	CPU: Pentium (R) D 3.00GHz Memory: 2GB
OS	Microsoft Windows XP Professional SP3
Programming tool	Microsoft Visual Studio 2005, MatLab 7

Fig. 3 shows a GUI window display to transfer packets for controlling motors of PTZ camera.

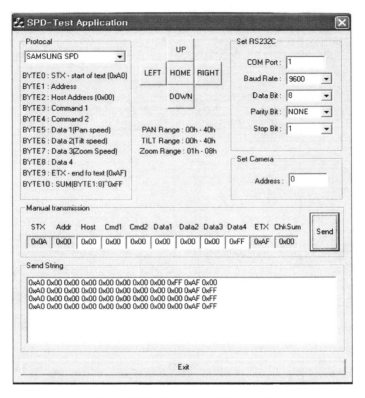

Fig. 3. GUI window for PTZ control

Fig. 4. Image frame data (5 frames interval)

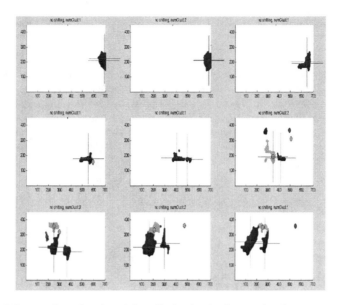

Fig. 5. Image clustering data (after eliminating background and preprocessing)

Fig. 4 shows input image frame data composed of 5-frame interval. We know that a man is walking toward nearly left side in the room. Fig. 5 shows image clustering data after eliminating background and performing image pre-processing. In this Fig.5, a large red cross mark means each cluster's centroid. In this system, we only use the cluster in which total number of pixels is greater than 2500. Fig. 6 shows the trace of moving object tracking.

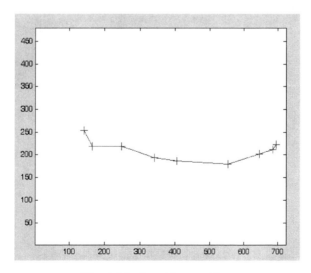

Fig. 6. Moving object tracking

5   Conclusions

In this paper, we implemented an image object tracking system for PTZ cameras. To detect moving object in the surveillance area, we used the mean shift algorithm. By using MatLab language we controlled the PTZ camera and showed the trace of the object moving routes in real-time. Mean shift algorithm has fast, stable and robust performance. For the future study, we will develop a new tracking system that can be utilized in the DSP (Digital Single Processor) 6400 series (especially, 6416) board in order to speed-up the total performance and to build that system in a stand-alone prototype using C++ , OpenCV and LabView tools. As a future research, we will develop a fast mean shift algorithm with robust kernel function and back ground elimination algorithm under the environment of changing lights and moving shadows. Further, we will also implement an integrated smart surveillance camera system having multiple cameras with networking and 3D concepts.

Acknowledgment. This work was supported by Security Engineering Research Center, granted by the Korea Ministry of Knowledge Economy.

References

[1] Sonka, M., Hlavac, V., Boyle, R.: Image Processing, Analysis, and Machine Vision, 3rd edn. Thomson (2008)
[2] Comaniciu, D., Meer, P.: Mean shift: A Robust Approach Toward Feature Space Analysis. IEEE Tr. on PAMI 24(5), 603–619 (2002)
[3] Comaniciu, D., Meer, P.: Kernel-based object tracking. IEEE Trans. PAMI 25(5), 564–577 (2003)
[4] Samsung Techwin, PTZ Dome Camera SPD-1000 Manual (2010)
[5] Lee, S.G., Hwang, S.K.: Implementation of an Object Tracking System using Mean shift Algorithm for PTZ camera. In: IPIU 2011, Jeju, Korea (February 2011)
[6] Nixon, M., Aguado, A.: Feature Extraction and Image Processing, 2nd edn. Academic Press, London (2008)
[7] Jain, A.K., Duin, R.P.W., Mao, J.: Statistical Pattern Recognition: A Review. IEEE Trans. Pattern Analysis and Machine Intelligence 22(1), 4–37 (2000)

Statistical Analyses of Various Error Functions for Pattern Classifiers

Sang-Hoon Oh

Mokwon University, Doan-dong, Seo-gu, Daejon, Korea
shoh@mokwon.ac.kr

Abstract. There are various error functions for pattern classifiers. This paper analyzes the error functions such as MSE(mean-squared error), CE(cross-entropy) error, AN(additive noise) in MSE, MLS(mean log square) error, and nCE(nth order extension of CE) error functions in a statistical perspective. Also, the analyses include CFM(classification figure of merit). The results of analyses provide considerable insights into the properties of different error functions.

Keywords: Classifier, error functions, statistical analysis, optimal solution.

1 Introduction

Pattern classifiers can be implemented using a discriminant function. The functional value corresponds to the degree of confidence that an input pattern belongs to a certain class and the decision of classification is done by selecting the class of maximal discriminant value [1]. Alternatively, the classifiers can be implemented based on the posteriori probabilities and this provides the Bayes classifier [2]. However, it is difficult to estimate the p.d.f.(probability density function) or the probability distribution of samples. The Parzen's window can estimate the p.d.f. of samples by locating the window function at each sample [3]. Still this method requires enough number of samples for the accurate estimation of the p.d.f.

In many cases, the discriminant function approach attains better performance than the posteriori probability approach and this supports the popularity of discriminant functions. Conventionally, MSE(mean-squared error) function is used to train the classifier whose outputs become discriminant values [4]. As a variant of MSE, Wang and Principe proposed the additive noise method in the desired signal of output [5]. In order to deal with outliers of samples, Liano proposed MLS(mean-log square) error function [6]. CE(cross-entropy) error is another error function for performance improvement [7] and nCE(nth order extension of CE) error is a more advanced formulation of CE [8][9]. All these error functions are tried to be minimized when training classifiers. CFM(classification figure of merit) function is another approach to be maximized during training of classifiers [10].

In this paper, the various error functions are analyzed in a statistical way in order to provide insights into the properties of them. Section 2 describes the mathematical analyses and the comparisons among them. Finally, Section 3 concludes this paper.

G. Lee, D. Howard, and D. Ślęzak (Eds.): ICHIT 2011, CCIS 206, pp. 129–133, 2011.
© Springer-Verlag Berlin Heidelberg 2011

2 Statistical Analyses of Various Error Functions

Let $\mathbf{x} = [x_1, x_2, \cdots, x_N]^T$ be an input sample and $\mathbf{y} = [y_1, y_2, \cdots, y_M]^T$ be an output vector of a classifier, of which desired vector $\mathbf{t} = [t_1, t_2, \cdots, t_M]^T$ is coded as follows:

$$t_k = \begin{cases} +1, & \text{if } \mathbf{x} \text{ originates from class } k \\ -1, & \text{otherwise.} \end{cases} \tag{1}$$

Classifiers are trained to minimize the distance between $\mathbf{t}$ and $\mathbf{y}$. MSE [4] defined by

$$E_{MSE}(\mathbf{x}) = \frac{1}{2} \sum_{k=1}^{M} (t_k - y_k(\mathbf{x}))^2 \tag{2}$$

is usually used as a distance measure. In the limit $P \to \infty$, the minimizer of E_{MSE} for all P training patterns converges (under certain regularity conditions, Theorem 1 in [11]) towards the minimizer of the function

$$E\{E_{MSE}(\mathbf{X})\} = E\left\{ \frac{1}{2} \sum_{k=1}^{M} (T_k - y_k(\mathbf{X}))^2 \right\}, \tag{3}$$

where $E\{\cdot\}$ is the expectation operator, T_k is the random variable denoting the desired value, and $\mathbf{X}$ is the random vector denoting the input sample. Since targets are coded as in (1), the square term in (3) can be written as

$$E\{(T_k - y_k(\mathbf{X}))^2\} = \int [(1 - y_k(\mathbf{x}))^2 Q_k(\mathbf{x}) + (-1 - y_k(\mathbf{x}))^2 (1 - Q_k(\mathbf{x}))] f(\mathbf{x}) d\mathbf{x} \tag{4}$$

where $Q_k(\mathbf{x}) = \Pr[\mathbf{X} \text{ originates from class } k \mid \mathbf{X} = \mathbf{x}]$ is the posterior probability. Let us seek the function $\mathbf{b} = [b_1, b_2, \cdots, b_M]^T$ minimizing the criterion (3) (in the space of all functions taking values in (-1,+1)). For fixed $Q_k(\mathbf{x})$, $0 < Q_k(\mathbf{x}) < 1$, the optimal solution is given by $\mathbf{b}(\mathbf{X})$, where the components of $\mathbf{b}$ are given by [8][11]

$$b_k(\mathbf{x}) = E\{T_k \mid \mathbf{x}\} = 2Q_k(\mathbf{x}) - 1, \ k = 1, 2, ..., M. \tag{5}$$

For performance improvement of MSE, additive noise in the desired signal was adopted as

$$E_{AN}(\mathbf{x}) = \frac{1}{2} \sum_{k=1}^{M} (t_k + n_k - y_k(\mathbf{x}))^2 \tag{6}$$

where n_k is the zero-mean white noise with variance σ^2 [5]. Then, since the random noise is independent of the desired and real output values,

$$E\{E_{AN}(\mathbf{X})\} = \frac{1}{2} \sum_{k=1}^{M} \int \left[\int (T_k + N_k - y_k(\mathbf{x}))^2 f(N_k) dN_k \right] f(\mathbf{x}) d\mathbf{x}$$

$$= \frac{1}{2} \sum_{k=1}^{M} \left[\int (T_k - y_k(\mathbf{x}))^2 f(\mathbf{x}) d\mathbf{x} + \sigma^2 \right]. \tag{7}$$

Here, N_k is the random variable denoting the noise. By applying (4), the minimizer of (7) has the optimal solution vector whose components are the same with (5).

CE [7] error is defined by

$$E_{CE}(\mathbf{x}) = -\sum_{k=1}^{M}[(1+t_k)\ln(1+y_k(\mathbf{x}))+(1-t_k)\ln(1-y_k(\mathbf{x}))]. \tag{8}$$

When we derive the optimal solution for minimizing

$$E\{E_{CE}(\mathbf{X})\} = -\sum_{k=1}^{M}\int[2Q_k(\mathbf{x})\ln(1+y_k(\mathbf{x}))+2(1-Q_k(\mathbf{x}))\ln(1-y_k(\mathbf{x}))]f(\mathbf{x})d\mathbf{x}, \tag{9}$$

the result is the same with the MSE case.

As an extension of CE, nCE error [8] is defined by

$$E_{nCE}(\mathbf{x}) = -\sum_{k=1}^{M}\int\frac{t_k^{n+1}(t_k-y_k(\mathbf{x}))^n}{2^{n-2}(1-y(\mathbf{x})_k^2)}dy_k. \tag{10}$$

The optimal solution for minimizing $E\{E_{nCE}(\mathbf{X})\}$ is derived and the kth component is given by [8]

$$b_k(\mathbf{x}) = g(h_n(Q_k(\mathbf{x}))), \tag{11}$$

where

$$h_n(q) = \left(\frac{1-q}{q}\right)^{\frac{1}{n}} \text{ and } g(u) = \frac{1-u}{1+u}. \tag{12}$$

Also, $g \circ h_n$ is strictly increasing.

In order to suppress the huge amount of weight updating by outliers, MLS error function [6] was proposed by

$$F_{MLS}(\mathbf{x}) = \sum_{k=1}^{M}\log\left(1+\frac{1}{2}(t_k-y_k(\mathbf{x}))^2\right). \tag{13}$$

Then,

$$E\{E_{MLS}(\mathbf{X})\} = \sum_{k=1}^{M}\int[Q_k(\mathbf{x})\log\left(1+\frac{1}{2}(1-y_k(\mathbf{x}))^2\right)$$
$$+(1-Q_k(\mathbf{x}))\log\left(1+\frac{1}{2}(1+y_k(\mathbf{x}))^2\right)]f(\mathbf{x})d\mathbf{x}. \tag{14}$$

Using the same procedure, the kth component of optimal solution vector for minimizing $E\{E_{MLS}(\mathbf{X})\}$ can be derived by

$$b_k(\mathbf{x}) = h^{-1}(Q_k(\mathbf{x})) \tag{15}$$

where

$$h(y) = \frac{y^3 - y^2 + y + 3}{-2y^2 + 6}. \tag{16}$$

It is easy to show that $h^{-1}(q)$ is a strictly increasing function of $q \in (0,1)$.

Contrary to the error functions which are minimized during training of classifiers, CFM is a criterion function to be maximized. CFM is defined by

$$CFM(\mathbf{x}) = \sum_{k \neq c} \frac{1}{1 + \exp(-\beta(y_c(\mathbf{x}) - y_k(\mathbf{x})))} \tag{17}$$

where y_c denotes the correct node and y_k denotes the incorrect node [10]. Therefore, the expectation is

$$E\{CFM(\mathbf{X})\} = \sum_{k=1}^{M} E\left\{ \int \frac{1 - Q_k(\mathbf{x})}{1 + \exp(-\beta(y_c(\mathbf{x}) - y_k(\mathbf{x})))} f(\mathbf{x}) d\mathbf{x} \right\}. \tag{18}$$

Since the fraction term in the integral is a monotonic increasing function of $y_c(\mathbf{x}) - y_k(\mathbf{x})$, $E\{CFM(\mathbf{X})\}$ is maximized when $y_c(\mathbf{x}) - y_k(\mathbf{x})$ is maximized. For a specific $\mathbf{x}$ in the class c, y_c is trained to be 1 and y_k ($k \neq c$) is trained to -1. That is,

$$E\{y_k(\mathbf{x})\} = (+1) \times Q_k(\mathbf{x}) + (-1) \times (1 - Q_k(\mathbf{x})). \tag{19}$$

As a result, the kth component of optimal solution is given by

$$b_k(\mathbf{x}) = 2Q_k(\mathbf{x}) - 1. \tag{20}$$

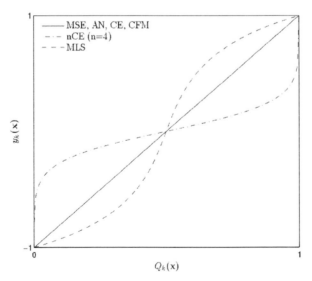

Fig. 1. The optimal solutions of $y_k(\mathbf{x})$ for minimizing the expectations of MSE, AN, CE, nCE, and MLS error functions and for maximizing the expectation of CFM. $y_k(\mathbf{x})$ denotes the kth output of classifier and $Q_k(\mathbf{x})$ denotes the posterior probability $\Pr[\mathbf{X} \text{ originates from class } k \mid \mathbf{X} = \mathbf{x}]$ when a random vector $\mathbf{X}$ is presented to the classifier.

Fig. 1 shows the optimal solutions of the various error or criterion functions. In the MSE, AN, CE, and CFM cases, the optimal solution is proportional to $Q_k(\mathbf{x})$. The optimal solution of nCE has a rapid slope when $Q_k(\mathbf{x})$ is near to 0 or 1. On the contrary, MLS has an optimal solution which is gentle when $Q_k(\mathbf{x})$ is near to 0 or 1. Although the curves of optimal solutions are different in some cases, all the error functions have optimal solutions which are strictly increasing functions of $Q_k(\mathbf{x})$. Therefore, the Bayes classifier can be defined with the decision rule "decide k, if $k = \max\ y_k(\mathbf{x})$".

3   Conclusions

This paper analyzed various error or criterion functions for classifiers in a statistical perspective. MSE, AN, CE, and CFM have the same optimal solution of classifier output, which is proportional to the posterior class probability. nCE and MLS have some rapid or gentle slopes when the posterior class probability is near to 0 or 1. For all error or criterion functions, the Bayes classifier can be defined with the "max" rule.

References

1. Park, W.J., Kil, R.M.: Pattern Classification with Class Probability Output Network. IEEE Trans. Neural Network 20, 1659–1673 (2009)
2. Fukunaga, K., Kessel, D.: Nonparametric Bayes error estimation using unclassified samples. IEEE Trans. Inf. Theory 19, 434–439 (1973)
3. Parzen, E.: On the estimation of a probability density function and mode. Ann. Math. Statist. 33, 1065–1076 (1962)
4. Rumelhart, D.E., McClelland, J.L.: Parallel Distributed Processing. MIT Press, Cambridge (1986)
5. Wang, C., Principe, J.C.: Training Neural Networks with Additive Noise in the Desired Signal. IEEE Trans. Neural Networks 10, 1511–1517 (1999)
6. Liano, K.: Robust Error Measure for Supervised Neural Network Learning with Outliers. IEEE Trans. Neural Networks 7, 246–250 (1996)
7. van Ooyen, A., Nienhuis, B.: Improving the Convergence of the Backpropagation Algorithm. Neural Networks 4, 465–471 (1992)
8. Oh, S.-H.: Improving the error back-propagation algorithm with a modified error function. IEEE Trans. Neural Networks 8, 799–803 (1997)
9. Oh, S.-H.: Error Back-Propagation Algorithm for Classification of Imbalanced Data. Neurocomputing 74, 1058–1061 (2011)
10. Hampshire II, J.B., Waibel, A.H.: A Novel Objective Function for Improved Phoneme Recognition Using Time-Delay Networks. IEEE Trans. Neural Networks 1, 216–218 (1990)
11. White, H.: Learning in Artificial Neural Networks: A Statistical Perspective. Neural computation 1, 425–464 (1989)

FPGA Implementation of Image Processing
for Real-Time Robot Vision System

Hayato Hagiwara, Kenichi Asami, and Mochimitsu Komori

Department of Applied Science for Integrated System Engineering,
Kyushu Institute of Technology,
1-1, Sensui, Tobata, Kitakyushu 804-8550, Japan
hagiwara@es.ise.kyutech.ac.jp, asami@mns.kyutech.ac.jp,
komori@ise.kyutech.ac.jp

Abstract. This paper presents a real-time robot vision system integrating adequate image processing and pan-tilt motion control which is implemented by FPGA reconfigurable logic device. The digital image processing acquired by CMOS image sensor is performed on the embedded FPGA board and Linux real-time video communication module. The integrated robot vision system aims to achieve a suitable platform available from both hardware and software. In addition, we also present spatial recognition by edge detection and tracking function of moving objects by determining color which are necessary for autonomous mobile robots.

Keywords: Robot Vision System, Autonomous Mobile Robots, FPGA, Real-time Linux, Device Driver.

1 Introduction

In recent years, research and development of autonomous mobile robots has been active in industry and academia. The autonomous mobile robots need the ability for spatial recognition and various detective functions to carry out the mission work. Because of the aging society with fewer children in Japan in the future, it is expected to be available for security applications such as watching and observation by autonomous mobile robots. The image processing system for autonomous mobile robots is necessary for the advancement of robot intelligence and the enhancement of processing performance. In the traditional robot vision systems, PCs and USB-attached cameras are equipped for the image processing. However the large power consumption and uncertain real-time response would be insufficient properties for the robot vision purposes [2], [3].

In this paper, a real-time video communication system using the networked FPGA board and real-time extension of Linux OS is introduced. The image capture circuit by hardware description language on the FPGA with FIFO and state machine is implemented. The stepper motor control circuit is integrated with the image recognition. In addition, the dedicated Linux OS with hard real-time extension

G. Lee, D. Howard, and D. Ślęzak (Eds.): ICHIT 2011, CCIS 206, pp. 134–141, 2011.
© Springer-Verlag Berlin Heidelberg 2011

manages the peripheral devices on the FPGA. Furthermore, CMOS camera module is attached to the pan-tilt mechanism for tracking moving target objects.

2 System Configuration

The mechanism of this robot vision system is constructed from two layers (Fig. 1). They are configured with two axes to rotate the total body from side to side and to turn the camera vertically. The main components include a FPGA board, a CMOS camera module, and two stepper motors shown in Fig. 2. The first layer fixes two stepper driver boards for controlling the pan-tilt motion and the second layer fixes the FPGA board. Functionally, the CMOS camera module is attached to the horizontal shaft borne at the middle between the two layers. The vertical shaft pierces the two layers in the center and suspends the body of the robot vision system.

The FPGA board SUZAKU-V (Model: SZ410-U00) [1] as an embedded Linux running environment is designed by Atmark Techno, Inc. This FPGA board installs Xilinx FPGA (Virtex-4 FX) device [6] that can be implemented user logic circuits. The FPGA board specifications include a PowerPC 405 processor, SDRAM 32MB x 2, flash ROM 8MB, 10BASE-T/100BASE-TX network interface and general purpose I/O 86 pins. The CMOS camera module NCM03-S is designed by Asahi Electronics Laboratory Co., Ltd. The image sensor specifications include sensing size of 1/4 inches, pixel size of 640x480, frame interval of maximum 30fps, horizontal angle of 105 degree and power supply of 2.8V. The stepper motors PA233PA are manufactured by Oriental Motor Co., Ltd. and the driver ICs SLA7052M are developed by Sanken Electric Co., Ltd. respectively. We built the stepper driver boards for two phase unipolar stepper motors, which consisted of resistors, capacitors, and the driver ICs for positioning control application. The signal pins for the CMOS image sensor and the stepper drivers are connected to the FPGA terminals.

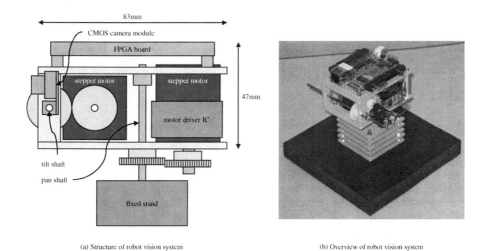

(a) Structure of robot vision system   (b) Overview of robot vision system

Fig. 1. Structure of robot vision system: (a) assembly view from the left side, and (b) crane shot of the whole system

(a) CMOS camera module   (b) FPGA board   (c) Motor drivers

Fig. 2. Main parts of robot vision system: (a) camera module and wire harness, (b) FPGA embedded board, and (c) stepper motors and driver ICs

3   Image Processing System

Image processing systems for robot vision based on PC are implemented by multithread software, so that hard real-time response can be difficult for the integration of image recognition and behavioral usage. Moreover many camera systems available with USB interface have a disadvantage for large-size data transmission of visual information. Therefore, the robot vision system uses image processing by FPGA directly connected to the CMOS image sensor. Xilinx ISE development environment supports designing process for embedded devices and enables the developers to implement specific logic circuits by using abstract hardware description language such as VHDL and Verilog HDL. Xilinx EDK integrates the development for embedded systems including processors with original IP cores that realizes the user specific function.

The image capture circuit for this robot vision system supplies MCLK (master clock) signal to the CMOS camera module as input, which is generated from DCM (digital clock manager) module. The CMOS camera module responds PCLK (pixel clock) signal to the output terminal, and serves PDATA (pixel data) in 8 bits as sequential image data together with HSYNC (horizontal synchronization) and VSYNC (vertical synchronization) signals. When the maximum-speed MCLK sets 27MHz, the CMOS camera module outputs image data at 30fps. The default pixel format of YUV422 at 640x480 VGA resolutions is produced in the PDATA terminals. The YUV pixel format reduces image data size compared to RGB format and is convenient to recognize objects by color without influence of brightness. The YUV model defines a visual representation for one luma (Y, brightness) and two chrominance (UV, color difference of blue and red) components. The YUV422 format gives luma (Y) for each pixel in 1byte, and shares chromaticity (U, V) for adjacent two pixels in 2bytes respectively. Namely the total amount of data size per pixel uses 2bytes. The visual receptors for human eyes are sensitive to changes in brightness more than in chromaticity, so that the image processing for the robot vision system can achieve efficient object recognition with suitable compressed data.

The block diagram for image processing circuit shown in Fig. 3 has been developed for the robot vision system. The image processing circuit includes a state machine circuit and a FIFO memory buffer that controls the timing for pixel data transmission from the image capture circuit to the embedded processor.

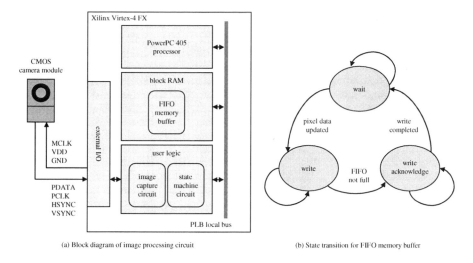

(a) Block diagram of image processing circuit   (b) State transition for FIFO memory buffer

Fig. 3. Structure of image processing circuit

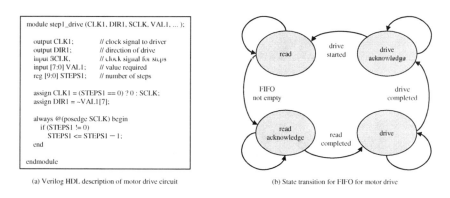

(a) Verilog HDL description of motor drive circuit   (b) State transition for FIFO for motor drive

Fig. 4. Implementation of motor drive circuit

The image capture circuit of this robot vision system receives one pixel data (1byte) at 27MHz from the CMOS camera module and collects four pixel data (4bytes) to deliver to the state machine circuit in compliance with the width size of 32bits in the FIFO memory buffer. The state machine circuit adjusts timing between the image capturing and the buffering to insert the 4bytes data for 2 pixels to the FIFO. Therefore there are three states of 'wait', 'write' and 'write acknowledge' for pixel data transmission. The device driver software on PowerPC processor can read pixel data at a suitable speed.

In addition, the motion control circuits of the stepper motors have been developed by Verilog HDL shown in Fig. 4. There are four states of 'read', 'read acknowledge', 'drive' and 'drive acknowledge' for driving two stepper motors according to stepping values from the processor that image recognition software provides. Small gained operations can be successfully managed in the state machine transitions.

4   Image Recognition Application

In this robot vision system, the application software for tracking moving objects using color information obtained from the CMOS camera module has been developed. This image processing application enables the robot vision system to turn the visual field to the target direction where the object has or is given to a specific color sign. The simple and primitive image processing library can be utilized in many cases for the robot vision system.

The method of object recognition by color sets thresholds for pixel components and all pixels are judged to be the target color. Fig. 5 shows the results for color judgment by using RGB and YUV formats respectively. Here the target color is assumed to be set to orange. The white pixels indicate the target color and the black non-target. Although the method using RGB format can not grasp the whole colored ball, the thresholds for YUV components are independent of brightness from the lighting. Therefore the object recognition by color adopts YUV format in order to grasp accurate region of the target object.

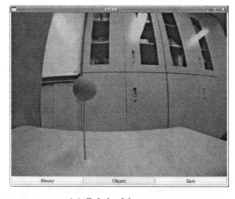

(a) Original image

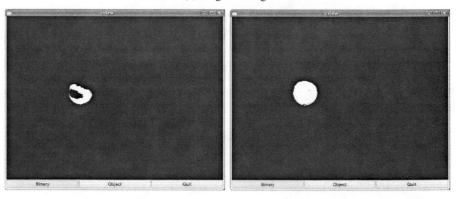

(b) Binary image by deciding RGB   (c) Binary image by deciding YUV

Fig. 5. Results of image binarization for detecting the target color object

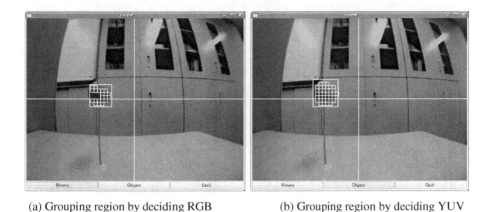

(a) Grouping region by deciding RGB   (b) Grouping region by deciding YUV

Fig. 6. Results of region detection for target object recognition

In order to eliminate the surrounding noises around the target object in the color detection, the region detection methods by segmentation and grouping are applied to the binary image by color detection. The method of region segmentation divides the whole image data into certain small segments and judges whether the segment indicates the effective region where many pixels with target color exist. The method of region grouping finds clusters of effective regions which are contiguous with each other and determines the biggest cluster with effective regions. Finally the central coordinates are decided for tracking the target object position. As shown in Fig. 6, the accurate central coordinates can be decided by using YUV format compared with using RGB format.

The autonomous mobile robots need the ability for spatial recognition and the function to correct their position and orientation. Specifically it is important to recognize the outline of obstacles or the perimeter of floors and walls in order to confirm their proper traveling route. Moreover it is useful to calculate the deviation from the position and orientation which are in advance taught during their traveling. In this robot vision system, the application software for image recognition has been developed for detecting passage boundaries between floors and walls in typical indoor environment. The methods of edge and line detections are applied for the image data from the CMOS camera module. The edge detection finds the continuous boundaries with contrast between light and shade in the image. First, the image data of YUV format is converted to the grayscale format. Second, it is smoothed by applying Gaussian filter. Third, gradient strengths and directions for all pixels are calculated by applying Sobel filter. Finally, the continuous boundary with the maximum strength and the same direction of gradient is traced. The line detection uses the Hough transformation algorithm where the image is divided into the left and right sides and the target directions are restricted to the assumed angles for boundaries between floors and walls. The result for boundary detection is shown in Fig. 7. This application software successfully recognized the boundaries of typical passage environment, and the ability of boundary detection was proved to be robust for some pedestrians and obstacles. The robot vision system is able to judge the position and orientation in the passage environment for self-localization using this boundary detection method.

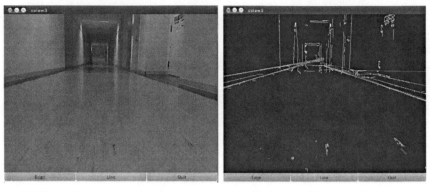

(a) Typical passage environment   (b) Result of boundary detection

Fig. 7. Result of boundaries detection in typical passage environment

5   Real-Time Module

This robot vision system uses embedded technology for real-time response and low power consumption. In embedded systems, real-time capabilities means that target processing can be ensured to start and finish within a given time or to repeat at a predictable time period [4]. In a multithread program, delays called latency in the start and end time can be caused by various factors such as device I/O capability and timing disparity between devices. This robot vision system needs to achieve hard real-time response to effective physical movement by image processing and to integrate high-level services such as image communication among user terminals and robots. Therefore, this robot vision system requires a real-time operating system on the embedded system. For this requirement, the hard real-time extension for general GNU/Linux operating system has a suitable capability, because of the real-time multithread scheduler, networking protocol stack, various device drivers, and useful development tools on the open platform environment for almost embedded systems.

Xenomai [5] real-time subsystem was ported to the embedded environment for this robot vision system. The Xenomai project develops micro kernel (co-kernel) with interrupt controller and preemptive scheduler tightly integrated with the GNU/Linux operating system. The embedded kernel and device driver for image processing was developed using GNU cross tool chain 4.4.1 for PowerPC, Linux kernel source 2.6.18, and Xenomai 2.4.10 for the FPGA board SUZAKU-V. The target kernel and device driver with real-time APIs were built and downloaded to the on-board flash ROM. The Xenomai operating system and the device driver for image processing run on the Virtex-4 FX device environment appropriately. The real-time device driver is mapped to the image capture circuit by using system API ioremap() function and captures image data from the FIFO memory buffer in the FPGA device through the peripheral local bus. The robot vision software integrates the real-time image capturing, functions for image recognition, pan-tilt motion control, and TCP/IP communication for remote manipulation.

6   Conclusion

In this paper, the real-time image processing system for robot vision system is presented by the FPGA implementation. The robot vision system detects and tracks a moving object smoothly. Moreover this robot vision system detects the corridor boundaries between floors and walls, and judges its position and orientation in the passage environment. This robot vision system is small enough and requires low power consumption for various robot applications. For example, a patrol service robot for finding a suspicious person and indoor surveillance of elderly people would be considered useful applications. The integration of hardware and software platform for real-time image processing could be applicable for various purposes. For the next work, image data communication system using the wireless LAN module will be developed. Furthermore, the FPGA implementation for image processing to detect stereo disparity is desired for measuring the distance to objects.

References

1. Atmark Techono, Inc.: SUZAKU-V Hardware Manual, SZ410-U00 (2011)
2. Kudo, K., Myokan, Y., Than, W.C., Akimoto, S., Kanamaru, T., Sekine, M.: Hardware Object Model and Its Application to the Image Processing. IEICE Trans. E87-A(3), 547–558 (2004)
3. Shimizu, K., Hirai, S.: CMOS+FPGA Vision System for Visual Feedback of Mechanical Systems. In: Proc. IEEE Int. Conf. on Robotics and Automation, pp. 2060–2065 (2006)
4. Yaghmour, K., Masters, J., Ben-Yossef, G., Gerum, P.: Building Embedded Linux Systems, 2nd edn., pp. 351–386. O'Reilly Media, Inc., Sebastopol (2008)
5. Xenomai Project, http://www.xenomai.org
6. Xilinx Inc., Virtex-4 FPGA User Guide (2008)

Security Requirements Prioritization Based on Threat Modeling and Valuation Graph

Keun-Young Park, Sang-Guun Yoo, and Juho Kim*

Sogang University, Department of Computer Science and Engineering, Seoul, Korea
{kypark,sangguun,jhkim}@sogang.ac.kr

Abstract. Information systems manage assets that are critical for the business processes of organizations. Therefore, it is imperative that information systems be guaranteed and secured from the beginning of their development life cycle. Several approaches such as misuse cases, attack tree, and threat modeling have been proposed by way of security requirements. However, these approaches do not prioritize security requirements, though it is necessary in many cases. For example, when the security budget is insufficient, security requirements need to be prioritized to decide what will be developed and what will not. In this paper, we propose an extension to threat modeling by creating a process that allows the prioritization of security requirements via the valuation of assets, threats, and countermeasures modeled in a tree-like structured graph that we refer to as a "valuation graph."

Keywords: Security Requirement Prioritization, Threat modeling.

1   Introduction

Recently, there has been an increase in the number of attacks on information systems because these systems manage important organizational assets. Moreover, the increasing complexity of applications and services implies a greater probability of security breaches [1]. This is why it is essential to guarantee and secure information systems from the beginning of their development life cycle.

Although the inclusion of security into the early phases of the development of information systems is cost effective, in majority of software projects, security is dealt with only after the system has been designed, developed, and put into operation. Quite often, this is due to inappropriate management of system security requirements. Many approaches have been proposed to solve this problem, such as the Trustworthy Computing Security Development Lifecycle [2] used by Microsoft, which includes security concepts spanning the whole development process. Other approaches such as misuse cases [3], attack tree [4], and threat modeling [5] help in the elicitation and specification processes of security requirements through an analysis of threats, vulnerabilities, and countermeasures. However, these approaches exclude the prioritization process, which is necessary in many cases. For example, when there is

---

* Corresponding author.

G. Lee, D. Howard, and D. Ślęzak (Eds.): ICHIT 2011, CCIS 206, pp. 142–152, 2011.

insufficient security budget, it is necessary to prioritize security requirements to decide which requirements will be developed and which will not. Another reason for prioritization of security requirements is when security functionalities of existing systems need to be developed following attacks; in this case, it is very important to first mitigate the most harmful attacks, beginning by developing the modules that have more priority. Prioritization of requirements is often used in software engineering processes, but the same methodologies cannot be used with efficiency when dealing with security requirements because there are additional elements that are unique to security.

In this paper, we propose an extension to threat modeling, creating a process that allows for the prioritization of security requirements via the valuation of assets, threats, and countermeasures modeled in a tree-like structured graph referred to as a "valuation graph." The rest of the paper is organized as follows. In section 2, we briefly explain some terms that are important in security requirements engineering and we detail our approach. Then, in section 3, we illustrate our proposal using a case study. Finally, by way of a conclusion, we review the main points of the paper and discuss possible future works.

2 Proposed Methodology for Security Requirements Prioritization

2.1 Terminology

To specify security requirements, it is critical to first understand the concepts underlying security engineering. The following list defines the security-oriented terms that will be used during the remainder of this paper [10].

- Asset: this is a resource of value such as the data in a database or on a file system, or a system resource.
- Threat: this is a potential occurrence (malicious or otherwise) that may harm an asset.
- Vulnerability: this is a weakness that makes a threat possible.
- Attack (or exploit): this is an action taken to harm an asset.
- Countermeasure: this is a safeguard that addresses a threat and mitigates a risk.

2.2 Security Requirements Prioritization Based on Threat Modeling and Valuation Graph

Several publications such as [5], [6], and [7] have analyzed the relationship between existing security requirements and assets to create support mechanisms that aid the software development process. For example, the main components that influence and are influenced by security requirements are detailed in [7]. These authors agree that security requirements are created to protect the assets of organizations, and that between these two elements (assets and security requirements), many other concepts play a role, such as vulnerability, risk, threats, and harm. The main idea of our approach is to value the common elements that comprise the relationship between security requirements and assets, and to then draw them in an easy to understand model. Our approach comprises eight steps, organized in three layers (see Fig. 1).

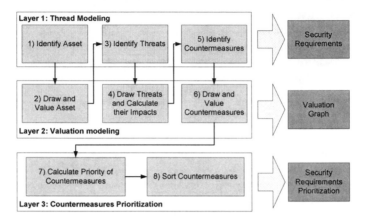

Fig. 1. Steps of security requirements prioritization methodology organized per layers

The first layer follows the traditional threat modeling approach, which deals with security requirements elicitation. The second layer, called valuation modeling, enables the measurement of the main components involved in the security requirements (assets, threats, and countermeasures), thus generating the valuation graph for prioritization purposes. A valuation graph is a diagram that stakeholders and developers can use to easily visualize the relation between the importance of an asset, the degree of harm (impact) associated with threats, and the cost of developing and using the modules that protect the assets from threats. Finally, the third layer, called countermeasures prioritization, allows for the calculation of the priority and the sorting of security requirements according to importance. The eight steps composing these three layers are described below.

- **Step 1: Identify assets:** In this step, a list of all critical assets of the system is created. Assets are very important to the business, so this step is not a stakeholder-only decision. The business users must also participate in determining which assets cannot be compromised and which can be without serious negative consequences. Every piece of information about the organization needs to be collected, categorized, organized, and analyzed. These information assets can comprise databases, data files, archived information, continuity plans, operational and support procedures, etc.

- **Step 2: Draw and value assets:** Once the list of assets requiring protection is recognized, the initial part of the valuation graph can be created. This graph begins with a node containing the name of the system; the assets of the system are drawn in new nodes that are joined to the initial node via edges (see Fig. 2).

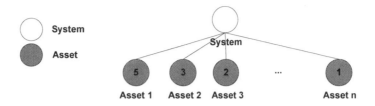

Fig. 2. Assets in valuation graph

Once the assets have been drawn, it is necessary to value them. Valuation can be performed using different valuation models. The calculated value of each asset must to be expressed as a number between 1 (low importance) and 5 (high importance). This value is very important during security requirements prioritization; therefore, this step needs to be taken seriously. In the case of small organizations, it can be difficult to value assets; in these situations, a simple security oriented asset valuation approach can be adopted (Table 1).

Once the value of each asset is recognized, these values are entered into the valuation graph, and the value is written in the node of each asset.

Table 1. Default asset valuation

Type of asset	Description	Value
Restricted	Most sensitive business information. The unauthorized disclosure of this information could have a serious negative impact on the organization.	5
Confidential	Less sensitive business information. The unauthorized disclosure of this information could have a negative impact on the organization.	3
Public	Information approved for public disclosures.	1

- ***Step 3: Identify threats per asset:*** In this step, the most important task is to adopt the mindset of an attacker, identify possible points of attack, and analyze the vulnerabilities of exposed assets. Existing publications and approaches can be used in this step, such as vulnerabilities lists (e.g. Web Application Security Consortium Threat Classification V1.0) or STRIDE. STRIDE [6], which is an acronym for Spoofing identity, Tampering with data, Repudiation, Information disclosure, Denial of service, and Elevation of privilege, is one the most popular threat modeling methodologies.

- ***Step 4: Draw threads and calculate their impacts:*** Once threats have been identified, they are drawn in the valuation graph using one node per threat and joining them to the assets via edges (see Fig. 3).

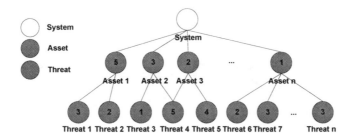

Fig. 3. Threats in Valuation Graph

The *impact* is the level of harm that an attack can cause to an asset when the threat is exploited. This is a number between 0 and 5, where 5 is the highest level of harm a threat can cause. The *impact* of the threat is valued using (1):

$$Impact= \frac{Damage+Recoverability+Likelihood}{3} \qquad (1)$$

Damage is a value that represents the harm caused by an attack when the threat is exploited, and this value is assigned according to Table 2. *Recoverability* is a value that represents the difficulty in recovering from a successful attack that exploited a given threat. This value is assigned according to Table 2. *Likelihood* is a value between 1 and 5 that expresses the possibility of the threat being exploited. This value is assigned according to Table 3, which evaluates the effort needed to exploit the threat, and the statistical level of occurrence of this threat in real implementations.

Table 2. Damage and recoverability valuation

Value	Damage	Recoverability
0	No damage is caused by the threat	-
1	Individual user data is compromised or affected	Easy to recover
3	Group data is compromised or affected	Hard to recover
5	Complete system data is compromised or affected	Impossible to recover

Table 3. Likelihood valuation

Effort / Occurrence	Very hard	Several steps required	Easy to perform
Low frequency of occurrence	1	2	3
Acceptable frequency of occurrence	2	3	4
High frequency of occurrence	3	4	5

Once the *impacts* of the threats are calculated, these values are entered into the valuation graph (see Fig. 3).

- **Step 5: Identify countermeasures:** Once the possible threats have been recognized, it is necessary to identify countermeasures. In this step, approaches like misuse cases and attack trees can be used to diagram and analyze the relation between threats, vulnerabilities, and countermeasures. Industry standards and case studies of other similar systems also can be used.

- **Step 6: Draw and value countermeasures**: Once the countermeasures are identified, they are drawn in the valuation graph via new nodes between assets and threats, joining the assets that are protected and the threats that are mitigated. When a threat is related to various countermeasures, it is necessary to clarify whether one of the countermeasures is enough to mitigate the threat, or whether all or parts of the countermeasures are necessary to mitigate the threat. In the valuation graph, logical operations (e.g., *AND* and *OR*) are used. *AND* is represented by a line, and *OR* is represented by a curve in the middle of the edges (see Fig. 4).

  Once the countermeasures are identified, their cost must be valued. When this value is measured, it is not enough to calculate only the development cost of the countermeasure because this cost is only partial. It is also necessary to calculate the user training cost, maintenance cost, and other important indirect costs of the countermeasure; this implies a calculation of the total cost of ownership (TCO). The TCO, which is a financial estimate designed to help consumers and enterprise managers assess direct and indirect costs, can be calculated using techniques such as those proposed in [8] and [9]. Once each countermeasure's TCO is calculated, this value is transformed into a number between 1 and 5, where 5 represents the highest cost of ownership. After this transformation, the values are entered into the valuation graph.

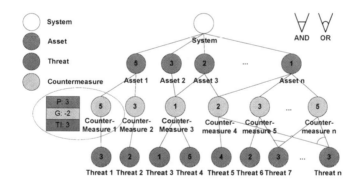

Fig. 4. Countermeasure assessment in the valuation graph

- **Step 7: Calculate priority of countermeasures**: As a result of the first six steps of our proposed methodology, the security requirements (countermeasures) and valuation graph are generated and, using this information, it is possible to prioritize the security requirements. First, the *total impact (TI)* of threats that a countermeasure mitigates is calculated. *TI* is the summation of the impacts of each threat related to the countermeasure. If a threat is related to more than one countermeasure, its impact is divided by the number of edges if the relation is an *AND*; the impact is summed if it the relation is an *OR*. Next, the *gain (G)* of each countermeasure is calculated by subtracting the *countermeasure TCO* from its *TI*. Finally, the *priority (P)* of the countermeasure is calculated by summing the *G* of the countermeasure and the value of assets that the countermeasure protects. Calculated values (*P*, *G*, and *TI*) are organized in Table 4, and these values also are drawn in the valuation graph (see Fig. 4); it is recommended that this option be incorporated into the tool that manages the valuation graph.

Table 4. Priority-gain-impact table

Countermeasure	Priority	Gain	Total Impact
Countermeasure 1	3	-2	3
Countermeasure 2	4	-1	2
Countermeasure 3	10	5	6
...			
Countermeasure n	-2.5	-3.5	1.5

• ***Step 8: Sort countermeasures:*** Once the *priorities* of each countermeasure have been calculated, they are sorted according to *priority*. If two or more countermeasures have the same *priority*, the second and third fields used to determine the order are the *asset value* and the *countermeasures TCO*. With this prioritization, risk management becomes easier, permitting visualization of the importance and impact of assets, countermeasures and threats using a simple model.

3   Case Study

In this section, we demonstrate how our proposed method can be applied to real systems through the specific example of an e-commerce web application that controls the bill-payment data of customers and the catalog of products sold on this site.

After eliciting the functional requirements of the system, three main logics were identified: bill payment, administration interface, and product catalog (see Fig. 5). The eight steps of the security requirements prioritization process for this example are noted below.

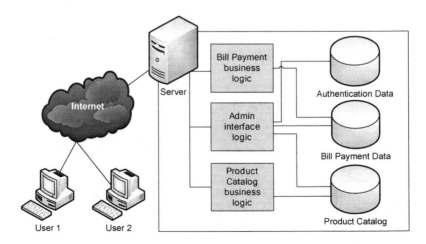

Fig. 5. Diagram of an e-commerce site

- **Step 1: Identify assets:** The assets that the organization wants to protect were identified based on functional requirements. The identified assets are authentication data, bill payment data, and the product catalog.

- **Step 2: Draw and value assets:** Once the assets have been recognized, they are drawn into the valuation graph. Next, the assets are valued using default asset valuation:

  - Bill payment data is restricted information, thus the assigned value is 5.

  - The product catalog is public information, thus the assigned value is 1.

  - Authentication data is restricted information, thus the assigned value is 5.

  Finally, these values are entered into the valuation graph (see Fig. 6).

Fig. 6. E-commerce application assets valuation

Note: In this part of the paper, we will only analyze the bill payment data asset because this is sufficient to understand the use of our approach.

- **Step 3: Identify threats per asset:** Threats for bill payment data were identified using the threat classification for Web application version 1.0, published by the Web Application Security Consortium. The identified threats are as follows: denial of service, brute force, spoofing, SQL injection, eavesdropping and repudiation.

- **Step 4: Draw threats and calculate their impacts:** The identified threats are drawn in the valuation graph and the impact of each threat is calculated using (1):

Threats	Damage	Likelihood	Recoverability	Impact
Denial of service	5	4	1	3.3
Brute force	1	5	1	2.3
Spoofing	3	2	5	3.3
SQL injection	3	3	3	3
Eavesdropping	3	3	1	2.3
Repudiation	1	3	3	2.3

Finally, the calculated impact values are entered into the valuation graph (Fig. 7).

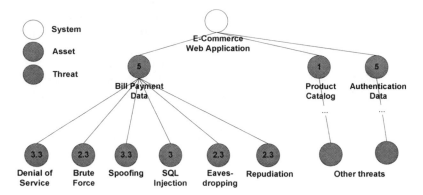

Fig. 7. E-commerce application threats valuation

- *Step 5: Identify countermeasures*: The countermeasures to threats were taken from vulnerabilities lists solutions and previous, similar cases. The identified countermeasures for bill payment data were as follows: an authentication system, query validation, a secure channel, and an auditing system.

- *Step 6: Draw and value countermeasures*: Countermeasures and relationships are entered into the valuation graph. Repudiation requires authentication and auditing systems in order to be completely mitigated, so the logical operator AND is used (see Fig. 8). The TCO of each countermeasure was calculated using data from previous, similar projects. TCOs were converted to a value between 1 and 5 and entered into the valuation graph.

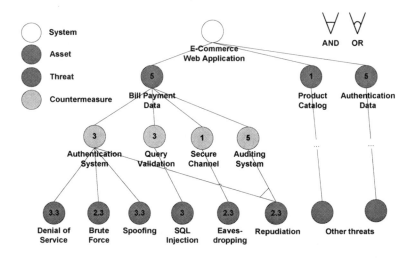

Fig. 8. E-commerce application countermeasures valuation

- **Step 7 and 8: Countermeasure Prioritization:** The total impact, gain and priority of countermeasures are calculated and the priority-gain-impact table is created (Table 5). These values are also then stored in the valuation graph (see Fig. 9).

Table 5. E-commerce application priority-gain-impact table

Countermeasure	Priority	Gain	Total impact
Authentication system	12.05	7.05	10.05
Query validation	5	0	3
Secure channel (SSL)	6.5	1.3	2.3
Auditing system	1.15	-3.85	1.15

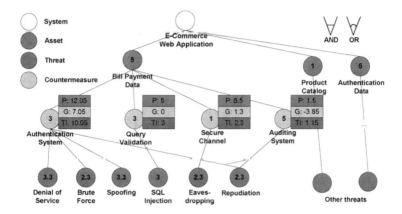

Fig. 9. E-commerce application valuation graph with prioritization information

Finally, the countermeasures are sorted according to priority, asset value, and countermeasures' TCOs (Table 6).

Table 6. E-commerce application prioritization list

Countermeasure	Priority	Asset value that protects	Counter-measure TCO
Authentication system	12.05	5	3
Secure channel (SSL)	6.5	5	3
Query validation	5	5	1
Auditing system	1.15	5	5

4   Conclusion

Not every project has sufficient budget to protect itself from all threats; therefore, it is not possible to ensure absolute security. However, we can deal with absolute risk acceptance, managing a balance between what is possible and what is acceptable via the risk management process (accept, transfer, remove, and mitigate the risk). For the purposes of this risk management, the prioritization of security requirements is indispensable. In this paper, we have proposed a methodology for prioritizing countermeasures that significantly aid the risk management process. The use of a valuation graph helps to show stakeholders why countermeasures are necessary, visualizing the danger relationship between threats and assets. On the other hand, a valuation graph also gives the system developer a model to easily understand the threats that a system is exposed to. Additionally, this graph enables the visualization of all the important values of threats, countermeasures, and assets, allowing stakeholders and developers to analyze and prioritize security requirements. In the future, we plan to polish this modeling process and develop a tool with which to manage it.

References

1. Walton, J.P.: Developing an Enterprise Information Security Policy. In: Proc. 30th Annual ACM SIGUCCS Conference on User Services, pp. 153–156. ACM, New York (2002)
2. Lipner, S.: The Trustworthy Computing Security Development Lifecycle. In: Proc. Computer Security Applications Conference, pp. 2–13. IEEE Press, Tucson (2004)
3. Sindre, G., Opdahl, A.: Capturing Security Requirements through Misuse Case. In: Proc. 14th Norwegian Informatics Conference (NIK 2001), Tromso, pp. 26–28 (2001)
4. Diallo, M.H., et al.: A Comparative Evaluation of Three Approaches to Specifying Security Requirements. In: Proc. International Working Conference on Requirement Engineering: Foundation for Software Quality(REFSQ 2006), Luxembourg (2006)
5. Myagmar, S., Lee, A., Yurcik, W.: Threat Modeling as a Basis for Security Requirements. In: Proc. Symposium on Requirements Engineering for Information Security SREIS, Chteseer, Paris, pp. 94–102 (2005)
6. Swiderski, F., Snyder, W.: Threat Modeling. Microsoft Press (2004)
7. Firesmith, D.: Specifying Reusable Security Requirements. Journal of Object Technology 3, 61–75 (2004)
8. Smith, J., Schuff, R., Louis, R.: Managing your IT Total Cost of Ownership. Communications of the ACM 45, 101–106 (2002)
9. MacCormack, A.: Evaluating Total Cost of Software Platforms: Comparing Apples, Oranges and Cucumbers,
   http://ideas.repec.org/p/reg/wpaper/168.html
10. Threats and countermeasures,
    http://msdn.microsoft.com/en-us/library/aa302418.aspx

An Efficient Hardware Countermeasure against Differential Power Analysis Attack

Amlan Jyoti Choudhury[1], Beum Su Park[1], Ndibanje Bruce[1], Young Sil Lee[1], Hyotaek Lim[2], and Hoon Jae Lee[2]

[1] Department of Ubiquitous IT, Graduate School of General, Dongseo University, Busan, 617-716, South Korea
[2] Division of Computer and Engineering, Dongseo University, Busan, 617-716, South Korea
choudhuryamlanjyoti@gmail.com, beumsupark@hotmail.com, ndibabruce@gmail.com, attract35@hotmail.com, htlim@dongseo.ac.kr, hjlee@dongseo.ac.kr

Abstract. Extensive research on modern cryptography ensures significant mathematical immunity to conventional cryptographic attacks. However, power consumption in cryptographic hardware leak secret information. Differential power analysis attack (DPA) is such a powerful tool to extract the secret key from cryptographic devices. To defend against these DPA attacks, hiding and masking methods are widely used. But these methods increase high area overhead and performance degradation in hardware implementation. In this aspect, this paper proposes a hardware countermeasure circuit, which, is integrated hardware module with the intermediate stages in S-Box. The countermeasure circuit utilizes the dynamic power dissipation characteristics of CMOS and provides countermeasure against DPA attacks.

Keywords: SPA, DPA, crypto-processor, cryptography, hamming weight, hamming distance, CMOS.

1 Introduction

Cryptographic algorithms may be hypothetically secure; however these algorithms can be easily broken by leakage of information through power consumption in the physical environment. These kind of sideways attacks on cryptographic hardware are called side channel attacks. Side channel attacks are broadly divided into three categories, which are invasive attacks, semi invasive attacks and non invasive attacks [1]. In invasive attacks, different hardware modules of cryptographic hardware are accessed directly with depackaging of the device (e.g. reverse engineering methods). Non invasive attacks are performed by analyzing the leakage power signals without physical damage to the device, such as, timing attack, power analysis attacks etc. The semi invasive attack is also performed by depackaging the device, however, the passivation layer is kept intact; (e.g. fault analysis attacks). The power analysis attacks fall under non invasive attack type.

G. Lee, D. Howard, and D. Ślęzak (Eds.): ICHIT 2011, CCIS 206, pp. 153–159, 2011.
© Springer-Verlag Berlin Heidelberg 2011

The study of power consumption in cryptographic hardware is the basis of power analysis attack. The power analysis attack was introduced by Paul Kocher in 1999 [2]. It is found that the power consumption mostly depends upon the number of bit transitions at a given time or hamming distance [3 - 6] between the bit streams. The consumption of power also depends on the hamming weight i.e. the number of 1s in the data bits [2], [7 - 9]. Power analysis attacks exploit these two properties to extract the secret data stored in cryptographic-hardware. There are mainly two types of power analysis attacks, viz. simple power analysis attack, SPA and differential power analysis attack, DPA.

In simple power analysis attack, the attackers examine the power consumption of a single power trace of the crypto-processor and try to deduce the key or a part of the key. As SPA exploit the key dependent characteristics of the power trace, these attacks are feasible for asymmetric key cryptography [10]. On the other hand, the power consumption of symmetric key encryption such as AES does not depend on the secret key or more specifically, the dependency is very less.

DPA however uses statistical method to deduce the secret key. In differential power analysis attack, around 1000 different random plaintext are used as sample and a part of the key is guessed. This non invasive method of attacking crypto-algorithm is a powerful way to extract the secret data from cryptographic hardware.

The existing hiding and masking methods significantly defends DPA attacks, but these two methods increases the area overhead by 50% (approx.) and degrades the performance by 50% (approx.) [10]. This paper proposes a hardware countermeasure circuit against DPA attacks that exploit the dynamic power consumption characteristics of a CMOS transistor.

2 Understanding DPA Attack

DPA is a very powerful method to extract secret information from secure cryptographic algorithm. The dependency of power consumption with the device as well as the intermediate values of the cryptographic algorithm (e.g., AES, DES etc.) is the key to DPA attack. In addition to that, DPA can be conducted in a very noisy environment as well, compared to SPA attacks. One of the unique advantages of DPA is that the knowledge of cryptographic algorithm is not required to perform DPA attack.

DPA attack breaks the key guess and brute force algorithm is applied. Hence the algorithm search is reduced to $\frac{n}{8} * 2^8$ instead of 2^n, where n denotes the number of bits in the key [11]. In the process of DPA attack, power analysis traces for each plaintext are collected and stored in computer. Subsequently, the adversary models the power consumption within the internal nodes, specifically the power consumption at the attack point within the circuit for each possible key value. After that, the attacker finds correlation between the measurement and the model. For wrong key value there should be no correlation. Nevertheless, for the correct key there should be some correlation.

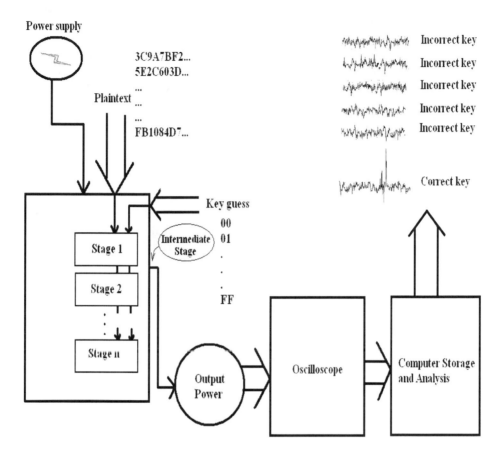

Fig. 1. DPA flow diagram

In figure 1, the DPA attack flow is shown, where some random plaintext inputs are fed by the adversary and a part of key guess is also given as an input to the crypto-module. Subsequently, the intermediate cipher text values are collected and converted into equivalent power values, which are further sampled by the oscilloscope and analyzed by the computer. As we can see in the upper right-hand side of figure 1, we can find sharp spikes for each part of a correct key guess.

3   CMOS Power Consumption

There are two kinds of power consumptions in CMOS circuits; viz. static power consumption and dynamic power consumption. CMOS device have very low static power consumption which is ideally assumed to be zero. But during switching at a high frequency, the dynamic power consumption contributes to significant power dissipation in the device. In 90nm CMOS technology, the dynamic power dissipation of a CMOS inverter at 1GHz frequency is 2.5μW [12].

4 Motivation and Architecture Design

As already discussed in section 1, most of the hiding and masking methods increases the area overhead and degrades the performance of the crypto-hardware-module. Hence, an area and performance efficient hardware countermeasure against DPA attack is a necessary requirement. In DPA attack, the key guess is correct if the correlation coefficient is close to ± 1 which is calculated as

$$r_{ij} = \frac{\sum_{d=1}^{D} (h_{d,i} - \overline{h_i}).(t_{d,j} - \overline{t_j})}{\sqrt{\sum_{d=1}^{D} (h_{d,i} - \overline{h_i})^2.\sum_{d=1}^{D} (t_{d,j} - \overline{t_j})^2}} \tag{1}$$

Equation 1 shows that the correlation of $h_{d,i}$ and $t_{d,j}$ will be zero if these two terms are independent. With reference to this concept, Liu et al [10] proposed ring oscillator based countermeasure circuit against power analysis attack with 121 gate area overhead. The design is not appropriate against DPA attack as the power consumption of the countermeasure circuit is not dominant over the S Box power consumption. An

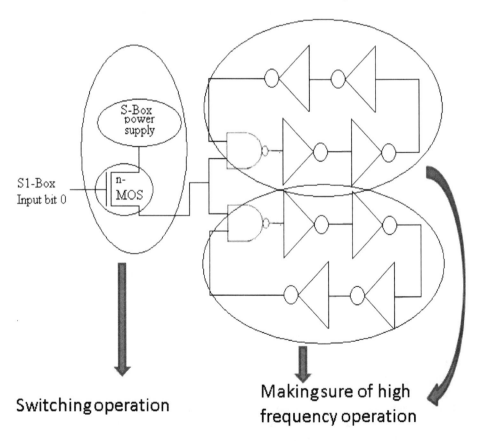

Fig. 2. The countermeasure module for one input of one S-Box is shown

LUT based S box consumes about 150μW in 90nm technology and for a hardware module to efficiently counter DPA attack, it must dominate in power consumption over S Box power consumption.

Hence, design an efficient hardware countermeasure against DPA attack is the prime concern for the proposed paper. The fundamental idea of our design is that, the dynamic power dissipation of a CMOS inverter is significant which apparently shows the spikes in every key guess. The basic design of the proposed hardware is shown in figure 2, which works only when the input logic to S-Box is 1. The circuitry will be idle if the input logic to S-Box is 0.

The operation of the above circuitry is explained below

1) In the switching operation block, the n-MOS transistor will act as an electronic switch, which will be "ON" when logic 1 will be on the S-Box input bit.

2) If the MOS switch is on, the second block will be in operation. The block will be operated at maximum possible practical frequency and consume maximum dynamic power.

3) When the S-Box input will be in logic 0 states, the MOS transistor will be in "OFF" state and input to "Making sure of high frequency operation" block will be in high impedance state. Hence, the second block will not consume any power when S-Box input bit will be in logic 0 state.

4) Hence in one S-Box, 8 similar circuitry will be placed parallel to the S-Box operation.

The power consumption of the proposed circuitry considering 1 GHz bit rate is around 200μW and hence, the power consumption is dominant over S-Box power consumption. As a result, in every attempt of DPA attack, the proposed countermeasure will be expected to show the following waveform shown in figure 3.

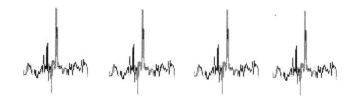

Fig. 3. Spikes in every key guess

5 Analyses and Conclusion

Table 1 shows comparisons among different countermeasure techniques based on area overhead and number of gates. Our design consumes approximately 200μW power parallel to one S-Box which dominates S-Box power consumption of 150μW. In addition to that, area overhead (in %) of our design in LUT and finite field methods S-Box are 12.56% and 35.08% which is quite better than the existing designs.

Table 1. Comparison with some existing designs

Procedure	Design type	Technology	Area (μm^2)	Area Overhead (%)	Gates
Lookup Table	LUT based S-Box w/o countermeasure	90nm	1784	-	637
	LUT based S-Box with countermeasure	90nm	2008	12.56%	717
	Liu et al. [10]	90nm	2123	19%	758
Finite Field	Composite S-Box	90nm	638	-	228
	Composite S-Box with countermeasure	90nm	718	35.08%	308
	Liu et al. [10]	90nm	977	53.13%	349
	Paramstaller [13]	90nm	N/A	200% (estimated)	N/A
	Akkar et al. [14]	90nm	3017	372.88%	1078
	Trichina et al. [15]	90nm	1821	185.42%	650

The DPA attack is considered to be one of the most critical threats to cryptographic hardware, and hence, several researches are going on to prevent power analysis attack. Upon most of the cryptographic devices the smartcard is most vulnerable to side channel attacks as the possibility of loss of smartcard are high. This paper proposes an efficient countermeasure to DPA attack which acts parallel to S-Box and which can be easily mounted on different S-Box implementation to resist DPA attacks.

Acknowledgment. This research was supported by Basic Science Research Program through the National Research Foundation of Korea (NRF) funded by the Ministry of Education, Science and Technology (grant number: 20100010488).

References

1. Quisquater, J.J., Rizk, M.: Side Channel attacks. Information-technology promotion agency, Japan technical report (October 2002) http://www.ipa.go.jp/security/enc/CRYPTREC/fy15/doc/1047_Side_Channel_report.pdf
2. Kocher, P.C., Jaffe, J., Jun, B.: Differential power analysis. In: Wiener, M. (ed.) CRYPTO 1999. LNCS, vol. 1666, p. 388. Springer, Heidelberg (1999)

3. Brier, E., Clavier, C., Olivier, F.: Correlation Power Analysis with a Leakage Model. In: vol. 1. Springer, Heidelberg (1973); vol. 6697 (2011), 0302-9743 (Print) 1611-3349 (Online)
4. Clavier, C., Coron, J.-S., Dabbous, N.: Differential power analysis in the presence of hardware countermeasures. In: Paar, C., Koç, Ç.K. (eds.) CHES 2000. LNCS, vol. 1965, pp. 252–263. Springer, Heidelberg (2000)
5. Coron, J.-S., Goubin, L.: On Boolean and arithmetic masking against differential power analysis. In: Paar, C., Koç, Ç.K. (eds.) CHES 2000. LNCS, vol. 1965, pp. 231–237. Springer, Heidelberg (2000)
6. Coron, J.-S., Kocher, P.C., Naccache, D.: Statistics and secret leakage. In: Frankel, Y. (ed.) FC 2000. LNCS, vol. 1962, pp. 157–173. Springer, Heidelberg (2001)
7. Mayer-Sommer, R.: Smartly analyzing the simplicity and the power of simple power analysis on smartcards. In: Paar, C., Koç, Ç.K. (eds.) CHES 2000. LNCS, vol. 1965, pp. 78–92. Springer, Heidelberg (2000)
8. Messerges, T.S.: Using second-order power analysis to attack DPA resistant software. In: Paar, C., Koç, Ç.K. (eds.) CHES 2000. LNCS, vol. 1965, pp. 238–252. Springer, Heidelberg (2000)
9. Messerges, T., Dabbish, E., Sloan, R.: Investigation of power analysis attacks on smartcards. In: Usenix Workshop on Smartcard Technology (1999), http://www.usenix.org
10. Liu, P.-C., Chang, H.-C., Lee, C.-Y.: Low Overhead DPA Countermeasure Circuit Based on Ring Oscillators. IEEE Transactions on Circuits and Systems 57(7), 546–550
11. Danis, A.U., Berna, O.: Differential Power Analysis Attack Considering Decoupling Capacitance Effect. In: European Conference on Circuit Theory and Design, ECCTD 2009, pp. 358–362 (October 2009)
12. Semenov, O., Vassighi, A., Sachdev, M., Ali, K., Hawkins, C.F.: Burn-in Temperature Projections for Deep Sub-micron Technologies. In: Proceedings of International Test Conference, ITC 2003, pp. 95–104 (2003)
13. Pramstaller, N., Oswald, E., Mangard, S., Gürkaynak, F.K., Häne, S.: A masked AES ASIC implementation. In: Proc. Austrochip, pp. 77–82 (2004)
14. Akkar, M.-L., Giraud, C.: An implementation of DES and AES, secure against some attacks. In: Proc. CHES 2001, pp. 309–318 (2001)
15. Trichina, E., Seta, D.D., Germani, L.: Simplified adaptive multiplicative masking for AES. In: Proc. CHES 2002, pp. 71–85 (2002)

A Study on Smartphone APP Authoring Solution Design for Enhancing Developer Productivity

Young-Hyun Chang and Dea-Woo Park[*]

Dept. of IT application technology, Hoseo Graduate School of Venture, Korea
baewhaoa@paran.com, prof1@paran.com

Abstract. This paper is to propose an authoring solution design for the development of smartphone applications with optimized productivity, by reducing cost and time period of development. Especially, more emphasis is placed on design methodology for authoring solution for developers to develop a business application program running on smartphones with easiness and speediness. The final objective of the thesis is to propose a method on how to develop a daily work process program by industry on a real time basis, using an authoring solution for smartphone application. In other words, this thesis covers a high-tech authoring solution design for the smartphone applications to be used by experienced developers with enough working knowledge in office.

Keywords: Smartphone, Smartphone Application, Authoring Solution, Developer Productivity, Business Application.

1   Introduction

Smartphones have an immense popularity initiated with the release of Apple's iPhone all over the world. A new industrial revolution has already started for applying IT technology on the basis of Smartphones and the wireless Internet in all industries and fields.

In general, job programs have been developed by developers who are used to program languages, e.g., Java, Visual Basic, C, etc.[1][2] However, in most cases, programs developed by developers who do not have relevant job knowledge and experience in business cannot meet the request in business and both of immense time and expenses are consumed.

Current APP programs show the tendency of requesting all members' capability for IT technology in each team even for unimportant things. Therefore, it is not possible for a small number of staffs in charge of computing to handle the work at all [3][4]. It is the most efficient that app is developed by the people who have a thorough knowledge of the relevant work for app development in a company which has lots of requirements and continues to change. That is, this is the time when the concept is needed to be introduced that the subject of development is transferred from IT development experts to staffs skillful in the relevant job.

---

[*] Corresponding author.

G. Lee, D. Howard, and D. Ślęzak (Eds.): ICHIT 2011, CCIS 206, pp. 160–166, 2011.
© Springer-Verlag Berlin Heidelberg 2011

Development of app editors is to apply and realize the concept. Namely, it is that staffs in charge of relevant work who are not experts in IT development develop and use desired app.

New innovative app editors for realizing such a request in business should be implemented by the technology that enables results to be immediately produced as if we use the word processors, Word or Excel which are the easiest word processors to learn with a computer.

Editors are required which allow actual app implementation to be processed automatically by an editor engine of artificial intelligence when an office worker designs desired app windows and functions with state-of-the-art GUI tools in the development environment that ordinary office workers can easily learn how to use it.

Mow, the most important element in app development is actual work experience or ideas about how to make which program. That is, this is the user-oriented time in which actual work experience or knowhow and the idea about what to make is important.

2 Design Methodology of Smart APP Editor for Efficient Development

2.1 Changes in Mobile IT Environment

Changes in computer processing in the environment with the infrastructure of mobile IT, firstly, evolve into automated field processing in real time where the job is needed to be processed, and, secondly, result from expansion in the object and scope for information due to high-tech input/output device and portability of smart phones [5].

With respect to changes in the objective of promoting IT technology, firstly, the enterprise system already established, e.g., ERP, CRM, GW, etc., is excluded for promoting new IT technology[6][7]. Secondly, general jobs of staffs in all job types in a company are automated to be compatible for the smart environment. Thirdly, since it is predicted that mobile IT technology will be explosively requested from all jobs and all teams of all companies, it would be impossible to meet such request with conventional software production systems and a new development environment is thus required.

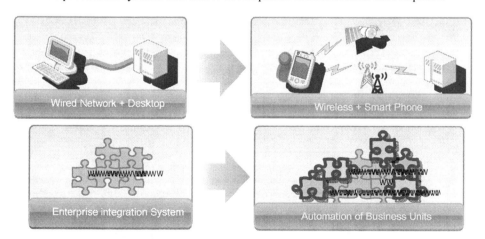

Fig. 1. Changes in mobile IT environment

2.2   Configuration and Connection of Mobile Expansion and Integration System

The configuration of mobile expansion and integration system applicable to business is described in 4 steps: the step of operating and managing 3 core resources of finance, material and manpower; the step of decision transfer between teams and works and decision making; and the step of supporting all processing in all types of jobs classified into supporting external users for services, using the app for processing external customer's orders and supporting in-house users for services, using the app for managing in-house member's request. The important 2 steps are classified in details in the following.

Table 1. Compare On-Site Working Solution and Event Handling Application

On-site working solution	Categories	Event handling application
Systems of lump-sum processing of job process of staffs in charge for each of in-house jobs	Overview of solution	Event processing app program, e.g., request and commission of job processing to a specific staff
The first key objective is automation of and application of IT technology to daily job processing and all processes of staffs in charge of each job. The auxiliary objective is to collect management and accounting data obtained in the process, connect it to and integrate it with the backbone system.	Objective	App program for processing events and behavior, e.g., request of processing, commissioning jobs to specific job and service providers by external customers or in-house staffs from the standpoint of using the job services, reviewing, and approving the work.
On-site real-time processing where the job is requested, e.g., customer sites, markets, moving vehicles, factories, offices, etc.	Where to use	Processing all sorts of jobs in a spot where they are requested, e.g., offices, homes, streets, etc.
(In field) salespersons, service persons, inspection of materials, storage in warehouses, issuing materials, process management, product release, transportation management, transportation personnel, guards, etc. (In office) marketing, ad PR, time and attendance, welfare, personnel affairs, facility management, reception of consumables, quality testing personnel, secretaries, etc.	App items to be developed	(In-house customer) simple job processing, e.g., holiday application form, report of absence, request of consumables, request of materials, request of business trip, request of guarding, request of certificate issue, settlement of travel and transportation cost, settlement of expenses, request of purchase, using offices, diary of using vehicles, job diaries, etc.
(Small and medium sized companies) 20~30 types (Large companies) jobs of 150~220 types	Job type and functions	(Small and medium sized companies) 50~60 types (Large companies) jobs of 300~400 types

3   Designing an Editor for Establishing Mobile Expansion and Integration System

3.1   Key Functions of the Editor

- GUI for designing app program windows and functions
- Implementation of business forms of high functions
- The function of engine of artificial intelligence for automatic implementation of analyzing document windows and then automatic design of DB
- Automatic creation of DB according to the design map
- Packing app modules for direct installation and running in smart phones
- Uploading packed app products and selling them in an open market
- Development, using high-tech technology which does not require coding
- Development of business SW of various types
- Development even by non-software experts
- Establishing an integration system of new concept fully integrating the wired and wireless networks
- Easy migration of existing systems
- Automatic implementation of data layers in designing program input/output windows
- Automatic creation of event and process control functions
- Wizard type database command operation.

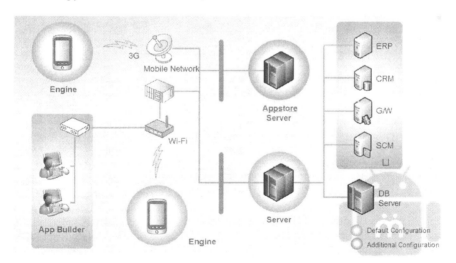

Fig. 2. Configuration of Smartphone APP authoring solution design

3.2   Applicable Jobs for Each Editor

Jobs that can be established with editors are described below.

- Develop app for unit job: Marketing, purchase, inventory, production, personnel management, finance and accounting, sales, campaign, A/S, customer management, transportation, storage, distribution, release, manager information, etc.

- Development of backbone systems, office work based systems, knowledge based systems, field work based systems, and transaction based systems
- Applications to each industry: Manufacturing, construction, wholesale and retail business, service industry/ banking, communication, hospital, intellectual   service industry/ public institutions, e.g., administration, defense, and schools
- Self-employed users and professional users: Business type, A/S, intellectual service, personal service, social care service, investigation and analysis, conducting public affairs.

3.3   Engine for Operating the Editor

Key functions of the engine include the following.

- Provides all functions required for developing and operating general app as universal elements.
- Activate, run and stop app programs and manages conversion of running and priorities.
- Systematically manages menus of all of installed app products and program modules.
- Controls server and DB access and communication, e.g., Wi-Fi, 3G, Bluetooth, etc.
- Performs real-time testing by direct connection to PC and Wi-Fi for app development.

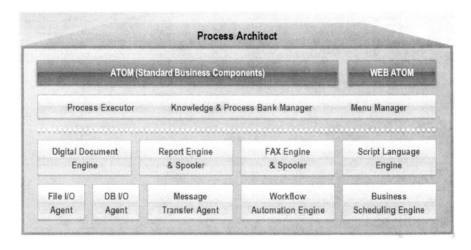

Fig. 3. Engine for operating the editor

3.4   Platform for Operating the Editor

Main functions of the operation platform are described below.

- Universal elements required for running all of app are provided to the server operation platform in a lump sum.

- Transaction processing monitor for safely processing immense amount transaction data.

- Management of activating, running and stopping app programs and priorities in the simultaneous multi-user environment.

- Registration and management of servers, DB, company information, organization system, users, access right, etc.

- Log information management for monitoring and analyzing access and processing by all users.

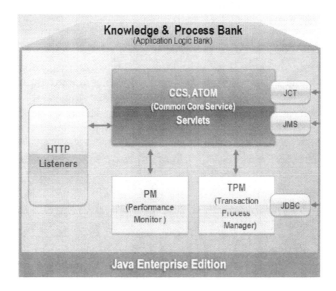

Fig. 4. Platform for operating the editor

3.5   APP Store for Editor

Main functions of app store are as follows.

- Provides license so that all members can use the powerful editor which enables office uses to make app for themselves.
- Distributes the produced app to the app store by means of easy key operation so that the entire organization can use it.
- Prior verification and control for functional connection and data sharing between the existing systems, e.g., ERP, G/W, etc., and app
- Mileage is support for systematically monitoring, encouraging and supporting app development and result of use.
- Systematic upgrade and version management of servers, platforms, editors, and app programs.

4   Conclusions

In most industry, there is an explosive demand for Biz App today. The second App heyday following Apple's success came. App business can be successful without capital and sales competence. The world market can be mine including domestic consumers if I efficiently utilize the open market established in the endless wireless Internet space.

There is an urgent need to improve university curriculums to meet the current conditions, especially to reorganize curriculums of departments closely related to industries, e.g., engineering, industrial engineering, management, management information, accounting, etc., to meet the request from the smart society. It is necessary to learn the capability of job automation and informationization for commercial implementation of the theories and knowledge learnt in universities as app programs, and for using mobile computing devices, e.g., Smartphones and tablet PC, etc.

For the education environment based on Smartphones and the wireless Internet, the basic infrastructure includes software and hardware, e.g., firstly tools for business process analysis and system design, secondly tools for mobile app program design and implementation edition, thirdly the engine for executing application programs, fourthly the app program operation platform server for business.

The app editor to be designed in this study is an app development solution that enables easy and quick development of applications for automating all business activities and tasks. Everyone including staffs in business, managers, students and the like without prior programming experience, and experts who is good at programming language can develop applications by means of minimum learning 30 times faster than using the development language, e.g., C or Java.

References

1. Kim, Y.J.: Research of web design management regarding to web accessibility, Design Management. SungKyunKwan University (2010)
2. Kim, S.J.: A Study for Selecting Web Accessibility Evaluation Tool in Suitable of Homepage Characteristics. Computer Science Incheon University (2010)
3. Chang, Y.H.: Design and Implementation of a Large Scale Qualification Management System for Performance Improvement Through the Use of a WCBT(Web and Computer based Test). Journal of The Korea Society of Computer and Information 13(2), 67–78 (2008)
4. Chang, Y.H., Seo, J.M.: Development of National Qualification Management System for Performance Improvement based on Real-Time Data Sharing. Journal of The Korea Society of Computer and Information 13(4), 214–220 (2008)
5. Park, J.Y.: Design and Implementation of Application system for Mobile Internet. Computer Science Dongguk University (2001)
6. Yoon, S.S., Won, R.Y., Bae, P.S.: Real-Time Evaluation System for Acquisition of A Computer Certificate of Qualification. Journal of The Korea Society of Computer and Information 14(1), 221–228 (2006)
7. Yun, J.K., Tak, J., Baek, Y., Lee, S.: Web Site Building System Using Templet. Journal of The Korea Society of Computer and Information 15(1), 83–87 (2010)

A Study of Disaster Preparedness Systems Operations Analysis and Financial Security Measures of Large Banking Network

Jong-Il Baek and Dea-Woo Park[*]

Dept. of IT application technology, Hoseo Graduate School of Venture
jibaig101@empal.com, prof1@paran.com

Abstract. The recent server system failure, which happened in April 2011, of Nonghyup bank left it in disgrace that the overall management of server system in the banking network has been in insolvent operation. It is one of the most urgent issues that the computer system administration and security should be properly managed as IT technology develops at speed of lightning. In this thesis, it will be comprehensively investigated how the financial institutions deal with the management system against the computer failure and the optimum solution to take care of their financial computer systems and information protection policies. For the optimized operation of banking network system, it is necessary to come up with the improvement plan through the quantitative evaluation of centralized and distributed processing system. The stability in the information protection system needs to deal with the role-based access control and distributed security management system against the access of internal & external users or illegal intruders. In this survey, it will be highlighted that the next generation of banking network infrastructure, which has been reinforced in terms of its stability.

Keywords: RBAC (Role Based Access Control), Centralized System, Distributed Processing System, Financial Security, Banking network.

1   Introduction

Paralysis of Nonghyup's computer network that broke out in Korea in April 2011 showed the weakness of information security. The accident was caused by the lack of awareness of information security, the absence of security policy on subcontractors, poor operation of intrusion prevention system, lack of system that can cope with failure and absence of CSO (Chief Security Officer) in charge of information security.

Availability and confidentiality are indispensible for effective system operation and security system design. In order to improve information security, the importance of information protection should be emphasized and strict security policy should be established and measures that can cope with unexpected problems should be prepared. On this end, it is necessary to study information protection by establishing an organization in charge of information protection.[1]

---

[*] Coresoponding author.

G. Lee, D. Howard, and D. Ślęzak (Eds.): ICHIT 2011, CCIS 206, pp. 167–173, 2011.

In chapter I, the background and need of this study were described. In chapter II, actual state of financial information network was examined. In chapter III, the weakness of information protection system was drawn by analyzing Nonghyup's operation of system to cope with network breakdown. In chapter IV, a measure to protect financial information network was proposed. In chapter V, this study was summarized and direction to future research was proposed.

2   Related Research

2.1   Financial Institution Operating System

A. Centralized System
Centralized system enables users to access through on-line and in remote terminal. Banks have used centralized system to support automatic teller machine (ATM: Automatic Teller Machine) that acts as terminal which enabled users to deposit or withdraw money in remote terminals. System under main frame that enterprises used in early stage depended on processing of highly efficient mainframe computer and in the system, mutual communication was impossible.

B. Distributed Processing System
Distributed computing is a field of computer science that studies distributed systems. A distributed system consists of multiple autonomous computers that communicate through a computer network. The computers interact with each other in order to achieve a common goal. A computer program that runs in a distributed system is called a distributed program, and distributed programming is the process of writing such programs.[2]

2.2   Disaster Recovery System

Disaster recovery is the process, policies and procedures related to preparing for recovery or continuation of technology infrastructure critical to an organization after a natural or human-induced disaster. Disaster recovery is a subset of business continuity. While business continuity involves planning for keeping all aspects of a business functioning in the midst of disruptive events, disaster recovery focuses on the IT or technology systems that support business functions.[3]

2.3   RBAC (Role-Based Access Control)

In computer systems security, **role-based access control(RBAC)**is an approach to restricting system access to authorized users. It is used by the majority of enterprises with more than 500 employees and is a newer alternative approach to mandatory access control (MAC) and discretionary access control (DAC). RBAC is sometimes referred to as role-based security.[4][5][6]

2.4   Actual State of Security System Operated by Financial Institutions

All security solutions that financial institutions are using were based on software method which made it difficult to protect information network from data hacking. The accident which Nonghyup suffered from was difficult to prevent though the improvement of security solution and monitoring system was made. There is limit in judging whether command from ROOT authority is normal or not in S/W level and only ROOT manager who implemented delete command knows whether command from ROOT authority is normal or not even in artificial intelligence system. Security measure that can prevent attack or force delete of data even when security fails due to virus or malicious code should be arranged.[7][8]

3   Analysis of Nonghyup's System Prepared to Cope with Network Failure

3.1   Actual State of Nonghyup's Network Operation

ROOT authority is powerful in all computer systems and can change and control setting of all applications operated in OS level. In the case of Nonghyup, it is assumed that file delete command was made by ROOT authority. A malignant hacker can make ill use of file delete command. File delete command can be caused by a variety of variables such as manager's negligence or mistake or malignant data attack by a malicious manager. Nonghyup's network breakdown was caused by the poor management of information protection system about accessible terminals. In addition, allowing subcontractors to have ROOT authority was responsible for paralysis of Nonghyup's computer network.

3.2   Analysis of Nonghyup's Network Breakdown

Nonghyup's network breakdown was caused by cyber terrorism. According to inspection , computer was infected while an employee who belonged to IBM that was in charge of server maintenance was downloading a movie in a notebook computer which was used for server management by using free download coupon obtained from a coffee shop on September 4, 2010. This notebook computer was infected with malicious code a hacker planted and suspects planted various malicious programs and stole important information on network management such as password and installed eavesdropping program for about seven months. It was revealed that attack command was installed at a notebook which degenerated into zombie PC at 8:20 a.m. April 12, 2011. Hackers implemented attack program through remote control that used internet at 4:50 p.m. and since then system destruction has been made. When delete command was made through notebook computer at 4:50 p.m. April 12, 2011, half of servers were destructed in only 30 minutes.

3.3   Analysis of Weakness in Operation of Nonghyup's Computer Network

Weakness found by analyzing actual state of operation of Nonghyup's computer network was classified into category of information security, backup system, management and table 1 shows problems.

Table 1. Analysis of weakness in operation of Nonghyup's computer network

Section	Problem
Information Security	- The capability that can cope with new and varietal malicious codes such as reverse connection that evolve rapidly is very weak. - Nonghyup's computer network system depends on packet/pattern analysis method which made Nonghyup's computer network system vulnerable to bypass attack. - Nonghyup's computer network system does not have any measures that can prevent data destruction after computer security incident broke out. - Information protection solution of Nonghyup's computer network system was operated at OS level which made it impossible to protect information when ROOT authority was exposed.
Backup System	- When infected file is backed up, backup itself causes data loss and resulting in adverse effect. - Securing physical availability is possible but securing data stability and data preservation is impossible.(securing recovery is impossible) - Securing updated data after computer breakdown is impossible. (data loss)
Management	- Allowing employees of subcontractors to have ROOT authority. - High budget was required to renew licenses and pay subcontractors.

4   Measures to Financial System Security

The effort to establish security measure for each area by analyzing weakness in actual state of Nonghyup's computer network and to apply the measure properly should be made.

4.1   Measures to Improve Weakness in Information Security

Measures to improve weakness in information security include implementation of role based access control policy and authority management system.

A. RBAC Enhancements

Within an organization, roles are created for various job functions. The permissions to perform certain operations are assigned to specific roles. Members of staff (or other system users) are assigned particular roles, and through those role assignments acquire the permissions to perform particular system functions. Since users are not assigned permissions directly, but only acquire them through their role (or roles), management of individual user rights becomes a matter of simply assigning appropriate roles to the user; this simplifies common operations, such as adding a user, or changing a user's department.

Three primary rules are defined for RBAC:

1. Role assignment: A subject can execute a transaction only if the subject has selected or been assigned a role.

2. Role authorization: A subject's active role must be authorized for the subject. With rule 1 above, this rule ensures that users can take on only roles for which they are authorized.

3. Transaction authorization: A subject can execute a transaction only if the transaction is authorized for the subject's active role. With rules 1 and 2, this rule ensures that users can execute only transactions for which they are authorized.

Additional constraints may be applied as well, and roles can be combined in a hierarchy where higher-level roles subsume permissions owned by sub-roles.

B. Account management system and authority management system

Account management system and authority management system involved account of retirees, shared account, password management, user status, absence of consistency in account policy tracking and system approach as shown in table 2.

Table 2. Method of improving security through account management system and authority management system

Section	Measures
Account Management	- Measure that blocks account right after employees leave a company should be taken - In the case of temporary accounts, automatic blocking should be made
1 person 1 account	- Use of shared account should be removed - Super account should not be used in APPS
Access enforcement	- Direct login of Super users should be controlled - Issuance of role based system account should be systemized
Audit	- Issuance of user's account and change history should be inspected - User's system access history should be tracked
Password Policy enforcement	- Policy that requires periodical password change should be implemented - Improvement of password management through self service

4.2   Measures to Improve the Weakness of Backup System

Measures to improve the weakness of backup system to cope with financial service breakdown involve real time dual and prevention of data loss to secure data availability and data preservation.

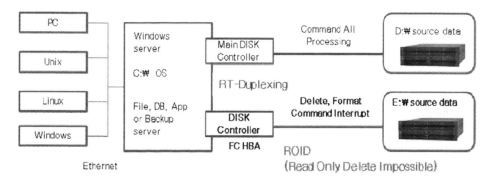

Fig. 1. System diagram for real time duplexing backup and prevention of data loss

Data preservation can be secured by adopting the function of "Delete Impossible" and supporting real time duplexing backup without extra application through real time duplexing backup at device level. Restoration is possible through Copy & Paste function and regeneration of backup data is also possible. Recovery can be made by simply using server reboot after changing connection route through auxiliary backup device.

4.3   Measures to Improve Weakness in Management

Internal guidance that can inspect information protection management system in which CSO plays a central role should be made to improve weakness in management. Thorough management about access to important system and control over subcontractors are necessary. Table 3 shows the management of internal human resources this study proposes.

Table 3. Method of managing account life cycle management for improvement of personnel security

Time	Method of account management
Incident	Creation of account and authority for employee in accordance to access rule that company established
Move department	Automatic change in authority when user is transferred to another department - VPN system account - Registration of fingerprint identification system account - Change in authority according to security levels - Account creation for internal application - OS account about management system
Promotion (Team manager)	User entitlement and authority delegation, authority acquisition to access company's internal data as team manager - Registration as manager - Identification and change of authority - Creation of business management system account
Project commitment	Acquisition of system authority and application authority according to role and task given in Task Force Team - Database access authority - Access to internal TFT File server - Access authority to applicable projects in group wares
Project termination	Automatic removal of authorities given after project is finished - Removal of access authority to internal database - Removal of access authority to file server - Removal of groupware authority
Retirement	When an employee leave a company, the authority, user ID that the employee had should be automatically removed - ID card and computer should be returned to a company - Access to system should be prohibited

5   Conclusions

This study aimed to find a measure to improve financial security and to analyze operation of system to handle computer network breakdown. Nonghyup's computer network breakdown shows how weak Korea's information security. The most important thing to enhance information security is the improvement of security awareness though investment in information security is important. This study analyzed actual state of Nonghyup's computer network and proposed a measure to improve the weakness of Nonghyup's computer network. Access policy and authority control policy should be established to establish information protection system. The effort to create system that can restore breakdown efficiently in the case of breakdown should be made. Routine security inspection should be made. Efficient security management policy should be made by competent experts.

Further research that aims to find effective active information security system that can properly cope with malicious codes that evolves rapidly and to design detection system for computer virus or malicious codes should be conducted in the near future.

References

1. Baek, J.-I., et al.: A Study on Traceback by WAS Bypass Access Query Information of Database. Journal of The Korea Society of Computer and Information 14(12), 181–190 (2009)
2. Park, S.-C., et al.: Barriers to information systems analysis, operational process improvement through research. The Journal of Korean Operations Research and Management Science Society, 136–140 (October 2008)
3. Chi-Ming, C., Macwan, A., Rupe, J.: Network disaster recovery [Guest Editorial]. The Journal of IEEE Communications Magazine 49(1), 26–27 (2011)
4. Jeong, S.-M.: Effective role for financial applications and secure role-based access control extract Applying, Korea Institute of Information Security and Cryptology. KIISC thesis Journal 18(5), 49–61 (2008)
5. Mun, H.-J.: Sensitive Personal Information Protection Model for RBAC System. Journal of The Korea Society of Computer and Information 13(5), 103–110 (2008)
6. Strembeck, M., Mendling, J.: Modeling process-related RBAC models with extended UML activity models. The Journal of Information and Software Technology 53(5), 456–483 (2011)
7. Mogull, R.: Database Activity Monitoring Is a Viable Stopgap to Database Encryption for the Payment Card Industry Data Security Standard (and Beyond), Gartner (July 2006)
8. Seol, M.-S.: Construction of Financial Networks based on Virtual Private Networks. The Journal of the Korea Contents Association 9(8), 41–48 (2009)

Malware Variant Detection and Classification Using Control Flow Graph

Donghwi Shin, Kwangwoo Lee, and Dongho Won*

Information Security Group,
School of Information and Communication Engineering, Sungkyunkwan University,
300 Cheoncheon-dong, Jangan-gu, Suwon, Gyeonggi-do 440-746, Korea
{dhshin,kwlee,dhwon}@security.re.kr

Abstract. The number of malware increases steadily and is too many. So a malware analyst cannot analyze these manually. Therefore many researchers are working on automatic malware analysis. As a result of these researches, there are so many algorithms. The representative example may be a behavior based malware automatic analysis system. For example, these are the Bitblaze [1], Anubis[2], and so on. However these behaviors based analysis result is not enough. So for more detail analysis and advanced automatic analysis feature, the automatic static analysis engine is necessary. Then some projects apply an automatic static analysis engine and the research on automatic static analysis is working. These analysis methods use the structural characteristic of malware, and that is the reason the malware is also software, there is a toolkit for a malware generation, and a malware author reuse some codes. For automatic static analysis, it is so useful that the static analysis engine uses the structural characteristic of malware. However previous researches have some problem. For example, these are a performance, false positive, detection ratio, and so on. Therefore we'll describe another method that used the structural characteristic of malware.

Keywords: Malware, Malicious Software, Control Flow Graph, Structural Analysis, Profiling, Signature, Security.

1 Introduction

The malware is not just simple threat. As a malware become a means for a profit, there is the malware industrialization. As a result, so many malware appeared and the number of malware increased rapidly. It made more difficult to a response. Therefore a research is working on automatic response. So an automatic analysis technology was developed and developing. However to guarantee high profits, many intelligent malware is come out and this is making more and more difficult to a response. In recently, a malware is becoming modularized and form a family. So it is an important issue to understand and analyze a relation with between executable files. Just a

---

* Corresponding author.

G. Lee, D. Howard, and D. Ślęzak (Eds.): ICHIT 2011, CCIS 206, pp. 174–181, 2011.

behavior based analysis system is not enough for the relation analysis. So this system needs a static analysis engine for a more accurate analysis. So a research is working on an automatic static analysis method. However there are so many difficulties. Nevertheless many researches proposed many static analysis algorithms. But, these algorithms had some problems. So in this paper, we propose an algorithm that considers a structural characteristic and an instruction semantic of executable files. This leads to solve some problems of previous researches and to prepare an algorithm application in field. This paper is structured as follows. Section 2 describes related works. Section 3 describes CFG based analysis and classification algorithm in detail. Finally in section 4 we conclude.

2   Related Works

CFG is the abbreviation of "Control Flow Graph". The component of CFG is a node and edge. In CFG, the node means a basic block. And the basic block is a continuous instruction sequence that is divided by jump instructions. And an edge means the path between basic blocks by jump instruction. Specially, there is an entry node and exit node. We call the entry node is a basic node at the start of CFG. And the exit node is a basic block at the end of CFG.

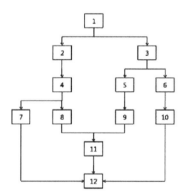

Fig. 1. CFG Diagram

The Fig. 1 is a CFG. The CFG includes 12 nodes and 14 edges. And the entry node is number 1 basic block and the exit node is number 12 basic blocks. According to circumstances, there may be many exit nodes. In the CFG structure, we are able to know that CFG has a flow characteristic of program.

In the Silvio Cesare research, he used that a CFG is representative of the program flow. He identified a variant of malware using CFG characteristic. At first, he disassembled a malware and the disassemble result was translated into an intermediate language. So the disassembled code is translated into the intermediate language depending on Fig. 2. And then he represented the intermediate language as a string. He calculated a similarity using edit distance algorithm [3]. I think that the main contribution of Silvio Cesare research is that he utilized instruction semantic, that a basic block include, as a factor of variant classification on a CFG based analysis. However the research did not present an enough experiment result in the evaluation part for a variant classification because the evaluation worked with little samples.

Fig. 2. The intermediate language grammar

As another research, there is the Christopher Kruegel research using a structural characteristic of executable file for the polymorphic worm detection. And the structural characteristic of executable file is extracted from CFG. The Christopher Kruege brought the k-subgraph, graph coloring technique, and adjacent matrix concept in a structural information extraction from CFG. And through graph coloring technique, the Christopher Kruege research used the instruction semantic for a classification. So the main contribution of the Christopher Kruege research is an effective algorithm for polymorphic worm detection. Another contribution is that the Christopher Kruege specified requirements of CFG based signature. The requirements are as following.

1. Uniqueness
2. Robustness to insertion & deletion
3. Robustness to modification

The signature for CFG based analysis must follow these three requirements. Especially though robustness parts are able to be ignored easily, the Christopher Kruege described these in detail. However the problem of Christopher Kruege research is also performance. His algorithm must identify each instruction semantic. These feature cause long analysis time. And it did not analyze a function that had basic blocks under number k because of k-subgraph concept. Except above-mentioned researches, Goldberg [4], Wehner [5], Carrera [6], Karim [7] analyzed malware mutual relation or variant and classified these by a structural characteristic. But, the major problem of these researches is a false positive, applicability, and performance including time complexity.

3   CFG Based Analysis and Classification

In this paper, we propose the new analysis method based on CFG. We can calculate the similarity ratio or value of malwares and identify that something is the most similar by the proposed method. The CFG based analysis algorithms have five elements as Fig. 3. These elements are divided as function.

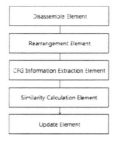

Fig. 3. CFG based analysis and classification system diagram

The function of each element is as follows.

Table 1. Element Description

Element	Input	Description
Disassemble Element	Target Malware	Disassemble executable file
Rearrangement Element	Instruction Sequence	Arrangement instruction sequence by function and create CFG
CFG Information Extraction Element	Instruction Sequence by Function	Extract CFG information
Similarity Calculation Element	Extracted CFG Information, Malware CFG Information on DB	Calculate a similarity about input malware
Update Element	Similarity Information	Update malware database

We disassemble a text area of executable file for an analysis. The "Disassemble Element" performs this function. The role of disassemble element is to get an instruction sequence. The output of disassemble element enters into "rearrangement element" as an input. This is a preparation process for a similarity calculation. After the classification process of rearrangement element, the basic preparation process is finished for CFG information extraction. After the rearrangement element arranges instruction sequence by function, the output is an input of the "CFG information extraction element". The CFG information extraction element extracts information from CFG for a similarity calculation. The extracted information is as followings.

- Number of basic block in CFG, Number of edge in CFG
- Number of call instruction in CFG, Number of push instruction in CFG

To remedy previous research's problem and to get advantage, we extracted these information from CFG. As we mentioned in section 2, the Silvio Cesare translated instruction sequence into intermediate language by instruction semantic. This method held an advantage that this method includes an instruction semantic. And Marius Gheorghescu mentioned that disadvantage of his research is what have no consideration for an instruction semantic. Therefore we extracted information from CFG for an instruction semantic and a structural characteristic. And this is to satisfy 3 requirements of the Christopher Kruegel about signature.

Let's look into the extracted information. The first information among extracted information from CFG reflects the structural characteristic of CFG. After we consider a structural characteristic by a number of basic block and edge, now we have to consider an instruction semantic that a CFG includes. So we have to select the most effective instruction. We select "call" and "push" instruction as the most considerable instruction.

We need to check a theoretical effectiveness about this information that is extracted from CFG. We experiment the information that is extracted from malware "Induc" variant and "Kido" sample.

The extracted CFG information is expressed in the coordinate value as following form.

$$[\text{number of basic block, number of edge,} \atop \text{number of call instruction, number of push instruction}] \qquad (1)$$

If we compared the extracted information from Induc variant samples, these values are very similar.

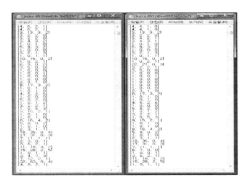

Fig. 4. Induc CFG Information Comparisons

However, a problem came up during a simply comparison. It is because that there are many values such as [1, 0, 0, 0] or [1, 0, 0, 1]. So these patterns are not enough to represent the malware as a signature. Therefore we excluded this information, for example [1, 0, 0, 0] and [1, 0, 0, 1] and so on, in a similarity calculation step. So we need to be a filter for excluding some meaningless information. In this research, we determine the criteria of filter as followings

- The number of basic block > 2
- The number of edge > 2
- The number of call instruction > 2
- The number of push instruction > 5

Of course, we cannot say that this criterion is certainly suitable. For a suitable or practicable criterion, we'll need more experiment. And you should set a criterion for your situations.

Until now, we extracted a CFG information for a comparison from the disassemble element to the CFG information extraction element. And then the output is a sequence of set that is composed of 4 values.

$S = \{x_i \mid x_i = [p, q, r, s]$, p = number of basic block, q = number of edge, r = number of call instruction, s = number of push instruction, $1 \leq i \leq$ number of function$\}$

The S is a set of point on 4 dimensional spaces. So if these points are very similar, the distance between these points is too close. And if these are the same point, the distance is zero. Now, the Similarity Calculation Element calculates should calculate the distance between points in set A and B on 4 dimensional space. For example, there are two points on 4 dimensional space and these points are m(p1, q1, r1, s1), n(p2, q2, r2, s2). Then the distance d is as following

$$d = \sqrt{(p_1 - p_2)^2 + (q_1 - q_2)^2 + (r_1 - r_2)^2 + (s_1 - s_2)^2}$$

Fig. 5. Distance between m and n

The Similarity Calculation Element calculated the distance and collected these result.

$C = \{d_i \mid d_i = $ distance between x_i and y_i, $x_i = $ CFG information from malware X, $y_i = $ CFG information form malware Y$\}$

The element set C is real number and there may be the same real number in set C. For example, if the extracted information from two functions is the same, the distance is zero. Therefore there is able to be the same element in set C. So we set section of distance and calculate ratio (%) of the section. And we assign a weighted value like the following table.

- Area 0 Ratio R0 = entity # of area 0 / n(C) * 100
- Area 1 Ratio R1 = entity # of area 1 / n(C) * 100
- Area 2 Ratio R2 = entity # of area 2 / n(C) * 100
- Area 3 Ratio R3 = entity # of area 3 / n(C) * 100
- Area 4 Ratio R4 = entity # of area 4 / n(C) * 100
- Area 5 Ratio R5 = entity # of area 5 / n(C) * 100

We just assigned the weighted value to the section that had a high similarity ratio. For example, there are malware X, Y, and Z. When we compared with these, there was a high similarity in section "0" in case X and Y. Then we assigned weighted value 50 to the section 0 between X and Y. And if there was a high similarity in section "1", we

Table 2. Weighted Value

Section	Area	Weighted Value
0	0	50(W0)
1	1 ~ 5	25(W1)
2	6 ~ 10	13(W2)
3	11 ~ 15	6(W3)
4	16 ~ 20	4(W4)
5	21 ~ 25	2(W5)
6	26 ~ 30	0
7	31 ~ 35	0
8	35 ~ 40	0
9	40 ~	0

assigned weight value 25. The following equation is the simply expression about weight assignment and similarity calculation algorithm.

$$\text{if } R0 == MAX(R0) \text{ then } S0 = W0 \text{ else } S0 = 0$$
$$\text{if } R1 == MAX(R1) \text{ then } S1 = W1 \text{ else } S1 = 0$$
$$\text{if } R2 == MAX(R2) \text{ then } S2 = W2 \text{ else } S2 = 0$$
$$\text{if } R3 == MAX(R3) \text{ then } S3 = W3 \text{ else } S3 = 0$$
$$\text{if } R4 == MAX(R4) \text{ then } S4 = W4 \text{ else } S4 = 0$$
$$\text{if } R5 == MAX(R5) \text{ then } S5 = W5 \text{ else } S5 = 0$$
$$\text{Similarity} = S0 + S1 + S2 + S3 + S4 + S5$$

Through a similarity calculation we can know what the most similar malware is. Finally we should update the malware database information for the next analysis. This update helps with a variant identification at the next analysis and helps with a variant detection performance. The "Update Element" performs this update.

4   Conclusion and Further Study

A fast-growing malware and a tool for generating malware creation the market, is related to malware, is developing. But, the progress of response human resource and equipment is not keeping up with a speed of market development. So there are so many researches about a malware response. Especially, the research about an automatic malware analysis is so active. As a result, a malware dynamic analysis is automated. However, the automatic dynamic analysis is not enough for a detail analysis. So the research about an automatic static analysis is working now. As previous mentioned in section 2, there algorithms for automatic static analysis had some problems. Therefore we proposed a new algorithm that used a structural characteristic for a variant identification. And this algorithm considered both a structural characteristic and instruction semantic. We designed this algorithm that focused on the detection performance. So if we need to improve an algorithm speed with keeping this algorithm basic concept, this algorithm will be used in the field.

Acknowledgments. This research was supported by the MKE (The Ministry of Knowledge Economy), Korea, under the "ITRC" support program supervised by the NIPA (National IT Industry Promotion Agency) (NIPA-2011-C1090-1001-0004).

References

1. BitBlaze: Binary Analysis for Computer Security, http://bitblaze.cs.berkeley.edu/
2. Anubis: Analyzing Unknown Binaries, http://anubis.iseclab.org/
3. Edit distance, http://en.wikipedia.org/wiki/Edit_distance
4. Goldberg, L.A., Goldberg, P.W., Phillips, C.A., Sorkin, G.B.: Constructing computer virus phylogenies. Journal of Algorithms 26(1), 188–208 (1998)
5. Wehner, S.: Analyzing worms using compression (2004)
6. Carrera, E., Erdélyi, G.: Digital genome mapping – advanced binary malware analysis. In: Proc. Virus Bull. Int. Conf., pp. 187–197 (September 2004)
7. Karim, E., Walenstein, A., Lakhotia, A., Parida, L.: Malware phylogeny using maximal p-patterns. In: Proceedings of the EICAR 2005 Conference, pp. 167–174 (April-May 2005)
8. Kang, M.G., Poosankam, P., Yin, H.: Renovo:A hidden code extractor for packed executables. In: Workshop on Recurring Malcode, pp. 46–53 (2007)
9. Gheorghescu, M.: An Automated Virus Classification System. In: Virus Bulletin Conference (2005)
10. Cesare, S., Xiang, Y.: Classification of Malware Using Structured Control Flow. In: Proc. 8th Australasian Symposium on Parallel and Distributed Computing (2010)
11. Krügel, C., Kirda, E., Mutz, D., Robertson, W., Vigna, G.: Polymorphic Worm Detection Using Structural Information of Executables. In: Valdes, A., Zamboni, D. (eds.) RAID 2005. LNCS, vol. 3858, pp. 207–226. Springer, Heidelberg (2006)
12. Trinius, P.: Visual Analysis of Malware Behavior Using Treemaps and Thread Graphs. In: Vizsec 2009, pp. 33–38 (2009)
13. Quist, D.A.: Visualizing CompiledExecutables for Malware Analysis. In: Vizsec 2009, pp. 27–32 (2009)
14. Zubair Shafiq, M.: PE-probe: leveraging packer detection and structural information to detect malicious portable executables. In: VB 2009 (2009)
15. Kaczmarek, M.: Architecture of a Morphological Malware Detector. Journal in Computer Virology (2008)
16. Vinod, P.: Static CFG analyzer for metamorphic. In: PIN 2009 (2009)
17. Song, D., Brumley, D., Yin, H., Caballero, J., Jager, I., Kang, M.G., Liang, Z., Newsome, J., Poosankam, P., Saxena, P.: BitBlaze: A new approach to computer security via binary analysis. In: Sekar, R., Pujari, A.K. (eds.) ICISS 2008. LNCS, vol. 5352, pp. 1–25. Springer, Heidelberg (2008)
18. Dullien, T., Rolles, R., Bochum, R.-U.: Graph-based comparison of executable objects (2005)
19. Sabin, T.: Comparing Binaries with Graph Isomorphisms (2004)

A Study on the Twofish Security Processor Design Using USB Interface Device

Seon-Keun Lee[1], Sun-Yeob Kim[2], and Yoo-Chan Ra[2]

[1] Faculty of Materials & Chemical Engineering,
Chonbuk National University Jeonju, Korea
[2] Department of Information and Communication Engineering,
Namseoul University Cheonan, Korea
caiserrisk@googlemail.com, sykim0599@nsu.ac.kr, ycra@nsu.ac.kr

Abstract. Since most cryptographic systems are PCI type, it is not suitable for users except experts to handle security system. In particular, most security programs used have not been verified against crack and are exposed to attack of virus or hacker. In this regard, this thesis designed Twofish cryptographic algorithm that uses USB, which can be easily used by users at general computers. Users can use security system easily using USB. In addition, Twofish cryptographic algorithm with various variable key lengths can be applied to various security system. Such Twofish cryptographic algorithm can improve performance for encryption and decryption, and increase efficiency of mobile security system in addition to USB.

Keywords: USB, Twofish, AES, PCI, Virus.

1 Introduction

Due to development of information and communication technology, we have at last 1 unit of PC for one person. To protect external invasion and damage from virus and spyware, various security programs are used for PC where personal information of users are saved.

However, unsecure configuration of IPv4/6, increase in PC user's dependence on internet information, hacking accident due to expansion of various hacking tools and DDOS have increased. Personal and corporate critical information are infringed due to various virus and regular system format at the office where security is important.

Accordingly, users should install security system at their PC but most security system are the PCI interface type that can be installed inside PC so that it is not suitable for the users unfamiliar with PC use[1]. Therefore, this thesis designed security processor of Twofish cryptographic algorithm that can prevent external invasion using USB interface technology, successful interface technology of PC.

2 The Transmission Types for USB

USB communication can be separated into the communication used for device enumeration and that used by application to perform the intended purpose of device.

G. Lee, D. Howard, and D. Ślęzak (Eds.): ICHIT 2011, CCIS 206, pp. 182–189, 2011.
© Springer-Verlag Berlin Heidelberg 2011

Transmission type in this communication is divided into control transmission, bulk transmission and interrupt transmission and isochronism transmission[2].

First, control transmission is used when host and device replace information on device function and when host reads and sets device information. It consists of setup stage, data stage and status stage, and each stage has over 1 transaction that consists of token phase, data phase and handshake phase. In setup stage, host starts setup transaction by sending information on request and token packet contains PID that recognizes itself as control transmission.

And data packet has information on request number, data stage status and information on request of data flow direction. Status stage consists of 1 IN or OUT transaction. In status stage, device reports success or failure of previous stage. Data packet source of status stage receives the data from data stage and if there is no data stage, device sends status stage data packet. At this time, status stage has the code regarding whether the data or handshake packet sent by the device is success or failure.

Accordingly, at the control transmission, setup stage and status stage are required and the data stage are optional, but required when it is specific request. Since all control transmission delivers information in both directions, message pipe uses endpoint address of same IN/OUT.

Second, bulk transmission is used for data transmission where transmission timing is not important. That is, since it gives way the bus to other transmission type and waits until it is available, it can send lots of data without interrupting other transmission of bus. The structure consists of over 1 IN or OUT transaction and has single-directional transmission. For bi-directional transmission, bus should be added and transmitted for each direction.

Third, interrupt transmission is used when data should be transmitted within given time. When the device sends data, it immediately has hardware interrupt on host and guarantees sending or transmission of data with the minimum delay. the interrupt structure is similar to bulk and has single-directionality.

At last, isochronism transmission can allow errors that occurs once in a while and is suitable for the streaming and real-time transmission that is meaningful only when the data arrives at certain rate or designated time. While it has an advantage of sending more data per frame compared to interrupt transmission, it has a shortcoming in that it is not ready to send again even if there is an error in the received data. As for structure of isochronism transmission, each frame has IN/OUT transaction at certain interval. Isochronism is single-directional like interrupt transmission and bulk transmission and requires separate pipe for each direction when bi-directional data is transmitted.

Host requests descriptor that has information on device using control transmission after enumeration, and requests another descriptors for the smaller components in the order of overall device, configuration, interface and endpoint during enumeration status. Upper-level descriptor gives information on additional lower-level descriptor to the host. As there exists only one descriptor that has information on overall device for each device, we can judge the number of configuration supported. Each device has over 1 configuration descriptor containing information on use of power and number of interface supported by configuration. Each interface descriptor of configuration has the number of endpoint descriptor containing information required to communicate with endpoint. Each endpoint descriptor has information on the method for the endpoint to transmit the data. In addition, interface without endpoint descriptor should use control endpoint.[3,4].

3   The Design of Twofish Cryptographic Processor for USB

Design of cryptographic system applied with Twofish cryptographic algorithm to prevent external invasion has the USB type for interface with general PC. Since general PC of user is always accessed to network and is vulnerable to external invasion or attack, the system was configured to protect the personal data and PC from external threat by inserting to PC of users anywhere anytime using the USB interface between general PC with cryptographic processor with Twofish cryptographic algorithm. Fig. 1 is the overall configuration of security processor that has Twofish cryptographic algorithm using USB.

Fig. 1. A block diagram of processor

Twofish cryptographic algorithm is one of 5 candidate algorithms of AES, and developed by USA Minneapolis research group. It is the block cryptographic algorithm that performs encryption using encrypted structure[5]. Encryption process of such Twofish cryptographic algorithm classifies 128 bit plaintext into 4 32 bit words using little-endian convention as shown in Fig. 2, and performs the process of input whitening with 4 32 bit subkeys and exclusive OR on each word. At each round, left 2 words are used as input of 2 g functions inside F function and 1 input word is entered through 8 bit left rotation. In addition, g function consists of S-boxes belonging to 4 8-by-8 bit keys and MDS matrix multiplier and its output was combined using PHT and 2 subkeys complete one round function by 32 bit modulo-2 addition.[6].

And 32 bit 2 word outputs of round function go through Exclusive-OR and 1 bit rotation of bit unit with left 2 words; left 2 words, right 2 words gone through Exclusive-OR with output of round function changed places to be used as input for next round, completing 1 time round. This round repeats 16 times repeatedly and the results of last round change places again, go through output whitening of Exclusive-OR with 4 32 bit subkeys in bit unit and creates 128 bit ciphertexts in the same method as little-endian convention applied to 128 bit plaintext. Decryption process is the reverse of the order of 40 subkeys applied in encryption process, and reverse of the process where outcome of F function went through Exclusive-OR with right 2 words and became one bit.

Fig. 3 shows the F function modified to transplant MDS-M2 blocks from single round. It is the circuit that can rearrange M0, M1, M2, M3 and S function values at q0 and q0 of h and g function in order to send the data to the simplified MDS-M2 block[7].

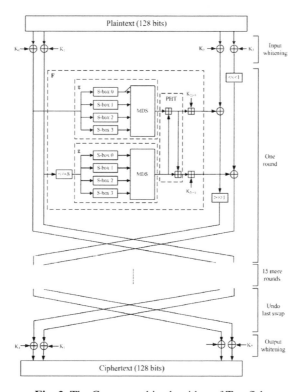

Fig. 2. The Cryptographic algorithm of Twofish

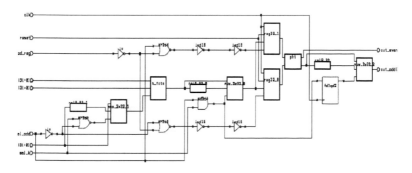

Fig. 3. Modified single F-fuction

Fig. 4 is the PHT transformation block that mixes the data of h and g function to create one data. Since PHT performs the function of combining h and g function into one function, original data cannot be restored once 2 data are mixed. Therefore, data is mixed using pseudo orthogonal function of PHT.

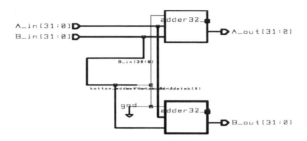

Fig. 4. PHT transformation block diagram

Fig. 5 shows S-box that processes data in h and g function with non-linear function. Since non-linear function is difficult to predict, it can be regarded as the main block in the cryptographic processor and basic operation is permutation in S-box.

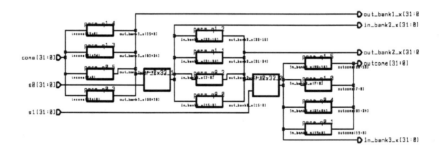

Fig. 5. S-box block diagram

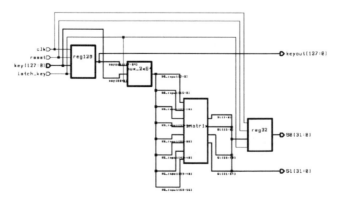

Fig. 6. Key Scheduler block diagram

Fig. 6 is the key scheduler block that creates encryption key used at security processor. Created keys provide separate data at each round.

Fig. 7 is the MDS-M2 block of cryptographic processor using Twofish cryptographic algorithm. To reduce bottleneck phenomenon due to multiplication, functions for multiplication were reduced as much as possible and common functions are indexed so that modulo-2 operation for overall multiplication is possible.

Fig. 8 is the control block that can control encryption/decryption of cryptographic processor using MDS-M2. Since encryption/decryption is possible simultaneously at Twofish cryptographic algorithm, security processor was configured using control transmission type of USB.

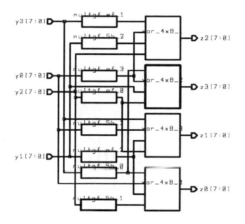

Fig. 7. MDS-M2 block diagram

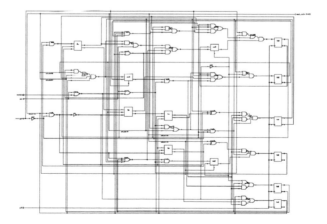

Fig. 8. Control block diagram for encryption/decryption

Fig. 9 is the overall block diagram of security processor where USB can be applied. By calculating the 128 bit output using 128 bit input and key information, it has 1:1:1 relation and has advantage of enabling encryption/decryption at the same time[8].

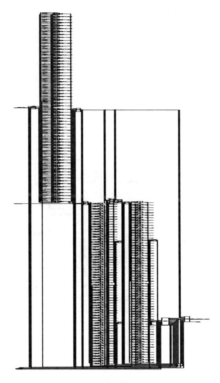

Fig. 9. Overall block diagram of cryptographic processor

4   Conclusion

This thesis designed security processor of Twofish cryptographic algorithm that can protect information against external attack using USB in order to keep information more safe and to protect users unfamiliar with PC use.

For security processor design efficiency of Twofish cryptographic algorithm, Top-down-VHDL was used, and Synopsis Ver. 1999. 10 used for synthesizing as CAD tool. QuartusII7.1v, device targeted Cyclone EP1C6Q240C8 for FPGA. The security processor using USB has an advantage of easy carrying and can be inserted into the PC of user. It is considered to protect PC and personal data from external attack.

Acknowledgments. Funding for this paper was provided by Namseoul University.

References

1. Schneier, B.: Applied Cryptography: Protocols, Algorithms, and Source Code in C. John Wiley & Sons, Inc., New York (1994)
2. `http://www.intel.com/technology/usb/`
3. `http://www.usb.org/developers/estoreinfo/`
4. `http://www.nxp.com/acrobat_download2/various/USBSTAND.pdf`
5. AES3 Proceedings of Third AES Candidate Conference, , pp. 44–54 (April 2000), `http://csrc.nist.gov/encryption/aes/round2/conf3/papers/`
6. `http://www.jetico.com/bcve_web_help/html/02_standards/02_algorithms.htm`
7. `http://software.intel.com/sites/products/documentation/hpc/ipp/ippcp/ippcp_ch2/ch2_two-fish_functions.html`
8. `http://www.encryptionanddecryption.com/algorithms/symmetric_algorithms.html#Twofish`

A Cyber-Security Implementation Framework for Nuclear Power Plant Control Systems

Cheol-kwon Lee[1], Jae-gu Song[1], Dong-young Lee[1],
Hyun-mi Jung[2], and Gang-soo Lee[2]

[1] I&C Human Factors Research Division, Kaeri
1045 DaeduckDaero, Daejeon, 305-353, Korea
jgsong@kaeri.re.kr
[2] Dept of Computer Engineering, Hannam University
Daejeon, 306-791, Korea
gslee@eve.hannam.ac.kr

Abstract. Control systems of nuclear power plants have been faced with the risk of cyber-security attacks from inside or outside agents. Thus control systems should efficiently and strongly account for the attacks. We propose a cyber-security implementation framework by integrating conventional concepts and paradigm such as CC, PP, ST, operational system evaluation, certification and accreditation, risk management, and etc.

Keywords: cyber-security, nuclear power plant, instrumentation & control system, security control, risk evaluation, common criteria, system protection profile, system security target.

1 Introduction

Instrumentation and Control Systems (CS) of Nuclear Power Plants (NPP) are isolated from external communication systems. Nevertheless, particularly the computers used in safety and safety-related systems must be effectively protected from possible intrusions. But other computers must be protected as well. The computers used to control the NPP are essential to assure the continuity of power production. The computers used to control access to sensitive areas are needed to prevent both unauthorized access that might be part of an attack, and to assure authorized access for safety and security reasons. Computers that store important and sensitive data have to be protected to assure that those data are not erased or stolen.

According to Title 10 of the *Code of Federal Regulations* (10 CFR) 73.54, "Protection of Digital Computer and Communication Systems and Networks", NPP CS's developers and Licensees should implement a cyber-security plan and program to protect digital computer communications systems and networks associated with the following functions from those cyber attacks, up to and including the design-basis threat[1]: safety-related and important-to-safety functions; security functions; emergency preparedness functions, including off-site communications, and support systems and equipment which, if compromised, would adversely impact safety, security, or emergency preparedness functions.

G. Lee, D. Howard, and D. Ślęzak (Eds.): ICHIT 2011, CCIS 206, pp. 190–195, 2011.

To cope with the problem, we study on cyber-security implementation framework by integrating conventional concepts and paradigm such as Common Criteria (CC), protection profile (PP), security target (ST), operational system evaluation, certification and accreditation, risk management, etc. [14]

In this paper, we propose a cyber security implementation framework for control system of NPPs. These results will be useful for stakeholders, such as developers, operators, and evaluators as well as certifiers, who develop and manage the development, operating, supervise and maintenance of a NPP control system.

In Section 2 related criteria for cyber security controls for NPP control systems are described. In Section 3, we propose a cyber-security implementation framework which is a CC based approach toward NPP cyber-security. Analysis and conclusions are presented in Section 4.

2   Related Criteria for Cyber-Security Controls for NPP Control Systems

FIPS 140-2 is an evaluation criterion for cryptographic module verification program (CMVP) is the U.S.A and Canada, as well as Korea and Japan [2].   **CC/CEM** is security functional evaluation criteria and evaluation scheme, respectively, which is also ISO/IEC 15408. CC includes a full set of security functional class and assurance criteria. A PP is not only a subset of CC, but also a typical type of security product such as a firewall. ST is an instance (or customizing) of a PP, which is a security requirement specification for specific security products such as X-firewall [3].

ISO/IEC 19791 is an extended evaluation criteria and guidance for assessing both the IT and the operational aspects of such systems. It is primarily aimed at those who are involved in the development, integration, deployment and security management of operational systems, as well as evaluators seeking to apply CC to such systems. System PP (SPP) and System ST (SST) correspond to SS and ST in CC, respectively [4]. **ISO/IEC 27001** (ISO/IEC 17799) is the criteria of information security management system (ISMS) that originated from the BS 7799 [5].

NIST SP 800-53 is a guideline for selecting and specifying security controls for information systems supporting the executive agencies of the federal government to meet the requirements of FIPS 200 [6]. **NIST industrial control system PP** is a system PP, targeted an extended EAL-3 level of assurance for CS. This SPP is intended to provide an ISO 15408 (i.e., CC) based starting point in formally stating security requirements associated with ICS. This SPP includes security functional requirements and security assurance requirements that extend ISO 15408 to cover issues associated with systems [7]. **SP 800-53 Appendix I**: Security control candidates for tailoring ICS and supplements [6].   **SP 800-82** is a guide to ICS security - including SCADA systems, DCS, and other control system configurations such as PLC [8].

RG 5.71 is a regulation guide for cyber-security program for NPP control system, which is developed by extending (customizing) 800-53 and 800-82. Appendix includes comparison of security controls between 800-53 and RG 5.71[1]. **IEC 62465** is the same as ISO/IEC 27001, except in replacing term 'Information systems' to 'I&C Systems' [10]. **NERC - CIPs** are cyber security standards for critical infrastructures from the North American Electric Reliability Corporation [11].

3   A Cyber-Security Implementation Framework

This framework is based on the CC (i.e., the concepts of common security function, security assurance, PP, ST) and the C&A (i.e., the concepts of common security control, certification, accreditation), and the RMF (risk management) paradigms [12]. The aim of the framework is supporting security development and operation of CS of NPPs which is becoming vulnerable to cyber-security attacks. Fig. 1 presents steps in the framework. Fig. 2 presents relationships among CSPP (control system PP), CSST (control system ST) and NPP's control systems.

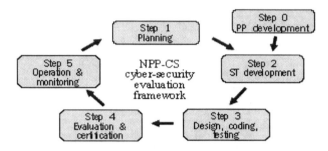

Fig. 1. Steps of NPP-CS cyber-security evaluation framework

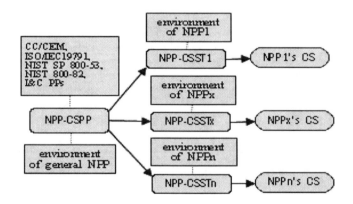

Fig. 2. Relationships among CSPP, CSST and NPP's control systems

3.1   PP Development (NPP-CSPP)

NPP-CSPP, that is a control system PP for a general NPP, developed by using the following standard, regulation, guidance: CC/CEM, ISO/IEC19791, NIST SP 800-53, NIST SP 800-82, NIST Industrial control system PP, and RG 5.71.

NPP-CSPP developed by means of the following steps: 1) environment analysis (analysis of typical NPP CS operation environments); 2) asset analysis (identifying & value estimation of asset in NPP CS (classification schema, value estimation scheme)); 3) attack analysis (identifying and likely measurement of attack to asset (attack agent - how - why - what - when - which)); 4) vulnerability analysis (vulnerability of operation environment); 5) risk analysis (risk is function of asset-value and attack-likelihood);

6) security object derivation; 7) countermeasure derivation (security control, security function are select from security control and functional DB); 8) setting assurance level (security strength).

NPP-CSPP includes the followings: 1) CSPP introduction (i.e., CSPP identification, NPP CS overview, domain organization); 2) conformance claims: only applicable if the CSPP claims compliance with one or more PP, PPs or security requirements packages; 3) security problem definition: risk identification (including asset, threat, vulnerability), organizational security policies; 4) security objectives; 5) extended components definition; 6) security requirements (i.e., selecting security controls and functions from ISO/IEC 19791, CC, SP 800-54, etc.). For each security domain, including the following: introduction; conformance claims; security problem definition; security objectives; security requirements.

The NPP-CSPP should be *evaluated* by an independent evaluating organization, and *certified* by a certification authority such as NRC, IAEA, KINS, or CC evaluation organizations. The evaluation is conducted by using PP and SPP evaluation criteria in CC and 19791, respectively.

3.2   Cyber-Security Planning

A chief-manager of NPP CS should establish a *cyber-security plan (CSP)* that has a *cyber-security program* for NPP CS. The CSP establishes a license base for a cyber-security program which should be complied with RG 5.71 Appendix A in community of NPP CS [1]. The cyber security program is aimed to implement security controls or security functions for a secure NPP CS. CSP is developed under and complied with *cyber-security policy* of an NPP.

3.3   ST Development (NPP-CSST*x*)

NPP-CSST*x*, that is a control system ST for specific NPP*x* (e.g., Wolsung NPP#1, Gori NPP#2, Hukushima NPP#1) is created by means of tailoring, instantiation and customizing the NPP-CSPP, with considering real operational environment (i.e., operational profile, security policy, specific security requirements and hardware and network flat-form of the NPP*x*).

Firstly, discovering and analyzing the real operational environment of control system from real NPP*x*. Secondly, customizing and instantiation the NPP-CSPP, and generating NPP-CSST*x*. It is important to note that this step is a *requirement analysis* phase in the life-cycle. NPP-CSST is regarded as a *security requirement specification* for NPP*x*'s control system. It will be used as a part of the *request for proposal* of control system in developing the system. NPP-CSST should be evaluated by using ST evaluation guidelines in CC and ISO/IEC 19791[3, 4].

3.4   Design, Coding and Testing

The NPP-CSST*x* is a security requirement specification for the NPP*x*'s control systems. Developers should design (architectural and detailed), implement and test (unit, integrate, system) the control system according to the NPP-CSST*x* and *security engineering* methodologies such as the security architecture, secured programming, security testing, penetration testing.

3.5   Evaluation and Authorization

The developed control system for NPP*x* should be evaluated by independent evaluating organization, and authorized by authorizing official (or licensee) such as NRC, IAEA, KINS (in Korea).

The activity of evaluation includes *security verification and validation* such as manual checking, static/dynamic security analysis, penetration testing, and vulnerability scanning. All deliverable (e.g., source code, documents) of evaluating system are submitted to the evaluating organization. Final result of this step is an authorization package that includes *authorization to operate* or *denial of authorization to operate*.

3.6   Operation and Evaluation

The developed and authorized control system will be operating XNPV. During the operation, the system conducts *information security management* including risk management [5]. If current security risk-level is greater than the predefined allowable risk-level, the system conducts the corresponding security control such as emergency power, network disconnect, fail in known state, flaw remediation etc.

3.7   Supporting Tool

The tool is useful for developers and evaluators of NPP-CSPP and NPP-CSST, as well as developers, evaluators, certifiers, licensee of a new NPP CS. The currently developing supporting tool has the following functions: searching the related statements of standard/guide/regulation; processing checklists; documentation; risk management (including asset, threat, and vulnerability analysis). The CPSAM analysis tool is a good reference tool [13].

4   Analysis and Conclusion

In this research, we are: (1) surveying related criteria for cyber-security controls for NPP control systems; (2) developing a NPP-CSPP that will be useful for development and standardization of NPP control systems; (3) proposing a *cyber-security evaluation framework* which is based on the CC, RMF paradigm, as well as complied with RG 5.71.

This work is an on-going research project. Thus, refinement and certification of NPP-CSPP, development and evaluation guide of NPP CS specific security controls and functions, and refinement of the supporting tool remain as further studies.

Acknowledgments. This work was supported by the nuclear Technology Development Program of the Korea Institute of Energy Technology Evaluation and Planning(KETEP) grant funded by the Korea government Ministry of Knowledge Economy (No. 2010161010001E).

References

1. Regulatory Guide 5.71, Cyber security programs for nuclear facilities, U.S. Nuclear Regulatory Commission (2010)
2. FIPS 140-2, Security Requirements for Cryptographic Modules, NIST (2001)
3. CCMB-2009-07-002, Common Criteria for Information Technology Security Evaluation, Version 3.1 (2009)
4. ISO/IEC TR 19791, Operational system protection profiles (2010)
5. Jayawickrama, W.: Managing Critical Information Infrastructure Security Compliance: A Standard Based Approach Using ISO/IEC 17799 and 27001. In: Meersman, R., Tari, Z., Herrero, P. (eds.) OTM 2006 Workshops. LNCS, vol. 4277, pp. 565–574. Springer, Heidelberg (2006)
6. NIST SP 800-53, Rev 3, recommended security controls for federal information systems (2009)
7. System Protection Profile - Industrial Control Systems, Version 1.0, NIST (2004)
8. NIST SP 800-82, Guide to Industrial Control Systems (ICS) Security, NIST (2008)
9. Catalog of Control Systems Security: recommendations for Standards Developers, Control systems security program, National cyber security division, Homeland security (2009)
10. IEC 62465 CD1 ed. 1.0, Nuclear Power Plants - instrumentation and control important to safety - requirements for security programmes for computer-based systems (2011)
11. CIP–002–3 ~ CIP–009–3 —Cyber Security (2011)
12. NIST SP 800-37, Rev.1, Guide for applying the risk management framework to federal information systems (2010)
13. DePoy, J., et al.: Critical Infrastructure Systems of Systems Assessment Methodology. SANDIA REPORT (2006)
14. Polk, W., Malkewicz, P.: Jaroslav Novak, Industrial Cyber Security From the Perspective of the Power Sector. Revision 1, DEFCON 18 (2010)

Design and Implementation of Mobile Forensic Tool for Android Smart Phone through Cloud Computing

Yenting Lai[1], Chunghuang Yang[1], Chihhung Lin[2], and TaeNam Ahn[3]

[1] Graduate Institute of Information and computer Education National Kaohsiung Normal
University, Taiwan
[2] Network and Multimedia Institute, Institute for Information Industry
[3] Security Engineering Research Center
Hannam University, Korea
f0963217595@hotmail.com, chyang@nknucc.nknu.edu.tw,
chlin@iii.org.tw, taenamahn@hotmail.com

Abstract. As time progresses, smart-phone features and wireless availability highlight the inner-mobile security issue. By detailed process of inner-mobile acquisition, analyzed result and reporting will be regarded as significant proof on the court. In this paper, researcher forensics implements system of Android smart-phone and delivers the acquisition data through cloud computing to get the forensic analysis and reporting. According to the forensic procedure of National Institute of Standards and Technology (NIST), forensics examiner acquires the inner-data when mobile turns on and then instantly sends it through the clouds. Results will be displayed immediately.

Keywords: Smart-Phone, Mobile Forensic, Android, Cloud Computing.

1 Introduction

According to marketing research from NPD Group, the smart phone marketing research report in third quarter in 2010 shows that Android's OS has up to 44 % beyond all smart phones compared to the increase of 11 % in the second quarter, and Android smart phone market share is now beyond the RIM phone (Blackberry) and iPhone4, and has reached up to the top of smart phone in the U.S. [14].

In Taiwan, the Personal Information Protection Law will renovate its regulation making it more detailed and specific. Later, various enterprises will encounter clashes toward this new policy [9]. Therefore, digital evidence becomes an important issue of innocence proof when enterprises have suspicions of personal information leakage.

Because of smart phone surfing feature, the availability of Internet and other advanced features, the mobile surfing availability rate is increasing annually, moreover, inner-mobile private data security is noteworthy. Mobile surfing rate has increased to 69.5% [5], according to the survey of National Communication Commission and Institute for Information Industry. By detailed forensic procedure, this paper represents the original scene restoration.

G. Lee, D. Howard, and D. Ślęzak (Eds.): ICHIT 2011, CCIS 206, pp. 196–203, 2011.
© Springer-Verlag Berlin Heidelberg 2011

2   Related Work

According to the forensic procedure of NIST, researcher applies live forensic of Android mobile implement system and receives data and instantly connect to the clouds. This chapter will cover detailed explanation of mobile forensic, live forensic, Android smart-phone, and cloud computing.

2.1   Mobile Forensic

There are four steps of standard procedure including preservation, acquisition, analysis, and reporting according to NIST standard mobile forensic procedure. Besides, digital evidence requires four characters that are testability, acceptance, error rate, credibility, and clarity [8]. Here is the explaining of preservation, acquisition, analysis, and reporting.

Preservation. Digital evidence is easily damaged by external factors, such as packaging, transportation, storage etc.; therefore, the evidence must be well-preserved to ensure its accuracy.

Acquisition. That acquired digital evidence by examiner from target mobile must be complete and analyzable and able to be presented by reporting.

Analysis. After forensic examiner receives the evidence result, analysis will be done in different way depending on the situation. The main purpose is to look for potential data connected to digital evidence to enhance the reality of evidence.

Reporting. Detailed conclusion will be presented in reporting including detailed acquisition procedure and result to reveal the original scene.

2.2   Live Forensic

Live forensic, one of the methods of digital forensic, provides foreseer the volatile evidence acquisition in random access memory with implement system. Volatile evidence volatilizes as mobile turns off [15]. So far, common forensic software executes acquisition when mobile turns off that causes the unavailability of volatile evidence in random access memory. To ensure the quality of volatile evidence, foreseer must confiscate the target mobile and turn off after its acquisition is done.

2.3   Android

Android is a mobile operating system developed by Java language and Android SDK application based on Linux operating system. There are four layouts in Android operating system, from the Linux 2.6 Kernel in the bottom layout, libraries and Android Runtime on the third layout, application framework on second layout and Android applications on the top layout (shown in figure 1) [1].

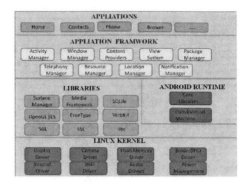

Fig. 1. Android Architecture

2.4   Cloud Computing

The network is now highly developed. It finally enters the cloud-computing era, after computer operation passed several stages like handset operation, parallel system, distributed system and grid computing. Cloud computing is a concept instead of an instrument [16]. Its definition is that without requiring any professional background knowledge, user easily gets the result through the network accessing other computers [4]. There are five characters of cloud computing according to NIST [10].

On-demand self-service. It automatically provides service according to user's need.

Broad network access. Service is available once user passes standard servo through cloud service on whatever instrument (e.g. PDA, smart-phone etc.).

Resource pooling. Cloud computing provider collects data for users. It doesn't require user to know in which servo his data stores but to be familiar with the application.

Rapid Elasticity. When the result returns back to the user, the operation capability rapidly surges on a large scale once it turns off.

Measured Service. Data usage is monitored while user processes to cloud service.

3   Comparison between Cloud Computing and Non-cloud Computing

Applications in this research apply Google cloud service to implement mobile forensic system under Android smart-phone. The idea of forensic through cloud computing is based on high storage capability and high operation of cloud computing [17]. Here is the comparison of mobile forensic through cloud computing and non-cloud-computing (shown in table 1).

Table 1. Table of Cloud Forensic and Conventional Forensic Comparison

	Forensic through Cloud Computing	Conventional Forensic
Data Protection	By Cloud Computing Provider	By Forensic Examiner
Storage Site	Clouds	Storage Medium
Reporting	Browser	Third Appellations
Procedure	Less Complicate	Complicate
Advantage	No Time, Location and Equipment Limitations	No External Effects of Digital Evidence
Limitation	Connection to Network is Required.	Connection to Network is not Required.

3.1   Security and Protection of Digital Evidence

The purpose of mobile forensic is to acquire digital evidence from mobile, analyze the result and present the reporting. Security and protection of digital evidence becomes a significant issue in mobile forensic. Here's the comparison of digital evidence security and protection. At cloud computing mode the forensic procedure in this study is to upload digital evidence instantly to Google Cloud Service, which transmits data by HTTPS that supports RSA public key cryptosystem to ensure data security. The main difference between cloud computing mode and Standard mode is that no excess human resources waste and extra procedure are required [12]. Google Cloud Service is maintained by several cloud providers to ensure data security [7] and on conventional mode according to standard NIST mobile forensic procedure; It needs to pack, transport, and store digital evidence well by forensic examiner to ensure the comprehensive and the accuracy [8].

3.2   Storage and Site of Digital Evidence

Standard mobile forensic tool reserves digital evidence into SD card, hard disk, disk, USB and other storage medium, and the storage space depends on user. Cloud mobile forensic tool store digital evidence onto the cloud and the storage space depends on provider. Take Google for instance, each account has about 7 GB storage space and it requires extra fee to get a larger storage space [6].

3.3   Reporting

Common non-cloud mobile forensic tool includes Oxygen Forensic Suite [13], and MOBILedit! Forensic [11], besides, PDF, CSV, XML, HTML, XLS and so on are common types of output data form. Third appellation is required to open forensic report, and the installation procedure varies when using different operation system. However, reporting of cloud mobile forensic tool requires only the connection to network to view the result.

4   System Implementation and Architecture

This study applied Android smart-phone and Google cloud service to implement forensic application under the turning on condition. Because of the volatile data stored in the random access memory, mobile forensic must be done under turning on condition Data in random access memory will volatize once mobile turns off, therefore, the data stored in random access memory must forensic under turning on condition, such a forensic procedure is called live forensic [3].

4.1   System Architecture

This study applies the mobile phone forensic application constructed by Android API and Google API. Before forensic, forensic examiners download the application from the cloud platform, install the system, and then execute the mobile forensic. The results will instantly upload to Google's cloud services, and then generate forensic report (shown in figure 2); Google API provides Google Client Login [2] as cloud identification mechanism and Google Docs.

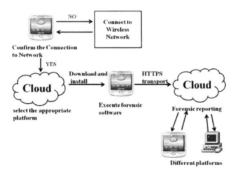

Fig. 2. System Architecture

4.2   System Implementation

As the development of technology, it improves mobile features, and smart-phone is now a hot issue in modern science and technology field, therefore, mobile usage security issue now attracts people's attention. In this study, the usage of Apache set up a cloud platform and provides forensic software for forensic examiner during the forensic. Before forensic begins and downloads the software, forensic examiner needs to identify the version and operation system of target mobile to ensure the software is suitable (shown in figure 4). After forensic examiners download, install and execute the forensic software, the inner-mobile data acquisition begins. Digital evidence instantly transmits to Google cloud platform after the acquisition. It provides reporting shown on cloud or mobile terminal. The procedure is shown in figure 3.

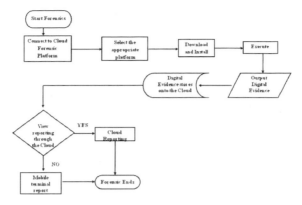

Fig. 3. System Forensic Flowchart

It requires forensic examiner having a Google account while executing the forensic software. According to NIST forensic procedure, it needs to establish forensic files, and forensic examiner must key in the name and date. If the date is incorrect, press the "Data Error" button to correct the date, and then key in the account ID and password to begin forensics (shown in Figure 4).

Fig. 4. Create Forensics File

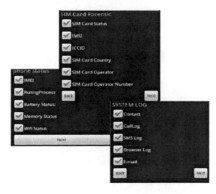

Fig. 5. Digital Evidence Collection

Enter Start; it begins with inner-mobile data acquisition choice which includes phone status, SIM card status, and system log files. Forensic examiner chooses options above (shown in figure 5), and then the acquisition begins with chosen options.

Forensic software in this study provides cloud and mobile terminal reporting. Cloud reporting presents reporting through browser and Google Docs (shown in figure 6) while the other provides instant viewing. Moreover, both reporting methods don't reserve any file inside the mobile.

Fig. 6. Cloud Reporting

5   Conclusion

The progress of information technology and widespread network promote smart phone features bringing the convenience but also threat, therefore, digital forensic becomes an important issue. Most of commercial forensic software process the acquisition when mobile turns off making it fail acquiring volatile data in the memory. In this study, researcher processes acquisition when mobile turns on and successfully acquires volatile evidence in mobile, meanwhile, instantly uploads the data onto the cloud. Without excess instrument, it provides forensic examiners process acquisition anytime, anywhere as long as examiners are able to upload the data through the Internet and browser. Instantly viewing reporting reduces extra time of installation of third appellation or operation system, and provides a more efficient acquisition. In the future, the aim will be the combination of cloud computing platform and mobile forensic that only requires browser to process mobile forensic.

Acknowledgments. This study was supported in part by research grants (NSC 98-2221-E-017-010-MY3) from the National Science Council of Taiwan and this study is also conducted under the "Wireless Broadband Communications Technology and Application Project" of the Institute for Information Industry which is subsidized by the Ministry of Economy Affairs of the Republic of China.

References

1. Android developers (2011), http://developer.android.com/guide/basics/what-is-android.html
2. Authentication and Authorization for Google APIs (2011), http://code.google.com/apis/accounts/docs/AuthForInstalledApps.html
3. Biggs, S., Vidalis, S.: Cloud Computing Storms. International Journal of Intelligent Computing Research, IJICR (2010)
4. Biggs, S., Vidalis, S.: Cloud computing: The impact on digital forensic investigations. In: International Conference for Internet Technology and Secured Transactions, ICITST 2009, pp. 1–6. IEEE, Los Alamitos (2009)
5. Foreseeing Innovative New Digiservices (2010), http://www.find.org.tw/find/home.aspx?page=many&id=275
6. Google Docs Help (2011), http://docs.google.com/support/
7. Google security and product safety (2011), http://www.google.com/corporate/security.html
8. Jansen, W., Ayers, R.: Guidelines on Cell Phone Forensics. NIST, SP, 800–101 (2007)
9. Law & Regulations Database of The Republic of China (2011), http://law.moj.gov.tw/index.aspx
10. Mell, P., Grance, P.: The NIST Definition of Cloud Computing. NIST (2009)
11. MOBILedit! Forensic (2011), http://www.mobiledit.com/forensic/
12. Molnar, D., Stuart, S.: Self Hosting vs Cloud Hosting: Accounting for the Security Impact of Hosting in the Cloud. In: Workshop on the Economics of Information Security (2010)
13. Oxygen Forensic Suite (2011), http://www.oxygen-forensic.com/us/
14. The NPD Group: Android Extends its Smartphone Market Share in the Third Quarter of 2010 (2010), http://www.npd.com/press/releases/press_101101.html
15. Thing, V.L.L., Ng, K.-Y., Chang, E.-C.: Live memory forensics of mobile phones. Digital Investigation, S74–S82 (2010)
16. Wang, P.: Walked into cloud computing, pp. P1-17–P1-20. Top Team Information Co.,Ltd., Taipei (2009)
17. Wolthusen, S.D.: Overcast: Forensic Discovery in Cloud Environments. In: Conference Proceedings 5th International Conference on IT Security Incident Management and IT Forensics, IMF (2009)

A SSL VPN Design Method for Enterprise FMC Security System Development

Hyun-mi Jung, Kyung-su Han, and Gang-soo Lee

Dept of Computer Engineering, Hannam University
Dae-jeon, 306-791, Korea
mihj@se.hannam.ac.kr, psksmail@hnu.ac.kr, gslee@eve.hannam.ac.kr

Abstract. Fixed mobile convergence service is an integrated environment which requires the function such as integrated authentication procedure, billing management, wire and wireless inter-networking, location management, and provides independent services of the connection technology. The connection technology in the integrated environment should provide fusibility, security, and elaborated connection control functions to the user, and also should provide convenience to the manager. Therefore, in this paper, we proposed a secure sockets layer virtual private network design method for enterprise Fixed mobile convergence security system.

Keywords: FMC, Enterprise FMC, Convergence, SSL VPN, Security system.

1 Introduction

Enterprise fixed mobile convergence system (FMC) is the system that can provide real-time business without delay to employees anytime, anywhere using private intra-net service, such as groupware, CRM, and ERP, in the enterprise. This system provides better fusibility, security, convenience for the remote mobile user than existing remote access solution devices. Especially, it could realize the smart office environment which can be performed business services like e-mail, instant messaging, e-payment and etc. based on current business calling features. FMC service is developing in the direction of implementing unified communication environment that means integrating a variety of existing wired and wireless technology including mobile communication technology, UMA technology, and wired internet technology [1]. For this reason, the design matching the characteristics of the companies must be take precedence in order to implement enterprise FMC service system according to analyze security threats of the physical environment, such as FMC controller, client, and wired and wireless communication technology, such as PSTN, VoIP, Wi-Fi.

In terms of access control for the FMC system, IPsec VPN system using at the existing remote access solution is expensive and also it is complicated to setup and manage. This device is difficult to provide fusibility, security, elaborated connection control functions for the user, and convenience for the manager. Hence, we need the

G. Lee, D. Howard, and D. Ślęzak (Eds.): ICHIT 2011, CCIS 206, pp. 204–211, 2011.

security control design matching the requirement of the companies through design of access control device using secure sockets layer virtual private network (SSL VPN). Therefore, in this paper, we proposed a SSL VPN design method for enterprise FMC security system.

In section 2, analysis for features and examples of FMC service is presented. In section 3, SSL VPN design requirements for designing FMC security system are analyzed and we proposed the efficient SSL VPN design for enterprise FMC system in section 4. Finally, conclusion is presented in section 5.

2.   FMC Service Features and Examples [2]

2.1   FMC Service Features

All FMC services should be designed to fit followed features.

- In the user's viewpoints, we should have to receive the same service even if we were moving between wired networks and wireless networks.
- In the service provider's viewpoints, they should have to provide the same QOS even if users were moving between wired networks and wireless networks.
- In the viewpoint of the mobility, different mobility functions depending on FMC scenarios should have to be provided.
- The user, in the mobile viewpoint, should have to be able to use the FMC service anytime, anywhere.
- Security-enhanced user ID and authentication techniques should have to be supported to users.

2.2   FMC Service Examples and Features

The followings show service examples and features required at FMC service system.

- The feature that could be selected access networks by users should have to be provided if providing various access networks at the same time. The service which is independent on access technology is provided to users simultaneously and, from this point of view, the convenient access service which could provide the same service independent on access technology, such as PSTN, 3G, Wi-Fi, should be provided anytime and anywhere.
- Through FMC service, users should have to use the improved service which could be possible to use the same VPN service independent on access network technologies and various terminal devices.
- Through FMC service, unified messaging service should be provided by integrating messaging services, such as short message services (SMS), multimedia message services (MMS), instant messaging, e-mail, which provided separately at the existing system.

3.   SSL VPN Design Requirement Analysis for Designing FMC Security System

3.1   The Purpose of SSL VPN Design

On the basis that the ultimate purpose of Enterprise FMC system is to increase productivity and efficiency of the companies, we would need to develop technology which could easy to add users like the remote mobile user in the untrusted environment.

For efficient security network connection using internet between the mobile user and the enterprise, virtual private network device is needed. Table 1 compares with IPSec VPN and SSL VPN in order to choose an efficient VPN device when applied to the enterprise FMC system.

Table 1. Comparison with IPSec VPN and SSL VPN

Type	Characteristic
IPSec VPN	• Installs the VPN Client software on all computers at a long range and requires management and maintenance • Requires patch and security management for clients • Problems of compatibility about various hardware on clients • Difficulties of suggestion of the correct solution when failures occurred • Needs the additional configuration at the firewall stage for accessible resources in the enterprise
SSL VPN	• If there was default web browser, VPN access would be possible anywhere easily. • Most browsers support SSL, thus the flexible act for implementation and management is possible. • Integrated security features: Encryption service using SSL at the application layer is provided and also TLS(Transport Layer Security[3]) is applied. • A suitable encrypted remote access method for mobile users • Provides the security connection for file sharing, e-mail, and the other applications • Enables SSO(Single Sign On) by using authentication token when accessing the portal

The limitation of current IPSec VPN is that only accurate authenticated users are able to access. Also, the elaborate management for whole IP network access is difficult. IPsec VPN is a suitable solution only at the particular environment as head offices and branches of the bank where each communication parties could trust each other.

However, SSL VPN is based on transmission protocol which is designed to ensure the safety of the data. Owing to this feature, SSL VPN is being introduced at the field, which required confidentiality and security of the communication, such as online banking and e-commerce. Also, using immediately by the web without the client is a major characteristic, unlike traditional VPN technology which should be installed IPSec client at the user device in advance. Hence, there is no restriction on the time and location, and various user environments are possible to accept [4, 5, 6].

In the convergence service environment like FMC service, the security technology is required when tasks and services through the internet are performed without changing the existing operational environment, and also convenience of use and management should be provided. Therefore, there need to introduce SSL VPN in FMC service environment.

3.2   SSL VPN Design Requirements for Enterprise FMC System

The followings show SSL VPN design requirements for efficient FMC system.

- Due to the characteristics of the convergence service environment, security technology which could eliminate disclosure of personal information and the exploitation of important information should be needed.
- The device which could defend effectively from security threats, such as identity theft, forgery of documents, personal information leaks, for secured e-commerce and business to business (B2B) which guaranteed the security without changing the current system should be needed.
- By being realize the remote working which is the ultimate purpose of FMC system, there should need the access technology, such as extra-net, that could connect between branches and partners or intranets in order to extend enterprise network to out of the local network which have purpose of interior information sharing.
- The solution that could provide convenient management, fusibility, access control functions by remote access technologies should be needed.
- VPN technology applying domestic standard encryption algorithm adequate for the domestic circumstances should be needed.

4.   SSL VPN Design for FMC Security System

In this section, we designed SSL VPN suitable for FMC security system. Figure 1 shows a designed structure of SSL VPN. This SSL VPN is composed of SSL Secure Gateway (SSG) system, SSL Secure Client (SSC) module, and Secure Management Tool (SMT) module as presented in following figure 2.

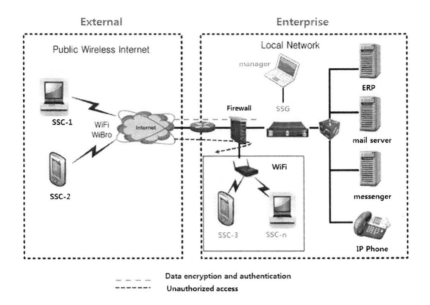

Fig. 1. Structure of SSL VPN system

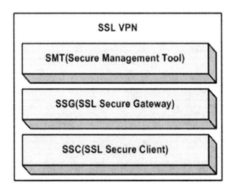

Fig. 2. Composition of SSL VPN

4.1 SSL Secure Gateway (SSG)

SSG system performs the role of security tunnel and security gateway and provides access control service according to the security policy about connection of clients. In order to build security tunnel, certificates and CA certificates, and configuration information are needed and they are provided by the SMT.

SSG is classified into two blocks that are composed of several modules. VPN control block which performs functions like security negotiation, building and maintaining of security tunnel is composed of practical behavior-related module of SSG. Manager performs additional management functions such as VPN start and shutdown.

I. VPN Control: VPN control block is composed of practical behavior-related modules such as VPN authentication, virtual network management, session management, SSL communication, for VPN.

   A. Authentication Module: Device authentication using a private certification and cross-certification between SSG and SSC are performed.
   B. Virtual Network Module: Virtual network drive installation and assigning virtual IP to the client are performed and virtual network drive, which is required to build security tunnel, is managed.
   C. Session management Module: CA, server, and client certificates, that are need at SSL VPN authentication, are managed.
   D. SSL Module: SSL protocol, handshake, and record protocol are performed.

II. Manager: Manager performs start and Shutdown of SSG, agent module for SSC, log management.

   A. Gateway Module: Requirements of client service are reflected and communication with internal server is performed.
   B. Start/Shutdown Module: The Function of SSG start and Shutdown is performed.
   C. Agent Module: Agent program which is needed to users in order to receive SSL VPN service is managed.
   D. Log management Module: System status and access information are managed.

4.2   SSL Secure Client (SSC)

User program installed at client performs functions such as building of security tunnel, management of user access information. In order to build security tunnel, certificates and CA certificates, and configuration information are needed and these information could be get to connect SMT web server after user login.

SSC is composed of VPN, Manager, and Active-X block and their features are as followings.

I. VPN Control: VPN control block is composed of practical behavior-related modules such as VPN authentication, virtual network management, session management, SSL communication, for VPN.

   A. Authentication Module: Device Authentication based on a private certification is performed.
   B. Virtual Network Module: Virtual network drive which is required to build security tunnel is installed and managed.
   C. Session Management Module: Session which is required to build and to maintain security tunnel is managed.
   D. Certificate Management Module: CA and Client certificates that are needed for SSL VPN authentication is managed.
   E. SSL Module: handshake and record protocol for SSL protocol are performed.

II. GUI Manager: Tray icon is generated and SSC is controlled by the generated icon.

   A. Tray Icon Generation Module: Tray icon for SSC control is generated at the client.
   B. SSC Control Module: SSC shutdown is controlled and the proxy is configured.

III. Active-X Control: Using Active-X, SSC start and shut-down functions are performed and client access information is removed when logout is executed.

   A. Start/Shutdown Module: Start and Shutdown function of SSC are performed.
   B. Access Information Management Module: when SSC is shutdown, Access information remained in the client is removed.
   C. Drive Check Module: The function, that checks whether the virtual network module is installed, is performed.

4.3   Secure Management Tool (SMT)

SMT, which is a management tool of SSL VPN, works as a web server at SSG and it provides certificates and access information to SSG and SSC and also provides functions such as the user and the operator management, security policy configuration, server and group management.

   SMT is classified into three blocks that are composed of several modules. System block controls system-related configuration information and security policy block manages SSL VPN security policy configuration information. Configuration block performs functions like system manager management and configuration information management.

I.   System-related configuration information is controlled in system block.

   A. System Management Module: System operation such as start, stop, shutdown and restart of SSG is controlled.
   B. Device Management Module: Duplication and load balancing is provided for several SSL VPN devices.
   C. Network Configuration Module: Internal IP and external IP of SSG are configured.
   D. Access Information Configuration Module: Configuration information and environmental information that are required to build the security tunnel is configured.

II. Security policy block manages SSL VPN security policy configuration information.

   A. Routing Configuration Module: When SSL VPN manages several internal sub networks, routing for each sub networks is configured.
   B. Server Management Module: Management functions such as internal server registration, modification, deletion are performed.

C. Group Management Module: Management functions such as access group registration, modification, deletion are performed for internal resource access.

D. User Management Module: Management functions such as client user registration, modification, deletion are performed.

III. Configuration block performs functions like system Operator management and configuration information management.

A. Operator Management Module: Management functions such as SSL VPN operator registration, modification, deletion are performed.

B. Backup File Management Module: SMT configuration information and revisions are managed.

5   Conclusion

The connection technology in the integrated environment should provide fusibility, security, and elaborated connection control functions to the user, and also should provide convenience to the manager. Designed SSL VPN system, in this paper, can be installed simply without changing the existing system and can be managed easily through familiar GUI to the user.

In the integrated environment like FMC service, independent services of the connection technology should be provided and functions such as integrated authentication procedure, billing management, wire and wireless inter-networking, location management are required. After designing according to these requirements, personal and business information protection should have to be guaranteed certainly. Therefore, the study of the specialized security technology for eliminating personal and business information leaks, critical information abuses is needed in the future.

References

1. The Start of Full-fledged Enterprise FMC Service, ATLAS Research (2009)
2. Ko, S.-j.: Standardization on FMC in the ITU-T. Telecommunications Review 18(4) (2008)
3. Thomas, S.A.: SSL and TLS Essentials. Wiley, Chichester (2004)
4. Barracuda Networks: VPN Technologies, Technical NThe Start of Full-fledged Enterprise FMC Service, ATLAS Research (2009)
5. Barracuda Networks, VPN Technologies, Technical Notes (2004)
6. Array Networks, SSL VPN vs. IPSec VPN, White Paper (2004)
7. VPN Security, Government of HKSAR (2008)

Malicious Software Detection System in a Virtual Machine Using Database

Hyun-woo Cheon[1], kyu-Won Lee[1], Sang-Ho Lee[2], and Geuk Lee[1,*]

[1] Dept of Computer Engineering Hannam University, Korea
[2] Dept of Military Studies, Daejeon University
csjhahaha@naver.com, kwlee@hnu.kr,
kilo2500@kornet.net, leegeuk@hnu.kr

Abstract. Malicious behavior detection using emulator or virtual machine is becoming an interesting issue in information security field because it is easy to re-initialize the system and execute codes in independent separate spaces which do not give any bad influence to the system. This paper proposes fast malicious codes detection system by using database of previous malicious codes.

Keywords: virtual machine, database, malicious software detection.

1 Introduction

Recent threats to computer resources are becoming various ranging from simple attacks using scripts to multifunctional malicious codes, and the extent of the damage is becoming serious as well. In addition, it is challenging to cope with the malicious codes with existing detection technology of vaccine programs or attack detection systems as they are evolving into multifunctional ones which have characteristics of virus and hacking tools.[1]

Furthermore, detection is not efficient after the system is infected or malicious code is activated because the system could have already been ruined. Therefore, malicious code detection technology which executes suspicious codes in emulators or virtual machines is a current interesting field of research.

Technology using a virtual machine directly runs the target file inside the virtual machine and detects malicious behaviors. Using a virtual machine makes code operate in an independent space and does not give any bad influence to the user system.

In addition, it does not take a long time to reinstall the system in case the user system or the host system is infected or broken by malicious codes. Virtual machines allow fast re-initialization and continuous trials of many codes.[2]

In this paper, malicious behaviors are detected quickly by adding behavior databases on the existing system as the system searches for certain patterns and the codes that match with the patterns.

---

* Corresponding author.

G. Lee, D. Howard, and D. Ślęzak (Eds.): ICHIT 2011, CCIS 206, pp. 212–218, 2011.
© Springer-Verlag Berlin Heidelberg 2011

2   Related Studies

2. 1   Detection Technology Using Emulator and Virtual Machine

Malicious behavior detection technology using emulator does not require extra hardware on the actual user's system and runs simulator on the independent area from the user space. Any suspicious files will be loaded on the virtual hardware and activated in a simulation environment In the simulation environment, it is possible to execute various malicious behaviors such as file infection, file delete, IRC server connection, email transmission, and listening port open. Emulator-based malicious behavior detection technology executes the target file in the emulator and detects malicious codes. Malicious behavior detection technology using a virtual machine runs the target file within the virtual machine which is logically separated from the user space. This makes it possible to execute codes without giving bad influence to the user systems.   Furthermore, target files are executed in each virtual machine so it is possible to re-execute and analyze the results of the target files on a new system by re-initializing the virtual machine.

In this system, system restoration time is reduced dramatically than the actual system which actually executes codes to decide malicious behaviors. VMware, in particular, can execute a lot of virtual machines on one system. Figure 1 is the concept of virtual machine. [3]

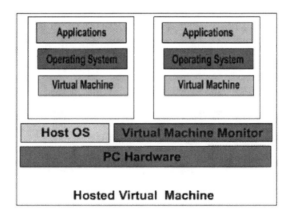

Fig. 1. Virtual Machine Key Map

2.2   Malicious Software Propagation

Malicious codes are executed to infect files or systems or propagate themselves to other systems or files. There are many ways to do so and understanding such ways is necessary to understand malicious codes. The propagation ways are "email-using spread", "network-using spread", "malicious codes activation methods with rebooting", "DLL injection", "Polymorphism", and so on. [4]

According to the report of AhnLab, Inc. in July, 2010, email takes up 70%, Trojan Horse and virus each takes up 10% of malicious code propagation way. The percentage of mass-mailer has been increased by 20% compared to the previous month owing to bagle worm, Netsky worm, and Mytob worm, which caused huge damage in June. [5]

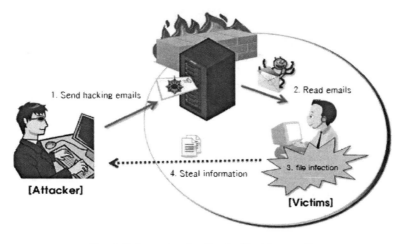

Fig. 2. Information Stealing by Hacking emails

As figure 3 shows, the title of hacking email attracts the receiver's attention and the attached files are various document files from MS Office(Word, Excel) to HWP, PDF, and so on. When the attachment files are opened, the malicious codes infect the system and transmit the internal data of the PC to midway points such as xxx.3322.org, xxx.youngkoala.com, and xxx.ods.org.[6]

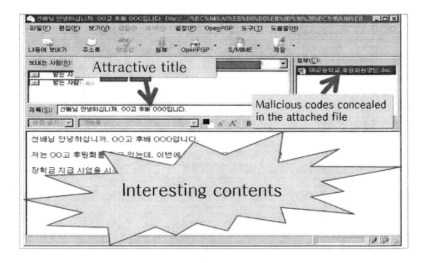

Fig. 3. An example of Malicious codes concealed in the Attached File

2.3   Malicious Software Detection System Using Virtual Machine

Malicious behavior detection technology using a virtual machine detects and executes malicious behaviors of each file in an independent virtual machine in the user's system. The virtual machine is initialized in case malicious software is executed and gives bad influence to the system. Therefore, it is possible to quickly re-execute malicious codes and measure the bad influence of each target file in a new virtual machine.[7]

System using a virtual machine reduces the system restoration time than the system which actually executes codes and decides malicious behaviors. Using virtual machines makes it easy to detect malicious behaviors of a lot of target files. A virtual machine like 'VMware' can activate several virtual machines simultaneously in a system.

Figure 4 shows overall architecture of malicious software system using a virtual machine. When an e-mail arrives, the system checks whether any attached file exists or not and sends the e-mails with attachment to Virtual Machine Cluster.

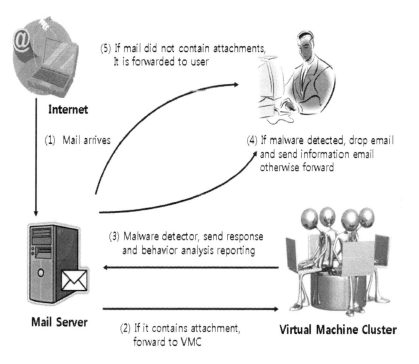

Fig. 4. Diagram of Malicious software Detection System Using Virtual Machine

In Virtual Machine Cluster, the detection tool separates a target file from email, executes the target file, and detects malicious behaviors of the target file. The attached files, which turn out to be malicious behaviors are not sent to the user, are discarded by the mail server and a notice email is sent to the user.

2.4   VMC(Virtual Machine Cluster)

The detection tool of VMware is G-DATA, which detects malicious behavior of target files and filters suspicious URL link. VMC generates behavior analysis report through Norman SnadBox Analyzer.

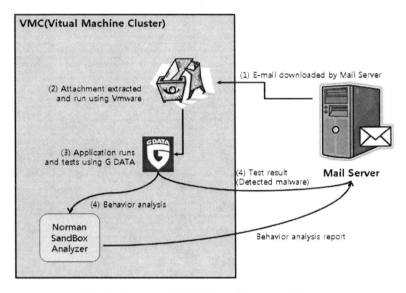

Fig. 5. Diagram of VMC(Virtual Machine Cluster)

3   Design of Malicious Software Detection System

3.1   Implementation of Prototype

The specifications of the test system are as follows.

Table 1. Test System

	Mail Server	VMC(Vitual Machine Cluster)
OS	Windows Server 2003 Standard Edition SP2	Windows XP Professional SP3
CPU	Intel Xeon 3.0GHz	Intel Core2Quad Q6660 2.4GHz
RAM	2GB	2GB
S/W	IIS	VMware Workstation 6.5.1 Build 126130 for Windows

AVK2009(G-DATA) whose detection rate is 99.8% was chosen as a detection tool according to the results of vaccine products test by Virus Bulletin of Germany.[8] G-DATA Internet Security 2009 is the official name for ACK 2009 which increases the detection rate by using BitDefender engine of BitDefender SRL, Rumania and Avast engine of ALWIL Software, the Czech Republic. G DATA Internet Security 2009 not only detects malignant codes but also blocks suspicious URL.

3.2   Decision Making Using Database

It is possible to detect malicious behaviors more rapidly by adding database to the existing VMC. Detecting methods are as follows. First of all, save the size, name and behavior analysis report of the previously detected malicious codes on the database. In case the e-mail has attachment, save the sender as well.

Then, if two of the size, name, and/or sender of the attachment match to the database, consider it as malicious codes. This way, there is no need to execute every attachment. VMC compares the attachment with the database and if the attachment is similar to any file that has already been detected as malicious code, the malicious code can be detected without execution.

The information of the size, name, and/or sender of previous malicious codes is saved in the database. If the information of the attached file do not match with more than two of the three (size, name, and/or sender), VMC decides whether it is malicious code or not by comparing its behavior to the existing behavior analysis report in database.

3.3   Test Results

The test results show that it takes 20 seconds on average to download the attachment, execute it on VMware, check if it is malicious behavior, and update Message Queue. It also takes approximately 20 seconds on average to reboot and initialize another VMware. The execution time increases when processing two OS images in VMware simultaneously compared to processing one by one, which is estimated due to system resource depletion.

Table 2. Prototype Test Results

	1 OS image	2 OS image
VMware Image upload	20 second	52 second
Message Queue Update after Target Files Execution	20 second	53 second
Total	40 second	107 second
Throughput / day	2,160	807

Table 3. Test Results Using Database

The Number of Executed Files	21,60
The Number of not Executed Files	160
Total	2320

Consequently, the test results show that 2,160 emails per day can be processed in one virtual machine, whereas, 2,320 emails can be processed by using database so as to sort malicious codes out without executing codes. Only about 160 emails are sorted out because database information is not sufficient, however, more attached files of emails could be processed if the information of malicious codes are added to the database.

4   Conclusion

In the existing VMC, it decides malicious behavior by executing the attached files of emails with malicious behavior detection tools in a virtual machine. In this system, we created previous malicious code database and compare the attachment of emails with the database. According to the results, VMC executes the selected attachments by sending them to detection tools. The virtual machine is re-initialized when malicious software gives bad influence to the system and compares the behavior of malicious software to the existing behavior analysis report. The system is faster than the existing VMC which executes every code and detects malicious software and makes it easy to detect malicious behaviors of a large number of target files.

For further studies, it is necessary to put more information of the existing malicious codes into database for faster and more accurate decision

Acknowledgments. This work was supported by a grant from Security Engineering Research Center of Ministry of Knowledge Economy.

References

1. Bacel, R., Mell, P.: Intrusion Detection Systems. NIST (2003)
2. Endorf, C., Schultz, E., Mellander, J.: Intrusion Detection & Prevention. McGraw-Hill, New York (2004)
3. Debar, H., Dacie, M., Wepsi, A.: A Revised Taxonomy for Intrusion- Detection Systems. IBM Report (1999)
4. Weaver, N., Paxson, V., Cunningham, R.: Taxonomy of Computer Worms. In: ACM CCS Workshop on Rapid Malcode (2003)
5. http://pc.ahnlab.com/bluebelt_pcdic/
   pcdic_view.do?BBS_SEQ=94759
6. National Cyber Security Center, Cyber Security, Personal authentication service (2010)
7. Seo, J.T.: Malicious Code Detection Technique in Virtual Environment (2007)
8. http://www.virusbtn.com/news/2008/09_02

Enhanced Tag Identification Method for Efficient RFID System

Bong-Im Jang[1], Yong-Tae Kim[1,*], Yoon-Su Jeong[2], and Gil-Cheol Park[1]

[1] Department of Multimedia Engineering, Hannam University
133 Ojeong-dong, Daeduk-gu, Daejeon, Korea
[2] Chungbuk National University Seongbong-ro, Heungdeok-gu Cheongju, Korea
Janggll@nate.com, bukmunro@gmail.com, gcpark@hnu.kr,
ky7762@hnu.kr

Abstract. Recently, many studies are made about RFID system which is spread in many areas such as logistics and distribution, medical treatment. For it is necessary to identify many objects during short time, RFID system needs the method to minimize the tag collision and to shorten the processing time for tag identification. Therefore, this paper proposes the method to identify the tag by using two readers, instead of conditions of only one reader which is generally used for the improvement of tag identification speed in RFID system. The proposed method minimizes tag collision and improves tag identification speed by identifying each of limited tag ID through the use of two readers.

Keywords: RFID, Tag Identification, Anti-collision, Framed query tree Algorithm.

1 Introduction

RFID system is the technology to recognize objects through radio frequency. It has merit to identify many objects in short time in contrast to bar-code. RFID system is made up of tag attached to object, reader to identify tag information, and back-end database to confirm tag information sent from reader[1]. Reader makes query through the way of broadcast and tag makes response to the query. In this process, it often happens that many tags within range of reader identification send information all together. Because of this characteristic, problem of tag collision happens in the process of tag identification. So, effective anti-collision algorithm for speedy identification of many tags is needed to solve this problem.

Various methods of prevention of tag collision[2-5] have been studied to improve RFID system until now, but the study is lack to improve speed and ratio of tag identification by increasing the number of reader. Therefore, this paper proposes the method of tag identification to improve speed and ratio of tag identification by minimizing tag collision through the use of two readers, instead of the general environment of one reader.

---

* Corresponding author.

G. Lee, D. Howard, and D. Ślęzak (Eds.): ICHIT 2011, CCIS 206, pp. 219–224, 2011.
© Springer-Verlag Berlin Heidelberg 2011

This paper is made up as follows. Chapter 2 introduces the existing algorithm for prevention of tag collision, and chapter 3 proposes the method of tag identification by using two readers. Chapter 4 describes the way of performance analysis of the proposed method, and Chapter 5 makes conclusion.

2   Related Work

This chapter compares and examines various methods for prevention of tag collision in RFID system.

2.1   Tree-Based Algorithm

Tree walking algorithm identifies tag by using the binary tree search[6]. At first, reader sends prefix B of k-bit to a tag. The tag compares the received prefix with its own prefix. If the prefixes coincide, k+1-bit is sent to the reader. If a collision does not take place, the reader makes a query with a new prefix B again. However, in case a collision occurs due to the transmission of "0" and "1" at the same time, reader makes a query with new prefix B ∥ 0 and prefix B ∥ 1 again. Tree walking algorithm has merit of identifying all tags, but has drawback that frequent collisions cause the delay of tag identification in case of many tags within range of reader identification.

Query tree algorithm is an improved method of tree walking algorithm and decides queries of tags depending on tag responses. Tag sends k+1-bit to the last n-bit to reader if the prefix sent from the reader coincides with the existing prefix[7]. In case a collision occurs, reader stores the prefixes in queue. Figure 1 indicate the process of query tree algorithm.

	Round	1	2	3	4	5	6	...
➢	Reader: query	ε	0	1	00	01	10	...
➢	Tag : response	Collision	Collision	Collision	Idle	101	100	
➢	Tag1(000)	000	000					
➢	Tag1(001)	001	001					
➢	Tag1(100)	100	100	100			100	
➢	Tag1(101)	101	101	101		101		
➢	Queue	0 1	1 00 01	00 01 10 11	01 10 11 000 001	10 11 000 001	11 000 001	

Fig. 1. Example of Query tree Algorithm Process

2.2   ALOHA-Based Algorithm

Slotted ALOHA algorithm, as shown in Figure 2, is the method to divide tag responding time to a few of slots and to identify tag just when selected slot sends ID without collision in the way that tags send ID to the randomly selected slot.

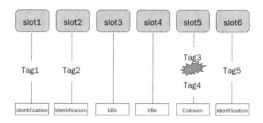

Fig. 2. The Process of Slotted ALOHA Algorithm

Framed Slotted ALOHA algorithm, as the modification of Slotted ALOHA, is the algorithm in which frame size used in communication between reader and tag is fixed. Frame means the period from reader's order-sending to the next order -sending. One frame consists of many slots. Framed Slotted ALOHA algorithm is easy to be realized, however, its disadvantages result from the fixed frame size. In an environment where there are a lot of tags but the size of frame is small, some tags can be not identified, whereas in an environment where there are a small number of tags but the frame size is too large, there can be useless waste of slots.

2.3   Hybrid Algorithm

Hybrid algorithm, a mixture of ALOHA algorithm and tree algorithm, is aimed to decrease the collision by dispersing tags in group. There are two representative Algorithms. One is framed query tree algorithm in which tags are divided randomly in unit of frame and each frame uses query tree algorithm[8-10]. The other is query tree ALOHA algorithm in which the real process of tag identification uses dynamic framed slotted ALOHA in the base of query tree algorithm[8]. Figure 3 indicate the process of framed query tree algorithm.

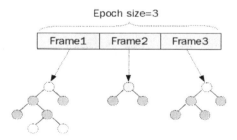

Fig. 3. The process of framed query tree Algorithm

3   Proposed System

This chapter proposes the method to classify and search tag ID by using two readers to minimize the tag collision in the process of tag identification of RFID system, to improve the speed of tag identification and to guarantee the performance of system.

But there is assumption that collision between readers is not available in the proposed system.

3.1   Tag Identification Using Two Readers

The process of tag identification using two readers, as shown in Figure 4, unlike the environment of identifying tag through one reader, decreases the possibility of tag collision and improves the ratio and speed of tag identification, for it identifies the tags appointed by each reader. One of two readers identifies only ID in which the number of "1" is odd among character string of tag ID. The other reader identifies only ID in which the number of "1" is even.

And, each reader performs the process of tag identification by using the framed query tree algorithm which is the speedy and stable in the ratio of identification among various collision-prevention algorithms to minimize the tag collision.

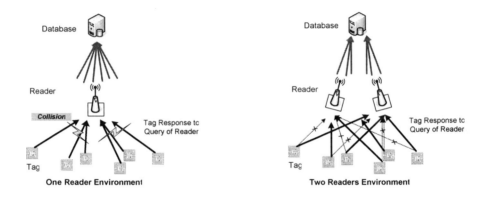

Fig. 4. Systems Comparison based on reader environment

3.2   Process of Tag Identification of Proposed System

If tag is identified in the range of reader identification, reader makes query to tag and receives tag ID. Subsequently, reader analyzes the character string forming received tag ID. After it classifies the number of "1" into even or odd, it identifies only ID assigned to it. If it's not assigned ID, for communication stops and the assigned ID is not included hereafter, the problem of overlapped identification of tag ID which can happen when using two readers is prevented.

4   Performance Evaluation

To evaluate the performance of the proposed system, this chapter compares and analyzes processing time about two system, that is, RFID system which uses one reader and proposed system which uses two readers.

The list of time required for tag identification is as follows.

• F_{sum} : whole processing time for tag identification
$$F_{sum} = E_{ft} + F_t + E_{lt}$$
  - E_{ft} : sending time of first epoch size to determine the adequate epoch
  - F_t : frame test time for estimation of the number of tag
  - E_{lt} : sending time of epoch size for tag identification after determining last epoch size
• R_t : searching time of relevant ID by each reader
• RT_{sum} : communicating time between reader and tag
$$RT_{sum} = P_t + T_t + BC_t + C_t + NP_t$$
  - P_t : comparing time between tag ID and prefix
  - T_t : time sending the rest tag ID from tag to reader
  - BC_t : sending time before collision
  - C_t : collision detecting time
  - NP_t : sending time of new prefix

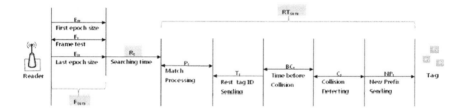

Fig. 5. Structure of Identification Time

To compare performance between two systems, it is necessary to analyze the communicating time between reader and tag under the environment of same number of tag identification. At first, when one reader is used, processing time should be calculated in proportion to the whole number of query and response between reader and tag. Secondly, when two readers are used, processing time should be calculated including analysis time of character string to search the relevant ID by each reader. After all, If the result is "$F_{sum} + R_t + (RT_{sum}/2) < F_{sum} + RT_{sum}$", the proposed system is supposed to have the better performance.

As above, the application of the proposed system to identify the tag ID limited by each reader by using two readers leads to improved performance of RFID system by decreasing the tag collision in the process of tag identification and decreasing the speed of identification between reader and tag.

5   Conclusion

This paper proposed the method of tag identification which uses two readers applying framed query tree algorithm among various tag collision prevention algorithms proposed to prevent the tag collision in RFID system.

The existing method of tag identification led to the decline of system performance by using one reader, identifying tag, and increasing the ratio of tag collision.

Therefore, so as to solve the problem as above, this paper prevented the overlapped identification of tag which can happen in the environment of multi-reader by using two readers, identifying tag ID selected by each reader, and minimized the tag collision by applying framed query tree protocol in the process of tag identification of reader.

In the future, it is necessary to study the system environment which arranges many readers for the optimum application of the number of reader in RFID system.

Acknowledgments. This work was supported by the Security Engineering Research Center, granted by the Korea Ministry of Knowledge Economy.

References

1. Finkenzeller, K.: RFID Handbook, 2nd edn. John Wiley& Sons, England (2003)
2. Chen, W.-T.: Performance Comparison of Binary Search Tree and Framed ALOHA Algorithms for RFID Anti-Collision. IEICE Transactions on Communications 91(4), 1168–1171 (2008)
3. Lee, J.-K., Kwon, T.-Y., Choi, Y.-H.: Improving RFID Anti-Collision Algorithms with Multi-Packet Reception. The Journal of the Korean Institute of Communication Science 31(11A), 1130–1137 (2006)
4. Weicheng, J., Chengfang, S.: An Anti-Collision RFID Algorithm Based on the Dynamic Binary. Journal of Fudan University. Natural science 44(1), 46–50 (2005)
5. Songsen, Y., Yiqu, Z., Weidong, P.: An Anti-collision Algorithm Based on Binary-tree Searching of Regressive Index and its Practice. Computer Engineering and Application 40(16), 26–28 (2004)
6. Juels, A., Ivest, R.L., Szydlo, M.: The Block Tag: Selective Blocking of Tags for Consumer Privacy. In: Proceedings of the 10th ACM Conference on Computer and Communication Security, pp. 103–111 (2003); ISBN:1-58113-738-9
7. Jacomet, M., Ehrsam, A., Gehrig, U.: Contactless identification device with anti-collision algorithm. In: Conference on Circuits, Systems, Computers and Communications, CSCC 2000. IEEE Computer Society, Athens (1999)
8. Shin, J.-D., Yeo, S.-S., Kim, T.-h., Kim, S.K.: Hybrid Tag Anti-collision Algorithms in RFID Systems. In: Shi, Y., van Albada, G.D., Dongarra, J., Sloot, P.M.A. (eds.) ICCS 2007. LNCS, vol. 4490, pp. 693–700. Springer, Heidelberg (2007)
9. Cho, J.-S., Shin, J.-D., Kim, S.K.: RFID Tag Anti-Collision Protocol: Query Tree with Reversed IDs. In: ICACT 2008, vol. 1, pp. 225–230 (2008)
10. Kim, S.-C., Yeo, S.-S., Kim, S.-K.: A comparative Study on the passive RFID tag anti-collision. Korea Information Technology Convergence Society 3(2), 163–177 (2010)

A Study on the MSNR Cryptographic Processor Design Appropriate for the RFID/USN Environment

Seon-Keun Lee[1] and Sun-Yeob Kim[2]

[1] Faculty of Materials & Chemical Engineering,
Chonbuk National University Jeonju, Korea
[2] Department of Information and Communication Engineering,
Namseoul University Cheonan, Korea
caiserrisk@googlemail.com, sykim0599@nsu.ac.kr

Abstract. Creation of RFID/USN environment has increased very rapidly due to activation of ubiquitous. Security environment suitable for RFID/USN environment, however, failed to comply with the speed of security thread increase in reality. Therefore, this thesis presented MSNR suitable for RFID/USN. MSNR presented showed increase in processing rate of 1.3 times compared to existing AES and showed 2 times of improvement in performance in terms of overall system efficiency. Therefore, MSNR is considered as the cryptographic algorithm suitable to overcome conditions of environmental resource conditions such as RFID/USN.

Keywords: RFID/USN, AES, High Speed, network management, Condition State.

1 Introduction

At present, the importance of security for the node network has been emphasized more and more with increase in the use scope of RFID/USN. While diverse security techniques have been applied in RFID/USN, however, it is still not sufficient as the protection technique against various attacks[1].

Therefore, this thesis aims to develop AES algorithm standardized in 2000 further, to increase processing speed and to keep security of node network in order to improve performance of cryptographic algorithm used in RFID/USN.

The presented method is the MSNR(Management of Speed and Network for Rijndael) technique for increasing processing rate and network management convenience for AES. Performance analysis was performed in comparison with AES and other symmetric cryptographic system.

Proposed MSNR cryptographic algorithm has faster processing speed than existing symmetric cryptographic algorithm and uses PRN(Pseudo-Random Number) that divides one cycle into N while mixing bit and byte operation for excellent network management. Therefore, MSNR is considered to become one of cryptographic system that is very required for RFID/USN environment with ever-complicated node network.

G. Lee, D. Howard, and D. Ślęzak (Eds.): ICHIT 2011, pp. 225–234, 2011.
© Springer-Verlag Berlin Heidelberg 2011

2   The MSNR Cryptographic Algorithm Including PRN

While various techniques have been presented to reinforce security in RFID/USN, however, it is still difficult to apply ordinary cryptographic algorithm to RFID/USN environment due to feature of restricted environment.

In this regard, this thesis presented MSNR technique for efficiency of security resources due to increase in node and for solution of problem in terms of realization, taking into account processing time, power consumption and size, and applied this to Rijndael cryptographic algorithm, AES.

Proposed MSNR technique uses exclusive OR as the basic operator similar to existing AES and uses byte for processing unit. Byte operation has excellent feature even with increase in secrete rate since it is difficult to tract back along with advantage of very high processing speed.

In addition, since it mixes existing Feistel and SPN structure, encryption/decryption can be executed at the same time, causing no decline in efficiency generated in Rijndael[2,3] or Serpent[4] cryptographic algorithm. Such feature can present the contents for the version following AES while satisfying sufficient precondition for AES. Due to such feature, cryptographic algorithm added with MSNR technology can solve problems of management due to increase in node, problem in implementation and secrete rate as well as real-time processing.

2.1   The MSNR Cryptographic Algorithm

Size of input/output/key block used for MSNR cryptographic algorithm is 128 bits and the size of plaintext, ciphertext and key is 1:1:1.

MSNR cryptographic algorithm contains following 4 function blocks as shown below like AES cryptographic algorithm and performs encryption/decryption using the round formed with transformation of byte unit while it passes through each stage.

  i) Inv/SubByte : Byte Transformation function using S-box with the feature of a CS
  ii) Inv/ShiftDiagonal : Shift function in row direction about CS
  iii) Inv/MixColumn : Mix function for each column of CS
  iv) AddRoundKey : Add function about CS and round key

Input of plaintext/ciphertext 128 bit is set as CS block initially. CS means the status with non-linear feature for forming MSNR cryptographic algorithm. CS determines the current status with the parameter given outside as shown in formula (1) and figure 1 and the determined current status becomes the reference value that forms the uncertain future status.

$$CS_{next} \leftarrow CS_{present}(prn_{ceed} \bmod 8) \tag{1}$$

CS performs the function to control substitution/transformation during encryption/decryption while making the data required for creation of key value using several parameters used as input. In Formula (1), mod 8 means the modular operation for input data and is the process to transform the format for input value to byte. In addition, PRN[5] SEED creates using the input data and key value as shown in Formula (2).

$$SEED = INPUT \oplus KEY_{7,10,\cdots,111,119,127} \tag{2}$$

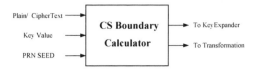

Fig. 1. A Characteristic of the Condition State

SEED created by Formula (2) calculates the output value of 2 bytes through PRN like Formula (3)

$$\text{PRN} = \text{PRN}_{odd} \parallel \text{PRN}_{even}$$
$$\text{PRN}_{odd} = x^{16} + x^{13} + x^{12} + x^{11} + x^7 + x^6 + x^5 + x^4 + 1 \tag{3}$$
$$\text{PRN}_{even} = x^{16} + x^{14} + x^{10} + x^9 + x^8 + x^6 + 1$$

Among 2 bytes created, odd information is used for controlling internal byte substitution of MSNR and even information is used for controlling key expansion. Odd and even information has the function to create the unpredictable function and has low possibility that internal substitution information will be leaked outside. Since it is created with input data and key information only, separate processing is not required.

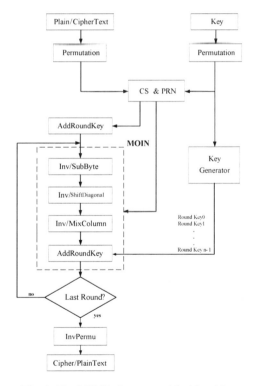

Fig. 2. The MSNR Cryptographic Algorithm

Formula (3) is the same formula as PRN used at mobile communication network and calculates output value, considering 2 byte as one equation.

CS performs n times of rounds after initial round key and exclusive OR. When all rounds are executed, encryption/decryption is finished. Overall flow of MSNR cryptographic algorithm is as shown in Fig. 2. The biggest feature of MSNR cryptographic algorithm is the fact that encryption/decryption is performed at the same time as shown in Fig. 2. Encryption/decryption mode is determined by control signal and processing operation works in normal/reverse order.

Data value for Inv / SubByte, Inv / ShiftDiagonal, Inv / MixColumn, Inv / Add RoundKey works separately by CS & PRN for each round. Therefore, secrete rate is determined depending on the number of round.

Defining these 4 transformation as MOIN(Minimum Operation for Inverse/Non-inverse) results in formula (4)

$$MOIN_{n-round} \leqq Inv/SubByte(odd)_n + Inv/ShiftDiogonal(odd)_n +$$
$$Inv/MixColumn(odd)_n + AddRoundKey(odd)_n$$
$$MOIN_{(n-1)-round} \leqq Inv/SubByte(even)_{n-1} + Inv/ShiftDiogonal(even)_{n-1} + (4)$$
$$Inv/MixColumn(even)_{n-1} + AddRoundKey(even)_{n-1}$$

Existing AES varies the length of data and key to 128, 192, 256 bits and determines the optimized number of rounds depending on the changing length. Therefore, the number of round is determined depending on the length of block of data. With this, if the data symbol size is recognized, the number of round can be identified and cracking is possible by DC and LC with data as well as key and number of round. In case of MSNR, however, the contents of data changes depending on the number of round internally even if it has fixed block and key size. Therefore, it has the advantage in that the number of round cannot be identified with size of data symbol only.

2.1.1   Inv/SubByte Transformation

Inv/SubByte transformation saves external status value by CS in each register using S-box composed of byte unit, and the saved values non-linearly transforms the bytes independently existed to create the byte group non-linearly transformed. S-box used for Inv/SubByte transformation allows inverse transformation and the inverse for the product exists at finite field $GF(2^8)$. It is the group of functions with non-linear transformation that can apply affine transformation defined as formula (5) to $GF(2^8)$.

$$[ab]_i = [ab]_i \oplus [ab]_{(i+4)mod\,8} \oplus [ab]_{(i+5)mod\,8} \oplus \qquad (5)$$
$$[ab]_{(i+6)mod\,8} \oplus [ab]_{(i+7)mod\,8} \oplus [ac]_i$$

Where, ab means the bit block of a parts, the part that transform the internal information among output information of CS & PRN. When it is $0 \leq i \leq 7$, it is the i-th bit of bytes that operates separately disconnected from each other. $[ab]_i$ is the i-th bit unit of relevant operation. mod 8 performs in byte unit for internal operation, but actual operation is bit unit. Therefore, by representing the modular operation for separated expression of bit and byte on the finite field GF, MSNR cryptographic algorithm represents that 1 byte operation becomes the reference. Representing this transformation in matrix type becomes equation (6), where ab' means new status layout transformed and $s_{00} \sim s_{77}$ organize the value for S-box. PN_0 is randomized numbers and a(alpha) means SEED value for the PRN.

$$
\begin{bmatrix} ab'_0 \\ ab'_{01} \\ ab'_2 \\ ab'_3 \\ ab'_4 \\ ab'_5 \\ ab'_6 \\ ab'_7 \end{bmatrix} = \begin{bmatrix} s_{00} & s_{00} & \cdots & s_{07} \\ s_{10} & s_{10} & \cdots & s_{17} \\ & & \cdots & \\ & & \cdots & \\ & & \cdots & \\ & & \cdots & \\ s_{60} & s_{60} & \cdots & s_{67} \\ s_{70} & s_{70} & \cdots & s_{77} \end{bmatrix} \begin{bmatrix} b_0 \\ b_1 \\ b_2 \\ b_3 \\ b_4 \\ b_5 \\ b_6 \\ b_7 \end{bmatrix} \times \begin{bmatrix} PN_0 \\ PN_1 \\ PN_2 \\ PN_3 \\ PN_4 \\ PN_5 \\ PN_6 \\ PN_7 \end{bmatrix} + \begin{bmatrix} a_0 \\ a_1 \\ a_2 \\ a_3 \\ a_4 \\ a_5 \\ a_6 \\ a_7 \end{bmatrix} \tag{6}
$$

2.1.2 Inv/ShiftDiagonal Transformation

Inv/ShiftDiagonal transformation is performed in diagonal direction simultaneously based on row and column unit in data status array.

$$
\begin{aligned}
S'_a(03,12,21,30) &= S_a(00,11,22,33) \\
S'_b(13,22,31,00) &= S_b(01,12,23,30) \\
S'_c(23,32,01,10) &= S_c(02,13,20,31) \\
S'_{da}(33,02,11,20) &= S_d(03,10,21,32)
\end{aligned} \tag{7}
$$

Such transformation can be expressed in equation like Formula (7). For each status array in Formula (7), row and column are transformed diagonally. Such diagonal transformation is not only easy for reverse substitution and calculation, but also guarantees randomness. Therefore, Inv/ShiftDiagonal transformation has the feature of 1/2 times of calculation speed of Rijndael or Serpent cryptographic algorithm and secrete rate has doubled.

In Inv/ShiftDiagonal transformation that moves diagonally like formula (7), the first row and first column with row number 0 will be shifted No. 3 and No. 0 and the last row and column are moved to No. 0 and 3. Such diagonal transformation brings the results of moving the bytes in the same rows to those with low rows and the bytes with low row number move to the position of upper row.

Since Inv/ShiftDiagonal transformation is transformed using diagonal transformation, inverse transformation is also transformed in the same process.

2.1.3. Inv/MixColumn Transformation

Inv/MixColumn transformation creates new transformation by multiplying fixed polynomial expression of $a(x)$. At this time, multiplication operation becomes the basic in total transformation formula. That is, transformed function $s'(x)$ will have the form of multiplying $a(x)$ function before transformation with $s(x)$. Such multiplication type is expressed as formula (8).

$$
\begin{bmatrix} S'_0 \\ S'_1 \\ S'_2 \\ S'_3 \end{bmatrix} = \begin{bmatrix} 02 & 01 & 03 & 03 \\ 01 & 03 & 03 & 02 \\ 03 & 03 & 02 & 01 \\ 03 & 02 & 01 & 03 \end{bmatrix} \begin{bmatrix} S_0 \\ S_1 \\ S_2 \\ S_3 \end{bmatrix} \tag{8}
$$

Result value of formula (8) are 4 bytes columns in status array and are transformed in Formula (9).

$$S'_0 = (\{02\} \cdot S_0) \oplus (\{01\} \cdot S_1) \oplus (\{03\} \cdot S_2) \oplus (\{03\} \cdot S_3)$$
$$S'_1 = (\{01\} \cdot S_0) \oplus (\{03\} \cdot S_1) \oplus (\{03\} \cdot S_2) \oplus (\{02\} \cdot S_3)$$
$$S'_2 = (\{03\} \cdot S_0) \oplus (\{03\} \cdot S_1) \oplus (\{03\} \cdot S_2) \oplus (\{01\} \cdot S_3) \qquad (9)$$
$$S'_3 = (\{03\} \cdot S_0) \oplus (\{02\} \cdot S_1) \oplus (\{01\} \cdot S_2) \oplus (\{03\} \cdot S_3)$$

In formula (9), there is a term requiring continuous calculation for the same coefficient value. Since such same operation is removed using exclusive OR, formula (9) can be summarized and transformed to b e like formula (10).

$$S'_0 = (\{02\} \cdot S_0) \oplus (\{01\} \cdot S_1)$$
$$S'_1 = (\{01\} \cdot S_0) \oplus (\{02\} \cdot S_3)$$
$$S'_2 = (\{02\} \cdot S_2) \oplus (\{01\} \cdot S_3) \qquad (10)$$
$$S'_3 = (\{02\} \cdot S_1) \oplus (\{01\} \cdot S_2)$$

By performing Inv/MixColumn transformation using Formula (10), only parts of status array can be transformed as shown in Fig. 3. While it performs diagonal transformation corresponding to Inv/ShiftDiagonal transformation, Inv/MixColumn transformation performs diagonal transformation corresponding to y = x only. That is, in diagonal transformation of MSNR cryptographic algorithm, diagonal transformation is separated as shown in formula (11).

Overall diagonal can be transformed by performing diagonal transformation x direction and -x direction for one status array like formula (11). Such diagonal transformation can reduce the inconvenience of performing multiplication operation like existing cryptographic algorithm. Presented MSNR cryptographic algorithm performs multiplication for other values only as shown in Formula (10) so that shortening and carry time in performing speed do not occur, while diagonal transformation is performed in opposite direction at the same time, improving secrete rate.

$$y = -x : \text{Inv/ShiftDiagonal}$$
$$y = x : \text{Inv/MixColumn} \qquad (11)$$

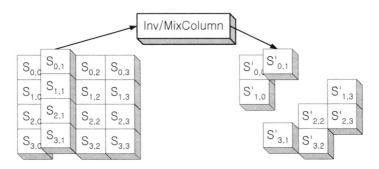

Fig. 3. A Inv/MixColumn Tranform

2.1.4   AddRoundKey Transformation

AddRoundKey transformation performs the operation of adding round key and status array(CS). Each round key calculates independent value whenever transition status occurs from key schedule and the calculated values perform the operation like Formula (12).

$$[S_0', S_1', S_2', S_3'] = [S_0, S_1, S_2, S_3] \oplus [W_{round} \cdot PN_{round}] \tag{12}$$

Where, W_{round} is the key schedule word for the execution of round and PN_{round} means the scope of round execution for CS & PRN. The number of counts for calculation differs depending on the scope of round of AddRoundKey transformation, the amount of calculation performs simple addition function only.

2.2   MSNR Key Scheduling

General cryptographic algorithm performs key creation algorithm and key scheduling works. Such key scheduling work is the mean to create more complicated key and performed absolute role for safety of algorithm.

MSNR cryptographic algorithm does not use the method of changing key scheduling depending on round, but simply creates it using CS & PRN method. PRN uses parts of plaintext to create ciphertext from, as SEED of PRN, and whenever each event occurs, PRN recognizes it as round course and creates key required for encryption. Such key creation enables effective creation of separate key required for each round and is not required to separately set the reference point for differentiating the round when performing encryption using the created key.

Key creation uses the value that changes whenever event of PRN occurs and value of PRN adjusts transformation while internal ciphertext changes at MOIN block, that is, whenever performing 4 stage transformation. Due to this, different cryptographic algorithms go through expansion and reduction process of key length to adjust size with key and ciphertext, but it is not required to separately operate key length in MSNR cryptographic algorithm.

3   Design of MSNR Cryptographic System and Simulation

MSNR cryptographic algorithm was implemented in top-down method using VHDL and Synopsys Design Analyser Ver. 1999.10, QUARTUS 7.0 were used for system synthesis. Synopsys VHDL Debegger, ModelSim 5.8C were used as tools for simulation. ALTERA Cyclone EP1C6Q240C8N device was used as testbed for implementation.

CS & PRN processing part is the core area of proposed MSNR cryptographic algorithm and actually performs MSNR cryptographic algorithm. CS block calculates condition status based on the status condition for input data and MSNR cryptographic algorithm, and CS, calculated condition status array, performs byte substitution. CS status changes through PRN and mod operation as shown in Formula (1). At this time, data used as input, is the encrypted sentence or plain sentence, key value and SEED value of PRNG(PRN Generator)[5].

For input data 128 bit, key value is also 128 bit and some data, among key value of 128 bit, is used as PRNG SEED value after performing input data and binary multiplication.

MOIN block inside MSNR cryptographic system performs the function of diagonal transformation in $y = x$, $y = -x$ directions at the same time.

MOIN block performs sequential functions for Inv/SubByte, Inv/ShiftDiagonal, Inv/MixColumn, AddRoundKey blocks. It is the part where actual data transformation is performed after AddRoundKey function at condition status. It transforms the bytes of data at Inv/SubByte block, performs diagonal transformation at Inv/ShiftDiagonal blocks and diagonal transformation again at Inv/MixColumn blocks.

Even, odd of PRNG is determined depending on encryption/decryption mode and the determined PRNG receives key information as SEED value and prints out the random key information. Key information printed out this time performs part of plain/ciphertext and performed pre-processing SEED operation. Key information after SEED format process is used as input of AddRoundKey inside MOIN block.

At this time, round of MSNR cryptographic system is determined to be 10 times. After 10 round operation is performed, plaintext or ciphertext are created after round determination process and last substitution process. keyscheduler block is the block that creates the random key information while entering key information for MOIN block.

MSNR cryptographic system yields output 128 bit for input 128 bit and prints out plaintext or ciphertext for ciphertext or plaintext.

Key information, required for encryption/decryption, is not actually required when encryption/decryption is actually performed, but only used as the basic materials for creating new key information.

Table 1 compares and analyzes existing encryption system and proposed MSNR cryptographic system[6-9].

Table 1. The evaluation of performance for MSNR

@50MHz	Structure	Round Number	Data Length(bits)	Key Length(bits)	Processing Rate(Mbps)
DES	Feistel	16	64	56	31.6
3DES	Feistel	48	64	112/168	15.6
SEED	Feistel	16	128	128	313.7
Serpent	SPN	32	128/192/256	128	197.3
AES	SPN	10	128/192/256	128	387.9
MSNR	Feisel & SPN	10	128	128	532.0

In the table 1, it was checked that processing rate of MSNR cryptographic system increased 130% compared to existing symmetric block cryptographic system. In addition, in terms of frequency of rounds and encryption efficiency per secrete rate, in MSNR cryptographic system, key information used for encryption is created by internal CS and PRN, and encryption/decryption can be executed at the same system at the same time so that MSNR encryption has 2 times of efficiency as existing block encryption system in terms of encryption/decryption. Therefore, in terms of overall system efficiency, MSNR encryption system has 2 times of performance as existing block cryptographic system.

As shown above, MSNR cryptographic algorithm is confirmed to be effective in environment with resource constraints such as RFID/USN compared to existing block cryptographic algorithm.

Figure 4 is totally cryptographic processor.

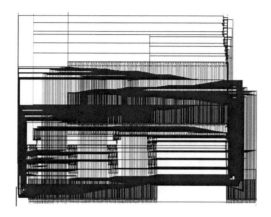

Fig. 4. The circuit of MSNR cryptographic processor

4   Conclusion

Modern society witnesses fast expansion to various applications by node network such as RFID/USN. However, it is very difficult to select cryptographic algorithm to keep the security due to constraints of environmental resource such as RFID/ USN.

This thesis presented MSNR block cryptographic algorithm, suitable for this RFID/USN resource constraints.

Proposed MSNR cryptographic system is the resource performing encryption and has its own information only. In addition, processing time and secrete rate have increased 130% in terms of processing compared to existing symmetric cryptographic system due to simultaneous operation. In addition, since it can process encryption/decryption as one system, it has performance 2 times of existing one in terms of overall system efficiency.

MSNR cryptographic algorithm proposed at this thesis has high transmission rate and system efficiency than existing cryptographic algorithm and is the structure-based algorithm requiring no determination of specific length of key so that system complexity is very low and processing time is fast. Therefore, proposed new MSNR cryptographic algorithm is considered as the cryptographic algorithm suitable for overcoming environmental resource constraints such as RFID/USN.

Acknowledgments. Funding for this paper was provided by Namseoul University.

References

1. Information protection in ubiquitous environments (November 2008),
   http://www.eic.re.kr
2. Ferguson, N., Kelsey, J., Lucks, S., Schneier, B., Stay, M., Wagner, D., Whiting, D.L.:
   Improved Cryptanalysis of Rijndael. In: Schneier, B. (ed.) FSE 2000. LNCS, vol. 1978, p.
   213. Springer, Heidelberg (2001)
3. NIST, AES Algorithm (Rijndael) Information,
   http://csrc.nist.gov/archive/aes/rijndael
4. Damaj, I., Itani, M., Diab, H.: Serpent Cryptography on Static and Dynamic Reconfigurable
   Hardware. In: IEEE International Conference on Computer Systems and Applications,
   aiccsa, pp. 680–684 (2006)
5. NIST: pseudo-random number generator, http://www.itl.nist.gov/div897/
   sqg/dads/HTML/pseudorandomNumberGen.html
6. Microelectronic Systems Laboratory, Implementation of DES Algorithm Using FPGA
   Technology (2002), http://www.alagger.com/des-vhdl/report.pdf
7. DES and 3DES cores, http://www.heliontech.com/des.htm
8. SEED block cryptographic algorithm,
   http://service2.nis.go.kr/pw_certified/seed.jsp
9. A candidate block cipher for the advanced encryption standard,
   http://www.cl.cam.ac.uk/~rja14/serpent.html

A Study on Applying RFID Systems of Korean Pharmaceutical Industry

Kwang NamGung, Yongjung Choi, Youngho Kwon,
Chulwoo Jun, Seongjin Park, and Jaejong Han

Hanmi IT, 45 Hanmi Tower, Bangi-dong, Songpa-gu, Seoul 138-724, Korea
{light557,cyj740,duke92,gte483w,park.sj,jayjay_han}@hanmi.co.kr

Abstract. In order to streamline the distribution process, the Korean government is currently preparing for a policy that mandates introducing RFID systems to various industrial participants that belong to the supply chain. Different from the industrial products, medical products are subject to a strict regulation on activities such as manufacture, distribution and consumption since it is directly related to people's life and health. This paper outlines RFID item-level tagging systems, a world first implemented by H pharmaceutical company and demonstrates the strategies based upon its accumulated experience along with the effect of introducing RFID systems.

Keywords: Pharmaceutical Industry, RFID, Item level tagging, Packaging materials, Tag encoder, Shipment verifier.

1 Introduction

The RFID systems in the supply chain provide effective stock management, optimal read time and accuracy, enhanced security and interoperability. Since 2009, the Korean government is preparing new legislation not only against medical products or liquors that are directly related with the consumers' safety, but also for the sake of supply chain transparency. The Ministry of Health and Welfare mandates RFID tag or 128-bar code on medical drugs, which will go into effect on Jan 2012 for all designated drugs, and in 2013 for all prescription drugs. Similar to RFID tag, the serialization will be applied to the new bar code system such as 2D bar code by Jun 2011[5]. The Korean Food and Drug Administration and the Ministry of Strategy and Finance have already formulated a 7% tax cut when adopting the serialized code system such as the RFID, 2D bar code scheme[3][7]. In addition, the Ministry of Knowledge Economy has proposed "pharmacy & IT convergence development strategies." The proposal now plays a substantial role in the impetus of RFID adoption; estimates suggest that by 2015, 50% of all medical products in Korea will be RFID tagged[6].

This paper introduces a case study of implementing RFID item level tagging systems, a world first accomplishment by H Pharm. in Korea and looks into the challenges and solutions during the implementation as well as the effect of the RFID adoption.

G. Lee, D. Howard, and D. Ślęzak (Eds.): ICHIT 2011, CCIS 206, pp. 235–243, 2011.
© Springer-Verlag Berlin Heidelberg 2011

2 The Background of Research

2.1 The Circumstances of Korean Pharmaceutical Industry

In 2008 the Korean pharmaceutical market was estimated to be worth $17 billion with a 12.6% increase in growth compared with the previous year. Meanwhile the production capacity amounts to approximately $14 billion with exports of $1.3 billion and imports of $4.3 billion. According to a report in 2008[4], Dong-A pharm. topped the list of annual revenue with $7.5 billion followed by Hanmi Pharm. with $5.8 billion and Daewoong Pharm. with $5.6 billion, respectively. As for the medical products distribution line, over 45% of all medical drugs are shipped to the wholesalers and 24% of drugs are directly shipped to private and general hospitals. And 33.5% of those shipped to the wholesalers are redistributed to local pharmacies and 12.3% to hospitals. More than 57% of drugs purchased by end customers is through local pharmacies and others through private and general hospitals.

2.2 The Case Study of RFID Systems at the Various Industries in Korea

In Korea, after the establishment of IT839 in March 2004, RFID began to make a mark as the next-generation technology replacing the bar code system and it is now widely utilized in a variety of industry sectors such as the pharmaceutical, publication, and liquor industries. Based on the circumstances, over a period of 2004 and 2005 the National Information Society Agency selected 12 subjects as RFID pilot projects such as Public Procurement Service, Ministry of National Defense, National Veterinary Research and Quarantine Service, Ministry of Environment, Ministry of Knowledge Economy, and others, which spurred other industries into adopting RFID systems [8]. The prominent examples are "the RFID-based ubiquitous electronic distribution systems" by CJ GLS, "Homeplus RFID cart" by Samsung Tesco, "the RFID-based process control systems" by Hynix semiconductor, and "the RFID-based u-medical information sharing systems" by The Ministry of Health and Welfare.

3 RFID Systems Implementation by H. Pharm

H pharm. is one of the biggest pharmaceutical companies in Korea. Fig. 1 shows the forward logistics in general and the reverse logistics specific to return & recall process performed by H pharm. The completed products, after going through the packing process along with RFID tagging process, are sent to the warehouse. When orders are received from wholesaler or retailers, medical products are then taken out of the warehouse and shipped to distributors. This process up until this point is called forward logistics. In case the medical products are subject to disposal due to expiration or deterioration, those products are shipped to manufacturer by way of retailers and wholesalers. Returned products are discarded by a waste disposal company. Hitherto the procedure is defined as reverse logistics.

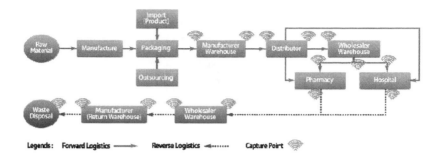

Fig. 1. Forward/Reverse Logistics of H Pharmaceutical Distribution in Korea

3.1 Measures to Improve the RFID Tag Reading Success Rate and the RFID Tag Data Security

H Pharm. currently produces 500 different medical products and uses a variety of packing materials such as paper carton packing, plastic or glass packing, metal packing, stretch packing and so on. The inner packs are also made with various materials for the purpose of preventing deteriorating the efficacy of a drug. However, RFID has an issue with penetrating some substances such as liquid and metal which significantly degrade the RFID tag read success rate. This section will seek for solutions to improve the read success rate against such as the liquid or metal-based materials.

3.1.1 A Space Divider: Adopting an Innovative Packing Structure to Raise the Readability

The box packing, except for glass packing products, is applied to 222 products out of the entire company products in a range of 450 different types. Aluminum foils for capsules or tablets and liquid substances such as collyrium, especially cause the diffusion of electro-magnetic wave. H Pharm. developed a unique wrapping technique placing a space divider and this technique is optimized for the individual box packing products.

Table 1. Prerequisite for applying space divider to box packing medical products

category	Prerequisites
Packing materials with space divider	• The read success rate against tablets, capsules, and collyrium products is required to exceed 95% • The read success rate is required to exceed 95% independent of the shapes of inner packing materials • As for item level tagging production, the read process speed is required to exceed 150 items per minute • As for item level tagging production, the read/write success rate is required to exceed 95% • As for tagging on a shipping box with unitary products and mixed products, the read success rate is required to exceed 95% • Stocktaking time in the retailers and pharmacies is required not to exceed 5 minutes

The following image demonstrates a paper packing material with a space divider, which may play a role as a standardized means for product packing in pharmaceutical industry. Presently, H Pharm. accomplished over 99% of read success rate on an average.

Fig. 2. Sample of paper based packing material equipped with a space divider

3.1.2 Development and Change in the Entire Range of Packing Materials Being Specific to the Properties of Each Medicine

In order to uniformly apply RFID technique into the entire line of company products, it is necessary to tackle the many obstacles to reading/writing tags which take place in the various environments such as manufacturing, shipping, stocktaking processes and so on. H Pharm. maximized the read/write success rate by developing and modifying the packing materials specific to each type of medical product such as tablets, capsules, powder, injection, collyrium, etc. Furthermore, this company established optimization, standardization and unification of applying packing technique and materials into the RFID item level tagging systems. Below are the prerequisites for reading the various types of medical products.

Table 2. Prerequisites for the various packing materials

category	Prerequisites for packing materials
Packing materials by each types of medical products	• The read success rate against tablets, capsules products is required to exceed 95% • The read success rate against glass products is required to exceed 95% • The read success rate is required to exceed 95% independent of the characteristics of ointments, paste-based products wrapped with aluminum tube, liquid medicines, chewable tablets and so on • As for item level tagging production, the read process speed is required to exceed 150 items per minute • As for item level tagging production, the read/write success rate is required to exceed 95% • As for tagging on a shipping box with unitary products and mixed products, the read success rate is required to exceed 95% • Stocktaking time in the retailers and pharmacies is required not to exceed 5 minutes

The table.3 shown below lists outer wrappers that H Pharm. currently utilizes. Each model has been modified over time in order to optimize RFID readability and set an industry standard.

In addition, this company has adopted a standard air interface protocol, EPC Class 1 Gen2, ISO 18000-6C, and leverages passive UHF 900Mhz RFID tags. The details are as follows.

Table 3. The details of outer wrapper models

[unit : mm]

Type/product name	Previous dimension (W*D*H)	Image	New dimension (W*D*H)	Image	Remark
ointments/ Echoron	120*23*33		120*23*40		divider on the upper side
hard capsule/ slimmer	120*23*33		110*30*75		divider on the lower side
collyrium/ Nunen	78*45*100		85*45*110		divider on the right side
Injection/ Triacson	-		-		Place a slipsheet
203 other models					

Table 4. The RFID tags leveraged by H Pharm

Category	Image	Dimension	Usage	Specification
Linear Tag		100mm×10mm	- carton boxes - shipping boxes	- ISO18000-6C (EPC Class1 Gen2)
Satellite Tag(1)		37mm×22mm	- bottle containers	- UHF (900MHz bandwidth) - 96bit or more (EPC sector)
Satellite Tag(2)		37mm×22mm	- glass containers	- Passive - reading/writing ability

3.2 RFID-Based Network Systems Architecture

The overall structure of RFID-based network systems is demonstrated in Fig. 3. The whole system is composed of 3-tiered system layers which allows the data to be seamlessly transferred to the upper layer regardless of network connection. Edge middleware associated with the shipping verifier collects EPC data from the EPC tag readers and batch-transfers them to the delegate middleware, which is stationed at the local server. Subsequently, the delegate middleware conveys the received data to the top middleware, which will later utilize those data for various business solutions.

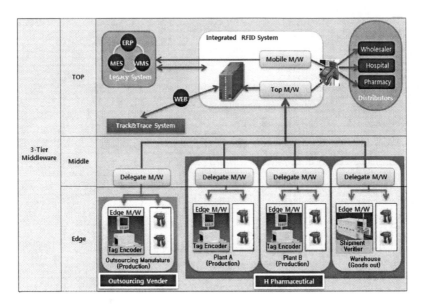

Fig. 3. The overall network systems structure

In case the data transmission fails, it keeps the data temporarily and batch-transfers them to the top layer when the network systems get back on track. The integrated RFID middleware at the top layer plays a bridge role for interoperating with the existing legacy systems via data communication. Such data communication methods reduce the complexity of combining with legacy systems, while increasing the maintainability by setting the tasks of legacy systems apart from the one of RFID systems.

3.3 RFID-Based Production/Shipment Process

3.3.1 Production Process

The relationship among each system during the production process is shown as Fig 4. When an order is received by the ERP system, a packing order (e.g. the information of product, manufacture, batch, shop floor, device, packing quantity, production date, and expiration date) is delivered to the MES. The reconfigured packing order (e.g. product code, manufacture info, batch info, packing quantity, production date, expiration date, work status, serial number) is then sent to Edge middleware of the tag encoder by way of RFID middleware. Packing process thereafter begins according to the packing order indicated through user interface.

The tag encoder first encodes an EPCglobal-standard SGTIN-96 EPC code[1][2], and shortly after, reads the code to verify any errors in encoding process. If no error is found, the tag information is sent to MES via RFID middleware and then stored into ERP as completed product info. Otherwise, the failed product is examined through encoding reverifier. The whole process is as illustrated in Fig. 5.

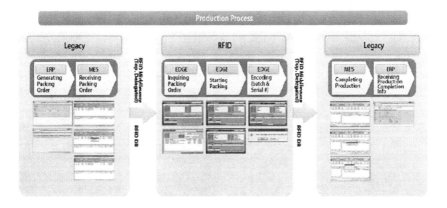

Fig. 4. Production process

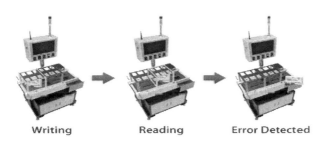

Writing   Reading   Error Detected

Fig. 5. Tag encoding/decoding process

Tag encoder not only accommodates the tag management but also promptly verifies the erroneous tags at the shop floor, which results from processing the tag data in real time.

3.3.2 Shipment Process

Fig. 6 demonstrates the relationship among each system at the shipping process. When an shipping order is received by the ERP system, the order info (e.g. the information of box quantity, completed box quantity, picking box quantity, client, release type, registrant, product code, box code and shipment quantity) is delivered to the WMS that later determines the shipping destination and stores the shipping data into RFID database. The stored data is sent to shipment verifier via RFID middleware. The shipment verifier reads the tag info of individual products as well as shipping box (that is a SSCC-based EPC code; denoting shipping code (13-digit decimal number), box code (5-digit decimal number) in terms of 62-bit code) and compares it with the order info by client, which enables the system to verify picking errors and associate the errors with shipping info.

Fig. 7 shows the shipment verifier. In order to scan to verify every single item in the shipping box, the device is equipped with multi-antenna in its scanning booth and shielded from reading tags outside of the booth by metal plates. This device not only reduced the time for manual examination but also rooted out erroneous shipments.

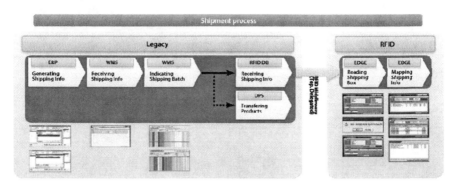

Fig. 6. Shipment process

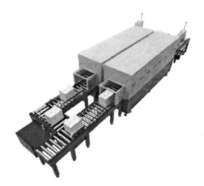

Fig. 7. Shipment verifier

4   The Effect on Adopting RFID Systems

The effect of adopting RFID systems by H Pharm. throughout its supply chain is, as show in Table.5, the optimal scale of management of stock, shipping cost, stocktaking time, workload of management staff as well as returning goods rate. This effect is expected to save 9,695,852.53 USD out of total annual expense.

Table 5. The expected effect on RFID systems of H Pharmaceutical company

Effectiveness	Amount	Computation Basis
Cuts down stock management cost	1,290,322.58	Average stock value * stock reduction ratio * financial cost = 129,032,258 USD * 25% * 4%= 1,290,322.58 USD
Cuts down shipping cost	2,101,382.49	Number of times being shipped to client per month : 4.2 times → twice 175,115.21USD/Month * 12 Months = 2,101,382.49USD
Reduces stocktaking time (pharmacy)	995,391.71	Before RFID : 30 min / client (6.26 days / month) After RFID : less than 5 min / client (1 day / month) = 82,949.31USD/Month * 12Months = 995,391.71USD
Reduces the workload of management staff	1,880,184.33	Daily average of visit more than twice → once 156,682.03USD/Month * 12Months = 1,880,184.33USD
Reduces the returning goods	3,428,571.43	OTC : 10.4% → 6%, prescription drugs : 2.9% → 2% = 285,714.29USD/month*12months = 3,428,571.43 USD
Total amount	9,695,852.53	USD

5 Conclusion

The RFID systems implemented by H Pharm. takes the lead in the transparency of the drug supply chain and plays a fundamental role in opening up a new chapter in the RFID industry in Korea. As a matter of fact, many other pharmaceutical companies are seeking advice about the experience and technique of H Pharm. and are following in their footsteps. One of the future projects of H Pharm. called "OASIS (On-site Applicable Smart Information Services)," provides a unified channel through which the any industrial participants share the various event data that takes place within the supply chain. Presently, the project has gone through the design phase and is now in development.

Acknowledgements. This work was supported by the Industrial Strategic Technology Development Program (No. 10035414, On-site Applicable Intelligent Software and Common Standard Platform Development) funded by the Ministry of Knowledge Economy (MKE, Korea).

References

1. EPC global, EPC Radio-Frequency Identity Protocols Class-1 Generation-2 UHF RFID Protocol for Communications at 860Mhz - 960Mhz Version 1.2.0 (2008)
2. EPC global, EPC Tag Data Standard Version 1.5 (2010)
3. Korea Food and Drug Administration, http://www.kfda.go.kr
4. Korea Health Industry Development Institute, 2008 Pharmaceutical Industry Report, Technical Report (2008)
5. Ministry of Health and Welfare, http://www.mw.go.kr
6. Ministry of Knowledge Economy, http://www.mke.go.kr
7. Ministry of Strategy and Finance, http://www.mosf.go.kr
8. National Information Society Agency (Korea), 2004-2005 RFID Demonstration Project Report, Technical Report (2006)

Authentication Platform for Provisioning in Cloud Computing

Hyokyung Chang, Changbok Jang, Hyosik Ahn, and Euiin Choi*

Dept. of Computer Engineering, Hannam University, Daejeon, Korea
{hkjang,chbjang,hsahn}@dblab.hannam.ac.kr, eichoi@hnu.kr

Abstract. Spreading of Cloud Computing draws users to ask for service provide that is more improved, faster, and securer. So, there come security issues in Cloud Computing continuously and there have been a lot of studies on authentication and accee control. Provisioning is a technology related to the process and action that prepare knowledge and resources and supply what needed in advance to find the opitimum among them. Provisioning is one of the technologies to make Cloud Computing more efficiently. This paper proposes an authentication for provisioing in Cloud Computing.

Keywords: Authentication, Provisioning, RBAC, Cloud Computing.

1 Introduction

Since Cloud Computing was introduced to IT market, Cloud Computing based products and services have been increased continuously. Enterprises related to IT industry in the world have been walking into Cloud Computing. Cloud Computing lets applications and data be taken care of by server and lessens a burden of server management by using virtualization technology. It also makes users use services with the price as they use IT resources through Internet. It tells that a variety of advantage of Cloud Computing has been being taken in the market. However, there are some complex problems to implement Cloud Computing such as data protection, management of resources, guarantee of availability, security of personal information in the field of service, platform, infra, and each service.

There are some security issues such as virtualization technology security, massive distributed processing technology, service availability, massive traffic handling, application security, access control, and authentication and password in the cloud computing service environment [4]. The issue of security of Cloud Computing cannot help being raised and it can be divided into two consumer areas. The first is the security from the point of view of individual user and the second is the one from the point of view of enterprise user [12]. In the point of individual user, all Cloud Computing should provide a proper method of security that includes user authentication, authorization, confidential, and a certain level of availability in order to confront the security problem of leak of personal information [10].

---

* Corresponding Author

G. Lee, D. Howard, and D. Ślęzak (Eds.): ICHIT 2011, CCIS 206, pp. 244–248, 2011.

Provisioning is a technology related to the process and action that prepare knowledge and resources and supply what needed in advance to find the opitimum among them. Provisioning is one of the technologies to make Cloud Computing more efficiently. So, this paper proposes the authentication platform, which considers security and provisioning that enables to draw resources optimization or service optimization in the concept of Pay-as-you-go in Cloud Computing.

This paper is composed of as follows. Chapter 2 describes related works, chapter 3 discusses authentication platform for provisioning, and finally chapter 4 draws conclusions and future work.

2   Related Works

2.1   User Authentication in Cloud Computing

There are written 10 of obstacles expected when taking Cloud Computing and 10 of solutions of them in the report announced at University of California at Berkeley in February, 2009. According to Gartner's analysis, Cloud Computing involves security dangers.

One of security challenges of cloud computing is user authentication with high guarantee. Cloud Computing provides IT resources as a service by using Internet technology and has a feature that users use resources like storage, server, and network and pay for only what they use. Services in Cloud Computing are mostly provided on the base of web. So, an authentication is necessary to confirm the identity of use. Authentication hosted in Cloud Computing, however, may include a possibility of authentication outside firewall and there are a lot of difficulties of security like low security of password, non-use, re-use, sharing, forgetfulness, theft, difficulty of input, key logging, weakness in man-in-the-middle attack, and so on [7, 10].

Users in the Cloud Computing environment have to complete the user authentication process required by the service provider whenever they use new Cloud service. Generally a user registers with offering personal information and a service provider provides a user's own ID (identification) and an authentication method for user authentication after a registration is done. Since then, the user uses the ID and the authentication method to operate the user authentication when the user accesses to use a Cloud Computing service [3]. Unfortunately, there is a possibility that the characteristics and safety of authentication method can be invaded by an attack during the process of authentication, and then it could cause severe damages. Hence, there must be not only security but also interoperability for user authentication of Cloud Computing [6].

User authentication in Cloud Computing is generally done in the layer of PaaS and representative authentication security technologies are Id/password, Public Key Infrastructure, multi-factor authentication, SSO (Single Sign On), MTM (Mobile Trusted Module), and i-Pin [12].

2.2   Provisioning

Provisioning technology is referred to the technology related to activities and the procedure that prepares knowledge needed in advance to find the optimized resources among multiple resources and supplies by requests. It makes users and enterprises use the system by allocating, arranging and distributing IT infrastructure resources according to need of user or business.

In particular, Storage provisioning identifies wasted or unused storage and drops it into a common pool. When the demand on storage is received, the administrator takes storage out form the common pool and provides to be used. It enables to construct infrastructure to heighten the efficiency of storage. Thin provisioning method and Hadoop provisioning method are the representative storage provisioning methods. Thin Provisioning method is the first announced technology by 3par storage which is a joint company of Sun, Oracle, and Symantec and is a management method of storage resources. It limited IT administrators to allocate the physical storage through virtualization. There are studies on Hadoop provisioning technique: the technique that sets up software required simply or generates basic configuration files by loading data in advance and operates Hadoop work automatically and the technique [proposed by Karthik Kambatla, et al.] that compares the similarity among resources groups and performs provisioning according to the optimized resources group which has the closest similarity [8]. Karthik proposed a provisioning technique based on signature. It is divided into RS Maximizer and RS Sizer. RS Maximizer calculates the optimized parameters for Hadoop work to use each resources group completely and RS Sizer decides the number of resources groups in order to maximize the performance and minimize the cost when each resources group is available. Signature based technique generates signatures of optimized resources group in advance, compares them to new signature group of resources required to operate, and allocates resources according to the most similar signature information.

2.3 RBAC (Role-Based Access Control)

Access control and user authentication are representative as security technologies used for platforms. Access control is the technology that controls a process in the operating system not to approach the area of another process. There are DAC (Discretionary Access Control), MAC (Media Access Control), and RBAC (Role-Based Access Control)[1, 9, 13].

DAC helps a user establish the access authority to the resources that he/she owns as he/she pleases. MAC establishes the vertical/horizontal access rules in the system at the standard of security level and area for the resources and uses them. RBAC gives an access authority to a user group based on the role the user group plays in the organization, not to a user.

RBAC is widely used because it is fit for the commercial organizations [12]. Technologies used to authenticate a user are Id/password, Public Key Infrastructure, multi-factor authentication, SSO (Single Sign On), MTM (Mobile Trusted Module), and i-Pin [12].

3   Authentication Platform for Provisioning

The authentication platform for provisioning in Cloud Computing proposed in this paper is composed of Pi (Personal Identification) Manager, Provisioning Manager, and Access Control Manager based on OSGi (open Service Gateway initiatives).

OSGi based on java is a middleware framework operated independently in the operating system or platforms. It is a open standard developed by OSGi Alliance, which defines the standard specifics for delivery, arrangement, and management of services in the network [2, 4, 5, 11].

PI manager operates authentication in the Cloud Computing environment and is composed of PI agent and PI Gateway. PI Agent is installed into client's PC that user program runs and it collects information of user authentication when a user logins. The information of user authentication from PI Agent is delivered to Authentication module loaded in authentication server and PI Gateway manages user authentication.

Provisioning Manager is composed of profile Registry and Profile Analyzer. Profile Registry registers profiles needed to find the optimum resources among a lot of resources in advance and Profile Analyzer analyzes registered profiles in order to provide the resources that will be provided and the procedure.

Access Control Manager is composed of RBAC (Role-Based Access Control) and Security Process Rule. RBAC defines the procedure of authentication and access control.

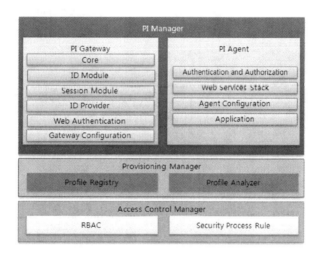

Fig. 1. Authentication Platform for Provisioning

4   Conclusions and Future Work

This paper looked at security technologies used in the platform in the Cloud Computing environment like access control and user authentication and provisioning technology that intends to optimize resources and service. Misuse of access authority to resources and leak of personal information which should be used to authenticate the user could affect faster and more powerful in the Cloud Computing environment compared to in the mono-system. And the use of inappropriate or un-optimized resources does not meet the concept of cost cutting provided by Cloud Computing, which you pay as much as you use. Hence, in this paper, the authentication platform is proposed for provisioning in the Cloud Computing environment. The proposed platform is expected to provide not only the security of user authentication, but also good quality of service by providing a provisioning for service optimization and resources optimization.

As future work, there shall be developed a user authentication service model which is suitable for Cloud Computing and a platform for provisioning.

Acknowledgments. This work was supported by the Security Engineering Research Center, granted by the Korea Ministry of Knowledge Economy.

References

1. Ferraiolo, D.F., Kuhn, D.R.: Role-Based Access Controls. In: 15th Computer Security Conference, pp. 554–563 (1992)
2. Wikipedia, the free encyclopedia, http://en.wikipedia.org/wiki/OSGi
3. Beritino, E., et al.: Privacy-preserving Digital Identity Management for Cloud Computing. Bulletin of the Technical Committee on Data Engineering 31(1) (January 2009)
4. Lee, H., Chung, M.: Context-Aware Security for Cloud Computing Environment. In: IEEK, vol. 47, pp. 561–568 (2010)
5. Lee, C., Hong, W.: Chang. H.: A Study on Authentication Technique based on OSGi Service Platform. In: KIISE, vol. 49(5), pp. 387–395 (October 2009)
6. Chang, H., Choi, E.: User Authentication in Cloud Computing. CCIS, vol. 120, pp. 338–342 (2011)
7. Chow, R., Golle, P., Jakobsson, M., Masuoka, R., Molina, J., Shi, E., Staddon, J.: Cloud Computing: Outsourcing Computation without Outsourcing Control, Palo Alto Research Center (2009)
8. Kambatla, K., Pathak, A., Pucha, H.: Towards optimizing Hadoop Provisioning in the Cloud. In: HotCloud 2009 Proceedings (2009)
9. Wikipedia, the free encyclopedia, http://en.wikipedia.org/wiki/RBAC
10. Kim, H., Park, C.: Cloud Computing and Personal Authentication Service. In: KIISC, vol. 20, pp. 11–19 (2010)
11. OSGi Alliance, OSGi Service Platform Core Specification, Released 4, Version 4.1 (April 2007)
12. Un, S., et al.: Cloud Computing Security Technology. In: ETRI, vol. 24(4), pp. 79–88 (2009)
13. Phillips, C.E.: Security Assurance for a Resource-Based RBAC/DAC/MAC Security Model. University of Connecticut (2004)

Grid Network Management System Based on Hierarchical Information Model

YoungWook Cha[1], KyoungMin Lee[1], ChoonHee Kim[2], Junguk Kong[3],
JeongHoon Moon[3], Woojin Seok[3], and HuhnKuk Lim[3]

[1] Dept. of Computer Engineering, Andong National University, Korea
[2] Dept. of Computer Management Administration, DaeGu Cyber University, Korea
[3] Korea Institute of Science and Technology and Information, Korea
ywcha@andong.ac.kr, tsoc@live.co.kr, chkim@dcu.ac.kr,
kju@kisti.re.kr, otello90@gmail.com, wjseok@kisti.re.kr,
hklim@kisti.re.kr

Abstract. We defined three-hierarchical information model and GNSI (Grid Network Service Interface) to reserve and activate network resources in grid environment. Management scheme of three-hierarchical information model is a little more complicated than that of existing two-hierarchical information model, but our model increases the utilization of network resources by allowing grid application to use several grid network resources over the same path. We designed and implemented grid network management system based on GNSI interface and three-hierarchical information model of network resources. We tested reservation and activation on grid network resources in the test-bed of GMPLS network overlaid over KREONET research network. We also successfully carried out international interoperability test on the control of grid network by deploying FENIUS-API over global research network.

Keywords: grid network, resource management, reservation, grid network interface.

1 Introduction

The grid technology that has been studying in America since 1998 enables high performance computation, mass data processing, and collaboration research in virtual space by connecting geographically distributed and various computing resources to high-speed network [1]. OGF (Open Grid Forum) specified open and integrated standardization, OGSA(Open Grid Services Architecture) to combine grid technology with web services. It is required to reserve, assign, and release resources over the network path among routers and switches in grid environment. NSI(Network Service Interface) working group [2] of OGF and Technical Issues working group[3] of GLIF are carrying out standardization issues for grid network resources.

Fig. 1 shows typical grid configuration with GRS (Grid Resource Scheduler), NRM (Network Resource Manager), and CRM(Computing Resource Manager) in grid environment, which has been adopted by several projects such as G-Lambda,

G. Lee, D. Howard, and D. Ślęzak (Eds.): ICHIT 2011, CCIS 206, pp. 249–258, 2011.

Phosphorus, and EnLIGHTened. The goal of G-Lambda project in JGN II is to define GNS-WSI (Grid Network Service-Web Services Interface) between GRS and NRM [4]. Phosphorus defines control plane and provisioning system of network resources, and it is offering relationship between core networks and grid resources in European GEANT2 test-bed [5]. EnLIGHTened project resolved cooperation issues between grid middleware and optical control plane for dynamic assignment and advanced reservation of network resources in National Lambda Railroad test-bed in America [6]. EnLIGHTened and G-Lambda have done the world's first cooperating project to resolve interconnection issues between different domains.

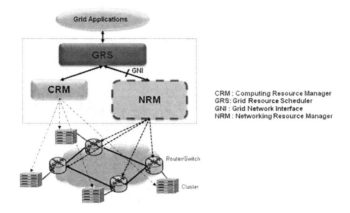

Fig. 1. Configuration of GRS and NRM

GRS requests NRM to reserve and assign network resources through GNI (Grid Network Interface). GRS selects one of network paths by referring an abstract topology created by NRM using physical topology of GMPLS. NRM performs path computation, resource admission, reservation, and assignment over the abstract topology under GRS's considerations of QoS, reservation time, reliability, and so on. GNI is a standard interface to manage grid network resources between GRS and NRM, and this interface is embodied by stateful web services of OASIS's WSRF (Web Services Resource Framework) [7] and resource management messages of GLIF (Global Lambda Integrated Facility) [3]. Existing projects [4-6] have adopted two-hierarchical information model with one to one relationship between grid network resource and network path.

In this paper, we defined three-hierarchical information model with GNP (Grid Network Path), GNR (Grid Network Resource), and GRR (Grid Resource Reservation) to provide grid application for various usage methods of network resources and maximize the utilization of network resources in grid environment. We designed and implemented NRM, dynamicKL which supports GNSI interface and reservation based resource admission control mechanism. GNI interface of GLIF is redefined as GNSI interface to support three-hierarchical information model of gird network resources. We newly defined GNR, notification and commitment messages in GNSI interface in addition to GNP and GRR messages of existing GNI interface. We measured activation and release delays on grid network resources in the test-bed of GMPLS network overlaid over KREONET research network. We also successfully carried out international interoperability test on the control of grid network by deploying FENIUS handler which mapped GLIF's interfaces into GNSI interfaces through global research network.

2 Information Model and GNSI Interface

2.1 Information Model of Network Resources

Fig. 2's (a) is two-hierarchical information model, in which network resource reservations are only possible on a grid network path. Fig. 2's (b) shows three-hierarchical information model, which consist of GNP, GNR and GRR, and describes their relationship. In three-hierarchical information model, there are more than one creation of GNRs on a single GNP according to QoS requirements such as bandwidth and delay. Several GRRs will be scheduled without the overlap of time zones on the same GNR. NRM will initiate activation or deactivation procedure of reserved network resources at the beginning or ending time of reservation. If one GNR is only mapped into one GNP, then the effect of three-hierarchical information model is almost same with that of two-hierarchical information model.

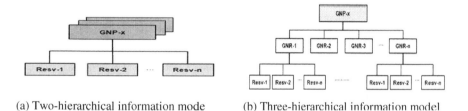

(a) Two-hierarchical information mode   (b) Three-hierarchical information model

Fig. 2. Information models

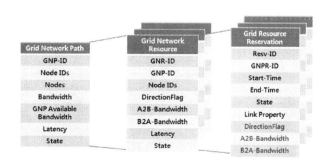

Fig. 3. Information elements of GNP, GNR, and GRR

Fig. 3 shows information elements of GNP, GNR, and GRR, and describes their relationship. GNP includes several information elements, those are GNP identifier

(GNP-ID), addresses of end nodes (Node IDs), addresses of intermediate nodes, total bandwidth, available bandwidth, latency, and state of GNP. DirectionFlag element represents whether this GNR supports unidirectional or bidirectional traffic, in which each direction is characterized by A2B-Bandwidth and B2A-Bandwidth elements. Start-Time and End-Time information elements of GRR represent starting and end times of reservation. More than one GRRs may be scheduled in a GNPR (Grid Network Path and Resource)-ID, which is concatenation of GNR-ID and GNP-ID.

Fig. 4 is a scheduling chart of grid network resources on three-hierarchical information model. GNP1's all resources (e.g., 1G bandwidth) are assigned to a GNR1, and several reservations for this GNR are scheduled in different time zones. If a single GNR is created over one GNP, then resource utilization of three-hierarchical informational model is the same with that of two-hierarchical informational model. GNP2's all resources are divided into three GNRs, those are GNR1 (500M), GNR2 (300M), and GNR3 (300M). Several reservations for each GNR will be scheduled without the overlap of time zones. Management scheme of three-hierarchical information model is a little more complicated than that of two-hierarchical information model, but three-hierarchical model increases the utilization of network resources by allowing grid application to use several GNRs over the same path.

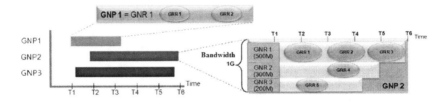

Fig. 4. Scheduling chart of grid network resources

2.2 Grid Network Service Interface

GNSI interface is base on GLIF specification [3] and it is extended to be aligned with three-hierarchical information model. Table 1 shows management messages for each service of GNSI interface.

Table 1. GNSI interface messages

Service	Message	Service	Message
GNP	CreateGridNetPath	**GRR**	CreateResourceResv
	DeleteGridNetPath		CancelResourceResv
	InitGNPR		ReleaseResourceResv
GNR	CreateGridNetResource		QueryAvailableResourceResv
	DeleteGridNetResource		UpdateResourceProperties
	GetAllGridNetResource		GetResourceProperty
Reliable Commitment	ResvCommit		GetAllReservedResources
	Abort	**Notification**	Notification

Grid network resource service is not defined in two-hierarchical information model of existing projects. Grid network resource service enables grid application to create several grid network resources by dividing resources of a grid network path. InitGNPR message is used to simultaneously create a GNP and its GNRs provisioned in the configuration file. We defined reservation service in GNSI interface to reserve grid network resource during the scheduled time period. Most of messages in reservation service are similar to them of GLIF's GNI interface except GetAllReservedResources message that is additionally defined in GNSI to query time and bandwidth information of existing GRRs between two specific end nodes.

Notification message is not used in existing GNI interface. In GNSI interface, this message is used to inform allocation or release event to GRS. Two-phase resource reservation is effective and reliable mechanism in which a GRS makes reservations with several NRMs. CreateResourceResv message is applied to the first phase, and ResvCommit message is used in the second phase. GRS checks whether all NRMs complete preparations of reservations in the first phase, and it commands each NRM to configure resource reservation in the second phase. On the other hand, Single phase mechanism is more adequate if a GRS reserves resources with one NRM. CreateResourceResv message in single phase mechanism is used to simultaneously prepare and configure reservation.

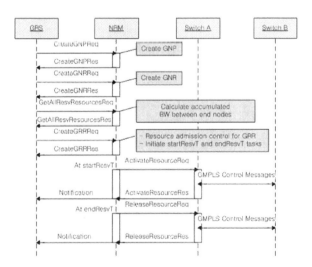

Fig. 5. Sequence diagram for managing GNP, GNR, and GRR

Fig. 5 shows a sequence diagram of grid network resource management. After creation of GNP and GNR, GRS will issue GetAllReservedResources to query time and bandwidth information of GRRs between two specific end nodes. GRS analyzes available reservation time zones and generates CreateResourceResv message with timing and bandwidth parameters which NRM will use to accept or reject new reservation. NRM let the ingress switch node activate or release network resources on the scheduled time of reservation, and it notifies the result of resource control to GRS through GNSI interface.

2.3 Reservation Based Admission Control for Grid Network Resources

Fig. 6 shows reservation based admission control algorithm for grid network resources, which requires starting and ending time parameters in addition to traffic and QoS parameters of existing admission control algorithms [8].

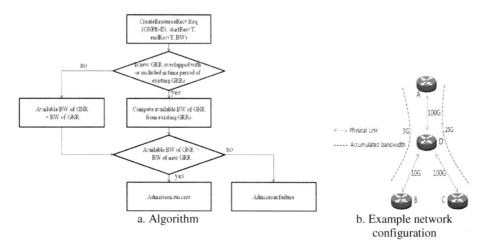

a. Algorithm   b. Example network
                                         configuration

Fig. 6. Reservation based resource admission control algorithm

GRS issues CreateResourceResv message with GNPR-ID, startResvT, endResvT, and bandwidth parameters. NRM searches existing reservations overlapped with or included in time period of new reservation. New reservation will be accepted if there are no existing reservations and configured bandwidth of the GNR is greater than bandwidth requirement of new reservation. NRM can compute available bandwidth for the GNR just using bandwidth information elements of existing GRRs overlapped with or included in time period of new reservation. Admission is successful only if available bandwidth is enough to support the bandwidth of new reservation. In the example network of Fig. 6, maximum configured bandwidth of path A-B with A-D and D-B links is 10Gbps. The available bandwidth of path A-B is 7Gbps if the accumulated bandwidth of this path for a specific period is 3Gbps.

3 Grid Network Management System

3.1 Design and Implementation of Grid Network Management System

Fig. 7 shows blocks and external interfaces of grid network management system, dynamicKL.

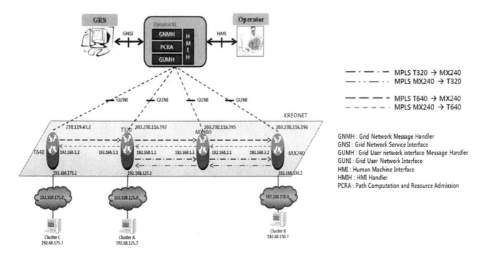

Fig. 7. Configuration of dynamicKL

GNMH (Grid Network Message Handler) block parses messages from GNSI or HMI interface, and delivers them to PCRA (Path Computation and Resource Admission) block. PCRA performs path computation and network resource admission for reservation. PCRA creates and deletes not only grid network paths to configure abstract topology, but also grid network resources on a grid network path. It makes more than one reservations over a grid network resource, and commands GUMH (Grid User Message Handler) block at scheduled time to activate or release network resources. GUMH delivers resource management information to GMPLS control plane for establishing or releasing LSPs (Label Switch Paths) over transport network through GUNI (Grid User Network Interface) interface. HMIH (HMI Handler) block provides web based user interface to handle operator's requests for managing grid network resources.

Each block of dynamicKL is coded with Java programming language on JDK 1.5 version. We adopts MVC (Model, View and Controller) model and implements web based user interface on Spring Framework 2.5. Web server function of dynamicKL is built up on Apache Tomcat 6.0 and server function for GNSI interface is based on web services core of Apache Muse. Dummy GRS was built up to do GNSI client function.

3.2 Test-Bed for Grid Network Resource Management

We built up test-bed with dynamicKL, dummy GRS, and commercial routers as shown in Fig. 7 to verify management functions such as creation, reservation, and allocation of grid network resources. GMPLS control and transport functions are provided by Juniper routers connected to KREONET in which T640 is installed in Seoul and other routers are in Taejon. MX 480 router operates as an intermediate node and other routers play a role of end nodes to which clusters are directly connected to exchange traffic of grid applications. ERO (Explicit Route Object) and FEC (Forwarding Equivalence Class) based LSPs are established by dynamicKL in scheduled time of reservation [9]. We used show-mpls-lsp-statistics command of Juniper router to check out statistics of traffic over the established LSP.

GUNI interface is used to control LSPs between dynsmicKL and Juniper routers through telnet based CLI (Command Line Interface). CLI processing and control plane delays are occurred when dynamicKL initiates activation or release of reserved resources on the scheduled time. Starting time of reservation (startResvT) is the point of exchanging traffic of grid application after allocation of network resources. Fig. 8 shows activation time, T_{acti} and three delays including D_{acti} (CLI processing and GMPLS control delays), D_{noti} (Notification delay from dynamicKL to GRS), and D_{app} (Grid application control delay). Considering these delays, dynamicKL initiates activation procedure a little earlier than startResvT. Activation time, T_{acti} is given by

$$T_{acti} = startResvT - D_{app} - D_{noti} - D_{acti}$$

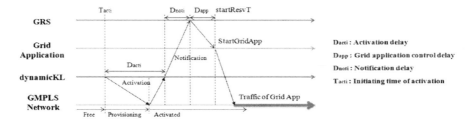

Fig. 8. Activation of reserved resources for grid application

D_{acti} may be the longest delay among delays related to activation time. As shown in Fig. 9, we measured activation and release delays between T320 and MX240 routers in the test-bed. Total activation delay takes about 3.15 seconds which consist of telnet session delay (0.65 seconds) and commitment delay to routers (2.5 seconds). Commitment delay includes CLI processing delay in the ingress router and GMPLS control plane delay among routers on the path. Telnet session delay for release is almost equal to that of resource activation. Total release delay takes about 1.95 seconds, which are the sum of telnet session and commitment delays (1.3 seconds).

Phase	siteA	siteB	Telnet	Comit
Activate	203.230.116.197	203.230.116.196	0.693	2.443
Activate	203.230.116.197	203.230.116.196	0.681	2.394
Activate	203.230.116.197	203.230.116.196	0.661	2.384
Activate	203.230.116.197	203.230.116.196	0.65	2.532
Activate	203.230.116.197	203.230.116.196	0.652	2.488
Activate	203.230.116.197	203.230.116.196	0.653	2.418
Activate	203.230.116.197	203.230.116.196	0.653	2.493
Activate	203.230.116.197	203.230.116.196	0.65	2.515
Activate	203.230.116.197	203.230.116.196	0.652	2.378
Activate	203.230.116.197	203.230.116.196	0.649	2.466

Phase	siteA	siteB	Telnet	Comit
Release	203.230.116.197	203.230.116.196	0.661	1.327
Release	203.230.116.197	203.230.116.196	0.657	1.327
Release	203.230.116.197	203.230.116.196	0.658	1.232
Release	203.230.116.197	203.230.116.196	0.679	1.268
Release	203.230.116.197	203.230.116.196	0.641	1.418
Release	203.230.116.197	203.230.116.196	0.653	1.349
Release	203.230.116.197	203.230.116.196	0.65	1.412
Release	203.230.116.197	203.230.116.196	0.651	1.523
Release	203.230.116.197	203.230.116.196	0.661	1.276
Release	203.230.116.197	203.230.116.196	0.654	1.217

(a) Activation delay   (b) Release delay

Fig. 9. Delays for activation and release of reserved resources

Fig. 10 shows super-agent, networks and their resource management systems, which were configured to test international interoperability in GLIF 2009 and 2010 meetings. Participating networks and their systems were dynamicKL in KREONET, IDC in Starlight, PHOSPHORUS in HPDMnet, and G-Lambda in JGN, which adopted FENIUS-API[3] for grid network management.

Mapping function between FENIUS API of super-agent and GNI of each resource management system is performed by FENIUS handler, which includes common API processing and translator modules. Super-agent requests resource management systems in each country to manage and reserve network resources through FENIUS-API. That is, user can make use of common web based interface to reserve network resources from Korea to Europe via Japan.

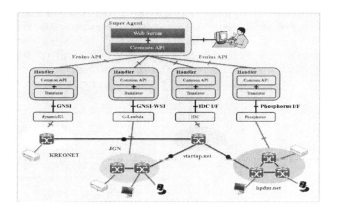

Fig. 10. Configuration of international interoperability test for grid network resource control

4   Conclusions

For grid tasks requiring guaranteed quality of service, we need to build converged infrastructure which manages network resources as well as computing resources. We defined three-hierarchical information model that is a little more complicated than that of two-hierarchical information model, but our model increases the utilization of network resources by allowing grid application to use several GNRs over the same path. For three-hierarchical information model, GNSI interface additionally includes GNR, notification and commitment messages in addition to GNP and GRR messages of existing GNI interface. We defined reservation based admission control algorithm, which requires starting and ending time parameters as well as traffic and QoS parameters. We designed and implemented, dynamicKL, which manages grid resources over GMPLS network. Activation and release delays were measured in the test-bed of GMPLS network overlaid over KREONET research network. We deployed FENIUS-API to carry out international interoperability test on the control of global research network.

Acknowledgements. This work was supported by the grant of KISTI.

References

1. Holmes, V.P., Johnson, W.R., Miller, D.J.: Integrating Web Service and Grid Enabling Technologies to Provide Desktop Access to High-Performance Cluster-Based Components for Large-Scale Data Services. In: ANSS 2003 (2003)
2. OGF, Network Service Interface Working Group,
   http://www.ogf.org/gf/group_info/view.php?group=nsi-wg
3. GLIF, Technical Issues Working Group,
   http://www.glif.is/working-groups/tech/
4. Takefusaa, A., Hayashib, M., Nagatsuc, N.: G-lambda: Coordination of a Grid scheduler and lambda path service over GMPLS. Future Generation Computer Systems 22, 868–875 (2006)
5. Markidis, G.: EU Integrated Project PHOSPHORUS: Grid-GMPLS Control Plane for the Support of Grid Network Services. Transparent Optical Networks (2007)
6. Battestilli, L.: EnLIGHTened Computing: An Architecture for Co-allocating Network, Compute, and other Grid Resources for High-End Applications. In: Grid Computing, High-performAnce and Distributed Applications, GADA 2007 (2007)
7. Banks, T.: Web Services Resource Framework. In: OASIS (2006)
8. Ali, Z.: A scalable call admission control algorithm. IEEE/ACM Transactions on Networking 16(2) (2008)
9. Yang, X., Lehman, T., Tracy, C.: Policy-Based Resource Management and Service Provisioning in GMPLS Networks. In: INFOCOM (2006)

SMCC: Social Media Cloud Computing Model for Developing SNS Based on Social Media

Myoungjin Kim and Hanku Lee[*]

New Millennium Bldg. 1309, Konkuk University, 1 Hwayang-dong,
Gwangjin-Gu, Seoul, Korea
{tough105,hlee}@konkuk.ac.kr

Abstract. Social media-based Social Networking Service (SNS) has grown dramatically over the last decade. Numerous users have strengthened their human networks by joining SNS. Although many service providers tend to develop SNS according to the needs of users, it is difficult to develop SNS based on large amounts of social media using traditional techniques with related to processing and transmitting it in large and DB system for storing it. Therefore, in this paper, we present a new concept of Social Media Cloud Computing model (SMCC) combined with cloud computing environment that provides elastic computing resources for processing and storing big social media data and platforms for providing SNS's development environments. Our services deal with the whole cloud computing services such as SaaS, PaaS and IaaS.

Keywords: Cloud computing, Social Networking Service, Cloud services, Cloud platform and Social media.

1 Introduction

Social media means media for social interaction, using highly accessible and scalable communication techniques [1]. Recently due to the dramatic proliferation of the use of smart phones and the advancement of network infrastructures, lots of users have easily used Social Networking Service (so-called SNS) based on social media contents such as Facebook and Twitter in either a mobile or PC environment. According to the Forrester Research, 75% of Internet suffers used social media contents in the second quarter of 2008 by joining social networks and reading blogs [2]. Overall, SNS based on social media contents based SNS plays a decisive role in activating the place of interchange where internet users are able to share their own thoughts, opinions, experiences and views with others [3].

Lots of users require SNS services based on multi-media data (video, audio and picture) including their own intuitive experiences, opinions, ideas and views rather than a text-based SNS. Although, many service providers have released social media

---

[*] Corresponding author.

G. Lee, D. Howard, and D. Ślęzak (Eds.): ICHIT 2011, CCIS 206, pp. 259–266, 2011.

based SNS containing social meanings, most of the present SNS have really broken away out of the existing micro blog-based services (the form of text).

Even though service providers are aware of the needs of users towards SNS, there are two reasons they have failed to provide social media-based SNS. Firstly, the traditional storage systems are faced with the limitation of storage capacity. In fact, the amount of data in Twitter users produce every day reaches up 7 TB, and even Facebook also produces 10 TB. There are lots of limitations in solving the storage problem through the traditional DB system. Secondly, service vendors also are confronted with difficulties in applying transmission techniques and data processing techniques to transmit large amounts of data to smart phone's users. As a matter of fact, to provide three different type terminals such as smart phone, TV and smart Pad, they should create and process three different video files suitable for those that are converted into different resolutions. New file system, DB system, processing techniques and computing power for processing large amounts of social media are required to achieve the procedures above.

For those, our paper are focusing on suggesting Social media Cloud Computing (SMCC) model that is able to provide element technologies, computing resources and cloud services needed to develop social media-based SNS. The main purpose of this paper is to study how to support cloud environment to develop SNS, multimedia services and social game based on social more easily over the whole cloud computing services such as SaaS, PaaS and IaaS. In this paper, we will introduce the concept of cloud computing as related work. Then, we will also present SMCC model as well as architecture over the whole cloud computing services such as SaaS, PaaS and IaaS.

2 Related Works

The Concept of Cloud Computing. The cloud computing was born in complex combination form distributed computing, grid computing, utility computing and so on [11]. Many people defines cloud computing[4,5,6,12] as an emerging computing paradigm where data and services reside in massively scalable data centers and can be ubiquitously accessed from any connected devices over the internet. That is to say, users are able to obtain useful information they want and use a variety of services and computing resources through the internet as a cloud without the constraints of time and place.

Also, the NIST[10] (National Institute of Standards and Technology) defines 'On-demand self-service', 'Broad network access', 'Resource Pooling', 'Rapid elasticity' and 'Measured Service' as the five essential characteristics of cloud computing. In addition, service models are classified into 3 service models: Cloud Software as a Service (SaaS), Cloud Platform as a Service (PaaS) and Cloud Infrastructure as a Service (IaaS). Lastly, deployment models are classified into 3 models: Private Cloud, Public Cloud and Hybrid Cloud [1,7].

Enabling Technologies in Cloud Computing. There are lots of enabling technologies in cloud computing environment. Table 1 shows the list of cloud computing techniques in terms of concept, meaning and element technologies.

Table 1. The enabling technologies in cloud computing

Technologies	Concept and Meaning	Element technologies
Virtualization	- A technology to operate a system beyond the limits of physical hardware	
Distributed processing in large	- A technology to distribute and process large amounts of data in large scalable server environment (over thousands of nodes)	Distributed processing technology
Open interface	- Applicable to expanding the existing services or changing their functions in cloud computing based SaaS and PaaS, etc.	
Service provisioning	- A technology to provide computing resources in real time - A concept to contribute to increasing the economic feasibility and flexibility of cloud computing by automating job tasks from applying service to service provision	Resource provisioning
SLA	Managing the operation quality in quantified form called service levels is needed according to the characteristics of cloud computing that utilize external computing resources	A system to manage the service levels
Security and privacy	- As companies or individuals' sensitive information is stored into external computing resources, security of the information has been emerging as a critical issue	Firewall, Prevention technology and access levels management
Multi-tenancy model	- An inevitable element to provide SaaS - A model that several groups of users utilize a single information resource instance in a completely distributed form	

3   Social Media Cloud Computing Model

The proposed SMCC model overcomes the limits in the conventional DB system, providing a new concept of cloud computing environment for service provider and developers to establish and offer social media-based SNS more easily. In addition, our model is a new service model combined with cloud computing environments so that it can support SNS services like Twitter or Facebook, social media services providing social media (pictures, music and video clips) like YouTube and social game services like social network games of Facebook. Figure 1 depicts the concept of our SMCC model.

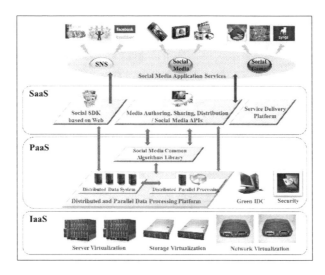

Fig. 1. Social Media Cloud Computing (SMCC) Model

4   Social Media Cloud Computing Architecture

The general idea of designing SMCC is to establish an environment supporting the development of SNS and addressing of numerous SNSs, to provide the approaches of processing big social media data and to provide a set of mechanisms to manage Infrastructure. SMCC is largely divided into 3 layers: SaaS layer, PaaS layer and IaaS layer. Furthermore, 3 layers are composed of 8 parts: Social Media Service Platform, Social Media Common Algorithms Library, Distributed Processing Platform, Cloud Security, Cloud QoS, Green IDC, Infra Management and Virtualization.

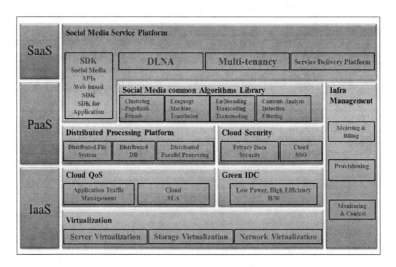

Fig. 2. Social Media Cloud Computing (SMCC) Architecture

5   Social Media Cloud Computing SaaS Platform

SMCC SaaS platform provides the function of interaction with social media. That is to say, users are able to create, distribute and share social media created by participants in SNSs using user-based SaaS platform. Our SaaS platform is composed of multi-tenancy, SDK social APIs, service delivery platform and DLNA (Digital Living Network Alliance).

Multi-tenancy. SMCC SaaS platform plays a role in providing SaaS service based on open solutions possible to guarantee a multi-tenancy service mode. The concept of multi-tenany is a critical issue in cloud computing because it is directly related to security and QoS in the aspect of companies and individual. Although it still faces a big challenge of sercurity and privacy problem[8], secured multi-tenancy should be appied in cloud computing enviriments to reduce cost correlated with builiding computing resources, especially storage resource and to effectively manage infrastructure. With the importance of multi-tenancy we apply secured multi-tenacy to our SaaS platform so that users can share and publish countless social media that is stored in a single cloud system.

Web Development Environment. Users or developers can make new services based on social media generated by participants in social sites using UI components, Service Components and a set of development tools such as social media APIs, SDK based on Web and SDK for application that SMCC provides.

Service Delivery Platform. It also contains service delivery platform that can deploy and develop new converged multimedia services quickly on a variety of smart devices such as Apple's iPhone and Google smart phones running on Android platform.

UPnP and DLNA. It includes UPnP (Universal Plug and Play) and DLNA (Digital Living Network Alliance) that deal with interoperability between networked

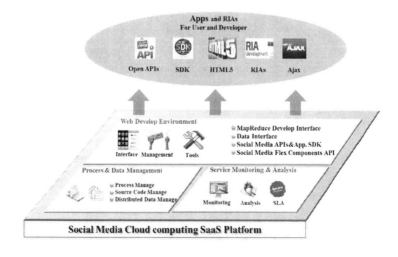

Fig. 3. Social Media Cloud Computing SaaS Platform

consumer devices permitting user-generated contents to be shared among household devices[9]. DLNA technology has a merit of easily sharing data among heterogeneous devices such as TV, home theaters, smart phones, DVD players, game consoles (PS3, Xbox360), cameras and set-top boxes. UPnP (User Plug and Play) technology is architecture to connect especially home PCs and smart devices or phones to a peer-to-peer network, and this technology makes various different smart devices automatically connect with each other and helps different users use various kinds of services through networking. With such technologies, SNS users can share, distribute and use countless social media data created by social media participants at hundreds of digital devices recently released.

In addition to these main functions, SMCC SaaS platform also provides MapReduce development environment to easily develop programing framework to distribute and process large amounts of social media in distributed processing systems.

6   Social Media Cloud Computing PaaS Platform and IaaS

This section explains PaaS platform likely to be the core of SMCC platform and IaaS functioning to provide physical computing environments. Figure 4 shows the whole architecture of PaaS platform and IaaS.

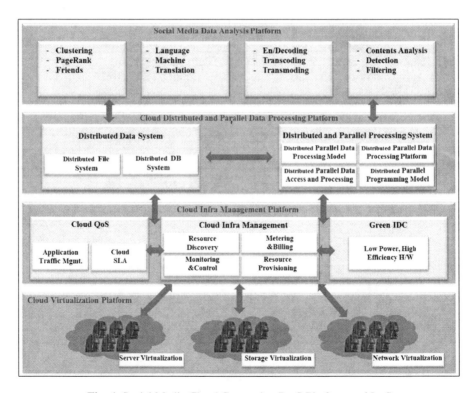

Fig. 4. Social Media Cloud Computing PaaS Platform and IaaS

SMCC PaaS Platform. PaaS platform consists of Social Media Data Analysis Platform (Social Common Algorithms Library), Cloud Distributed and Data Processing Platform and Cloud Infra Management Platform. Firstly, the main functions of Social Media Data Analysis Platform are to analyze usage pattern, relationship between users and social media data on demand, and to provide the functions of encoding, decoding, transcoding and transmoding as the form of libraries. Transcoding means converting one media file into files suitable for numerous digital devices in terms of file forms. Transmoding means converting one media file into files in terms of file size. Secondly, Cloud Distributed and Parallel Data Processing Platform as a core platform of our model is able to store, distribute and process social media data created by users by applying HDFS (Hadoop Distributed File System), MapReduce and HBase (Hadoop Database System) to its system. Lastly, Cloud Infra Management Platform contains the concepts of cloud QoS, Green IDC and Cloud Infra Management. Cloud Infra management manage and monitor computing resources that do not depend on specific OS or Platform. It includes the functions of resource scheduling, resource information management, resource monitoring and virtual machine management. These functions are provided on web service based on Eucalyptus. In addition, our IaaS is designed to offer elastic computing resources including servers, storage and bandwidth using virtualization techniques based on Xen[13].

7   Conclusion and Future Works

In this paper we introduce Social Media Cloud Computing model for developing SNS based on social media. The main objective of this paper is to present new cloud computing model that combines SNS based on social media with cloud computing environment, describing element technologies required in the proposed model

The SMCC architecture is designed to support the whole cloud computing services such as SaaS, PaaS and IaaS. Under 3-layers structure, SMCC consists of 8 sub-parts in total. We also describe all sub-parts in terms of service characteristics and technological characteristics in detail.

In the future work, we are planning to concretize the suggested model more technologically. Moreover, we will implement SMCC Model in an actual cloud computing environment, while verifying the superiority of our new model through benchmarking and performance evaluations. Finally, for our further study, we are focusing on suggesting various new models in the field of cloud computing, by combining SMCC Model with various different business models and other new models.

Acknowledgments. This research was supported by the MKE (The Ministry of Knowledge Economy), Korea, under the ITRC (Information Technology Research Center) support program supervised by the NIPA (National IT Industry Promotion Agency (NIPA-2011 – (C1090-1101-0008)).

References

1. Wikipedia (2011), http://en.wikipedia.org/wiki/Social_media
2. Kaplan, A.M., Haenlein, M.: Users of the world, unite! The challenges and opportunities of Social Medial. Journal of Business Horizons 53(1), 59–68 (2010)
3. Kim, M., Yoon, H., Lee, H.: An Intelligent Multi-Agent Model for Resource Virtualization: Supporting Social Media Service in Cloud Computing. SCI, vol. 365, pp. 99–111 (2011)
4. Foster, I., Zhao, Y., Raicu, I., Lu, S.: Cloud Computing and Grid Computing 360-Degree Compared. In: Grid Computing Environments Workshop GCE 2008, Article no. 4738445 (2008)
5. Armbrust, M., Fox, A., Zaharia, M.: A View of Cloud Computing. Magazine Communication of the ACM 53(4) (April 2010)
6. Raghavan, B., Vishwanath, K., Ramabhadran, S., Yocum, K., Snoeren, A.: Cloud control with distributed rate limiting. In: Conference on Applications, Technologies, Architectures, and Protocols for Computer Communications, pp. 337–348. ACM, New York (2007)
7. Wang, L., Von Laszewski, G.: Cloud Computing: A Perspective Study. New Generation Computing 28(2), 137–146 (2010)
8. Li, X.-Y., Shi, Y., Guo, Y., Ma, W.: Multi-Tenancy Based Access Control in Cloud. Computational Intelligence and Software Engineering, 1–4 (2010)
9. Ha, K., Kang, K., Lee, J.: N-screen service using I/O Virtualization thechnology. Information and Communication Technology Convergence, 525–526 (2010)
10. http://www.nist.gov
11. Song, S., Yoon, Y.: Intelligent Smart Cloud Computing for Smart Service. CCIS, vol. 121, pp. 64–73 (2010)
12. Rich, B., Thain, D.: DataLab: transactional data-parallel computing on an active storage cloud. In: 17th International Symposium on High Performance Distributed Computing, pp. 233–234. ACM Press, New York (2008)
13. http://www.xen.org/support/documentation.html

Development of Visual Navigation System for Patrol Service Robot

Kenichi Asami, Hayato Hagiwara, and Mochimitsu Komori

Department of Applied Science for Integrated System Engineering,
Kyushu Institute of Technology,
1-1, Sensui, Tobata, Kitakyushu 804-8550, Japan
asami@mns.kyutech.ac.jp, hagiwara@es.ise.kyutech.ac.jp,
komori@ise.kyutech.ac.jp

Abstract. The visual navigation system for a mobile patrol robot using image processing by FPGA and real-time device driver is presented. The CMOS image sensor and the stepper motor driver ICs are connected to external I/O ports of the FPGA. The image processing and motor drive circuits are implemented into the FPGA device together with state machine and FIFO memory buffer to adjust timing for pixel data transmission. The real-time device driver couples the flexible hardware circuits with software applications for the robot vision.

Keywords: Autonomous Mobile Robot, Patrol Service Robot, Real-time Image Processing, Visual Navigation System.

1 Introduction

Advanced fully and semi-autonomous mobile robots for life support purpose would acquire greater importance in order to adapt future social environment, such as the aging of society and declining number of children. Autonomous mobile robots need to recognize the surrounding by using image processing together with sensors and transducers. The robot vision systems require real-time response and low electric power consumption for the accurate and long time movement for mission tasks. The effective usage of visual information processing could expand the autonomy and the ability of mobile service robots. The adequate integration of hardware and software for image processing is designed for the robot vision system.

This paper describes the development of patrol service robot using FPGA (Field Programmable Gate Array) device for image processing and real-time device driver for the fast and deterministic time response to device I/O. A CMOS image sensor is connected to I/O ports of the FPGA device on the pan-tilt mechanism for tracking an object. The periodically-scheduled image data gets available to application software for object recognition, pan-tilt motors control, data communication, and so on. The visual navigation system provides the function of automatic correction of self-localization for the patrol service robot to recognize the position and the orientation in the typical passage environment. The patrol service robot enables to continue autonomous repetitive traveling without deviation from the route.

G. Lee, D. Howard, and D. Ślęzak (Eds.): ICHIT 2011, CCIS 206, pp. 267–270, 2011.
© Springer-Verlag Berlin Heidelberg 2011

2 System Configuration

The patrol service robot shown in Fig. 1 consists of a FPGA board, a CMOS camera module, two stepper motors for pan-tilt motion, two stepper motors for wheel drive, motor drivers, a wireless LAN module, and a Lithium-ion polymer battery. The FPGA board equips Xilinx Virtex-4 FX with PowerPC 405 processor at 350MHz as a hard core in the 12,312 logic cells. The platform of embedded systems has 64MB SDRAM, 8MB flash ROM, 86 external I/O ports, and UART serial communication interface, which realizes the running environment of Linux operating system and a variety of useful open source software. The CMOS camera module has a size of 12mm x 12mm x 11mm with a 1/4 inch receiving area in VGA resolution as digital output. For the electrical power consumption, the camera module requires 80mW at 2.8V in 12MHz. The standard video data output is served in the form of ITU_R RT.656 in YUV 4:2:2 pixel format, which compresses data size into 16 bits for a pixel. The two-phase unipolar stepper motor has a size of 20mm x 20mm x 30mm and a weight of 50g with the stepping angle of 1.8 degree, and the maximum stationary torque produces 0.02Nm. The stepper motor drivers are designed for simple requirement and made with easily available parts. The motor driver IC includes four power FETs to amplify electrical current, two PWM controllers to synchronize two phases, and a logic sequencer to generate excitation patterns for a full or half step.

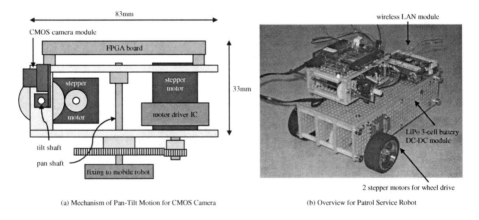

(a) Mechanism of Pan-Tilt Motion for CMOS Camera   (b) Overview for Patrol Service Robot

Fig. 1. The patrol service robot composed of a FPGA board, a CMOS image sensor, stepper motors, and a wireless LAN module

The constructed patrol service robot installs a large capacity battery of 1350 mAh, a DC-DC module and a wireless LAN module. The total mobile robot has a length of 180mm, a width of 140mm, a height of 130mm, and a weight of 960g. The stepper motors for wheel drive records and reproduces the traveling route according to the stepping values without feedback mechanism or rotary encoder. The pan-tilt mechanism enables to change the direction of visual field in the CMOS image sensor by itself. The wireless LAN module enables the user to manipulate the mobile robot from a distant terminal and to receive the image data through TCP/UDP/IP networks by IEEE802.11b/g/n standards.

3   User Logic Circuits by FPGA

PC-based image processing systems have a typical problem for unpredictable time response to critical tasks concerning for behavioral decision, which are achieved by multithread programs on uniprocessor in many cases. In addition, video data transmission by USB interface could become a bottleneck for the image processing speed. In the other cases, dedicated hardware systems based on DSP could fall into inflexible image processing structure and result in a lot of useless computing resources. In order to solve this problem, robot vision systems based on FPGA devices could have the advantage of configurable efficiency for large-scale and high-speed data processing.

The image capture circuit supplies the master clock to the CMOS camera module, which is generated from a digital clock manager that combines a frequency synthesizer, a phase shifter and a clock delay locked loop. Once the CMOS camera module starts the operation, the image capture circuit receives input signals: pixel data in 8 bits, pixel clock for data synchronization, horizontal synchronization for a line and vertical synchronization for a frame. According to the width of data bus of 32 bits between the processor and the peripheral IPs, the image capture circuit collects four pixel data that includes two luma components (Y, brightness) and two chrominance components (UV, color difference of blue and red) for two pixel elements, and then transfers the quadruplet to the state machine circuit and the FIFO memory buffer.

In the same way as the image processing circuit, the motor drive circuit has been built in the reconfigurable device as original logic. The FIFO receives an operation value in 4 bytes for motor drive from application software at the states of read and read-acknowledge. The FIFO transmits the value to the motor drive circuit at the states of drive and drive-acknowledge. According to the directive value, the sequential logic generates clock signal for steps to the driver ICs, and then the stepper motors start a drive. When the number of steps becomes zero, drive-end signal is returned to the state machine circuit in order to permit the next drive operation. Small-grained operations can be successively managed in the state transition.

4   Visual Navigation System for Self-localization

The visual navigation system provides the function of autonomous correction of self-localization with the patrol service robot. In order to revise the deviation from the right location of the mobile robot, image processing from the CMOS image sensor recognizes the suitable position and orientation for the traveling route. It is supposed that the patrol service robot is used in the typical indoor passage environment. The image processing for correcting self-location accomplishes the suitable straight passage movement for a certain distance, since a gap in the orientation for the straight passage results in large deviation from the proper traveling route. The image processing methods using edge and boundary detection judges the gap in the direction to the straight line of passage. The visual navigation system removes the difference between the absolute and the relative directions on the traveling route that the patrol service robot records in the teaching stage.

In a typical indoor passage, there must be left and right boundaries as lines with the strong features, when the patrol service robot moves in the straight direction of passage. The recognition of left and right boundaries is useful to navigate the suitable position and orientation on the traveling route for the mobile robot. In order to detect only left and right boundaries in passages, the boundary model would be applied in detecting lines in the image data. First, the edge image is divided into the left and right sides for finding the individual boundaries in the both sides. Second, the angle of boundary is restricted to the fixed range for finding the boundary between a floor and a wall. The visual navigation system selects the closest line to this side as the respective boundaries. The boundary detection was verified to be robust for some passing people and obstacles.

5 Conclusion

This paper presented the development of patrol service robot with real-time image processing using FPGA and device driver. Application software for correcting the self-localization by detecting boundaries of passage was developed. This patrol service robot enables to manipulate by remote control and provide image data transmission to a distant terminal based on wireless LAN. The system would be useful for applications of watching an aged person and detecting a suspicious person by indoor patrol service robot. The integration of hardware and software platform for real-time image processing could be applicable for various robot vision purposes. For the future development, common image processing by FPGA circuits, stereo vision measurement by trigonometry, and real-time image communication by using RTnet Ethernet controller driver would be needed.

References

1. DeSouza, G.N., Kak, A.C.: Vision for Mobile Robot Navigation: A Survey. IEEE Trans. on Pattern Analysis and Machine Intelligence 24(2), 237–267 (2002)
2. Kudo, K., Myokan, Y., Than, W.C., Akimoto, S., Kanamaru, T., Sekine, M.: Hardware Object Model and Its Application to the Image Processing. IEICE Trans. E87-A(3), 547–558 (2004)
3. Morita, T.: Tracking Vision System for Real-time Motion Analysis. Advanced Robotics 12(6), 609–617 (1999)
4. Okada, K., Inaba, M., Inoue, H.: Walking Navigation System of Humanoid Robot using Stereo Vision based Floor Recognition and Path Planning with Multi-Layered Body Image. In: Proc. IEEE Int. Conf. on Intelligent Robots and Systems, pp. 2155–2160 (2003)
5. Sawasaki, N., Nakao, M., Yamamoto, Y., Okabayashi, K.: Embedded Vision System for Mobile Robot Navigation. In: Proc. IEEE Int. Conf. on Robotics and Automation, pp. 2693–2698 (2006)
6. Shimizu, K., Hirai, S.: CMOS+FPGA Vision System for Visual Feedback of Mechanical Systems. In: Proc. IEEE Int. Conf. on Robotics and Automation, pp. 2060–2065 (2006)
7. Xilinx Inc.: Virtex-4 FPGA User Guide (2008)

Mobility Information and Road Topology Based Inter-vehicle Routing Protocol in Urban

Si-Ho Cha and Keun-Wang Lee[*]

Dept. of Multimedia Science, Chungwoon University
San 29, Namjang-ri, Hongseong, Chungnam, 350-701, South Korea
{shcha,kwlee}@chungwoon.ac.kr

Abstract. Vehicular ad-hoc networks (VANETs) are special cases of mobile ad-hoc networks (MANETs) where nodes are highly mobile, so network topology is changing very fast. In VANET, vehicles move non-randomly along roads. Inter-vehicle communication is wirelessly connected using multi-hop communication without access to some fixed infrastructure. Because of the rapid movement of vehicles and the frequent topology change of VANET, link breakages occur repeatedly and the packet loss rate increases. Geographical routing protocols such as greedy perimeter stateless routing (GPSR) are known to be very suitable and useful for VANET. However, they can select stale nodes for relay nodes and generate the local maximum problem. This paper presents an inter-vehicle routing protocol based on mobility information of vehicles and digital map on city roads to solve those problems in urban area. Mobility information includes the position, velocity, and direction of vehicles. Road topology is used to obtain the location of intersections and whether it is an overpass or a dead end road. Simulation results using ns-2 show performance improvement in terms of packet delivery ratio than existing routing protocols.

Keywords: VANET, IVC, V2V, Link Breakage, Local Maximum, Stale Node.

1 Introduction

VANET provides both vehicle-to-infrastructure (V2I) communication and vehicle-to-vehicle (V2V) communication. V2I can provide real-time the road traffic conditions, weather, and basic Internet service through the communication with backbone networks. V2V can be used for providing information about traffic conditions and/or vehicle accidents based on wireless inter-vehicle communication (IVC). In V2V communication environments, vehicles are wirelessly connected using multi-hop communication without access to some fixed infrastructure [1]. VANET has unique characteristics such as high node mobility and a rapidly changing network topology compared to MANET. Because of the rapid movement of vehicles and the frequent topology change of vehicles, link breakages occur repeatedly, as illustrated in Fig. 1. The frequent link disconnection is also cause by the characteristics of VANET such as

---

[*] Corresponding author.

G. Lee, D. Howard, and D. Ślęzak (Eds.): ICHIT 2011, CCIS 206, pp. 271–277, 2011.
© Springer-Verlag Berlin Heidelberg 2011

vehicle movements are constrained in roads and traffic lights have great influence on the vehicle movement [2]. The frequent link disconnection may increase the possibility of local maximum. Due to these problems, geographical routing protocols such as GPSR [3] are known to be more suitable and useful to VANET than existing routing protocols designed for MANETs. For instance, in Fig. 1, geographical forwarding could use node N2 instead of N1 to forward data to D.

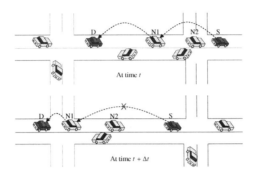

Fig. 1. Route (S, N1, D) that was established at time t breaks at time $t + \Delta t$ when N1 moves out of the transmission range of S

Geographical forwarding is one of the most suitable solutions for routing in VANET because it maintains only local information of neighbors instead of per-destination routing entries. GPSR selects the node that is the closest to the destination among the neighbor nodes as the relay node. However, GPSR may generate the link loss problem in urban environments. Because GPSR does not take into account road structure and the speed and moving direction of vehicles, it is able to select stale nodes as relay nodes. As vehicle movements are constrained by roads, GPSR without taking urban environment characteristics is not applicable for VANET [2]. To solve this problem, greedy perimeter coordinator routing (GPCR) [5] and greedy perimeter urban routing (GPUR) [6] are proposed as possible solutions. However, GPCR may cause transmission delay and path selection error because it identifies nodes that are on a junction by detecting coordinator nodes to select relay nodes. GPUR selects a node that has 2-hop neighbors for the relay node. It will cause serious transmission delay. GPUR does not resolve local maximum problem due not to considering road specifications such as dead end roads. Our target is to improve the routing protocol for V2V, based on vehicles' movement information like position, direction, velocity, and road topology. To do this, we assume that each vehicle knows its location through GPS like most of the related geographic routing protocols, and has a digital street map for road information. We also suppose the availability of a location service, so the source node can get the destination information. The position prediction based on digital map is more realistic if two reads are superposed with or without different running directions.

The rest of the paper is organized as follows. Section 2 discusses the related work, and Section 3 introduces the proposed inter-vehicle routing protocol. The performance evaluation is discussed in Section 4. Finally, in Section 5, conclusions are made including the future research.

2 Related Work

Traditional MANET routing protocols, such as AODV [7], DSR [8], are not suitable for VANET. To deal with the rapidly changing network topology of VANET, greedy forwarding protocols based on geographic information have been proposed.

GPSR [3] is one famous greedy forwarding protocol. GPSR makes greedy forwarding decisions using only information about the immediate neighbors in the network topology. GPSR may increase the possibility of getting the local maximum and link breakage because of the high mobility of vehicles and the road specifics in urban areas. This is because it just selects the nearest node to the destination as a relay node within its transmission region to make packet forwarding decisions. GPSR may also generate the link loss problem because it maintains stale nodes as neighbor nodes to select a relay node in greedy mode. The local maximum and link breakage problems can be recovered in perimeter mode forwarding, but packet loss and delay time may appear because the number of hops is increased in perimeter mode forwarding. This decreases the reliability of a VANET. Fig. 2 shows an example. Assume vehicle S wants to send a packet to D, and S has two neighbors: N1 and N2. GPSR will choose N2 to forward the packet because N2 is closer to D. But in common sense, we should choose N1 as vehicle movements are constrained in roads.

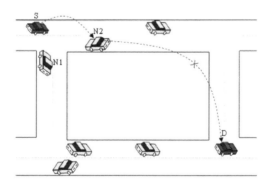

Fig. 2. GPSR chooses N2 instead of N1 to forward the packet because N2 is closer to D within the transmission range of S

GPCR [5] was proposed to improve the reliability of GPSR in VANET. The basic behavior of GPCR is similar to GPSR, but it selects a relay node by considering information about the road structure. GPCR makes routing decisions on the basis of streets and junctions instead of individual nodes and their connectivity. However, GPCR forwards data packets based on the node density of adjacent roads and the connectivity to the destination. Thus, if the density of nodes is low or there is no connectivity to the destination, then the delay time increase and the local maximum problem are still not resolved.

GPUR [6] selects a relay node based on information about the road characteristics, like as GPCR. However, unlike GPCR, GPUR selects a relay node among nodes that have 2-hop neighbors. It makes periodic beacon messages to estimate the presence of 2-hop neighbors about all relay candidates. The periodic beacon messages used to evaluate the presence of 2-hop neighbors will cause serious transmission delay. GPUR does not also resolve local maximum problem due not to considering road specifications such as dead end roads.

3 Proposed Routing Protocol

Our routing protocol adopts the road segments based routing approach with street awareness. Therefore, data packets will be routed between vehicles, following the street map topology and the road segments in the real area. Our target is to improve the routing protocol for IVC based on vehicles' movement information such as position, direction, and velocity as well as road topology. To do this, we assume that each vehicle knows its location through GPS like most of the related geographic routing protocols, and has a digital street map for road information. We also suppose the availability of a location service, so the source node can get the destination information. The position prediction based on digital map is more realistic if two reads are superposed with or without different running directions.

In our protocol, a sender should first look at all relay candidates similar to GPUR. Then, among these relay candidates, it selects the most stable one based on the future vehicle's position using a digital map. The advantage of the proposed routing protocol is that the predicted positions will be more realistic. If a vehicle is on the curved road without the help of a digital map, its position prediction will wrong. The positions of relay candidates for a sender can be predicted using formula (1).

$$\vec{n}_{iFP} = \vec{n}_{iCP} + \Delta V_i \times T \times \{ \frac{(\vec{S} - \vec{n}_{iCP})}{\vec{S} - |\vec{n}_{iCP}|} \} \tag{1}$$

where $\vec{n}_{iFP}$ and $\vec{n}_{iCP}$ are the future position vector and the current position of a relay candidate, respectively. ΔV_i is the relative velocity variation and $\vec{S}$ is the position vector of a sender. T is the elapsed time since the sender receives the beacon response from the relay candidate. If there is any relay candidate within the sender's transmission range, the velocity variation can be estimated with the relative velocity between the sender and the relay candidate using formula (2).

$$\Delta V_i = \frac{V \times n_i (x_j - x_i, y_j - y_i)}{S(x_i, y_i)} (\cos \theta + \sin \theta) \tag{2}$$

where V is the velocity of the relay candidate. n_i and S are the relay candidate and the sender, respectively. The sender selects a node that has the larger relative velocity than the others as a relay node.

The algorithm select relay nodes based on both mobile positioning information and road topology information. Followings are the algorithm used in our protocol to select relay nodes.

1. S sends beacons to $N = \{x|x$ is a node with the radio range of $S\}$.
2. S predicts all ΔV_x.
3. S estimates all $\overrightarrow{n}_{xFP}$ based on a Road Map.
4. S excludes nodes that on a dead end road and/or in other direction from relay candidates.
5. If S find out all relay candidates are on the road that possible to conjunct with the road on which destination node,
6. Then S selects a node nearest to the destination node among the relay candidates that was founded as a relay node.
7. Else S selects a node nearest to the destination node among all relay candidates as a relay node.

For instance, in Fig. 3, the proposed protocol can use node N3 instead of N2 to forward data to D. GPCR may send packets to N4 based on the node density at the junction roads and GPUR may forward packets to N1 and N2, following the moving direction and the shortest path from S to D. It cause local maximum and unnecessary traffic overhead and delay. Our proposed protocol therefore can avoid the dead end road by using digital map for realistic road information.

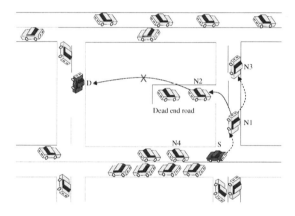

Fig. 3. The proposed routing protocol chooses N3 instead of N2 to forward the packet because N2 is on a dead end road even if N2 is moving to D and it has the shortest path to D

4   Performance Evaluation

We analyze and compare the performance of the proposed protocol and the existing GPSR, GPCR and GPUR using the ns-2 simulator. In this performance evaluation, we just considered the packet delivery rate at this point. The simulated area is based on a real map from Seoul with a 1000 m * 1000 m size. Table 1 summarizes our simulation parameters. The simulations were performed for 180 secs, and the number of nodes was increased from 10 to 100. The moving velocity of the nodes was increased from 20 km/h to 100 km/h. The experiments were performed thrice and average values were used. Maximum and minimum values were excluded.

Fig. 4 and Fig. 5 show the packet delivery rate according to the number of nodes and the packet delivery rate according to the velocity variation of the nodes, respectively. The proposed protocol represents the best packet delivery rate compared to the existing GPSR, GPCR, and GPUR.

Table 1. Simulation parameters

Parameter	Value
Topology size	1000m*1000m
Transmission range	250m
MAC protocol	IEEE 802.11
Node number	10 to 100
Node velocity	20km/h to 100km/h
Beacon time	1sec
Bandwidth	2Mbps
Packet size	1000bytes

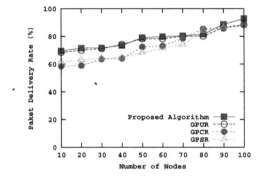

Fig. 4. Packet delivery rate according to number of nodes

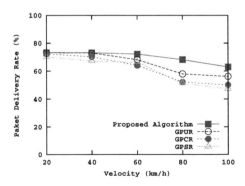

Fig. 5. Packet delivery rate according to velocity of nodes

Fig. 4 shows that the performance of the proposed protocol appears similar to that of GPUR. However, the proposed protocol shows better performance than GPUR while the number of nodes increases. This reason can be inferred from the error occurred in selecting relay nodes because GPUR considers only the location of 2-hop nodes in selecting relay nodes without considering road topology. On the other hand, the proposed protocol can be found showing a high delivery rate by selecting relay nodes based on road topology. As shown in Fig. 5, all protocols show the packet delivery rate drops with increasing node speed. However, the decreasing rate of the proposed protocol is lower than the other protocols. This reason can be expected that the delay time to determine the position of nodes is lower than that of the other protocols because the proposed protocol knows the position of nodes with a digital map.

5 Conclusion

This paper has presented inter-vehicle routing protocol based on mobility information and digital map in urban, which is a class of VANET routing protocols for city-based environments that takes advantage of the layout of the roads to improve the performance of routing in VANETs. The proposed routing protocol can reduce the possibility of link breakage by selecting the relay node based on mobility information as well as the road topology. Simulation results showed that our protocol has a high packet delivery rate compared with GPSR, GPCR and GPUR for VANETs.

As the future work, we will consider the probability of local maximum and the packet breakage rate in our simulation. And we will incorporate more realistic factors into our routing protocol, such as the density of vehicles and delay tolerant characters.

References

1. Tse, Q.: Improving Message Reception in VANETs. In: The International Conference on Mobile Systems. In: The International Conference on Mobile Systems, Applications, and Services (MobiSys), Krakow, Poland (2009)
2. Luo, J., Gu, X., Zhao, T., Yan, W.: A Mobile Infrastructure Based VANET Routing Protocol in the Urban Environment. In: The International Conference on Communications and Mobile Computing (CMC), pp. 432–437 (2010)
3. Karp, B., Kung, H.T.: GPSR: Greedy Perimeter Stateless Routing for Wireless Networks. In: The International Conference on Mobile Computing and Networking (MobiCom), pp. 243–254 (2000)
4. Nzouonta, J., Rajgure, N., Wang, G.: VANET Routing on City Roads Using Real-Time Vehicular Traffic Information. IEEE Transactions on Vehicular Technology 58(7), 3609–3626 (2009)
5. Lochert, C., Mauve, M., Fler, H., Hartenstein, H.: Geographic Routing in City Scenarios. ACM Mobile Computing and Communications Review 9(1), 69–72 (2005)
6. National Ryu, M.-W., Cha, S.-H., Cho, K.-H.: A Vehicle Communication Routing Algorithm Considering Road Characteristics and 2-Hop Neighbors in Urban Areas. The Journal of Korea Information and Communications Society, KICS 36(5), 464–470 (2011)
7. IETF: Ad Hoc On-Demand Distance Vector Routing. RFC3561 (2003), http://www.faqs.org/rfcs/rfc3561.html
8. IETF: The Dynamic Source Routing Protocol for Mobile Ad Hoc Network for IPv4. RFC 4728 (2007), http://www.faqs.org/rfcs/rfc4728.html

Communication Quality Analysis for Remote Control of a Differential Drive Robot Based on iPhone Interface

Hahmin Jung, Yeonkyun Kim, and Dong Hun Kim

Department of Electrical Engineering, Kyungnam University
Changwon-si, Gyeongnam, Korea
dhkim@kyungnam.ac.kr
http://www.kyungnam.ac.kr/~dhkim

Abstract. This paper presents a tangible interface based on iPhone for robot control. In this scheme, two axes of the acceleration sensor attached inside iPhone are used for remote control of a mobile robot. In particular, this study presents four possible communication methods for wireless communication between a mobile robot and iPhone. For each method, the communication quality of delay time and number of acquisition data is analysed and compared. The experimental results show that the proposed iPhone-interface control is effectively used for controlling a mobile robot remotely.

Keywords: iPhone, mobile robot, tangible interface, acceleration sensor, Blutooth communication.

1 Introduction

Many researchers have proposed various interfaces for a mobile robot or a biped robot. Xing-Han Wu et al. [1] proposed a hand gesture interface for robot control. The interface for navigation of a car-robot used a 3-axis accelerometer and recorded a user's hand trajectories. The received trajectories are then classified to one of six control commands for navigating a car-robot. Manigandan et al. [2] proposed a hand gesture recognition interface that is similar to [1]. However the interface [2] used a vision recognition based on perceptual color space such as HIS, HSV/HSB, and HSL where a robot finds the center of gravity (COG) of the hand region as well the farthest point from the COG. The robot identifies the sign based on recognized hand. Shah et al. [3] proposed a sketch interface for controlling robots. A operator of the interface uses a pen tablet to control a robot remotely by sketching commands, where the robot interprets and executes correctly. Barbosa et al. [4] proposed a brain computer interface that is applied to activate the movements of a mobile robot, associating four different mental activities to robot commands. Kerscher et al. [5] proposed a haptic handle to enable the user to steer interactive behavior operated shopping trolley (InBOT). The handle is composed with different sensors for detecting the forces

G. Lee, D. Howard, and D. Ślęzak (Eds.): ICHIT 2011, CCIS 206, pp. 278–285, 2011.

and torques applied by an user. Interfaces using a Personal Digital Assistant (PDA) or Smartphone have been proposed by many researchers. Valero et al. [6] proposed an desktop-based interface and a PDA-based interface for remote control of mobile robots, where situation awareness of the operator is reduced when using PDA interface than the desktop-based interface. Jong Hyun et al. [7] proposed a PDA-based user interface to control a mobile robot. The interface consists of an image for the robot camera, indicators for distance estimation, and sensor data for robot navigation. And it needs a server PC through WiFi router that requires restricted area. In [8] and [9], iPod was treated for robot control. [8] and [9] proposed a touch interface based on gesture navigation. Although it is similar to [1], iPod touch has small size and low computation capacity.

In this paper, we present an iPhone-based interface for control of a differential drive robot. The goal is to propose possible communication methods and analyse their communication quality of differential drive robot based on iPhone interface. Specifically the following four methods are attempted: *M1* method of sending a start byte for one packet, *M2* method of sending a start byte for a respective data, *M3* method of using data transmission by requesting, and *M4* method of controlling robot by the smallest size packet.

2   Robot and iPhone

2.1   Differential Drive Robot Model

The configuration of a differntial drive robot is completely described by three vectors that consist of current position and orientation angle from reference frame. By assuming that wheels of the robot do not slip on the plane, the motion of the point for robot is subjected to the following equation [10],

$$\dot{x}\sin\phi - \dot{y}\cos\phi = 0 \tag{1}$$

where (x, y) is the center of gravity of the robot in the inertially fixed coordinate system, ϕ is its orientation. This natural constraint is nonintegrable, i.e., nonholonomic. In other words, the robot does not directly move to a x direction or a y direction. The longitudinal velocity v and angular velocity ω[13] are given by

$$\dot{x}\cos\phi + \dot{y}\sin\phi = v$$
$$\dot{\phi} = \omega \ . \tag{2}$$

Hence, the kinematic model is given by

$$\mathbf{\dot{q}} = \begin{bmatrix} \dot{x} \\ \dot{y} \\ \dot{\phi} \end{bmatrix} = \begin{bmatrix} \cos\phi & 0 \\ \sin\phi & 0 \\ 0 & 1 \end{bmatrix} \begin{bmatrix} v \\ \omega \end{bmatrix} . \tag{3}$$

v and ω are given by

$$v = \frac{p_r + p_l}{2} \ \ and \ \ \omega = \frac{p_r - p_l}{d} \tag{4}$$

where p_l and p_r are pulse width modulation (PWM) signals of left and right motor, respectively. d is the distance of two wheels.

2.2   Measurement of Acceleration Sensor Value

The acceleration sensor of iPhone measures an acceleration quantity in gravity field. When the measured value is 1.0, it means that the $1g$ (gravity) power influences to certain direction. If the iPhone is stood up vertically, the value of sensor means that $1g$ force influences toward y-axis of sensor. However, in the case that the acceleration sensor of iPhone is used, gravity values have large chattering values. Thus, the study uses a low pass filter for reducing chattering values. The digital low pass filter is given by

$$\widetilde{A} = \alpha A + (1 - \alpha)\widetilde{A} \tag{5}$$

where $\widetilde{A} = [\widetilde{x_a}, \widetilde{y_a}, \widetilde{z_a}]$, $\widetilde{A}$ is the value of estimated acceleration sensor. And $A = [x_a, y_a, z_a]$, A is the value of measured acceleration sensor. $\alpha = s_t/(t_c + s_t)$, α is an weight constant. where s_t is a sampling time and t_c is a time constant. In this paper, the parameter values are used as follows: the value of s_t is 0.01 and t_c is 0.2.

2.3   Robot Hardware and Control

Figure 1 shows that the differential drive robot is controlled by changing gravity value in an acceleration sensor of the iPhone. In this study, the y and z-axis of the acceleration sensor for the robot control are used. The z-axis value changes to -1 when tilting forward the iPhone. If the iPhone tilts to left and right, y-axis value of acceleration sensor changes between -1 and 1. Left and right motor values are given by

$$\begin{aligned} p_l &= s_m(k_1(\widetilde{z_a} + z_i) - k_2\widetilde{y_a}y_r) \\ p_r &= s_m(k_1(\widetilde{z_a} + z_i) + k_2\widetilde{y_a}y_r) \end{aligned} \tag{6}$$

where k_1 and k_2 are the weight value given by $k_1 + k_2 = 1$. z_i is the initial value of z-axis and y_r is the value of left and right rate. s_m is the maximum speed of the robot, which is changed by selection mode.

3   Communication Method Between iPhone and a Robot

In the study, the maximum speed data of the robot and the acceleration data of y and z axis, and so on are communicated between iPhone and the differential drive robot. All this information is packed in one Bluetooth packet and then send out to the robot. In this paper, four methods are proposed. one packet in each method is shown in Fig. 2 where a piece of rectangular shape means 1 byte. *M1*: Method of sending a start byte for one packet.

The packet sent from iPhone in Fig. 2. (a) includes y and z axis values of acceleration sensor, which are packed into one 15-byte Bluetooth packet. This method has an advantage of sending large data using a '#' as start packet and a ',' as data division. However, the packet length is long, since the packet includes

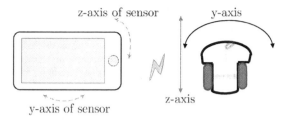

Fig. 1. Remote robot control using iPhone

a start letter and a division letter in the packet. The packet is saved in buffer while the robot is receiving the packet from iPhone. Then, the packet is analysed in the robot and the analysed packet is used for the control input of robot. While the robot analyses the packet, the robot does not process a next packet, which occurs a time delay. Thus, real time control is not easy since the method occurs time delay between the robot and iPhone.

M2: Method of sending a start byte for respective data

M2 method excludes the '#' letter differently from the *M1* method. Instead, a starting letter and an ending letter as a mode value, and y-axis and z-axis values are added. Figure 2. (b) shows the structure of a communication packet in the *M2* method where '*m*', '*y*', and '*z*' mean a start letter, respectively, and ',' means ending of data value. The packet received from the iPhone is analysed by comparing starting letters of the data. This method is faster than the *M1* method for robot control. However, in this method, the data length is still long and a delay time exists.

M3: Method of using data transmission by requesting

The *M1* and *M2* methods transmit packet from iPhone without regardless of robot condition. While, *M3* method sends a packet only when the robot requests as shown in Fig. 2. (c). First, the iPhone waits for request from the robot. When the request data is '*y*', the iPhone sends the y-axis value. Then, the iPhone waits for next request. In the method, data size of the *M3* method is smaller than the other methods of *M1* and *M2*. However, when communication timing of the *M3* method is not appropriate, the robot may not receive transmited data from iPhone. If the PWM value of motor is decided, then it is converted to different type of data from the saved data. Thus, the *M3* method occurs time delay between the robot and the iPhone.

M4: Method of controlling a robot by a small size packet

This method reduces the calculation time of actuating motors and the time delay between the robot and the iPhone. The PWM values of a robot are calculated in iPhone. Then the robot receives the PWM values from iPhone and uses these values. Figure 2. (d) shows the *M4* method , where the '#' represents the start of packet data. The PWM value consist of 8 bits. The highest bit means a direction of a rotating motor. For example, when the highest bit is 0 value, the motor rotates count clockwise. The other bits mean PWM value. In the paper, the *M4* method is used for controlling the robot where the sampling time is around $33ms$ that is shorter than the other methods.

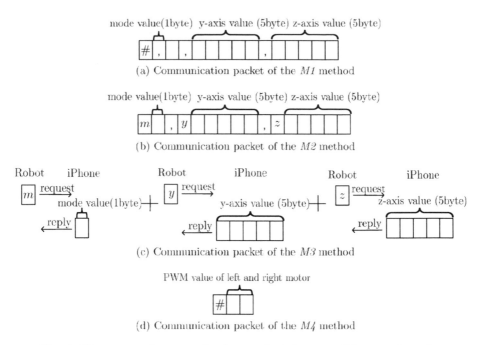

(a) Communication packet of the *M1* method

(b) Communication packet of the *M2* method

(c) Communication packet of the *M3* method

(d) Communication packet of the *M4* method

Fig. 2. The proposed communication mathods between iPhone and a robot

4   Analysis of Experimental Results

The communication between the iPhone and the differential drive robot is based on Bluetooth. In experiment, the overall structure of the proposed system is shown as Figure 3.

Figure 4. (a) shows the output values of robot after transfering data from iPhone through Bluetooth communication. The packet is saved in buffer while the robot is receiving the packet from iPhone. Then, the packet is analysed in the robot and the analysed packet is used for the control input of robot. Thus, while the robot analyses the packet, the robot does not process a next packet. Hence, a time delay occurs. As for the *M1* method in Fig. 4. (a), real time control is not easy since the method occurs time delay between the robot and iPhone. However this method may be appropriate for a slow responding system. *M2* method in Fig. 4. (b) does not response to fast section during [3, 3.2]. In Fig. 4. (c), the robot using the *M3* method is received data from iPhone by request. Thus, sampling rate is slower than the *M1*, *M2* method. Result of the *M4* method in Fig. 4. (d) is similar to that the robot directly uses the saved input data from buffer without wireless communication from iPhone. The robot responses to fast section during [3, 3.2]. In the paper, the *M4* method is adopted for controlling the robot where the sampling time is around 0.0334 second, which is shorter than the other methods.

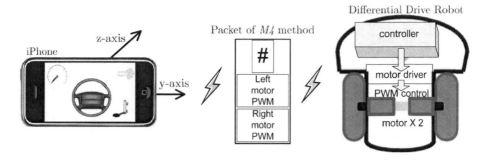

Fig. 3. Structure of the proposed system

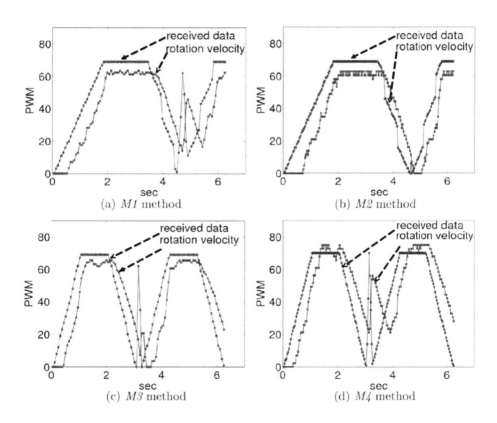

Fig. 4. PWM data and motor speed using the proposed methods

Figure 5 shows a sampling time and the number of transferred data for each method. In the experiment, the robot receives 620 data from the iPhone having sampling time of 10ms. The sampling time of the $M4$ method is lower than the other methods. The number of data in the $M4$ method that received from 620 data sent every 10ms is more than other methods. Figure 6 shows an experimental result of the robot control using iPhone. In the experiment, the $M4$ method is

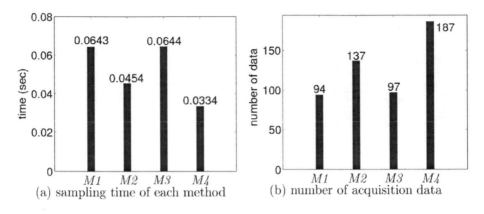

(a) sampling time of each method   (b) number of acquisition data

Fig. 5. Sampling time of each method and number of acquisition data

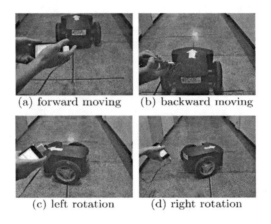

(a) forward moving   (b) backward moving

(c) left rotation   (d) right rotation

Fig. 6. Control of a robot using iPhone [11]

used, where $s_m = 40$, $k_1 = 0.6$, $k_2 = 0.4$, $z_i = 0$, $y_r = 0.1$ are used as parameter values. Figure 6 shows the experimental results that is shown as a movie file in [11].

5   Conclusion

In the paper, the iPhone interface for wireless controlling the differential drive robot is presented. The proposed iPhone controller uses Bluetooth communication that does not have place limit differently from use of WiFi communication. In the paper, the communication quality is analysed for remote control of a differential drive robot based on iPhone interface. Comparative results of four communication methods are proposed, and the $M4$ method is adopted for communication between the iPhone and the robot. The experimental results present that the proposed iPhone-interface control is effectively used for controlling a mobile robot remotely.

Acknowledgments. This work was supported by Kyungnam University Research Fund, 2011.

References

1. Wu, X.H., Su, M.C., Wang, P.C.: A hand-gesture-based control interface for a car-robot. In: 2010 IEEE/RSJ International Conference on Intelligent Robots and Systems (IROS), pp. 4644–4648 (2010)
2. Manigandan, M., Jackin, I.M.: Wireless Vision Based Mobile Robot Control Using Hand Gesture Recognition Through Perceptual Color Space. In: 2010 International Conference on Advances in Computer Engineering (ACE), pp. 95–99 (2010)
3. Shah, D., Schneider, J., Campbell, M.: A robust sketch interface for natural robot control. In: 2010 IEEE/RSJ International Conference on Intelligent Robots and Systems (IROS), pp. 4458–4463 (2010)
4. Barbosa, A.O.G., Achanccaray, D.R., Meggiolaro, M.A.: Activation of a mobile robot through a brain computer interface. In: 2010 IEEE International Conference on Robotics and Automation (ICRA), pp. 4815–4821 (2010)
5. Kerscher, T., Goller, M., Ziegenmeyer, M., Ronnau, A., Zollner, J.M., Dillmann, R.: Intuitive control for the mobile Service Robot InBOT using haptic interaction. In: ICAR 2009, International Conference on Advanced Robotics, pp. 1–6, 501–518 (2009)
6. Valero, A., Randelli, G., Botta, F., Saracini, C., Nardi, D.: Give me the control, I can see the robot! In: 2009 IEEE International Workshop on Safety, Security & Rescue Robotics (SSRR), pp. 1–6 (2009)
7. Park, J.H., Song, T.H., Park, J.H., Jeon, J.W.: Usability analysis of a PDA-based user interface for mobile robot teleoperation. In: 6th IEEE International Conference on Industrial Informatics INDIN 2008, pp. 1487–1491 (2008)
8. Kubota, N., Wakisaka, S., Yorita, A.: Tele-operation of robot partners through iPod touche. In: 4th International Symposium on Computational Intelligence and Intelligent Informatics, ISCIII 2009, pp. 75–80 (2009)
9. Kubota, N.: Sato. W.: Robot Design Support System based on Interactive Evolutionary Computation using Boltzmann Selection. In: 2010 IEEE Congress on Evolutionary Computation (CEC), pp. 1–8 (2010)
10. Siegwart, R., Nourbakhsh, I.R.: Introduction to Autonomous Mobile Robots. MIT Press, Cambridge (2004)
11. http://www.kyungnam.ac.kr/~dhkim

Remote Control and Monitoring of an Omni-directional Mobile Robot with a Smart Device

Yong-Ho Seo

Department of Intelligent Robot Engineering, Mokwon University,
Mokwon Gil 21, Seo-gu, Daejon, Republic of Korea
yhseo@mokwon.ac.kr

Abstract. New robot applications using smart devices are recently providing interesting and feasible solutions to longstanding problems. In this study, we propose a remote control and monitoring method using an omni-directional mobile robot with a smart device. We also developed remote control and monitoring application software for the smart device by using Microsoft's .Net Compact Framework and Bluetooth data communication to support various smart OS Platforms. We evaluated the feasibility and the effectiveness of the proposed method by conducting several experiments involving remote control of a mobile robot using a gravity sensor in a smart device and remote monitoring.

Keywords: Smart Device, Omni-directional Mobile Robot, Remote Control and monitoring, Bluetooth Communication.

1 Introduction

Smart device based markets are currently expanding as smart device use becomes increasingly popular. Laptop is a computer with mobility compared to desktop computer. Advanced types of laptops such as netbooks, UMPCs, and tablet PCs are smart devices that consume less electric power. Smart phones are distinguished from portable computers, which in turn evolved from desktop PCs. A smart phone is an advanced mobile phone that offers advanced computing ability and connectivity. A smart phone takes the place of a PDA while also offering various multi-media content.

Smart devices are new computing devices that use diverse contents. Examples include smart phones, smart books, and smart TVs. A smart book is a portable device that has advantages of a smart phone and a tablet PC. Smart phones have dominated the market recently, replacing feature phones. Smart books are also becoming popular for using various multimedia contents including e-books. Smart TVs are similarly attracting attention for use in the home [1].

Today, we can easily use various internet-based applications, as Wi-Fi hotspots are being installed in most public areas, including bus terminals, airports, libraries, and buildings. In ubiquitous environments with seamless wireless connection, smart

G. Lee, D. Howard, and D. Ślęzak (Eds.): ICHIT 2011, CCIS 206, pp. 286–294, 2011.

devices are becoming hub devices by using service market places and personal cloud computing.

In consideration of this recent trend in mobile computing technologies and with the recent rapid diffusion of smart devices, interconnection of smart devices and robots via wireless communication will attract public attention and will have a major influence on the markets of u-Healthcare, u-Home Service, and intelligent service robots. Fig. 1 shows a service architecture involving cloud computing, content, smart devices, and u-services that will be attainable in the near future.

In this study we propose a new smart device based robot application that can support remote control and monitoring of an omni-directional mobile robot. In the related experiments, gravity sensor steering, whereby the robot is controlled by the gravity sensor of a smart device, and remote monitoring, whereby the robot is controlled while monitoring a remote scene including the robot, are conducted to verify the effectiveness of the proposed system. We also evaluated the performance of remote monitoring.

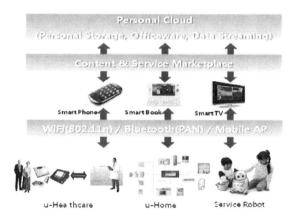

Fig. 1. Ubiquitous Service Architecture that involves cloud computing, content, smart devices, and u-Services

2   Application Software of Smart Device

2.1 .Net Compact Framework and Dalvik VM

There are two dominant operating systems, and set-top boxes developed by Microsoft. Windows and Andriod for smart devices nowadays [2]. .NET Compact Framework is a version of the .NET Framework that is designed to run on resource constrained smart devices such as personal digital assistants (PDAs), mobile phones, factory controllers. The .NET Compact Framework inherits the full .NET Framework architecture of the common language runtime for running managed code. It provides interoperability with the Windows CE operating system of devices so the user can access native functions and integrate his or her favorite native components into applications [3].

The UI development is based on Windows Forms, which is also available on the desktop version of the .NET Framework. User interfaces can easily be created with Visual Studio by placing .NET Compact Framework controls such as buttons, text boxes, etc. on the forms. Features such as data binding are also available for the .NET CF [4].

Android is a software stack for mobile devices that includes an operating system, middleware, and key applications developed by Google. The Android SDK provides the tools and APIs necessary to begin developing applications on the Android platform using Java programming language [5].

Dalvik is the process virtual machine (VM) in Google's Android operating system. Dalvik is thus an integral part of Android, which is typically used on mobile devices such as mobile phones, tablet computers, and netbooks. Before execution, Android applications are converted into the compact Dalvik Executable format, which is designed to be suitable for systems that are constrained in terms of memory and processor speed.

Fig. 2 shows the .Net Compact Framework (left) and Dalvik VM (right) as representative latest smart device development environments.

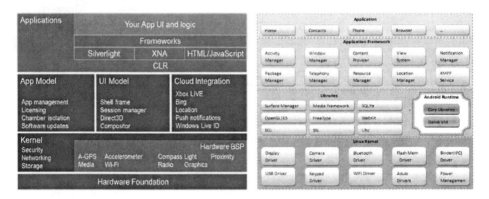

Fig. 2. .Net Compact Framework (left) and Dalvik VM (right) based Smart Device Development Environment

2.2 Application Software for Control and Monitoring of Omni-directional Mobile Robot

Software configuration of the proposed system is show in Fig. 3 below. The software of the mobile robot is configured with C compiler based firmware, which transmits the robot's sensor measurements to the smart device and receives motor commands from the smart device. The software in the smart device is configured with .Net Compact Framework, which drives the robot based on the received sensor information from the robot and a C# based mobile application.

The user interface of the smart device application for mobile robot control is show in Fig. 4. This UI was developed using a smart application interface based on C# and .Net Compact Framework 3.5. Various buttons are placed to manually control the robot and to make it easy to connect to Bluetooth, select which mobile mode to run,

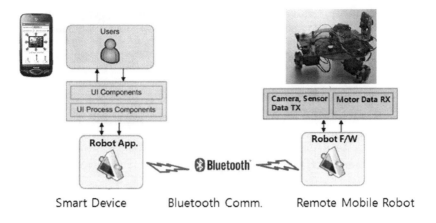

Fig. 3. Software Configuration of Smart Device based Mobile Robot

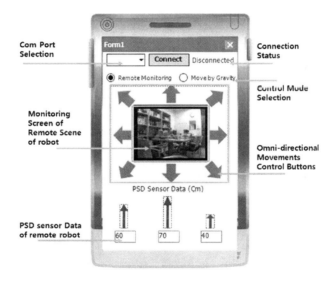

Fig. 4. User Interface of Smart Device Application

and verify current PSD sensor measurements. The developed robot software is available on various smart devices and Windows CE based embedded devices.

The developed software supports a user interface that has various control mode options for the user to select a specific mode among control mode and gravity steering mode. When the mobile robot moves according to the mobile mode when selected via the smart device, it collects information of the surrounding environment through IR sensors and transmits the information to the smart device via Bluetooth. The smart device then analyzes the data from the robot and generates new motor control commands and transmits the commands to the mobile robot.

Also, the mobile robot smoothly moves by adopting velocity control of the Dynamixel AX-12A robot servo motors using TTL-based data communication with the microcontroller. A smart device application is developed to control the robot according to the selected mobile mode after connecting with the mobile robot through Bluetooth.

3   Omni-directional Mobile Robot and Smart Device Configuration

3.1   Omni-directional Mobile Robot

The omni-directional mobile robot is composed of three wheel motors of Dynamixel AX-12A, a CMOS Jpeg camera, three PSD range sensors, and a 8-bit microcontroller with a Bluetooth module. An illustration of the omni-directional mobile robot and its configuration with a smart device is shown in Fig. 5. The images captured from a CMOS camera sensor of the remote robot are transmitted via Bluetooth to the connected smart device while a user of the smart device controls the robot through the control UI of application of the device.

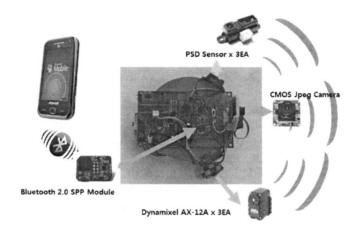

Fig. 5. Illustration of Omni-directional mobile robot and its Configuration with Smart device

The mechanical design and the graphical representation of motor distribution of the omni-directional robot are shown in Fig. 6. Each wheel is placed so that the axis of rotation points towards the center of the robot and there is an angle of 120 degree between the wheels. For this type of omni-directional configuration, the total displacement is achieved by summing up all the three vectors contributions.

Since the robot forward direction is represented by Yr, each motor contribution consists of the cosine of the angle of the desired direction projected on each wheel drive direction multiplied by the velocity. Considering that the three wheels driving directions of the robot are 150, 270 and 30 degrees respectively, the contribution for each motor for linear velocity is given by the following equations [6].

$$F1 = velocity \cdot \cos(150 - DesiredHeading) \qquad (1)$$

$$F2 = velocity \cdot \cos(270 - DesiredHeading) \qquad (2)$$

$$F3 = velocity \cdot \cos(30 - DesiredHeading) \qquad (3)$$

Where F1, F2 and F3 denote the motor vector contributions of the motors 1, 2, 3 respectively, velocity denotes the linear velocity the robot should move and DesiredHeading denotes the angle of the desired movement.

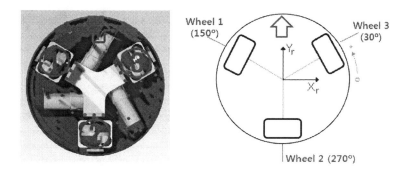

Fig. 6. Mechanical design and Graphical representation of motor distribution of Omni-directional robot

The mobile robot also has CMOS Jpeg serial port camera module from LinkSprite. It can capture pictures of various resolutions such as VGA, QVGA and 160*120, then outputs JPEG images through UART. The camera module can be easily integrated into existing robot design.

At the mobile robot, the microcontroller continually acquires information from other sensors such as PSD through the ADC and sends it to Bluetooth using a serial port. The firmware of the microcontroller was programmed to receive motor drive commands from the smart device. Table 1 shows the hardware specifications of the mobile robot and the smart device used in this paper.

3.2 Smart Device Configuration

We used a Windows Mobile based Samsung smart phone named T-Omnia2 to communicate with the microcontroller via Bluetooth for controlling the remote mobile robot. To retain portability on various Windows based smart device OS environments, .Net framework and Bluetooth are used for the software implementation. To develop a smart device user interface and handle the gravity sensor efficiently, Windows Mobile SDK and the corresponding smart phone's SDK are used [7].

Table 1. Hardware Specifications of Mobile Robot and Smart Device

H/W Item	Feature
Microcontroller	ATMEL ATMega164P-20MU
Motor	ROBOTIS Dynamixel AX-12A x 3EA
Sensor	CMOS JPEG camera x 1EA, PSD Sensor x 3EA
Bluetooth	SPP / BT2.0
Smart Device	Samsung SCH-M715(T-Omnia2)
Smart Device OS	Windows Mobile 6.5
Dev. Tool	.NET Compact Framework 3.5 with C# Language
Comm. Speed	115200 bps

4   Experimental Results

4.1   Remote Movement Control by Gravity Sensor

In remote control by gravity sensor mode, the user can control the mobile robot akin to a user driving a car with a steering handle. In this mode, the smart device simulates the physical car handle by estimating the current pose using a 3-axis gravity sensor. The user also can feel the speed of the mobile robot through vibration of the smart device driven by an embedded vibration motor [8].

To move the robot at a constant velocity, navigation functions such as forward acceleration and immediate left and right turn are implemented. Fig. 7 shows screenshots of the control experiment of the remote mobile robot involving gravity control mode with the user interface.

4.2   Remote Monitoring

In remote monitoring mode, the user can control the mobile robot while monitoring the remote scene captured from the camera on the robot. The user can also see the

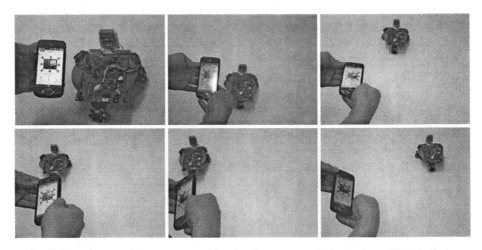

Fig. 7. Experiments of Remote control by Gravity sensor and Monitoring of Mobile Robot

fore-facing 3 PSD sensor measurements from the robot. Control functions in the remote monitoring mode are 8 desired directions for the robot to move without rotating the robot using the omni-directional three wheels mechanism and additional left and right turning for rotating the robot. Fig. 7 also shows images from the experiment of the remote monitoring mode, where the robot was controlled successfully according to the user's commands.

4.3 Performance Evaluation of Remote Monitoring

We performed a performance evaluation of remote control and monitoring with a smart device by measuring the wireless data transmission latency and the frame rate of the captured images at 160*120 resolution. In evaluating the wireless data transmission latency of the wireless smart device based robot control system, the average wireless data transmission latency was measured after 10 byte data packets were sent and received 20 times in a loopback manner. Table 2 shows the average data transmission latency and the fame rate according to the Bluetooth communication speeds.

Table 2. Data Latency andn Frame Rate according to Bluetooth Communication Speeds

Comm. Speed of Bluetooth (bps)	Average data latency (mSec)			Frame Rate (Frames/Sec)
	Near	5M	10M	
57,600	33.6	36.8	58.9	2~3
115,200	22.3	25.5	33.9	4~5

5 Conclusion

In this paper, we described the remote control and monitoring of an omni-directional mobile robot using Bluetooth with a smart device. We also conducted several remote robot experiments and a performance evaluation to verify applicability of the developed control and monitoring system using the smart device.

In the experiment, the gravity steering mode and the remote monitoring mode were tested. The control algorithms for the remote mobile robot were implemented at the smart device side and the smart device controlled the mobile robot by sending a motor control signal through wireless connection. To validate this wireless control system, data transmission latency was measured at different communication speeds and different distances between the robot and the smart device.

The proposed smart device based remote robot control system makes a significant contribution to the field in that it is a new type of robot application, successfully integrating and exploiting a traditional PC based mobile control method and the distinct features of a smart device, including mobility in particular.

When considering the rapid advancement of smart devices with various embedded sensors, we expect that smart devices will soon be able to be used instead of desktop PCs for controlling robots. We also hope that the service variability of intelligent service robots will be widened through their connection with smart devices and by using the various features and contents available on smart devices.

Acknowledgments. This work was supported by Basic Science Research Program through the National Research Foundation of Korea(NRF) funded by the Ministry of Education, Science and Technology(MEST) (2011-0013776).

References

1. Yoon, Ho, J., et al.: Smart Device, merging into '3S' market, Timely Report, ROA Consulting (2010)
2. Lin, F., Ye, W.: Operating System Battle in the Ecosystem of Smartphone Industry. In: Proc. of the International Symposium on Information Engineering and Electronic Commerce (IEEC 2009), pp. 617–621. IEEE Computer Society, Washington, DC (2009)
3. Microsoft.NET Framework Developer Center,
   http://msdn.microsoft.com/netframework
4. NET Compact Framework, http://msdn.microsoft.com/en-us/library/
   9s7k7ce5v=vs.80.aspx
5. Android developers, http://developer.android.com/guide/basics/what-
   is-android.html
6. Ribeiro, F., Moutinho, I., Silva, P., Fraga, C., Pereira, N.: Three Omni-directional Wheels Control on a Mobile Robot, Control. University of Bath, UK (2004)
7. Samsung Mobile Innovator, http://innovator.samsungmobile.com
8. de Souza, M., Carvalho, D.D.B., Barth, P., Ramos, J.V., Comunello, E., von Wangenheim, A.: Using Acceleration Data from Smartphones to Interact with 3D Medical Data. In: Proc. of the SIBGRAPI Conference on Graphics, Patterns and Images (SIBGRAPI 2010), Gramado, pp. 339–345 (2010)

User Following of a Mobile Robot
by Fusing Different Kinds of Sensors

Yeon-Gyun Kim[1], Hahmin Jung[1], and Dong Hun Kim[2]

[1] Department of Advanced Engineering
[2] Department of Electrical Engineering
Kyungnam University, Changwon-si, Gyeongnam, Korea
dhkim@kyungnam.ac.kr

Abstract. This paper presents a mobile robot system for an user-following by detecting the direction of an user based on a fusion of ultrasonic sensors. One of the most important characteristics of the ultrasonic sensor has a property of the straight within a constant angle range. In addition, the ultrasonic sensor is able to promptly detect the user with fast operating speed and resists influence of external environment. The mobile robot follows the user quickly through a simple hardware and algorithm. Extensive simulation presents to illustrate the viability and effectiveness of the proposed user-following mobile robot system.

Keywords: User-following, ultrasonic sensor, mobile robot.

1 Introduction

Recently, a mobile robot has been extensively studied based on various sensors due to development in sensor technology. The ultrasonic sensor among various sensors is suitable for a mobile robot since it is inexpensive and useful. The ultrasonic sensor is used for environmental recognition such as localization and obstacle avoidance. The localization based on ultrasonic sensors use several ultrasonic transmitters fixed to a ceiling to give a mobile robot location information. And an ultrasonic receiver is mounted on the mobile robot that does not know location information. The fixed ultrasonic transmitter uses two different transmission frequencies that are 40kHz and 25kHz. Two ultrasonic receivers are used for receiving 40kHz and 25kHz frequencies, respectively. The localization of a mobile robot is calculated by the time variation taken when an ultrasonic receiver detects the ultrasonic signal transmitted from the ultrasonic transmitter [1]. In [2], the localization of a mobile robot is measured based on a fixed transducer ultrasonic sensor and four receiving ultrasonic sensors that is equipped on the mobile robot. The ultrasonic transmitter immediately sends out an ultrasonic signal as soon as receiving a starting signal via the RF switch controlled by a central computer. The four ultrasonic receivers detect an ultrasonic signal transmitted from the ultrasonic transmitter at different time, respectively. Owing to different receiving time, the distance from each ultrasonic receiver to the ultrasonic transmitter is also different. The location of a mobile robot is calculated by the measured distance between each ultrasonic receiver and the ultrasonic transmitter.

G. Lee, D. Howard, and D. Ślęzak (Eds.): ICHIT 2011, CCIS 206, pp. 295–302, 2011.
© Springer-Verlag Berlin Heidelberg 2011

In many cases, a mobile robot recognizes the existence of obstacles using ultrasonic sensors. The mobile robot is equipped with several ultrasonic transmitter and receivers for avoiding obstacles. The distance from a robot to an obstacle is calculated by time variation taken between transmitting the ultrasonic signal and reflecting ultrasonic signal from an obstacle, which is used for obstacle avoidance [3]. In the study, a mobile robot system following an user through the ultrasonic sensors is presented.

Many approaches to track a human using cameras or laser range finders have been studied for recognizing a human. In [4], the mobile robot detects a human with a light emitting device by use of a camera. After calculating the relative distance and direction between the recognized human and the mobile robot, the robot follows the human. In [5] a CCD color video camera that is mounted on a pan-tilt unit on the robot recognizes skin color of a human. The CCD color video camera controlled by a pan-tilt control follows a human. The human recognition performance based on a camera is sensitive to camera setting such as a camera angle, view points and pixels of the vision. Moreover, the reliability and robustness of a vision system to an object depends on the change of illumination and climatic condition. On the other hand, the laser range scanner detects human legs to follow human in [6]. If the laser range scanner is mounted to a mobile robot, the mobile robot is become bigger and more expensive since the laser range scanner is rather big, even high-priced. In [7], the method that detects a direction of human position by a clapping sound and then recognizes human legs based on a camera and laser range scanner has been proposed.

This study proposes that the mobile robot follows an user by detecting the direction of an user based on a fusion of ultrasonic sensors. The ultrasonic sensor for a mobile robot is small in size and low in price. In addition, the ultrasonic sensor is able to promptly detect the user with fast operating speed and resists influence of external environment. One of the most important characteristics of the ultrasonic sensor has a property of the straight within a constant angle range. Based on such a characteristic of the ultrasonic sensor, it is possible that the mobile robot follows the position of an user.

2   Mobile Robot Model and Hardware Configuration

2.1   Robot Model

In this study, the mobile robot is a grounded vehicle model with four wheels. The mobile robot has two wheels on left and right side respectively. However the mobile robot is similar to an unicycle robot, since two wheels of each side are driven by the same signal [8]. Hence, the kinematical model of a mobile robot is given by

$$\Delta x_c = \frac{\cos \theta_c (v_r + v_l)}{2} = \cos \theta_c \cdot v$$

$$\Delta y_c = \frac{\sin \theta_c (v_r + v_l)}{2} = \sin \theta_c \cdot v \tag{1}$$

$$\Delta \theta_c = \frac{v_r - v_l}{2d_c} = \omega$$

The robot model is described by a three vectors $q_c = [x_c \quad y_c \quad \theta_c] \cdot [x_c \quad y_c]$ Indicates the center position of each wheel and θ_c indicates the orientation of the robot with respect to the x-axis. d_c is the distance between the point $[x_c \quad y_c]$ and each side wheel. The tangent and angular velocities are given by

$$v = \frac{v_r + v_l}{2} = \frac{r(\omega_r + \omega_l)}{2} \quad \text{and} \quad \omega = \frac{v_r - v_l}{2d_c} = \frac{r(\omega_r - \omega_l)}{2d_c} \tag{2}$$

where the velocities v_r and v_l are the tangent velocities of contact point between each side wheel and the ground, and ω_r and ω_l are the angular velocities of each side wheel. r indicates the radius of each wheel.

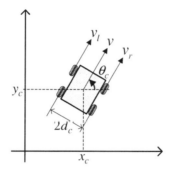

Fig. 1. Robot model

2.2   Hardware Configuration

In this study, the user-following mobile robot system is divided by the user part and the mobile robot part. The configuration of the user module is made as a shown in Fig. 2 (b). The user module is equipped with an ultrasonic transmitter HG-40C for transmitting ultrasonic signals to the mobile robot and a Bluetooth communication module ESD-110 for calling the mobile robot out. The module ESD-110 is able to communicate within 100 meters. The ultrasonic transmitter and a wireless communication through the Bluetooth communication module are controlled by 8bits microprocessor mounted on the user module, ATmega128A16u.

The configuration of the mobile robot is shown in Fig. 2 (a). The mobile robot is equipped with three ultrasonic receivers HG-40NTII for detecting the ultrasonic signal from the user, two infrared sensors for avoiding obstacles, two DC motors on left and right side respectively for user-following, two channel DC motor drivers for operating the four DC motors, and the Bluetooth communication module ESD-110 for receiving a call from the user. An 8bits microprocessor ATmega128A16u is used for controlling the whole mobile robot system.

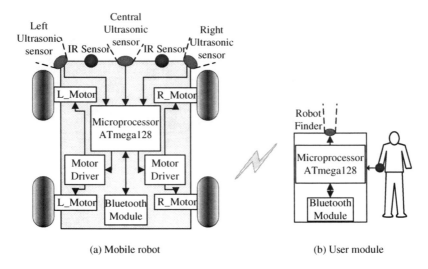

(a) Mobile robot   (b) User module

Fig. 2. Block diagram of the proposed user following system

3   The Proposed Method

The ultrasonic sensor has a property of the straight within a constant angle range, which is used for user-following in the study. Thus, due to the properties of straight, an ultrasonic sensor enables a mobile robot to detect a direction of the user where he or she is. The ultrasonic transmitter mounted on the user module transmits an ultrasonic signal to constant direction and its range. The mobile robot is equipped with three ultrasonic receivers for detecting ultrasonic signals from the transmitter. Three ultrasonic receivers attached in front of the mobile robot are able to detect the ultrasonic signal from separate directions within constant range. Figure 3 (a) and (b) show the separate range of receivers and the direction of transmitter, respectively.

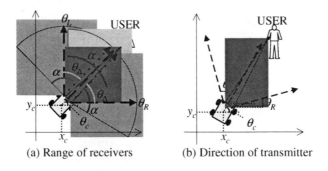

(a) Range of receivers   (b) Direction of transmitter

Fig. 3. The range and direction of the ultrasonic sensors

The directions of three ultrasonic receivers are given by

$$\theta_L = \theta_c + \theta_\Delta, \quad \theta_F = \theta_c, \quad \text{and} \quad \theta_R = \theta_c - \theta_\Delta \tag{3}$$

where θ_L, θ_F, and θ_R are the left, center, and right receiving direction, respectively. θ_c is the forward direction of the heading of a mobile robot. And θ_Δ is the constant angle tilted with respect to a heading of the mobile robot. The direction of the ultrasonic transmitter mounted on an user module is ψ. Each ultrasonic receiver on the mobile robot has a separate range for detecting the ultrasonic signal transmitted from the user. The ultrasonic transmitter on the user module transmits an ultrasonic signal with a constant range. Figure 4 shows the range of three ultrasonic receivers for detecting an ultrasonic signal.

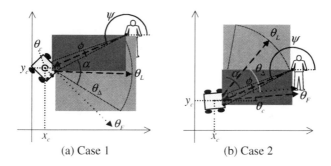

(a) Case 1   (b) Case 2

Fig. 4. The range of the left ultrasonic receiver

Figure 4 shows a constraint of a condition to detect on the left ultrasonic receiver. Figure 4 (a) and (b) show the range constraint of the left ultrasonic receiver attached on the mobile robot for detecting the ultrasonic signal that is transmitted from the ultrasonic transmitter on the user module. The angle conditions Fig. 4 (a) and (b) of a left receiver are given by

$$(\psi + \frac{\phi}{2}) - (\theta_L + \frac{\alpha}{2}) = \pi \quad \text{and} \quad (\psi - \frac{\phi}{2}) - (\theta_L - \frac{\alpha}{2}) = \pi \tag{6}$$

where $\frac{\phi}{2}$ is the constant transmitting angle respect to ψ, and $\frac{\alpha}{2}$ is the constant receiving angle on the left ultrasonic sensor with respect to θ_L. Two angle conditions are obtained according to the location of the user. Similarly to the range of the left ultrasonic receiver, from the three ultrasonic receivers on front of the mobile robot, the conditions for detecting a transmitted signal from the ultrasonic transmitter mounted on user module are given by

$$\pi - \frac{\alpha}{2} + \frac{\phi}{2} \leq \psi - \theta_L \leq \pi + \frac{\alpha}{2} - \frac{\phi}{2}$$

$$\pi - \frac{\alpha}{2} + \frac{\phi}{2} \leq \psi - \theta_F \leq \pi + \frac{\alpha}{2} - \frac{\phi}{2} \qquad (7)$$

$$\pi - \frac{\alpha}{2} + \frac{\phi}{2} \leq \psi - \theta_R \leq \pi + \frac{\alpha}{2} - \frac{\phi}{2}$$

The three ultrasonic receivers on the mobile robot face a different direction, respectively. Each ultrasonic receiver detects the ultrasonic signal within the angle with a constant range. And the ultrasonic transmitter on the user module transmits the ultrasonic signal within the angle with a constant range. Through the ultrasonic signal transmitted from the ultrasonic transmitter to each ultrasonic receiver, the mobile robot is able to recognize the relative direction of the user.

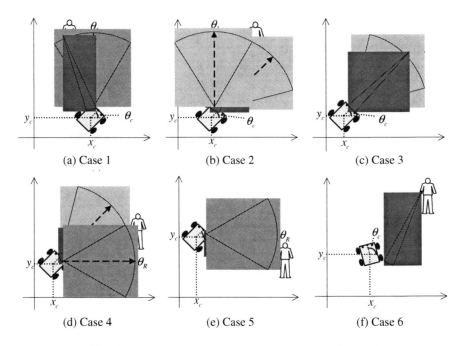

(a) Case 1   (b) Case 2   (c) Case 3

(d) Case 4   (e) Case 5   (f) Case 6

Fig. 5. Six cases to transmit and receive an ultrasonic signal

In this study, the user and the mobile robot represent U and R, respectively. The mobile robot follows the user based on three ultrasonic receivers and one ultrasonic transmitter that are mounted on the mobile robot and user module, respectively. The user-following by the ultrasonic sensor is carried out by three ultrasonic receivers. To receive the ultrasonic signal on each ultrasonic receiver mounted on the mobile robot and transmit ultrasonic signal from ultrasonic transmitter attached on user module is divided into six cases, as shown in Fig 5.

4   Simulation

The simulation of the algorithm presented in section 3 is performed using MATLAB. The simulation is performed, based on the hardware performance a real robot. The mobile robot has control constraints that are the limits of tangent velocity and angular velocity. The constraints of the tangent and angular velocity are given by

$$0 < v < v_{max} \quad \text{and} \quad -\omega_{max} < \omega < \omega_{max} \tag{8}$$

where v_{max} and ω_{max} are the fastest tangent and angular velocity of a mobile robot. In this simulation v_{max} and ω_{max} of the mobile robot are $0.6m/\sec$ and $\frac{\pi}{6} rad/\sec$, respectively. Also, in this simulation, it is assumed that the mobile robot stops when it reaches the user within one meter.

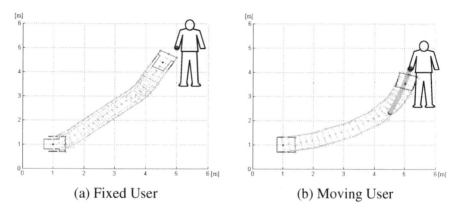

(a) Fixed User   (b) Moving User

Fig. 6. User-following for the static user

Figure 6 shows the scenario that the robot recognizes and follows the user who does not move. The small circles in front of the mobile robot are the three ultrasonic sensors mounted on the mobile robot. The heading of the left and right ultrasonic receivers are tilted $45°$ and $-45°$ with respect to the heading of the mobile robot, respectively. And the heading of the central ultrasonic receiver is the same as the heading of the mobile robot. In figure 6, the user is located at $(5,5)$. The mobile robot located at initial position $(1,1)$ is heading to the right and the user is tilted $45°$ with respect to the heading of the mobile robot. The ultrasonic transmitter equipped with the user heads to $240°$ with respect to x-axis. Figure 6 presents that the mobile robot follows the user well by the proposed user-following system.

5   Conclusion

In this study, we present that the mobile robot follows an user by detecting the direction of an user based on a fusion of ultrasonic sensor. The user is equipped with one ultrasonic transmitter and the mobile robot is equipped with three ultrasonic receivers for the user-following. The mobile robot follows the position of the user by operating its motors depending on detection of three ultrasonic receivers. If the user pushes the button, the mobile robot promptly finds and follows the user like the KITT (Knight Industries Two Thousand) broadcasted in American TV Drama [9]. A key development is that the mobile robot follows the user quickly through a simple hardware and algorithm. Extensive simulation presents to illustrate the viability and effectiveness of the proposed framework.

Acknowledgement. This work was supported by Kyungnam University Research Fund, 2011.

References

1. Kim, H., Choi, J.: Advanced indoor localization using ultrasonic sensor and digital compass. In: International Conf. on Control, Automation and Systems, pp. 223–226 (2008)
2. Tsai, C.: A Localization System of a Mobile Robot by Fusing Dead-Reckoning and Ultrasonic Measurements. IEEE Transactions on Instrumentation and Measurement, 144–149 (1998)
3. Borenstein, J., Koren, Y.: Obstacle Avoidance with Ultrasonic Sensors. IEEE Journal of Robotics and Automation, 213–218 (1998)
4. Ohya, A.: Human robot interaction in mobile robot application. In: 11th IEEE International Workshop on Robot and Human Interactive Communication, pp. 5–10 (2002)
5. Sidenbladh, H., Kragic, K., Christensen, H.I.: A Person Following Behaviour for a Mobile Robot. In: IEEE International Conf. on Robotics and Automation, Detroit, pp. 670–675 (1999)
6. Kim, H., Chung, W., Yoo, Y.: Detection and tracking of human legs for a mobile service robot. In: IEEE International Conf. on Advanced Intelligent Mechatronics, pp. 812–817 (2010)
7. Luo, R.C., Huang, C.H., Lin, T.T.: Human Tracking and Following Using Sound Source Localization for Multisensor Based Mobile Assistive Companion Robot. In: 36th Conf. IEEE Industrial Electronics Society, pp. 1552–1557 (2010)
8. Maalouf, E., Saad, M., Saliah, H.: A higher level path tracking controller for a four-wheel differentially steered mobile robot. Robotics and Autonomous System 54, 23–33 (2006)
9. KITT- Wikipedia, the free encyclopedia, http://en.wikipedia.org/wiki/KITT

A Remote Firmware Upgrade Method of NAN and HAN Devices to Support AMI's Energy Services

Young-jun Kim, Do-eun Oh, Jong-min Ko, Young-il Kim,
Shin-jae Kang, and Seung-Hwan Choi

S/W Center, KEPCO Research Institute, Daejeon, Korea
juny@kepco.co.kr

Abstract. As AMI system is applyed in Korea gradually, electric company can provide a lot of services to customers like DR(Demand Response), CSS (Customer Service System). In this case, field deployed devices which are AMI's basic component must be able to upgrade its firmware to support the services. In this paper, we introduce the remote firmware upgrade management system(FUMS) using distributed processing for field deployed device like DCU, and HAN Devices(Smart Meter, IHD, LCD,PCD) in Korea AMI. This article provides FUMS architectures and the upgrade process of communication between FUMS Server and DCU and between DCU and HAN Devices.

Keywords: Advanced Metering Infrastructure, Remote Firmware Upgrade System, Network Service Management, HAN (Home Area Network) Devices.

1 Introduction

AMI is an essential system for implementing Smart Grid and a principal means of realizing Demand Response based on Supply-Demand mutual recognition and AMI system can improve energy efficiency and accept various distributed energy resources for the electric industry to grow constantly for the next decades solving global energy security crisis and global warming problem form excessive CO_2 emission.
[Fig.1] shows Korea's AMI system and it consists of smart appliances, IHD, PCT, ESI, smart meter, DCU, ADCS, MDMS Server, and Utility application system from the end user to the electric company.

With the advent and the extension of AMI, the electric devices (ex: DCU, smart meter and IHD) needing maintenance will increase explosively and a lot of energy information services (DR, CSS) are being developed quickly. For that reason, the efficient method of management and upgrade electric devices to support AMI's various function is being discussed.

To solve this problem, Korea AMI is plans to adopt Network Service Management (NSM) system and Firmware Upgrade Management System (FUMS) are going to be installed by May 2011 and through these systems are able to manage and control more than 31,500,000 electric devices, spread throughout the country, effectively.

In this paper, we provide FAN (FUMS and NSM) architecture and describe the process of HAN devices and DCU's Remote Firmware Upgrade Method in AMI Environment.

G. Lee, D. Howard, and D. Ślęzak (Eds.): ICHIT 2011, CCIS 206, pp. 303–310, 2011.

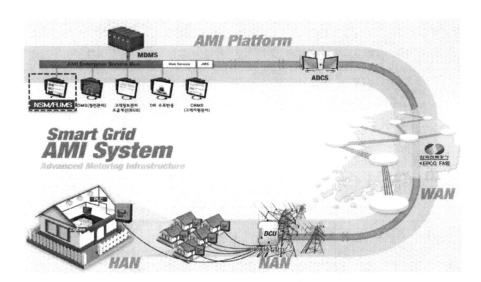

Fig. 1. This shows a AMI system architecture and AMI's electric devices (energy efficiency services, MDMS, DCU, ADCS, HAN devices)

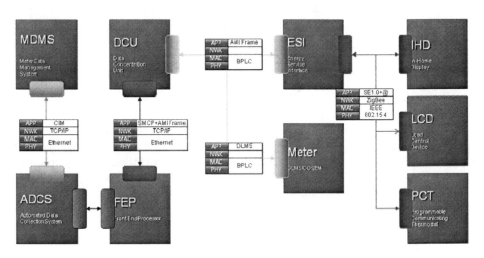

Fig. 2. Korea-AMI's protocols from MDMS to HAN devices(Smart Meter, IHD, LCD, PCT). Korea-AMI has a lot of data protocol format and it is using global standard to communicate between devices for compatibility and scalability.

2   Device Management System

As shown fig2, AMI system consists of many and various electric devices. And they are divided into electric company's asset (MDMS, ADCS, DCU and Smart Meter) and a customer's asset (HAN Device: ESI, IHD, PCT and smart appliances).

Especially, among these, DCU and HAN Devices, excluding smart appliances, are the devices in the middle of AMI system and one of the most important devices to support electric services which will be newly developed in the future.

These deployed devices need to be flexible in dealing with a change. But they are a lot and spread throughout the country, as shown [table1], so manual firmware upgrade by man is actually impossible.

Table 1. The number of electric devices managed by MDMS

Number of ADCS per MDMS : 10 ea
-     Number of FEP per ADCS : 150
-     Number of DCU per FEP : 100
-     Number of Meter per DCU : 100
Number of DCU and Smart Meter per MDMS
-     Meter : 15,000,000
-     DCU : 1,500,000
Number of Home Appliance per MDMS
-     Home Appliance: 15,000,000 (minimum)
Total of the managed devices: 31,500,000

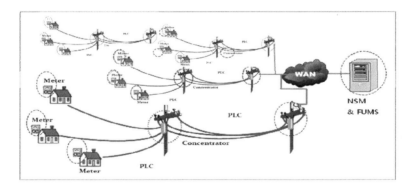

Fig. 3. Firmware Upgrade target devices: DCU, Smart meter and Home appliance

To solve this problem, we have to design AMI system for considering remote device management and maintenance from start to finish.

Furthermore, when more than 31,500,000 devices are version monitoring or F/W upgrading, to reduce the load of NSM and FUMS server, they should be distributed control system by giving some of their features and functions to an agent device. Through an agent device, the entire network management system can be configured to operate efficiently and reliably. In Korea AMI system, DCU is perfect as an agent device.

2.2 NSM and FUMS System

Fig4 shows the proposed NSM and FUMS's architecture.

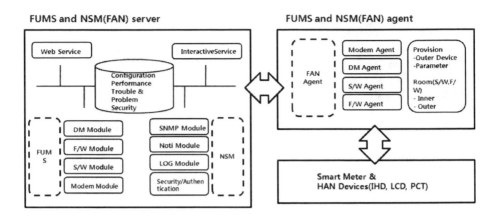

Fig. 4. FAN server and FAN agent's configuration

They are operated in electric company's headquarters or branch offices but their agent for distributed processing is installed in DCU. It performs sub-system's monitoring and f/w upgrade through communication with NSM and FUMS servers.

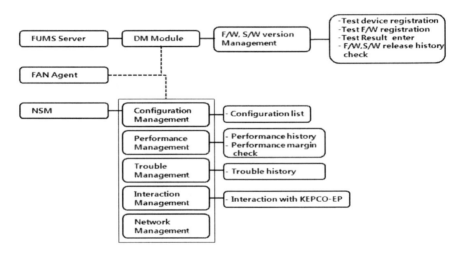

Fig. 5. FAN's inner composition modules

2.3 NSM and FUMS Main Functions

With NSM and FUMS, the three main functions can be implemented. First, Plug and Play (PnP) function for automatic registration for electric devices, then update function of NAN/HAN devices' configuration information, and finally remote firmware distribution and installation.

2.3.1 Automatic Registration for Electric Device

Smart meter or DCU send authentication messages to NSM server as soon as they are turned on. After completing authentication, the server sends automatically the devices' information to FUMS. Since the devices are registered as the managed devices, NSM and FUMS will monitor the devices' status and firmware version.

2.3.2 Device Configuration Update of Electric Devices

NSM and FUMS servers send their agent program installed in DCU the information of their managed devices' status and configuration so that DCU can manage NAN/HAN devices. The list of the managed devices is updated by NSM periodically.

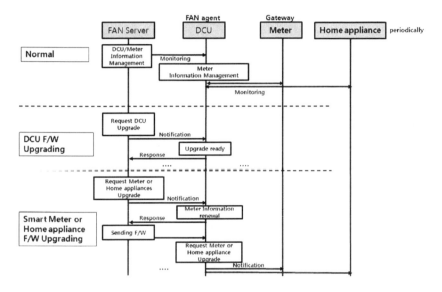

Fig. 6. Configuration update process

2.3.3 Upgrade of F/W and S/W

If DCU or HAN devices have some bugs or need to change its F/W or S/W for supporting new energy service, FUMS Server checks the devices information and then installs new version of firmware. In the case of upgrading DCU, as shown Fig7, FUMS and NSM server is communication with DCU directly. FUMS server performs monitoring DCU's status and version as well as controls device upgrade process directly.

But in HAN device case, to reduce the server's load, it uses distributed processing method. In order to do that, first of all, as you saw fig4 before, it needs to install 'FUMS agent' demon. It performs monitoring and upgrade control as FAN server's substitute.

And then FUMS server send the managed list of HAN devices to its agent and let the agent perform monitoring of version/status on behalf of the server.

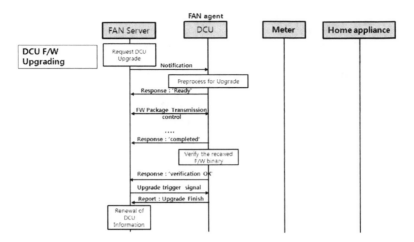

Fig. 7. DCU's firmware upgrade process

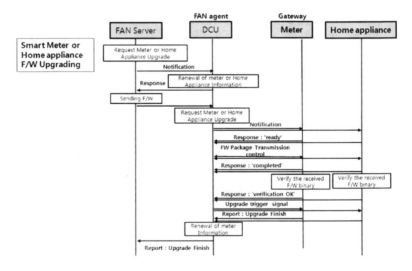

Fig. 8. HAN device's firmware upgrade process

For example, when the agent implements a meter upgrade process, first it divides a firmware binary received by the server into 128Kbyte sized blocks(the size limit is in DLMS/COSEM specification) and then sends it to the meter several times [Fig 8]. After the file transfer is finished, the agent sends the meter a trigger signal to install the new firmware in the meter itself.

For more details, refer to table 2.

Table 2. Font sizes of headings. Table captions should always be positioned *above* the tables.

Level	Process
Step 1	Get Image Block Size
Step 2	Initiate Image transfer
Step 3	Transfer Image Blocks
Step 4	Check completeness of the Image
Step 5	Check the Image before activation
Step 6	Verify Image
Step 7	Activate Image

The upgrade flow chart of DCU and HAN device is as follows

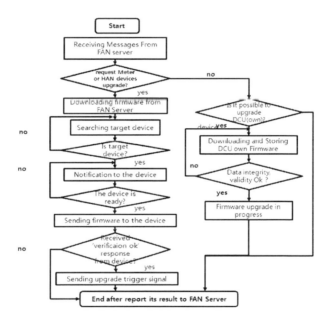

Fig. 9. FAN agent's upgrade flow chart

3   Conclusion

With the development of Advanced Metering Infrastructure (AMI), the new harmonious relations of power supply and application between the user and the grid company will be built up. In order to that, AMI system enables to be managed flexibly and to upgrade its electric devices with new version of firmware for supporting energy efficiency service provided to customer.

In this paper, we provide NSM and FUMS architecture and describe the process of HAN Device and DCU's Remote Firmware Upgrade Method in AMI Environment.

References

1. Korea Electric Power Research Institute, Development of AMI(Advanced Metering Infrastructure) system, pp. 4–17 (2009)
2. Farhangi, H.: The path of the smart grid. IEEE Power & Energy Magazine 8, 18–28 (2010)
3. Open Meter, Requirements of AMI (2009)
4. Utility AMI, High-Level Requirements (2006)
5. SCE, SCE Use case (2006/2009)
6. NIST, Framework and Roadmap for Smart Grid Interoperability Standards 1.0 (2010)
7. IEC, IEC SMB SG 3 Smart Grid (2009)

Real-Time Monitoring System for Watercourse Improvement and Flood Forecast

Cristina Sotomayor Martínez, Antonio J. Jara, and Antonio F.G. Skarmeta

Department of Information and Communication Engineering
University of Murcia, Murcia, Spain
{csotomayor,jara,skarmeta}@um.es

Abstract. This works presents the developed Monitoring System to carry out the processing, compilation, analysis and presentation of the data collected from the deployed monitoring platforms based on Future Networks for the semiarid zone of the "El Albujón" located in the South of Spain. This Future Networks-based system is used for the evaluation and improvement of the watercourses and flood forecast. The developed system is formed by the parts Complex Event Processing and a flexible architecture focused on Web Services and applications, which presents a new relevance with respect to the current solutions in aspects such as the real-time event processing of the collected information, and the capacity to offer the scalability, flexibility and ubiquity for the access and collection of the information, which are features required for the future watercourse improvement and flood forecast solutions based on Future Internet.

Keywords: Hydrological monitoring, Future Networks for Sensor-based Monitoring Systems, Sensor Web, Open Geospatial Consortium (OGC), Complex Event Processing (CEP).

1 Introduction

The southeast of Spain presents a semiarid zone, where an important part of the rainfall is infiltrated in the dry soil surface. This reduces the natural watercourse and makes more difficult the flood forecasting, since in case of be an uneven area with several highland and lowland zones such as the located in "El Albujón" at the south of Spain, where are located three highland zones, which are draining to a low area prone to flooding. Therefore, the problem arises for the reason that the sum of the volume of water from the highlands is not directly related to the risk of flooding in the lowlands because of the infiltration influence [1].

This problem is also located in other many countries, where it has been defined solutions to study of hydrological parameters in order to understand their natural environments, such as national projects in Sweden based on aerial photos [2], in UK with modeling grid-to-grid [3], or geological survey in several islands from USA [4].

In order to carry out the flood forecasting in this kind of areas, it is defined a flood forecasting model based on the hydrological information, such as rainfall or soil

G. Lee, D. Howard, and D. Ślęzak (Eds.): ICHIT 2011, CCIS 206, pp. 311–319, 2011.
© Springer-Verlag Berlin Heidelberg 2011

moisture. Watercourse presents a complex and nonlinear behavior due to the involvement of many physical parameters and the influence of among them in the semi-arid environments. For that reason it is required an exhaustive monitoring of the area, in order to reach a high and accurate level of data of the area. The collection of hydrological information in wide areas is a technique already applied for Water Quality Assessments in other related projects such as [5], but they present some differences in term of the kind of sensors and the features of the deployment with respect to the flood forecasting deployments,

The deployment and solution defined in this project has been carried out in the frame of the project ("Hydrological Modeling of Semiarid Zones"), which has been denominated MHZS. This project involves the monitoring of the hydrological process that occurs in the semiarid areas, where changes can be very harsh and cause flooding or other problems. This studies and monitories the evolution of phenomenon such as precipitation, evapotranspiration, infiltration recharge to aquifers, and surface runoff.

The mentioned monitoring has implied the development of a monitoring system composed of a hardware platform to connect the sensors and log the sensor's value, as such as a post-processing of the data system, which is presented in this work.

Specifically, the project is focused on an area south-east of Spain, marked by major drought periods, the "Campo de Cartagena", which covers an area of 550 km^2. Its main drainage basin is the "Rambla del Albujón", shown in Figure 1, where has been deployed a sensing system composed by 8 hardware platforms/data loggers to monitor a set of hydrological variables and environment information, which is linked with the Geographic Information System (GIS) for its subsequent treatment and analysis with the post-processing information System.

Therefore, in an initial step for this project has been developed a monitoring platform, which consists of a low-power microcontroller, a long-range radio module, a persistent memory and 12 interfaces to connect various sensors in order to be able to monitor different phenomenon and in different positions inside of a monitored region.

Monitored information is used, on the one hand, to visualize the current state of the zone in real-time, and on the other hand, to flood forecast, i.e. prevent the runoff formed by the water infiltrated. This system is required since the periodic and manual monitoring by a group of technicians, such as carried out in other similar projects, is not enough, because it does not generate relevant information, since the area status is

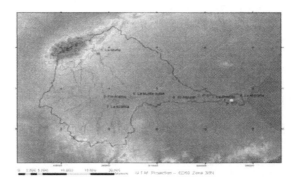

Fig. 1. Monitored area

dry for many months, and it is usual in this zone that suddenly arise an overflow due to sudden storms. This led us to seek the solution to a system that can monitor these areas continuously, and is able to react in time to an emergency situation. This paper is going to present the developed architecture in order to support this real-time monitoring solution to flood forecast and improvement of watercourse.

2 Real-Time Monitoring System for Flood Forecasting

Sensor networks enable automated collection of data that require large amount of sensors, monitoring platforms and control systems, this allows to build in the field the traditionally used in industrial environments Supervisory Control And Data Acquisition (SCADA) systems, which are capable of providing communication with field devices, in addition to filter and manage the information collected, as such as monitor and evaluate the status of the process and of the environment in real-time.

Additionally, it is extended the capabilities to collect data automatically and in an unattended way with the use of Wireless Sensor Networks. Thereby, it is significantly reduced the cost and environmental damage of these solutions. In addition, they are also able of measuring hydrological variables with high spatial and temporal resolution, allowing the use of models for integration of the distributed data, and the development of Intelligent Information Systems and Knowledge Based Systems for monitoring of the state in real time, transmitting the relevant information to the processing stations which will analyze all the collected data, in conjunction with other historical information to detect potential dangerous situations in a collaborative way.

Therefore, the most relevant stages of the process in the real-time monitoring system are, initially the data analysis and presentation, together with interpretation of the results, and finally the report with the conclusions about the status and probabilities of dangerous situations. These backend systems are what show how successful the monitoring activities have been in attaining the objectives of the assessment. It is also the part of the solution that provides the information needed for decision making, and assessing the environment state [6].

The next sections are going to present the solution defined to carry out the process of data analysis with Complex Event Process technology, and presentation of the information with Web Services and Sensor Web approach, which allows a scalable access to all the collected information and status of the environment.

3 System Architecture

The system architecture developed for monitoring the evolution of the watercourse and the flood forecast covers both hardware and software parts. The next subsections are going to show an overview of the hardware design, and a deep analysis and presentation of the software design. The architecture has been designed with modular and flexible principles, in order to allow that each part can be interchangeable, adaptable and scalable, independently of the other parts from the system.

Monitored information is used to visualize the current state of the zone in real-time, and to flood forecast, i.e. prevent the runoff formed by the water infiltrated. This system is required since the periodic manual monitoring such as carried out in other similar projects is not enough, since it does not generate relevant information, because the area status is dry for many months, and it is usual in this zone that suddenly arise an overflow due to sudden storms. This led us to seek the solution to a system that can monitor these areas continuously, and allow to react in time to an emergency situation. This paper is going to present the developed architecture in order to support this real-time monitoring solution to flood forecast and improvement of watercourse.

3.1 Hardware Design

The hardware architecture includes the sensors network, the telemonitoring system and the information system, which defines the 5 components shown in Figure 2.

- **Base stations:** are the most important hardware elements, they are formed by sensors for the humidity, water level, temperature etc. and the developed platform for datalogging. On the one hand, the sensors define the information captured from the environment, and on the other hand, the dataloggers are responsible for managing the sensors, pre-analyze data and transmit them to the processing station.
- **Acquisition station:** is part of the backend formed by a set of servers, which perform the main SCADA operations, i.e. filtering, storing data locally and control the state received from the sensors, dataloggers and finally report alarms.
- **Processing station:** is also part of the backend and performs GIS functions, such as look for behaviors, patterns and characteristics of the monitored environment through the processing based on the paradigm CEP (Complex Event Processing).
- **Spreading station:** is a web server that provides access to the information generated by the sensor network using GIS and the Open Geospatial Consortium (OGC) services. This web server also provides an interface to manage the sensor network through the SCADA system.
- **Monitoring stations:** are the users and systems, which are accessing to the data obtained by the sensors, as such as the information from the SCADA and the knowledge contextualized by the information systems.

As shown in Figure 2, the system architecture consists of different stations for which data will flow from its collection in the environment by the sensors to its presentation in a Web application for the end users. The data processing is carried out in the backend through different distributed information systems. This data processing is divided in different stages with different levels of abstraction of the data, information and knowledge. The main change of the data to the knowledge i.e. detection of flood risk is made in the processing station, which generate complex information through the analysis and fusion of the different data sources.

The environmental data is generated periodically by the sensors connected to the Base Stations. The Base Stations carry out an initial calibration to compensate for the imperfections of electronic components, as such as pre-analysis of the data. These corrected and pre-analyzed values from the sensors are temporarily stored and transmitted using GPRS communications to the Acquisition Station.

Acquisition Station performs an adaptation of the formats through the component called "PLC software", which converts the real time collected data to a format

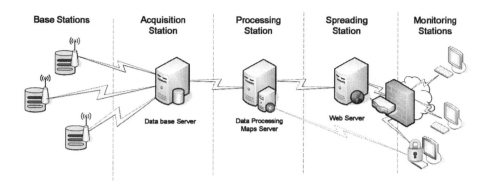

Fig. 2. Hardware architecture

compatible with the SCADA. After this conversion the data is stored in a PostgreSQL database. This "raw" information may be consulted by users with appropriate permissions using JNLP application via a web browser, and it will be also used for the mentioned processing stage.

The processing stage is carried out through the application of the defined hydrologic models to the data received from the acquisition station and the use of additional information from the GIS. This processing system is based on CEP in order to support the decision-making, and monitoring the watercourse in the highlands from the basin of "El Albujon". Thus, through the analysis of the water level and the time that remains in that level is studied the contribution from the highlands to the lowland, in order to detect overflows and raise alarms when required.

The main information of the system is obtained through the sensor networks which are connected with the base stations, which feed to the acquisition system. That is not all the information used by the system, such as mentioned it is used the GIS and the historical of the previously captured values. Thereby, it can be obtained enough information and knowledge about the real environment situation.

3.2 Software Design

The design of the software architecture is divided into several layers of abstraction, which at the same way that for the hardware components, they can be modified individually to enhance the capabilities of the final system.

The software architecture supports the heterogeneity of the data sources in order to include all the data collected by the sensors deployed in the environment, the information obtained through other external entities, which provides information through Web Services such as the Embedded Web Services based on RESTful in the Future Internet Architecture focused to Internet of Things and Web of Things [7], and finally other relevant environmental information from the GIS. The GIS contains information about the basins geometry and the geographical location of the sensors. The mentioned information in conjunction with the soil features are essential to correctly process the data and analyze the situation.

The Figure 3 presents the software architecture, where in a first layer is found the already presented Data Sources which are going to feed the Acquisition Layer, which is awareness of the sensors diversity and specific considerations, e.g. meaning of the voltage ranges with respect to the value, and communication interface considerations.

The collected values by the Base Stations are stored in the GIS in function of the location of the sensor. Thereby, this allows to the processing system works with the data from a certain area, since it knows the sensors which generate information related to that area.

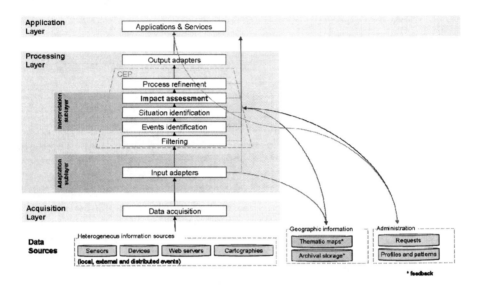

Fig. 3. Software architecture

The second stage is focused on the processing; it is what presents a higher complexity in the architecture, since this requires to perform several processing stages. Specifically, in an initial stage, this adapts the input data and stores the raw information, this processes the data and injects the events into the CEP engine, in order to classify, identify and finally generate new high-level events. After that, the information obtained is adapted to the output formats, which is stored in the generated thematic maps.

The last layer is focused on the interfaces and services offered to the end user applications and services provided to third parties. Specifically, it has developed a Web Application and a set of Web Services.

The web application developed for this project presents the information about the project, allows the access and management of the sensors features, as such as access to the sensor measurements, and finally the events generated by the processing layer.

This Web Application has been included in a map viewer, where it is shown the data acquired and generated by the system and those provided by external entities, making use of Web services, such as the previously mentioned based on RESTful.

The Figure 4 presents a set of screenshots from the Web Application, where it is shown in the left side the information and management for a specific sensor; in the middle the map view of the deployed sensors, and in the right an information page.

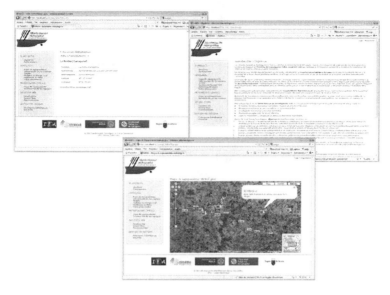

Fig. 4. Web application

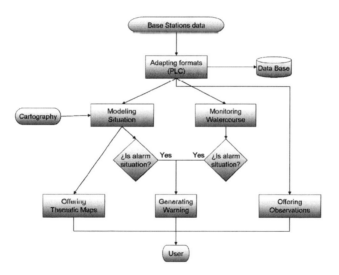

Fig. 5. Information flow diagram

The set of Web Services developed are compliance with the OGC open specification. Since, such as mentioned, the collected values are geographically located; they can be located on a map and shown the geographical image with its inferred characteristics. For this purpose, the system has a service based on the regulations from the WMS (Web Map Service), in order to provide the inferred information by our system along to the information from other WMS servers, and from the services based on WFS (Web Feature Service), which allows consultations of the properties shown in the map. On the one hand, WMS dynamically produces a map with the data located in function of its geographical reference, e.g. generation point. Therefore, it is a representation of the geographic information on an image which represents that area. The operations defined in the standard which are implemented to access to the service are: GetCapabilities, GetMap, and GetFeatureInfo. On the other hand, WFS allows to retrieve/update geospatial data using platform-independent functions. This defines interfaces to access and manipulate the geographical features. The operations defined in the standard to access to this service are: GetCapabilities, DescribeFeatureType, GetPropertyValue, GetFeature, GetFeatureWithLock, and LockFeature.

Finally, Figure 5 presents a flow diagram with the evolution of the information in the proposed system.

4   Conclusions

This work has presented the developed Monitoring System to carry out the processing, compilation, analysis and presentation of the data collected from the deployed sensors based on Future Networks. This offers a scalable and flexible hardware architecture based on Future Sensors Networks, and a software solution based on Web Services and Web Applications to allow a Sensor Web, which reaches the Future Internet concept of Web of Things. In addition, it has defined several abstraction and processing layers, from a simple pre-analysis until a Complex Event Processing. Finally, this solution allows a real-time access to the information, as such as a high scalability, flexibility and ubiquity, which are features required for the solutions based on Future Internet. The current platform has been deployed in "El Albujon" located at the Southeast Spain, and it is being now started to be used by the geographical department for the definition and evaluation of the geographical models.

Acknowledgment. This work has been sponsored by the Autonomous Region of Murcia (CARM, Spain), with funds for Science and Technology Program (PCTRM 07/10), through the project "*Modelización Hidrológica en Zonas Semiáridas (MHZS)*" (Hydrological Modeling in Semiarid Zones), this has been also supported by the Foundation Seneca, through the Excellence Researching Group Program (04552/GERM/06), and finally by the FPU program (AP2009-3981) from the Education and Science Spanish Ministry.

References

1. Siccardi, F., Boni, G., Ferraris, L., Rudari, R.: A hydrometeorological approach for probabilistic flood forecast. J. Geophys. Res. 110, D05101 (2005)
2. Ståhl, G., et al.: National Inventory of Landscapes in Sweden (NILS)—scope, design, and experiences from establishing a multiscale biodiversity monitoring system. Environmental Monitoring and Assessment (2010), doi:10.1007/s10661-010-1406-7
3. Hannaford, J., Lloyd-Hughes, B., Keef, C., Parry, S., Prudhomme, C.: Examining the large-scale spatial coherence of European drought using regional indicators of precipitation and streamflow deficit. Hydrological Processes (2010) (in press), doi:10.1002/hyp.7725

4. Carpuso, W., Busciolano, R.: Real-Time Hydrologic Monitoring by the U.S. Geological Survey on Long Island and in the Five Boroughs of New York City (2008)
5. Water Quality Assessments - A Guide to Use of Biota, Sediments and Water in Environmental Monitoring - published on behalf of WHO (1996) ISBN 0419215905
6. Freiberger, T.V., Sarvestani, S.S., Atekwana, E.: Hydrological Monitoring with Hybrid Sensor Networks. In: International Conference on Sensor Technologies and Applications, Valencia, pp. 484–489 (2007)
7. Guinard, D., et al.: Interacting with the SOA-Based Internet of Things: Discovery, Query, Selection, and On-Demand Provisioning of Web Services. IEEE Transactions on Services Computing 3(3), 223–235 (2010)

Material Information System According to Change of Building Type

Jung-Soo Han[1] and Gui-Jung Kim[2]

[1] Division of Information & Communication Baekseok University,
Cheonan-city, Chungnam-do, Republic of Korea
jshan@bu.ac.kr
[2] Dept. of Biomedical Engineering Konyang University,
Nonsan-city, Chungnam-do, Republic of Korea
gjkim@konyang.ac.kr

Abstract. This paper develops the virtual building technology that helps make building design effectively done by constructing building materials as components and assembling them as patterns. It also aims to develop the Flexible Building Design System that supports a technology which makes change of constructing easy and reduces the cost effectively through a simulation of building design like design, analysis, change information, etc., by grafting the virtual building technology into the process of building. It especially is possible for a designer and user to change a building easily by using patterns and according to the change, pattern information of materials needed and the plans of the building are made automatically. Kin-search that can be happened through personal proficiency or knowledge virtualization is also the reason why the technology should be embodied.

Keywords: Steel house, Pattern, Material, Building, Assembly.

1 Introduction

This paper develops the virtual building technology that helps make building design effectively done by constructing building materials as components and assembling them as patterns. It also aims to develop the Flexible Building Design System that supports a technology which makes change of constructing easy and reduces the cost effectively through a simulation of building design like design, structure analysis, offering change information, assembly, etc., by grafting the virtual building technology into the process of building. It especially is possible for a user as well as a designer to change a building easily by using patterns and according to the change, pattern information of materials needed and the plans of the changed building are generated automatically [1][2]. This paper also aims to embody Kin-search that can be happened in the industrial field through personal proficiency or knowledge virtualization according to the current situation of worker or the job context in the working place, education place, and other space and time. It expresses matched knowledge information to worker's context (working order, proficiency, experience,

G. Lee, D. Howard, and D. Ślęzak (Eds.): ICHIT 2011, CCIS 206, pp. 320–328, 2011.
© Springer-Verlag Berlin Heidelberg 2011

etc.) as a pattern type and supports working and learning together in real time. In order to do this, a repository based on pattern information should be constructed, coaching and advice should be taken, and multi-dimensional relations should be identified and searched easily.

The framework needs to be developed, which enables to do patterned assembly modeling. The repository helps use pattern information by making it library and knowledge. It also utilizes pattern-based component module and develops a pattern module like assembling composed patterns brought from the pool without primitive detailed modeling work which lines or figures are drawn in detailed in an existing way [3]. A virtual producing method is introduced, which a user enables to design, change, and mix interactively after modeling an actual sized building using VMU (Virtual Mock-up) technology in the virtual space. It is certain that a technology of assembly/generation/change of patterned building based on this technology is an innovative technology in the field of building design [4].

2 Related Works

2.1 Appearing Background of Steel House

Steel house helps build a stronger house with advantages of wood house because it is built of steel in a way of building wood house. There are some countries that build steel houses such as America, Japan, Australia, and some in Europe. These countries use similar methods, but the reason why they developed steel houses is different. For example, there are a lot of termites in America which gnaw wood in traditional wood houses. It causes a shortening of house' life, so that is why steel houses are built. In the case of Japan, there are lots of earthquakes, so steel houses are introduced to build strong houses in earthquakes. Another reason why steel house method is positively introduced to many countries in the world is that steel is 100% recyclable. It means it is an environmentally friendly method that enables to solve the environmental problem caused by closed materials [5][12].

It was 1996 when steel house was introduced in Korea. Since then, POSCO, RIST (Research Institute of Industrial Science & Technology), and KOSA(Korea Iron & Steel Association) have lead a research on steel house, localized materials, and enacted a standard of design. The background of steel house introduced is to deliver a counterproposal as a dry method into the building market where a wet method is the main and to draw a variety of materials. A dry method like steel house has an advantage to shorten the period of construction due to minimizing the field work and to enable to build houses with the same quality because end products from the factory should be assembled in the site. And Korea produces iron and steel as much as ranked as No. 1 or 2 in the world, so it is easy to get galvanized steel sheets used as a material of steel house at a stable price. They have high strength and an equaled quality compared to other materials, so a house could be built into minor and light materials. The strength of steel house is that it is functioned of steel house and environmentally friendly and it is possible to build in any season in a year and to design and innovate freely.

2.2 Steel House Method

In order to construct more effective design system, this paper tried a Flexible Building Design System based on steel house which is easy to assemble and noticed as a future building method. Steel house is a nondearing-wall-structured house as the same as a common wood house, but the difference of steel house from common wood house is to have a basic frame in steel. Steel stud, which the strength is heightened after an around 1mm-thick galvanized steel sheet is processed in the shape of letter C, is used for a structural frame of steel house. Steel is light, so it is easy to be dealt with and to be processed. The volume of wood, forestry resources, is being produced less and less, however, steel stud is produced from factories, so there is an advantage of making plenty of it in a good quality. When steel is used as a structural frame, the amount of used wood will be dropped a lot. Furthermore, steel is recyclable. About 70% of steel is being recycled currently and steel is stronger in financial point because it keeps the same quality after recycling. Steel house is a kind of green home housing that helps protection of forestry resources. Also steel house is free to structure a design and has a lot of advantages as a material of mobile assembly structural design. Steel house has shown that it is superior through several structural tests. Wall panel made of around 1mm-thick galvanized steel sheet is about 1.5 times stronger in natural disasters like typhoon than the existing 2x4 wall panel. The real value of steel house was proved in typhoon happened in Florida, USA and the Kobe earthquake in Japan in 1990s [6, 7, 8].

Steel house has a structure which high strengthened outer panel supports the entire house, so it is free to design inner structure of the house and to change the structure simply because the wall is to be moved easily. It makes it advantageous to remodeling the house. Steel house is the house that naturally deals with the change of family members like the change of housing style and children's growth. Therefore, this paper constructed a Flexible Building Design System by reflecting the steel house method.

3 Embodiment Scheme of Material Change

If the size of the material is input by using material pattern information, the information needed for this is designed to be shown. The information to be digitized is based on figures on utilities drawing. Size, standard, the number of material is expressed by using digitized figures (width/height of wall, etc). For that, used materials which are composed in advance (standard required when designing a building) are as follows: outer wall–stud, track, joist; room wall–stud, track, joist; window–component; gate–component, the material composed in advance (standard required when designing a building); roof–stud, ridge cap, ladder trust, figure data of materials input [11][12]. First, outer wall, that is, the wall is to be formed. Once formation of the wall is completed, rooms should be formed and windows and doors should be formed in addition. After this formation is completed, the roof should be formed, and then a design for a house is done. However, the formation of roof can be designed regardless of the order once the formation is designed. The formation of every function is designed as a pattern, so it will be variable and possible to be modified and changed if a user wants. Java was used as an implementation language for a system. javax.swing.* library, one of Java libraries, was used because of making a window, not a system consol. In addition, AWT library was used to complete Event which lacks in swing. Additional event generation is required to use swing, so action event is basically used a lot and Open library and Open source are used to implement because events used in each swing sentence are different. The algorithm written before using swing was used and shown on system consol and the analysis was done if the process of each function of algorithm went well. After that, the algorithm was changed and modified to process nicely. The algorithm was also implemented considering that there is required a different result output from basic system consol when substituted for swing by using the tested algorithm.

The values of figures input when a system was implemented were stored in arrays and used. addItemListener was registered to express each stored array as JComboBox and the each stored array enabled to be registered in ComboBox. ItemStateChanged was registered to choose the registered items in ComboBox and use. It used ItemEvent to refer. So as to choose items from ComboBox, it calls up each item and use by using ItemEevent .getItem().equals(Item). Each item has different contents. First, it choose an item form chosen ComboBox and it is expressed as JComboBox.getSelectedItem().equals(). JCheckBox was used for the composing function of window and door. Whether choosing JCheckBox, it used JCheckBox.isSelected().equals (Item). To do this, ActionEvent should happen or refer. It also used if-sentences and processed when an event happened. Each function generates dialogues and accesses. Main screen (initial screen) plays a role as a parent and each function dialogue plays a child. Parents' fields are generated and children store at parents' fields by using data. If it wound not happen in this way, data would not be transmitted. So there would be a system error and functions would not work properly. Data transmission would be unclear if the setting of global variables and local variables is uncertain or done well [9][10][13].

For Flexible Building Design of steel house, it in this paper was constructed to support not only a design using each pattern but also what materials change automatically when a user wants to change a type into another type among A, B, C, D made in advance. The standard of building type is $99\,\mathrm{m^2}$ and it stores each data at DB. It enables to change the type and also provide data such as more needed materials, leftover materials, and recyclable materials by using a recyclable function which is one of the advantages of steel house when the type is changed.

4 Steel House Data Automatic Transmission

Generally in the case of construction, there are hundreds of parts and it is impossible for a designer to choose parts one by one when he/she assembles them on the screen. Therefore, to operate this effectively, there should be functions as follows.

◦ Locational Composition

Design location of the product, for instance, in the case of house, it should compose all parts as patterns and assemble them automatically once it inputs the parts that go into the location after all the locations like roof, wall, pillar, and floor are fixed. Also according to requests, there should be a function that a designer enables to designate a 3D area and assemble all the parts belonged to the area.

◦ Functional Composition

Function information of each pattern, for example, once there input function for the roof, window function on the wall, and so on, it should have a function that assembles the parts of the function.

◦ Assembly Information of Pattern

It expresses assembly composition information of pattern as a graphic on the screen and chooses it, so it assembles by selecting assembly unit needed.

◦ Assembly by combining location, function and assembly information

It should compose a building by combining location information, function information, and assembly information. For example, in the case of wall, there should be a function that assembles the assembly unit of the other parts except from the glass of the window.

◦ Assembly Function of pattern

To combine patterns, there should be a function that it composes them as a type of template in order to connect patterns via interface and enables to assemble by letting them be connectable patters to make them right pattern for input/output.

◦ Automatic Change Function of Type

It shows types of building like A, B, C, D and displays the data of changed material in the case of changing A into B. Data provided will be displayed as currently recyclable material or additionally needed material or leftover material. A user needs to aware that which should be needed when changing the building.

◦ Wall Composition Function

In order to compose the wall material should be chosen and there are input values of height, total width, and interval. The displayed values are input values which are material, total width, height, and interval. The number of materials are calculated and displayed. Like Fig.1, it displays the number of materials after automatically calculating the values input on the complete screen when inputting the wall composition information.

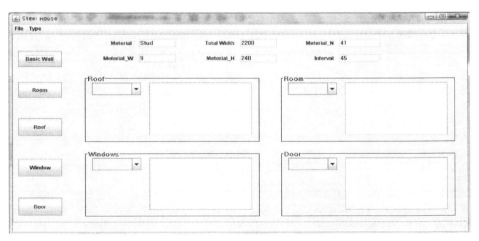

Fig. 1. Display of the Material Number

° Room Composition Function

The material composed as a wall is used basically to compose a room. To compose a room, the name for the purpose should be given and the size should be given by inputting width and length. The designated room displays the room information of the initial screen. Like Fig. 2, when inputting composition elements of the room, the size of width and length will be displayed.

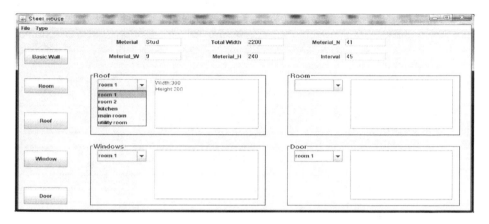

Fig. 2. Room Information

° Window Composition

Window composition will be done after room composition is done. That is, once the room composition has done, you can compose windows. The size of the window here is the size that was defined at the beginning. When composing a window in a size of 800*1300mm, you choose a room to put the window and input the number of windows. You may choose if there is a window when the wall of the room is made.

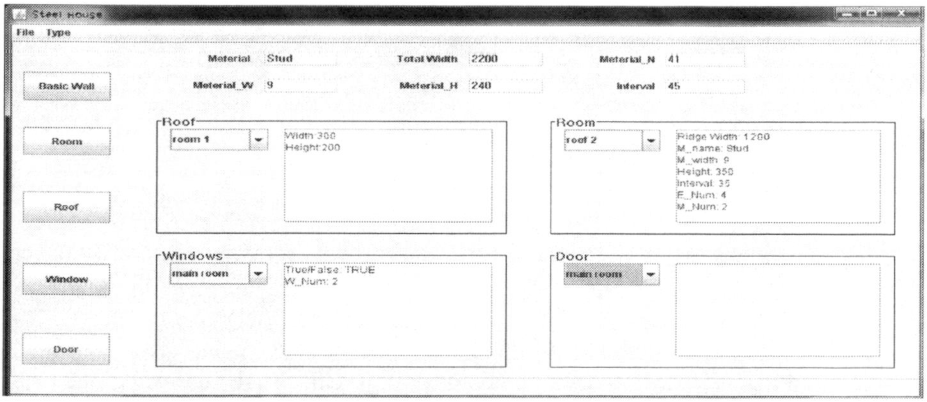

Fig. 3. Window Information

The room which the composition is completed will display the window information on the initial screen. Fig. 3 shows that the information related to window is displayed as a View after completing the window composition.

◦ Door Composition

Door composition will be done after room composition is done as well. That is, once the room composition has done, you may compose door composition. The size of the door here is the size that was defined at the beginning. When composing a door in a size of 1300*2100mm, you choose a room to put the door and input the number of doors. The room which the composition is completed will display the door information on the initial screen.

◦ Roof Composition

You input the name of the roof at the wanted place, choose the material you want, and input the height, the width of ridge cap, and the number of fascia cap. For the roof, there should be input the height, the width of ridge cap, the number of fascia cap, and intervals of materials. If you choose the roof on the screen after composing the roof, each composing element of the roof will be displayed like in Fig. 4. The number of materials for the roof may be calculated in the same way to get the number of materials for the wall.

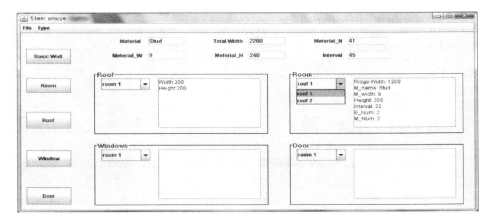

Fig. 4. Roof Information

◦ Choosing a Type

A type shows the entire information of each composition because wall, room, window, door, roof are input in advance. Hence, when the chosen type is changed to another one, it also provides information of the state of material change and recyclable materials, additionally needed materials, leftover materials according to the change. Fig. 5 displays the information of type B, tells the state of material change in addition when type A was changed into type B, and shows the information of type B.

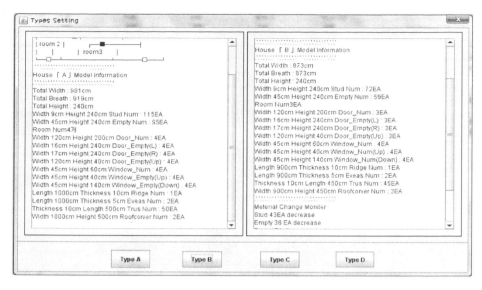

Fig. 5. Change Type A into Type B

5 Conclusions

This paper aims to develop a virtual building technology that enables to compose building materials as components and assemble the components as patterns in order to design a building effectively. It also aims to develop a Flexible Building Design System that supports a technology which changes a building easily and cuts the cost effectively through the simulation of building design such as design, supply with changed information, and assembly by using the technology above.

This paper is not for a lot of buildings. It chose a sample building and abstracted components from it, and then it composed metadata for each component and metadata for patterns from components. Each component has a semantic relation for pattern composition and it shall be connected with the relation. Once the part of building using pattern is designed, a new building design will be done while the components related to will be recombining according to pattern. In the case of general building, materials are composed of hundreds of parts, so it is impossible for a designer to choose part's name needed one by one and work when assembling on the screen. Therefore, to do this effectively, it tries to construct a patterned flexible building design system. Here the concept of steel house was introduced and automatic change system for material was constructed. Hence, if the size of the material is input, there will be figured information needed by using material pattern information in this paper. Information needed to digitize is based on the figures on the plans.

As a further research, building design shall be cone in a visual way through 3D modeling.

Acknowledgment. "This research was supported by Basic Science Research Program through the National Research Foundation of Korea (NRF) funded by the Ministry of Education, Science and Technology (No. : 2011-0003950)".

References

1. Kozaczynski, W., Booch, G.: Component Based Software Engineering. IEEE Software 5(9) (September 1998)
2. Gamme, E., Helm, R., Johnson, R., Vlissides, J.: Design Pattern: Elements of Reusable Object -Oriented Software. Addison-Wesley, Reading (1995)
3. Tonella, P., Antoniol, G.: Object Oriented Design Pattern Inference. In: Proceedings of the IEEE International Conference on Software Maintenance, pp. 230–238 (1999)
4. http://sketchup.google.com
5. Korea Steel House Technical Association, Steel House, Sikong Publishment (2007)
6. http://www.hyunsunghousing.kr/
7. Jung, H.-s.: North America Steel House Supply and Technical Trend. Korea Steel Structure Association, 48–49 (2001)
8. Editorial Department, Steel House using STEEL STUD. House Publishing Company (2001)
9. http://www.ewoodland.com/bom/zboard.php?id=jj22
10. Won, W.-y.: A Study on Module Design of Steel House Structure. KonKuk Univ. in Korea, Master's Thesis, 66–69 (February 2010)
11. Jung, H.S., Choi, K.: An Introduction to the Construction of Low-Rise Steel-Framed Housing in Korea, vol. 12(1). RIST Pohang Univ, Korea (1998)
12. Jung, H.S., Lee, P.G.: Construction Optimization through use of Prefabricated Light-gauge Steel Panels, vol. 13(4). RIST Pohang Univ., Korea (1999)
13. http://download.oracle.com/javase/1.5.0/docs/api

A Design of Execution Tool for an IPTV Service Deployment

Jung-Ho Kim, Jae-Hyoung Cho, and Jae-Oh Lee

Information Telecommunication Lab,
Dept. of Electrical and Electronics Engineering,
Korea University of Technology and Education,
Cheon Ahn, Korea
{jungho32,tlsdl2,jolee}@kut.ac.kr

Abstract. There are diverse terminal equipments like computer smart phone and tablet PC these days. According to the situation, service providers are developing and serving each Internet Protocol Television (IPTV) services depending on the user's terminal equipments. However, these developments make inconsistent services and duplicate services causing waste of resources. Therefore, we suggest an IPTV execution tool independent on service deployment at IP Multimedia Subsystem (IMS). The IPTV execution tool is composed with the part of user's terminal design and part of Application Server (AS)'s service design. And, this tool reduces the dependence of the terminal platform and allows easy to develop IPTV services.

Keywords: IPTV, IMS, IPTV Service.

1 Introduction

Nowadays, mobile network evolved from Circuit Switched (CS) network to Packet Switched (PS) network based on Internet Protocol (IP). Therefore, users can be provided high speed and high quality services. Furthermore, these networks make that users want various services like multimedia service based on IP, Voice of Internet Protocol (VOIP) or mobile IPTV. So, service providers need network architecture to feed user's needs. IMS has been proposed to satisfy these needs. The IMS can process IP based multimedia services and CS network based services [1]. The IMS provides a variety of IP-based services and applies the existing services without major change. Also, it does not depend on the type of network both wired and wireless. In contrast, the current IPTV terminal equipments are various according to the communication agencies and platforms. It is hard for service provider to develop IPTV service acting all equipments. By using the IPTV execution tool proposed in this paper, therefore, service developers can easily develop the services using interface unified by IPTV execution tool between application server and user's terminal equipment. IPTV manager placed at user's terminal equipment uses unified Application Programming Interface (API) to check the terminal equipment, and application manager placed at AS uses a unified API for serving function and procedure specifications. These specifications are only required for deploying IPTV service. Service developers don't need to consider the terminal equipment because application manager makes the

G. Lee, D. Howard, and D. Ślęzak (Eds.): ICHIT 2011, CCIS 206, pp. 329–336, 2011.
© Springer-Verlag Berlin Heidelberg 2011

function and procedure specification referring to unified device API. For this reason, service development time and development cost would be reduced.

In chapter 2, we research basic theory and structures about the IMS, IPTV and AS on the IMS. Then, we suggest an IPTV execution tool independent on service deployment and simple scenario in chapter 3. Chapter 4 shows conclusion and discuss future works.

2   Related Works

2.1   The Structure of the IMS

The IMS is developed to realize real-time or non-real-time multimedia service in the mobile environment. The service object of the IMS provides synthetic multimedia services using IP protocol such as voice, video and data services. Additionally, the IMS has advantages that are easy in service developments and modifications. The IMS can also improve a completive price as well as is easy to interlock with various 3rd party applications that are based on efficient session management. It enables mobile operator to support the structure based on Session Initiation Protocol (SIP) such as session management, security, mobility, Quality of Service (QoS) and charging [2,3].

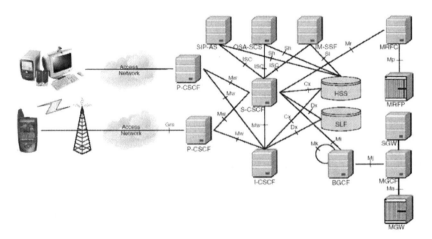

Fig. 1. IMS Structure and Interface

Figure 1 shows the IMS structure. In the IMS network, Call Session Control Functions (CSCFs) process signal. The CSCFs based on SIP implement basic functions to control multimedia session with infra-system, and are divided into P(Proxy)-CSCF, I(Interrogating)-CSCF, S(Serving)-CSCF by roles. The CSCFs do the function of subscriber registration, authentication, charging, service triggering, routing of relevant application server, collating the receiving of user location, and compressing and decompressing of the SIP message. And profile information, authentication and related location information data of a subscriber are stored in

Home Subscriber Server (HSS). AS has real service logic to provide services. For multilateral service, the Multimedia Resource Function (MRF) is used to control the signal and process media mixing. Additionally, the IMS can interact with Public Switched Telephone Network (PSTN) through one or more PSTN gateways. Media Gateway Control Function (MGCF) is responsible for inter-working with PSTN. The Breakout Gateway Control Function (BGCF) is the IMS element that selects the network in which PSTN breakout is to occur. If the breakout is to occur in the same network as the BGCF then the BGCF selects a MGCF. The MGCF then receives the SIP signaling from the BGCF [1, 4].

2.2 IPTV

IPTV provides digital television services over Internet Protocol for residential and business users at a lower cost. These IPTV services include commercial grade multicasting TV, Video on Demand (VoD), triple play, Voice over IP (VoIP), web/email access, and well beyond traditional cable television services. IPTV is a convergence of communication, computing, and content, as well as an integration of broadcasting and telecommunication. IPTV has a different infrastructure from TV services, which use a push metaphor in which all the content is pushed to the users. IP infrastructure is based on personal choices, combining push and pull, depending on people's need and interest [5, 6].

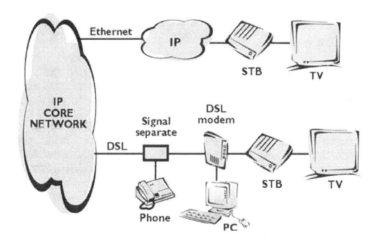

Fig. 2. IPTV Home Network Picture

2.3 Application Server

The IMS application provides a specific service to the end user. The IMS end-user services include multiparty gaming, videoconferencing, messaging, community services, presence, and content sharing. The IMS architecture lets the service provider deploy multiple application servers in the same domain. Different application servers can be deployed for different application types (for example, telephony or presence application servers) or different groups of users. The S-CSCF decides whether it should forward an incoming initial SIP request to a given application server. The decision it makes is based on filter information received from the HSS. The HSS stores and conveys this filter information on a per-application-server basis for each user. When the HSS transfers the name and address of more than one application server, the S-CSCF must contact each application server in the order provided. The S-CSCF uses the first application server's response as input to the second application server. The application server uses filter rules to decide which of the many services deployed on the server should handle the session. During the service logic's execution, the application server can communicate with the HSS to get additional information about a subscriber or to learn about changes in the subscriber's profile [7].

3   An IPTV Execution Tool and Simple Scenario

In this paper, an IPTV execution tool is separated client part and server part. First, the client part is IPTV manager. IPTV manager defines specifications of IPTV Set-Top Box (STB) and user's terminal equipments. Current IPTV equipments use their own API according to the type of equipment and platform. Also, current IPTV services require the different interfaces according to the equipments. When developing the services, service developers should develop differently depending on the equipments. These circumstances not only increase development time and cost, but also drop the scalability of services. The proposed IPTV manager proposed can solve these problems by offering unified device API. Unified device API make easy to develop IPTV service, because developer don't need to care IPTV terminal equipments only

Table 1. Example of unified device API

Device Functions	Unified API	Existing APIs
WebCAM Data Receive	CamRead(param)	aCamGetData(param) bCamRecv(param)
WebCAM Data Transmit	CamWrite(param)	aCamSendData(param) bCamTrans(param)
GPS Data Receive	GPSRead(param)	aGPSGetData(param) bGPSRecv(param)
Bluetooth Settings	BTSet(param)	aBTConfig(param) bBlueSet(param)
Bluetooth Data Receive	BTRead(param)	aBTGetData(param) bBlueRecv(param)
Bluetooth Data Transmit	BTWrite(param)	aBTSendData(param) bBlueTrans(param)

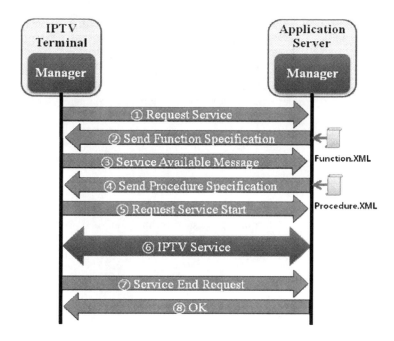

Fig. 3. IPTV Service Flow using IPTV Execution Tool

care IPTV manager's unified device API. Therefore, service development time is decreased and service provider can reduce service development cost.

Table 1 shows an example of unified device APIs. Existing API is mapped to unified API according to device functions. The column of Existing APIs field has two ones in the below Table. When IPTV manager judges whether the user can be served or not, this API is used.

The second part is application server manager. Application server manager generates two specifications using unified device API. One is function specification, the other is procedure specification. The function and procedure specifications are enough to deploy the service. Application server manager interacts with IPTV manager. When user requests IPTV service, application server manager checks the function for deploying the service and sets service procedure using function specification and procedure specification. Service developers only make the service scenario and feed scenario to application server manager. Then application server manager makes specifications using Extensible Mark-up Language (XML). These actions make that service developers don't need to develop whole application program. So, service development time and cost are reduced.

Figure 3 shows a simple sequence scenario between IPTV terminal manager and application server manager, when IPTV user requests an IPTV service. Detailed steps are explained below.

1. IPTV user sends the service request message to application server manager.
2. Application server manager sends function specification to IPTV terminal manager. A file, Function.XML, is a function specification including essential device list. Then, the IPTV terminal manager checks required functions and device libraries.
3. If user can execute the service, IPTV terminal manager sends response message.
4. After receiving the service available message, application server manager sends a file of Procedure.XML to IPTV terminal manager.
5. After setting the user's equipment to start the service according to the file of Procedure.XML, IPTV terminal manager sends request service start for starting the service.
6. The IPTV service is activated between user and application server.
7. When user wants to terminate the service, the user sends service end request message.
8. Application server responses OK message, then, service is terminated.

The file of Function.XML transmitted at step 2 is shown in the following.

```
<?xml version="1.0" encoding="euc-kr" standalone="yes" ?>
<managerinfo xmlns="urn:xml:servermanager1" version="1">
  <Devices id="1">
    <Name>Camera</Name>
    <RequiredLib>CamRead.DLL</RequiredLib>
    <APIReplace>
      <LibName>CamRead.DLL</LibName>
      <StandardAPI>CamRead(Parameter)</StandardAPI>
      <StdParamType>Media</StdParamType>
      <PrevAPI>CamDataGet(Parameter)</PrevAPI>
      <PrevParamType>Binary</PrevParamType>
    </APIReplace>
  </Devices>

  <Devices id="2">
    <Name>Bluetooth</Name>
    <RequiredLib>BtRead.DLL, BtWrite.DLL, BtSet.DLL</RequiredLib>
    <APIReplace>
      <LibName>BtRead.DLL</LibName>
      <StandardAPI>BtRead(Parameter)</StandardAPI>
      <StdParamType>Media</StdParamType>
      <PrevAPI>GetBtData(Parameter)</PrevAPI>
      <PrevParamType>Binary</PrevParamType>
    </APIReplace>

    <APIReplace>
      <LibName>BtWrite.DLL</LibName>
      <StandardAPI>BtWrite(Parameter)</StandardAPI>
      <StdParamType>Media</StdParamType>
      <PrevAPI>SendBtData(Parameter)</PrevAPI>
      <PrevParamType>Binary</PrevParamType>
    </APIReplace>

    <APIReplace>
      <LibName>BtSet.DLL</LibName>
      <StandardAPI>BtSet(Parameter)</StandardAPI>
      <StdParamType>Integer</StdParamType>
      <PrevAPI>SetBtData(Parameter)</PrevAPI>
      <PrevParamType>Integer</PrevParamType>
    </APIReplace>
  </Devices>
---------------skip---------------
```

Fig. 4. Example of Function.XML

```
<?xml version="1.0" encoding="euc-kr" standalone="yes" ?>
<managerinfo xmlns="urn:xml:clientmanager1" version="1">
  <Thread id="1" endcondition="ServiceEnd">
    <Procedure id="1">
      <RunType>FuncCall</RunType>
      <CallAPI>CamRead(Parameter)</CallAPI>
      <paramName>A_Buf</paramName>
      <paramType>Media</paramType>
    </Procedure>

    <Procedure id="2">
      <RunType>FuncCall</RunType>
      <CallAPI>Player1(Parameter)</CallAPI>
      <paramName>A_Buf</paramName>
      <paramType>Media</paramType>
    </Procedure>

    <Procedure id="3">
      <RunType>FuncCall</RunType>
      <CallAPI>SocketWrite(Parameter, Parameter)</CallAPI>
      <paramName>Socket_1, A_Buf</paramName>
      <paramType>Socket, Media</paramType>
    </Procedure>

  </Thread>
---------------skip---------------
```

Fig. 5. Example of Procedure.XML

Function.XML makes the terminal equipment use unified API and device library. It can have multiple *Devices id* elements. The *Devices id* has sub elements such as *Name*, *RequiredLib*, and *APIReplace*. The *Name* element identifies the type of device. And, the *RequiredLib* element describes the necessary device library. The *APIReplace* element is used to change existing device API to unified device API. The *APIReplace* also has multiple sub elements such as *LibName*, *StandardAPI*, *StdParamType*, *PrevAPI*, and *PrevParamType*. If one device refers to multiple existing libraries, the *APIReplace* element is appeared several times. The *LibName* element is existing library name and the *StandardAPI* is the unified API used by manager. The *StdParamType* element sets the type of Standard API's parameter. The *PrevAPI* element shows the existing API that is mapped to unified API. The *PrevParamType* element means the type of *PrevAPI*'s parameter. By rewriting the required functions like figure 4, it is easy to change existing libraries to unified device API. Moreover, it is easy for IPTV terminal manager to check the available services.

An example file of Procedure.XML transmitted at step 4 of figure 3 is shown in the figure 5. The *managerinfo* has the information of xml version and the generation number. The Thread element describes the actual behaviors of IPTV terminal manager. The *Thread* element is separated as *Thread id*. The *Procedure* element, the sub element of *Thread*, represents a sequence of services. The *RunType* element defines what the IPTV terminal manager does. The *CallAPI* element shows which function is called. The *paramName* element and the *paramType* element have the name of parameter name and the type of parameter that are used in the function of *CallAPI*. The *Procedure id="1"* calls *CamRead(Parameter)* function. The name of *Parameter* used in *CamRead* is *A_Buf*. The type of *Parameter* used in *CamRead* is *Media*. The *Procedure id="2"* is a procedure for playing the media data transmitted by *CamRead(Parameter)* in *Procedure id="1"*. The *RunType* is *FuncCall* and *CallAPI* is *Player 1(Parameter)* that is a declared function in advance for playing media at IPTV terminal. By using *A_Buf* which is read to the *Procedure id="1"* as the parameter of *Procedure id="2"*, the media data can be played on *Player 1*.

Thus, the "Function.XML" redefines the unified API from existing API and the "Procedure.XML" uses the unified API for performing services. In this way, an IPTV execution tool has some advantages. At the IPTV terminal side, IPTV terminal do not need to develop all application programs and can get a variety of services. At server side, it is easy to make a service development by taking the approach which divides its related service function and procedure.

4   Conclusions

In this paper, we suggest the IPTV execution tool based on application server. By providing unified format when IPTV service is deployed, users can be served various services regardless of the terminal equipment and communication agency and service developers can develop the service easier.

In current, we focus on the design of IPTV execution tool. So we will enhance the functionality of the IPTV execution tool in the future.

Acknowledgments. This paper was (partially) supported by the Education and Research Promotion Program of KUT.

References

1. 3GPP TS 23.228, IP Multimedia Subsystem (IMS) V10.1.0 (June 2010)
2. Camarillo, G., Garcia-Martin, M.A.: The 3G IP Multimedia Subsystem (IMS) merging the internet and the cellular worlds. Wiley, Chichester (2006)
3. Mikoczy, E., Sivchenko, D., Moreno, J.I.: IPTV Services over IMS: Architecture and Standardization. IEEE, Los Alamitos (2008)
4. Cho, J.-H., Lee, J.-O.: The IMS/SDP structure and implementation of presence service. In: Ma, Y., Choi, D., Ata, S. (eds.) APNOMS 2008. LNCS, vol. 5297, pp. 560–564. Springer, Heidelberg (2008)
5. Xiao, Y.: Internet Protocol Television (IPTV): The Killer Application for the Next-Generation Internet. IEEE Communications Magazine (November 2007)
6. Jain, R.: I Want My IPTV. IEEE Multimedia 12(3), 95–96 (2005)
7. Khlifi, H.: IMS Application Servers Roles, Requirements, and Implementation Technologies. IEEE Computer Society, Los Alamitos (2008)

Middleware Integration of DDS and ESB for Interconnection between Real-Time Embedded and Enterprise Systems

Yunjung Park[1], Duckwon Chung[1], Dugki Min[1], and Eunmi Choi[2]

[1] School of Computer Science and Engineering, Konkuk University, Seoul, Korea
{sm6280p,dwchung,dkmin}@konkuk.ac.kr
[2] School of Business IT, Kookmin University, Seoul, Korea
emchoi@kookmin.ac.kr

Abstract. As technology evolves, the border of embedded system and enterprise system is gradually meaningless. ESB is a well-known middleware for integration of enterprise services and systems, and DDS is a promising real-time communication middleware for embedded systems. This paper introduces structurally integrated architecture between DDS middleware and ESB middleware for interconnection between real-time embedded systems and enterprise information systems. An integrated architecture of DDS and ESB allows integration of embedded system environment and enterprise system environment. It supports integration of diverse embedded devices from ubiquitous systems with real-time communication and allows effective data processing using enterprise services, which are generated from embedded systems. We design DDS binding component for ESB to allow DDS communication to enterprise services. This DDS-ESB integration architecture could be applied in the field that generates huge of data for monitoring and processes data for providing user services in real-time, such as u-City or home network.

Keywords: DDS, ESB, Middleware Integration, Embedded and enterprise hybrid system.

1 Introduction

As evolving of IT infrastructure, diverse of embedded devices are connected by network for providing ubiquitous services. As these devices generate huge amount of data, real-time communication and rapid processing of data are required. In addition, communications among heterogeneous devices are needed for full integration of systems. Usually real-time communication was one of embedded system requirements and system integration and information processing were enterprise system requirements area. However, as the power of devices has evolved and as the volume of information and its complexity has increased, the border of embedded system and enterprise system is gradually meaningless.

G. Lee, D. Howard, and D. Ślęzak (Eds.): ICHIT 2011, CCIS 206, pp. 337–344, 2011.
© Springer-Verlag Berlin Heidelberg 2011

DDS (Data Distribution Service)[1] is a data-centric communication middleware for distributed real-time applications. DDS supports publish/subscribe communication model, which has strong point of time critical message delivery and many-to-many communications. Currently, DDS is adopted in real-time embedded systems such as defense, smart grid, complex telemetry (e.g. NASA Rocket Launch System), public transformation, and financial services[2].

ESB[3] is a SOA-based middleware that combines distributed services and organizes diverse applications based on web services, content-based routing and message transformation technologies. Core function of ESB is integration of services, applications, resources and functions in enterprise environment.

For interconnection between embedded systems and enterprise systems, we suggest ESB-DDS middleware integration architecture. DDS allows real-time and stable communication and ESB integrates DDS-based embedded systems and other legacy enterprise systems. Embedded devices such as sensors or robots deliver their data using DDS middleware in real-time and DDS delivers it to ESB (if necessary), and ESB routes data to its suitable processor enterprise service. This kind of architecture ensures both interoperability in high-level (enterprise system level) and real-time communication in low-level (embedded system level). This DDS-ESB middleware integration architecture could be applied in the field that generates huge of data for monitoring and processes data for providing user services in real-time, such as u-City or home network.

In this paper, we introduce architecture of ESB-DDS middleware integration, especially a design of DDS binding component for ESB and sample configuration for DDS-UDP message transformation. In addition, we implemented and evaluated a test-bed system to proof our architecture design concept in the end of this paper.

2 Related Works

DDS is a promising real-time communication middleware, and there are several approaches for hybrid DDS and other existing IT technologies to allow real-time communication in their domains. Among them, we introduce two outstanding works.

2.1 Integrating CEP with DDS

CEP(Complex Event Processing) consists of complex type of events from everywhere and CEP engine analyzes and identifies the meaningful events within the collected complex events and takes proper actions in real-time[4]. Current CEP engines are distributed and needs special management such scalability, load balancing and needs special mechanisms for integration and interoperability between systems.

DDS is logically data centric. However, physically DDS is a fully distributed system. DDS does not need broker and it guarantees high performance, scalability and configurable QoS. DDS-CEP integrated system allows these DDS features into distributed CEP engines. Figure 1 shows current CEP system model and DDS-CEP integrated system model.

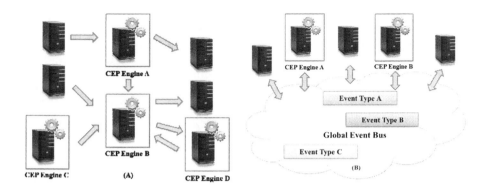

Fig. 1. (A) Current CEP system model. Each CEP engine and system is connected directly. (B) DDS-CEP integrated system model. Each CEP engine and system is connected indirectly. They communicate through DDS bus (Global Event Bus)[5].

One of DDS vendors already contains CEP as a part of their solution[5], and there is another research that integrates open source CEP solution and open source DDS solution[6].

2.2 Integrating HLA with DDS

HLA (High-Level Architecture) is a distributed simulation integration middleware. DDS and HLA have highly similar architecture. Both communicate based on publish/subscribe-model, support of diverse dissemination semantics and support application portability via a standardized API specification. There remarkable difference is that DDS is based on real-world data and HLA is based on virtual-world (simulation) data. Figure 2 shows equivalents between HLA and DDS.

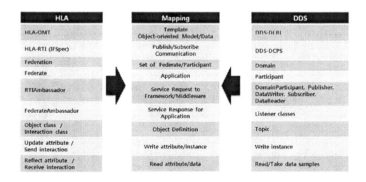

Fig. 2. Summary of HLA-DDS equivalents[2-4]

There is already structural mapping research between DDS and HLA in 2006[7], and its implementation research is processing by this paper authors in the point of the CPS (Cyber-physical System)[8] research view.

3 Architecture of ESB-DDS Middleware Integration

Based on JBI (Java Business Integration) 1.0 standard[9], ESB architecture is separated in service logics and communication processing. ESB architecture consists of 3 core parts. Service Engine (SE) is for service logic processing such as rule engines, BPEL engines, XSLT engines and scripting engines. Binding Component (BC) communicates using remote protocols and normalize/denormalize messages what it receives. Normalized Message Router (NMR) delivers messages between SE and BC.

DDS is a communication middleware and it is based on RTPS (Real-time Publish /Subscribe) protocol[10]. If DDS/RTPS data can be changed to other protocol data type, then DDS applications and other applications can be easily integrated. Protocol transformation is one of main function of ESB. Usually, DDS applications are for embedded systems and ESB integrates enterprise systems. It means that Binding Component (BC) for DDS/RTPS will interconnect DDS middleware and ESB, and other systems.

Figure 3 shows overall structural architecture of ESB-DDS middleware integration. DDS/RTPS Binding Component (BC) connects embedded devices with ESB services and enterprise applications. Data processing engine which processes data from embedded device is a Service Engine (SE) component and it delivers processed data to target enterprise applications such as information service system. ESB Framework (NMR and management modules) connects Service Engines and Binding Components.

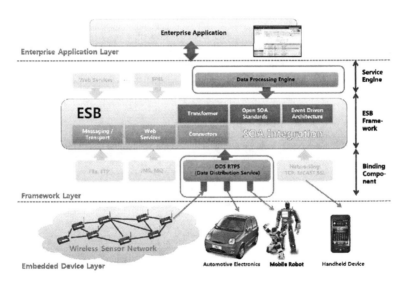

Fig. 3. Overall Architecture of ESB-DDS Middleware Integration

This architecture has benefits for existing enterprise services to get DDS abilities such as real-time messaging performance, scalability and QoS support. In the same way, DDS-based embedded systems could get some benefits with this architecture, such as easy integration with other protocol based enterprise services and applications.

4 Design of DDS Binding Component for ESB

This chapter contains detail design of DDS binding component for ESB, based on JBI standard. We optimized this component, especially for Apache ServiceMix v3.3.2[11] and OpenSplice DDS Community Edition v5.3[12].

4.1 DDS Binding Component

DDS BC acts a connector or adaptor between DDS middleware and ESB. Without DDS BC, DDS-based embedded systems cannot access to ESB. For interconnecting between DDS and ESB, DDS BC is only one that is needed for additionally.
  DDS BC has mostly same functions of other ESB BCs. It waits for message receiving from outside of system and sends it to suitable SEs. Otherwise, it sends message to outside of system when SE requests message sending. Binding component can control only message flows. DDS BC contains control logics for DDS communications. However, we designed and added additional message routing mechanism in DDS BC. It is introduced in next section.

4.2 Design of Message Routing

Most ESB BC supports two kind of message routing mechanisms: In-Only (one-direction communication) and In-Out (two-way communication). Data collection for processing will be delivered through normal route using NMR. Interconnection with other protocol applications is also available In-Only and In-Out message routing. However, NMR communication doesn't consider real-time requirement. To achieve two goals-interoperability and real-time communication, we add additional message routing mechanism, which is called 'delegation'. Every message in ESB should pass NMR. However, delegation message routing sends its messages to target device directly without using NMR for real-time processing. After sending its data to target device, then BC send its data to NMR for data processing. Delegation message routing is only available when both source and target devices are using DDS protocol. If protocol transformation is needed for interoperability, then delegation message routing will not be available. For protocol transformation is one of services, NMR message passing is mandatory.

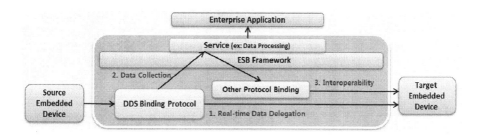

Fig. 4. Example Message Routing of DDS Binding Component

Delegation message routing is used in the case that DDS applications need to be real-time communication or it cannot communicate directly, for instance network coverage problems. Most of DDS middleware supports only LAN area, so delegation message routing will be useful when DDS data should across WAN. Figure 4 shows example message routing of DDS Binding Component.

4.3 Sample Configuration for DDS-UDP Message Transformation

There is sample configuration for message transformation between DDS-UDP applications. This example consists of Bean SE and EIP SE (Pipeline Pattern) for message transformation and routing, and DDS BC and UDP BC to connect DDS and UDP applications to ESB. Figure 5 shows design structure of this DDS-UDP message transformation in ServiceMix

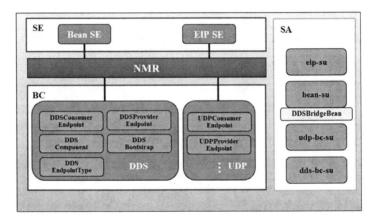

Fig. 5. Structure of Sample DDS-UDP Message Transformation in ServiceMix

For message transformation, core 'xbean.xml' configuration of each Service Unit (SU) is below:

(1) Service Unit of DDS BC (dds-bu-su)

```
<dds:consumer service="dms:dds-inonly-bean-service5"
          endpoint="dms:dds-inonly-bean-endpoint5"
          targetService="dms:dds-eip-pipeline1"
          mep="in-only" read_partition="MsgWrite"
          topicName="PP_bridgetoudp_topic"
          topicRegisterType="messaging::PP_bridgetoudp_msg"
/>
```

(2) Service Unit of UDP BC (udp-bu-su)

```
<udp:provider service="dms:udp-outonly-bean-service1"
          endpoint="dms:udp-outonly-bean-endpoint1"
        mep="in-only" targetPort="9000" targetIP="127.0.0.1"
   />
```

(3) Service Unit of EIP SE (eip-su)

```
<eip:pipeline service="dms:dds-eip-pipeline1"
    endpoint="dms:dds-eip-endpoint1">
  <eip:transformer>
    <eip:exchange-target
      service="dms:dds-bridge-bean-service" />
  </eip:transformer>
  <eip:target>
    <eip:exchange-target
      service="dms:udp-outonly-bean-service1" />
  </eip:target>
</eip:pipeline>
```

5 Implementation and Evaluation of Test-Bed System

In this section, we introduce our test-bed system to evaluate proposed architecture design. We made simple 'naval warfare system' using two heterogeneous embedded devices and two sensor devices. Each device acts as our army ship and a sensor acts as army ship's radar sensor for detecting enemy army ships. In this scenario, we have two our army ships. When enemy army ship is detected, then our army ship fires missiles to enemy army ship. During this scenario, every data is delivered using DDS, and their information is processed by information center using ESB. In other words, important and urgent command is delivered using DDS directly, and information such as results of missile fire and status of our army ship is processed by Information Center.

Figure 6 shows our physical experiment environment and software UI for our army ship missile control. Detected sensor data is delivered to an embedded device which acts as our army ship, and our army ship data is delivered to PC which acts as Information Center. We setup this test-bed system using OpenSpliceDDS Community Edition v5.3 as a DDS communication middleware and Apache ServiceMix v3.3.2 as an ESB middleware.

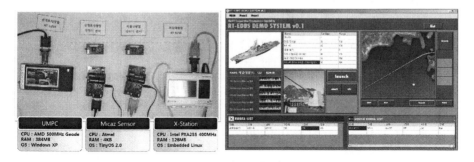

Fig. 6. Experiment Environment of Test-bed Physical System and Software UI

6 Concluding Remarks

For interconnecting real-time embedded systems and enterprise systems, we suggested DDS-ESB middleware integration architecture. We designed DDS (or

DDS/RTPS) binding component of ESB to allow communication between DDS and ESB. This BC supports 3 kinds of message routing mechanisms: In-Only, In-Out and Delegation. We made a test-bed system using 'naval warfare system' scenario to proof our architecture design concept. The DDS-ESB middleware integration architecture could be applied in the field that generates huge of data for monitoring and processes data for providing user services in real-time, such as u-City or home network.

In this paper, we focused on real-time message delivery between DDS and ESB. Therefore, it explains message transformation logics and routing mechanisms. However, we didn't concern structure of DDS dynamic topic type in this paper. It made lose half of DDS's strong point. Therefore, in the future research will include dynamic topic configuration and its performance evaluation. In addition, we have a plan to integrate our other research on DDS-HLA integration results with this DDS-ESB middleware integration research results.

Acknowledgments. Corresponding author of this paper is Dugki Min (e-mail: dkmin@konkuk.ac.kr). This paper was supported by OSS Community Supporting Program from National IT Industry Promotion Agency (NIPA) of the Ministry of Knowledge Economy of Korea and supported by the Konkuk University in 2010.

References

1. Schmidt, D.C., van't Hag, H.: Addressing the Challenges of Mission-Critical Information Management in Next-Generation Net-Centric Pub/Sub Systems with OpenSplice DDS. In: IEEE International Symposium on Parallel and Distributed Processing, IPDPS 2008 (2008)
2. Corsaro, A.: The Open Source Messaging Accelerating Wall Street. PrismTech. (2009)
3. Ziyaeva, G., Choi, E., Min, D.: Content-Based Intelligent Routing and Message Processing in Enterprise Service Bus. In: International Conference on Convergence and Hybrid Information Technology, ICHIT 2008 (2008)
4. Complex event processing (CEP), http://en.wikipedia.org/wiki/Complex_event_processing
5. Oberoi, S.: Complex Event Processing with RTI Data Distribution Service. In: RTI (2007)
6. Corsaro, A.: Stream Processing with DDS and CEP. PrismTech. (2011)
7. Joshi, R., Castellote, G.-P.: A Comparison and Mapping of Data Distribution Service and High-Level Architecture. In: RTI (2006)
8. Cyber-physical system (CPS), http://en.wikipedia.org/wiki/Cyber-Physical_Systems
9. JSR-000208 Java Business Integration 1.0 , http://jcp.org/aboutJava/communityprocess/final/jsr208/index.html
10. Pardo-Castellote, G.: Analysis of the Advanced Message Queuing Protocol AMQP) and comparison with the Real-Time Publish Subscribe Protocol (DDS-RTPS Interoperability Protocol). In: RTI (2007)
11. OpenSplice DDS, http://www.prismtech.com/opensplice
12. Apache ServiceMix, http://servicemix.apache.org/

Application of SOA in Safety-Critical Embedded Systems

Douglas Rodrigues[1], Rayner de Melo Pires[1], Júlio Cézar Estrella[1],
Marco Vieira[2], Mário Corrêa[3], João Batista Camargo Júnior[3],
Kalinka Regina Lucas Jaquie Castelo Branco[1], and Onofre Trindade Júnior[1]

[1] Institute of Mathematics and Computer Science - University of São Paulo,
São Carlos - SP, Brazil
{douglasr,rayner,jcezar,kalinka,otjunior}@icmc.usp.br
[2] University of Coimbra, Coimbra, Portugal
mvieira@dei.uc.pt
[3] University of São Paulo, São Paulo - SP, Brazil
{mario.correa,joao.camargo}@poli.usp.br

Abstract. Service-Oriented Architecture (SOA) are having a widespread use in enterprise computing applications, being Web services the most common implementation. The use of SOA has also been proposed for embedded systems, although very little could be found in the literature on the use of SOA for Safety-Critical Embedded Systems. This paper discusses the use of SOA for the development of this class of systems. Safety-critical embedded systems have specific requirements such as high reliability and real time response, making the use of SOA more challenging than for standard applications. To make concepts clear, a case study on Avionics for Unmanned Aerial Vehicles (UAVs) is presented. This is a complex application based on a reference model proposed by the authors. SOA shows to be a promising approach to implement parts of this reference model, especially in what concerns the missions played by the aircraft.

Keywords: Safety-Critical Embedded Systems, Unmanned Aerial Vehicles, Service-Oriented Architecture.

1 Introduction

Embedded systems are computing systems that are part of a larger system. They provide a predefined set of tasks, normally dedicated to a particular real time application, and present special requirements. In fact, they typically provide real-time monitoring and control for an entire system. These systems are considered to be safety-critical when failure events can lead to human live losses or high valued asset losses. In some applications, such as in aviation, safety-critical embedded systems must present failure rates as low as a serious fault every 10^5 to 10^9 hours of operation.

Embedded systems have becoming increasingly more complex in hardware and software. On the other hand, they are becoming more and more usual in domestic and professional environments for control or information management. Multicore and

G. Lee, D. Howard, and D. Ślęzak (Eds.): ICHIT 2011, CCIS 206, pp. 345–354, 2011.
© Springer-Verlag Berlin Heidelberg 2011

multiprocessor systems are becoming common, further increasing the complexity of the software [7].

The key concepts of Service-Oriented Architectures (SOA) have received significant attention from the community of software development, although there are some conflicting understanding on SOA concepts and usage. Various types of service-oriented architectures have emerged and Web services are the most common implementation [3].

The advantage of using the SOA paradigm is the interoperability achieved by the use of XML, which allows not only conventional communication in the Web, but also communication between devices ranging from a small sensor to sophisticate domestic, commercial or industrial equipment. Self-describing open components that support quick and seamless integration are characteristics of this kind of paradigm.

In this paper we discuss whether Web services are suitable or not for embedded systems. While the use of SOA in the business application domain is well established, several aspects must be considered in the embedded systems domain, mainly the availability of enough resources (processing Power and memory size). Embedded systems are attached to the fast growth of the Internet, communication technologies, pervasive computing and portable consumer electronics. System sizes range from the tiny to the big, complex multicore systems. In many of them there are not enough resources to make possible the use of Web services.

This paper encourages the use of SOA in the normally bigger non-critical sections of complex safety-critical embedded systems. This provides a simple approach to the problem, allowing the use of different paradigms to solve different parts of a complex system, avoiding the disadvantages already mentioned.

As a case study, an UAS (Unmanned Aerial System) reference model is proposed. The system was modeled (a layered reference model), showing its critical and non-critical sections. Services and protocols between layers are under development and should be presented in a future paper. The advantages of the use of SOA are discussed, easing the implementation of many high-level functions.

The remainder of this paper is organized as follows. Section 2 presents the related work especially on the use of SOA in Embedded System and Safety-Critical Embedded Systems. Section 3 proposes and discusses a Reference Model for UASs. Section 4 presents a Framework on how to use SOA in the target applications. And finally, in Section 5, the conclusions of the work are presented.

2 Related Work

This section presents a review on the use of SOA in Embedded Systems and in Safety-Critical Embedded Systems.

Service-oriented computing is a paradigm that uses services as basic blocks for application development. This paradigm allows applications to be built on a cooperative network that crosses the boundaries of universities and organizations. These blocks must comply with some standards and patterns. Web services are the SOA implementation that has achieved the highest market penetration. This paradigm enables interoperability of applications due to a series of standards that are based on XML (eXtensible Markup Language). However, Web services do not have predefined clients and therefore must be adapted to different contexts and are, somehow, a type of client/server system especially structured to make the best use of Web standards. Services are offered by providers and used by service consumers (clients). The main architectural units in service-oriented computing are service description, service discovery and service consumption.

Many complex embedded systems are coupled with a high-level information system. SOA can provide the integration of low-level embedded system services and high-level information system services. This integration is still an incomplete work, in spite of the many related works found in the literature [7] [5] [4] [6] [13] [8] [9] [12] [2] [1]. In practice, the use of SOA in Embedded Systems can provide a lot of benefits, such as: decoupling configuration from environment; improvement of reusability and maintainability; higher level of abstraction and interoperability; more interactive interface between devices and information systems; and easy use of resource-hungry services provided by more powerful internet servers.

Safety-critical embedded systems are computational modules integrated to physical devices and equipment, which have a predefined set of tasks, usually with special requirements. These systems must present very low failure rates. Furthermore, real time performance is almost mandatory and must be guaranteed in any circumstances. Several mechanisms of SOA, for example service discovery, have potential for non-deterministic behavior, not compatible with the basic requirements of safety-critical embedded systems. Complex safety-critical embedded systems can be almost always split into critical and non-critical sections. Despite being algorithmically more complex, the critical sections are normally much smaller than the critical sections of the code.

The use of Web services architectures for embedded systems in distributed automation applications is presented in [5]. Kakanakov shows some results of using the TCP Client/Server model in networks of embedded systems. The author establishes that the performance of the protocol is totally dependent on the operational system and on the kind of device.

A related work [6] extends the work presented in [5]. The authors discuss the possibility of adaptation of the Service-Oriented Architecture in a distributed embedded system. They provide a description of the architecture and a tool for creating services in Java and C/C++. They test two systems (java-enabled and SOAP-enabled) and both present the same results: they allow the system to be much more than an application, to be a service.

Thramboulidis et al. (2007) [13] propose in their work an approach to use service-oriented architectures in embedded systems. The paper proposes a framework that allows easier development of embedded systems using features available as services. They provide an easy way to integrate components, plug-and-playing the desirable features that should be provided by the service-oriented architectures.

Moritz et al. (2008) [9] presents an approach to add Web services to low cost microcontrollers, without losing real-time capabilities. The work described had just started and they did not present important results. The measurements necessary for convincing conclusions were missing.

Lee at al. (2008) [8] discuss and present an implementation of mobile applications using a service-oriented paradigm. In their work, services have optimized implementations for mobile devices having scarce resources (mainly memory size and processing power).

Other projects related to the development of SOA platforms that enable the implementation of embedded systems are eSOA [12], SODA [2] and the SIRENA project [1].

This work has a different focus than the works reviewed. Most of them do not address safety-critical systems. For this class of systems, the references on the use of SOA are poor and almost inexistent. This is due to the specific requirements that standard SOA does not address.

A typical application of a complex safety-critical embedded system is an UAS (Unmanned Aerial System). The term UAS (Unmanned Aircraft System) was adopted by both the FAA (Federal Aviation Administration) and the international academic community to designate systems that comprise not only the aircraft but all associated elements such as the payload, the ground control station and communications links [4]. An UAS can operate for a longer period of time without human pilot intervention.

There are different types of UAS presenting different capabilities. Some aircraft can fly autonomously following a pre-programmed flight path (grid or waypoint based) [15], while others fly receiving commands from pilot-operated ground stations. The aircraft size can range from the micro to the big, and the ground control station can be implemented in smartphones, tablets, notebooks and a network of workstations. Aircraft can vary not only in size, but also in shape, type of propulsion and performance. The human-computer interface can vary from a smartphone touchscreen up to a tangible user interface. The performance of the communication links and the type of payload are also very important to accomplish the intended mission for the system.

All the papers and roadmaps on UASs found in the literature typically presents UASs implemented using traditional approaches [11] [17] [18] [19] [20] [14]. There are roadmaps showing the expected advances in UASs that are periodically published by military organizations, such as the US Air Force [16].

3   Structure of an UAS - A Reference Model

3.1   Reference Model Architecture for Unmanned Aerial Systems

Architecture is a structure that identifies, defines and organizes components. The relationship and principles of design of the components, functions and the interface established between subsystems can also be defined by architecture.

The reference model architecture is an architecture where the entities, relationships and information units involved in interactions between and within subsystems and components are defined and modeled. In summary, it is a model of something that embodies the basic goal or idea and can be considered as a reference for various purposes.

UASs have been extensively used for precision agriculture, national security (military missions) and environmental monitoring. In this kind of systems it is necessary to make clear the complex structure of the system making easier the tasks of the system designers.

The NIST (National Institute of Standards and Technologies) provides a reference model for UAS [10]. In this specific standard, the reference model was proposed to specify the military rules, uses and commands in an understandable and intuitive way for a human commander.

Figure 1 presents a Reference Model Architecture for UASs proposed by the authors of this paper showing entities and theirs relationships, a different approach of that presented by NIST [10]. The components of an UAS can be split into an aerial segment and a ground segment. The aerial segment is hierarchically composed of the physical layer, the distributed RTOS layer, the system abstraction layer, monitoring & control layer, navigation & services layer and mission layer. The ground segment is divided into a physical layer and a ground control station layer.

3.2 Protecting the Critical Parts

The separation in layers allows the system to be divided into subsystems that can be implemented in different ways. This division helps to protect the critical parts that compose the entire safety-critical embedded system providing the best of the both worlds: protect and perform the critical parts complying with all the necessary requirements and take full advantage of the facilities offered by the Service-Oriented Architecture. The shaded parts in Figure 1 indicate the non-critical parts that can be easily implemented using SOA.

3.3 The Smart Sensor Interface and Protocol and In-flight Awareness

This reference model and the use of SOA support the development of the SSI (Smart Sensor Interface). The SSI is a concept that makes easier the payload integration to an UAS. In fact, the mission is isolated from the aircraft, being part of the sensor (smart sensor, normally a MOSA - Mission Oriented Sensor Array). The aircraft provides the capability of motion to the sensor. The sensor is in charge of the mission, directing the aircraft for its accomplishment. The SOA service discovery can to be used to allow the sensor to choose or compose the best service to perform the proposed mission.

Not always the aircraft can fulfill all the requirements necessary to accomplish a specific mission. When connected, the aircraft and sensor communicate using the SSP (Smart Sensor Protocol) for exchanging information in order to agree on the requirements for the mission feasibility. As a result, the mission can be completely feasible, partially feasible or not feasible. This step is made whenever a new sensor is connected. Missions can be adaptative and some configuration can change during the execution of a mission.

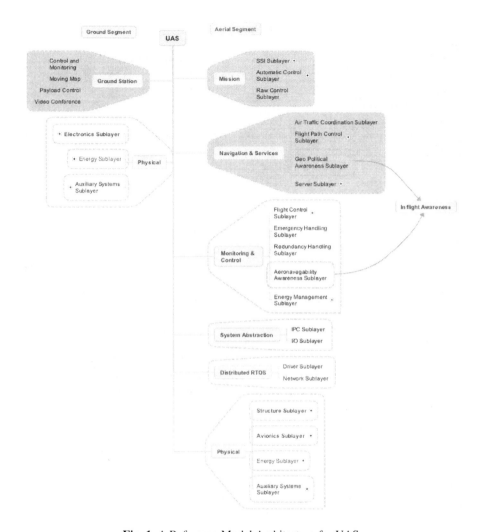

Fig. 1. A Reference Model Architecture for UASs

In the near future (10+ years), UAVs should dominate the skies over manned aircraft. In fact, the only manned aircraft that will make sense will be the ones that carry passengers. In this sense, all aircraft must coordinate the use of the airspace without human intervention.

Another important issue is the replacement of some functionality provided today by the human pilot. The authors call this replacement In-flight Awareness (IFA). The human pilot can smell alarming odors; evaluate cloud formations; listen to usual and to unusual noises; feel usual and unusual vibrations; stay aware of political boundaries and the ground being own over. All this knowledge can be used to avoid dangerous situations and select the best procedure in emergencies. SOA has great potential to support the implementation of IFA due to its naturally dynamic behavior.

4 A Knowledge Based Framework for Dynamically Changing Applications

4.1 The Framework

All communication in SOA is performed through the SOAP protocol in XML. The exchange of messages is done via plain text in the raw mode, without any concern for security.

The SOA basic operation is shown in Figure 2. The provider publishes the service. The client searches the UDDI repository to find the service and finally a message passing communication is established between the service provider and the client. A problem is that the chosen service could not be the best service for each moment in time during the operation of the system. To overcome this problem, this paper proposes the Knowledge Based Framework for Dynamically Changing Applications (KBF), presented in Figure 3.

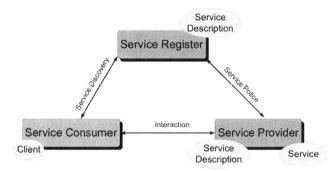

Fig. 2. SOA basic operation.

KBF extends the broker service discovery capability adding knowledge about the application domain. In this way, the application designer, or even the application itself, can choose or compose the best service based on a set of usage rules and some selection criteria such as: dependability, security, performance and real time response.

4.2 Reconfigurable Matrices

KFB maintains a Knowledge Database to store all the information and selection criteria established by the user and the application. Another key issue is the assembling of reconfiguration matrices. These data structures correlate available services, its functionalities and other selection criteria with the application procedures. They can be: static - defined off the system, manually or with the help of a supporting tool; semi-static - defined at system startup; and dynamic - defined during system operation, when a service status changes (availability, selection criteria).

Static configuration matrices never change. Semi-static configuration matrices change during the startup of the system and cannot be changed during normal system operation. Dynamic matrices can be changed or composed during the system operation. These matrices have potential for non-deterministic behavior and despite its flexibility, should be avoided in the critical parts of the system.

Using all information available in the configuration matrices, the KFB can choose or compose the best service to perform a mission defined by the user.

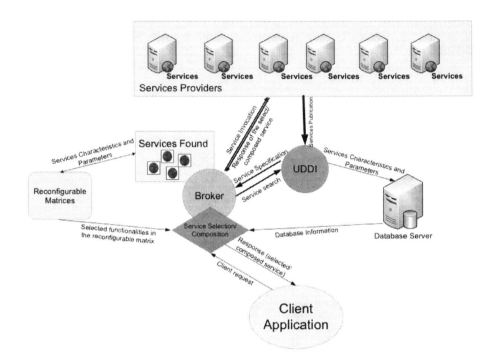

Fig. 3. Knowledge Based Framework on SOA

4.3   Reasons to Use SOA and the KBF in an UAS

UASs are complex systems that perform complex missions. Big UASs are distributed systems, with tens of different processor boards. In these later systems, processor and memory costs are not an issue. Therefore, performance is normally not an issue. The use of SOA can provide a more easy and quick development of the non-critical parts of such systems as discussed before. A variability of services with different functionalities and levels of performance (including security, reliability, safety and real time) can be developed and composed.

Different missions, defined by different MOSA subsystems and different UASs, can be integrated by the KFB that can choose the best service to fit in the scenario. This is the basis of implementation of the SSI and SSP. The mission can be adaptative. During a mission, based on a configuration matrix, the UAS can dynamically adapt the characteristics of the mission choosing services that fits better depending on the situation. In critical fault situations the use of the SOA paradigm can also help. Different strategies, based on configuration matrices, map default and the best procedures for its handling can be selected.

5 Conclusions

Web services, the cornerstone of the current SOA technology, are widely used for linking service providers and clients in different areas such as banking and financial services, transportation, manufacturing, to name a few. They are designed for interoperable machine-to-machine communication, allowing different systems to invoke remote methods or to exchange documents without mutual knowledge of internal implementation details.

Several studies show the application of Web services in embedded systems. Most of them focus an optimized version of Web services to avoid the lack of resources normally presented by such systems. Web services and safety-critical embedded systems often are seen as non-related areas.

This paper introduces the use of SOA in critical embedded systems, providing dynamic behavior and flexibility to this class of systems. However, the problem of choosing the parts of the system that can be implemented with this technology, without compromising its safety-critical nature, is not a trivial task. In this paper a reference model on UAS is presented to show how this technology can be used in a complex safety-critical embedded system. Complex systems of this kind normally have non-critical parts that can take advantage of all facilities that SOA, and in particular Web services, can provide.

A framework was proposed to allow the use of SOA in the non-critical parts of Safety-Critical Embedded Systems, helping the implementation of the SSI and SSP, as they are good first candidates for application of the ideas presented in this paper.

Acknowledgments. The authors acknowledge the support granted by CNPq and FAPESP to the INCT-SEC (National Institute of Science and Technology - Critical Embedded Systems - Brazil), processes 573963/2008-9 and 08/57870-9.

References

1. Bohn, H., Bobek, A., Golatowski, F.: SIRENA - Service Infrastructure for Real-time Embedded Networked Devices: A Service Oriented Framework for Different Domains (2005)
2. Deugd, S., Carroll, R., Kelly, K.E., Millett, B., Ricker, J.: SODA: Service-Oriented Device Architecture (2006)
3. Erl, T.: Service-Oriented Architecture: Concepts, Technology, and Design. Prentice-Hall, Upper Saddle River (2005)
4. GAO: Unmanned Aircraft Systems - Federal Actions Needed to Ensure Safety and Expand their Potential uses within the National Airspace System, GAO-08-511 (2008)
5. Kakanakov, N.R.: Experimental Analysis of Client/Server Application in Embedded Systems. In: Electronics, Sozopol, Bulgary (2005)
6. Kakanakov, N.R., Spasov, G.: Adaptation of Web Service Architecture in Distributed Embedded Systems. In: International Conference on Computer Systems and Technologies, CompSysTech 2005, pp. IIIB.10-1 – IIIB.10-6 (2005)
7. Kouloheris, J.: Future of System Level Design. In: Panel Discussion of First IEEE/ ACM/ IFIP International Conference on Hardware/Software Codesign and System Synthesis (2003)

8. Lee, M., Yoo, C., Jang, O.: Embedded System Software Testing Based on SOA for Mobile Service. International Journal of Advanced Science and Technology 1, 55–63 (2008)
9. Moritz, G., Pruter, S., Timmermann, D., Golatowski, F.: Web Services on Deeply Embedded Devices with Real-Time Processing. In: The Proceedings of the IEEE International Conference on Emerging Technologies and Factory Automation, ETFA 2008, pp. 432–435 (2008)
10. National Institute of Standards and Technologies: 4D/RCS: Reference Model Architecture for Unmanned Vehicle Systems Version 2.0 (2002)
11. OSD UAV Roadmap 2002-2027: Office of the Secretary of Defense, (Acquisition, Technology, and Logistics) Air Warfare (December 2002)
12. Scholz, A., Buckl, C., Sommer, S., Kemper, A., Knoll, A., Heuer, J., Schmitt, A.: eSOA – SOA Fuer Eingebettete Netze (2009)
13. Thramboulidis, K.C., Doukas, G., Koumoutsos, G.: A SOA-Based Embedded Systems Development Environment for Industrial Automation. EURASIP Journal on Embedded Systems 2008, 1–15 (2007)
14. Trindade Jr, O., Barbosa, L.C.P., Neris, L.O., Jorge, L.A.C.: A Mission Planner and Navigation System for the ARARA Project. In: ICAS - 23rd International Congress of Aeronautical Sciences, Toronto (2002)
15. Trindade Jr, O., Neris, L.O., Barbosa, L., Branco, K.R.L.J.C.: A Layered Approach to Design Autopilots. In: IEEE-ICIT 2010 International Conference on Industrial Technology, vol. 1, pp. 1395–1400. IEEE Press, Chile (2010)
16. United States Air Force: Unmanned Aircraft Systems Flight Plan 2009-2047, Headquarters, United States Air Force, Washington DC (2009)
17. Unmanned Aircraft Systems Roadmap 2005-2030: Office of the Secretary of Defense (August 2005)
18. Unmanned Systems Roadmap 2007-2032: Office of the Secretary of Defense (January 2009)
19. Unmanned Systems Integrated Roadmap FY2009-2034: Office of the Secretary of Defense (April 2009)
20. Valavanis, K.P.: Advance. In: Unmanned Aerial Vehicles: State of the Art and the Road to Autonomy. International Series on Intelligent Systems, Control, and Automation: Science and Engineering, vol. 33 (2007)

A Study on the Integration of Power System Operational Data and Application

Do-Eun Oh, Young-Il Kim, Young-Jun Kim,
Seung-Hwan Choi, and Il-Kwon Yang

KEPCO Research Institute 103-16 Munji-dong Yuseong-Gu Daejeon, Korea
hifive@kepri.re.kr

Abstract. Modern utilities monitor and control the power system via a vast network of communication-enabled devices. The data generated from power plants to consumers has been treated independently and applications monitoring and controlling the power system have been operated separately. However, operation of the power system is now completely dependent on the exchange of data among the applications to achieve a comprehensive view of end-to-end reliability of the power system. In this paper, the integration of power system operational data and application is described.

Keywords: Power system, Data integration, Application integration.

1 Introduction

The amount of data being collected is increasing exponentially in the power system. This rapid expansion of data retrieval results from the fact that more field devices are being installed and that these field devices are becoming more intelligent both in what power system characteristics they can capture, and also in what calculations and algorithms they can execute which result in even more data. As distribution automation extends communications to devices on feeders, as substation automation expands the information available for retrieval by substation planners, protection engineers and maintenance personnel, and as more power system asset information is stored electronically, even more varieties and volumes of data will need to be maintained and managed.

As the role of consumer through the two-way communications between utilities and consumers is increasing in the power system, utilities are required to build an integration infrastructure for operational application integration quickly to provide a base for adaptable business models. An enterprise-wide operational application integration offers a tremendous opportunity for improved operational efficiency, improved control of customer processes based on supply system conditions, use of customer-owned and operated generation and power quality improvement technologies as part of overall system management and to achieve the required levels of reliability and power quality at the end user level.

On the other hand, the requirements of transforming the centralized and one-way flow of control and information by utilities to the decentralized and two-way flow

G. Lee, D. Howard, and D. Ślęzak (Eds.): ICHIT 2011, CCIS 206, pp. 355–362, 2011.

structure enabling plug-and-play are rapidly increasing. Moreover, the importance of reliable, high-quality electrical power continues to grow as society enters into a new era of economics driven by digitally-based technologies. Enabling the digital economy of the future will require an intelligent gird that utilizes information generated from every application being operated in the power system. The future power system such as an intelligent grid ultimately will require an information infrastructure wherein the electric power system operates as an organic system through the integration of applications and data, decentralization of control, and interaction of functions. This paper describes the integration of power system operational data and application.

2 Power System Operational Applications

Korea's power system has various operational applications such as Energy Management System (EMS), SCADA, Distribution Automation System (DAS), and Automatic Meter Reading (AMR). EMS controls the power plants and monitors the situation of each substation. If there is something wrong during monitoring, the SCADA system takes charge of transmitting to the operator and EMS. DAS is an automation system that performs remote monitoring of abnormalities of the distribution lines and prompt search for and restoration of the outage area. Currently, SCADA and DAS are under a data link with a one-way communication wherein DAS mainly uses SCADA data.

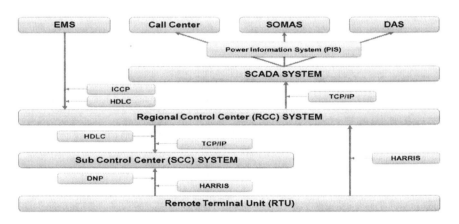

Fig. 1. Layered protocols of the power automation system

2.1 Energy Management System (EMS)

The elements of the EMS are shown in the figure 2. The followings are the major functions:

- Supervisory Control and Data Acquisition (SCADA)
- Automatic Generation Control and Economic load Dispatch (AGC & ED)
- Generation Scheduling

- Network Analysis
- Outage Scheduling
- Information Storage and Retrieval
- Dispatcher Training Simulator

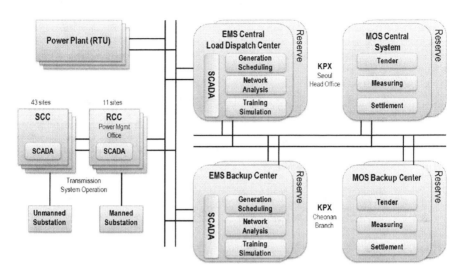

Fig. 2. Relation of EMS and SCADA

2.2 Supervisory Control and Data Acquisition (SCADA)

The SCADA system consists of SCADA in the Regional Control Center (RCC) and SCADA in the Sub-Control Center (SCC). It monitors the operation status such as line flow, MTR flow, bus voltage, circuit breaker, etc. of the power plants and substations.

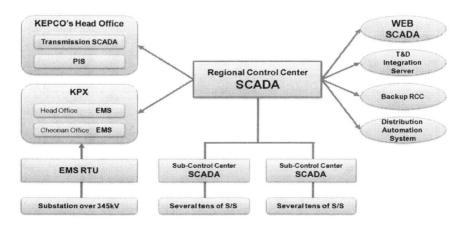

Fig. 3. Conceptual diagram of SCADA system

2.3 Distribution Automation System (DAS)

The DAS monitors and controls the distribution devices remotely. It provides advanced application functions including area load control, automatic troubleshooting, calculation of short-circuit faulty current, calculation of voltage sag, calculation of distribution loss, restructuring of power line, analysis of protection coordination, and preparation of operation sequence.

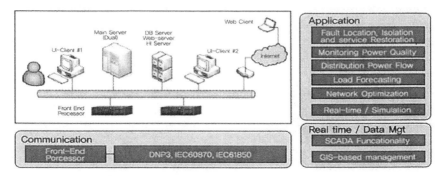

Fig. 4. DAS system

2.4 Automatic Meter Reading (AMR)

The AMR system collects meter reading data for the meter every 15 minutes. It also provides monitoring function of device failure, outage duration, etc.

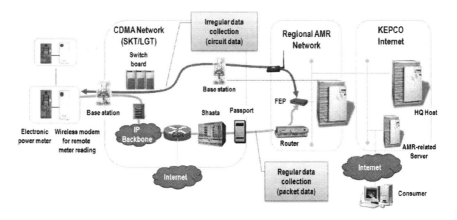

Fig. 5. AMR system

3 Need for Data and Application Integration

The power system operational functions are changing and growing more complex with exciting requirements of the digital economy [1]. Moreover, the importance of reliable, high-quality electrical power continues to grow as society enters into a new era of economics driven by digitally-based technologies. Enabling the digital economy of the future will require an intelligent gird that utilizes information generated from every application being operated in the power system [2]. The future power system such as an intelligent grid ultimately will require an information infrastructure wherein the electric power system operates as an organic system through the integration of applications and data, decentralization of control, and interaction of functions.

In case of distribution automation, advanced functions involve applications in the control center supplemented by applications implemented in field equipment. The control center applications provide the global analysis of the distribution system state and capabilities and are the overarching functions in control of distribution system operations, while the field equipment applications provide local information and control. The advanced distribution functions in the control center rely heavily on data from many different sources such as SCADA, Geographical Information System (GIS), etc. SCADA system provides real-time data acquired from field equipment and GIS gives power system facilities data. The primary requirements for advanced distribution functions are focused on data management. Correct, available, and timely data are crucial to the advanced distribution functions operating properly. However, since data comes from many different sources and since the systems acting as these sources usually are provided by different vendors, the coordination, synchronization, integration of systems, and mapping of data elements across these systems is a major problem.

4 Power System Common Information Model

A number of applications being operated in the electric power system differ in terms of operation system, data structure, etc. In case point-to-point connection is in use, $0.5 \times N^2$ adaptors are normally necessary for data exchange between systems when there are N applications.

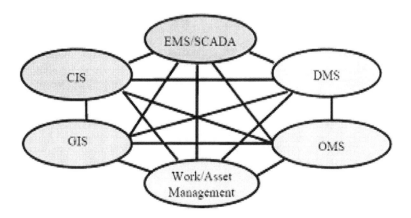

Fig. 6. Point-to-point connection

Accepting all adaptors imposes considerable burden on a system developer. In this case, if a common information model is in use, making only N adaptors for N systems solves the data exchange problem. Also, a common information model reduces the time and expenses required for adaptor development. Developers can then solely concentrate on the improvement of the applications.

At the IEC, standardization work is now in progress with regard to the common information model that can be applied to the energy management system and electricity distribution management system [3].

This common information model is the one modeling all components of the electric power system as the example shown in figure 7 using the object-oriented method; it is defined using Unified Modeling Language (UML).

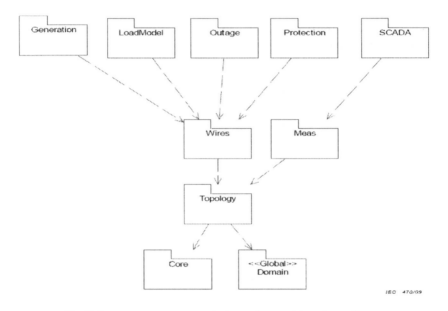

Fig. 7. Power system common information model package diagram

5   Integration of Power System Operational Application

For the integration of power system operational data and application, we use a message bus using the power system common information model. The message bus supports the management of service interaction occurring between participants and virtualization. It serves as the connection between service suppliers and requesters and enables interaction even if correct matching is not made. The power system common information model provides the unified semantic method to exchange data among heterogeneous systems.

The conceptual architecture of integration of power system operational data and applications is shown in figure 8. Applications for integration are SCADA, Transmission GIS, DAS, Distribution Information System (DIS), Customer Information System (CIS) and AMR. The CIM database manages the meta-data and

GIS database has GIS and facility information. The transmission and distribution status data are collected by SCADA and DAS and then are transferred to the data warehouse using high speed data access interface.

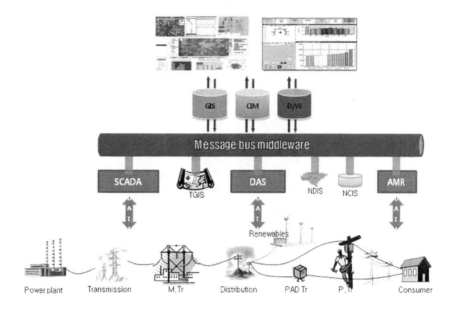

Fig. 8. Conceptual architecture of data and application integration

The existing applications like SCADA have their own data structures and types. To exchange data through power system common information model among the applications, the data conversion is required. Figure 9 shows an example of a process converting the types and contents of messages within the message bus.

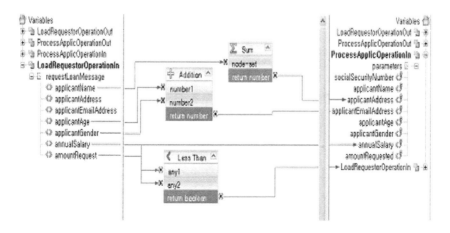

Fig. 9. Message conversion

6 Conclusion

In the electric power system, there are several applications for the power system operation; these applications perform their own intrinsic functions according to the purpose. The importance of reliable, high-quality electrical power continues to grow as society enters into a new era of economics driven by digitally-based technologies. Enabling the digital economy of the future will require an intelligent gird that utilizes information generated from every application being operated in the power system. In this paper, the integration of power system operational data and application are described.

References

1. Gellings, C.W.: The Power Delivery System of the Future. In: EPRI (2003)
2. Hughes, J., et al.: The Integrated Energy and Communication Systems Architecture. In: EPRI, vol. I (2004)
3. McMorran, A.: An Introduction to IEC 61970-301 & 61968-11: The Common Information Model, University of Strathclyde (2007)

KepriSNS: A Business SNS Platform Based on Open-Source

Dongwook Kim, Changhun Chae, and Namjoon Jung

S/W Center, Kepco Research Institute
Munji Road 105, Yuseong-Gu, Daejeon 305-760, Korea
{dongwook,chchae,njjung}@kepco.co.kr

Abstract. Social Network Service (SNS) is a service that enables internet users make human network and strengthen the existing human relationships. But, many companies, currently, are aware of the importance of SNS for business areas. The emergence of SNS has brought a lot of changes in the business environment of companies. At the first, companies have utilized the SNS service as the communication channels with their customers. But, they have recently been considering SNS as a way to strengthen their organizations. Our company has the prototype version, KepriSNS, before building an official business SNS platform. This paper addresses the considerations for building a business SNS platform and introduces the development of KepriSNS based on open-source StatusNet.

Keywords: SNS, Social network service, KepriSNS, StatusNet.

1 Introduction

Social Network Service (SNS) is a service that enables internet users make human network and strengthen the existing human relationships. At the beginning of SNS's emergence, SNS have focused on the support of internet users' communication. But, many companies, currently, are aware of the importance of SNS for business areas[1].

The emergence of SNS has brought a lot of changes in the business environment of companies. At the first, companies have utilized the SNS service as the communication channels with their customers[2]. But, they have recently been considering SNS as a way to strengthen their organizations[2,3]. Therefore, many companies are trying to use SNS for employees to build their own human networks and an idea bank for information exchange. Particularly, IT and Telecommunication companies are adopting SNS for conversations among their employees, as well conservative financial institutions[3].

Our company has built a prototype version, KepriSNS, before publishing an official SNS platform. This paper addresses the considerations for building a business SNS platform and introduces the development of KepriSNS based on open-source StatusNet[7].

G. Lee, D. Howard, and D. Ślęzak (Eds.): ICHIT 2011, CCIS 206, pp. 363–370, 2011.
© Springer-Verlag Berlin Heidelberg 2011

2   Strategy for the Establishment of a Business SNS Platform

So far, many compaines have been trying to accumulate their own knowledge through the collection and the manipulation of personal vaulable information. To this end, from the past, the companies have built their own knowledge management systems to accumulate knowledge spreaded within their own companies. And then they built up information search systems to find more correct knowledge information from the huge accumulated information. In the attempts, there are many successful cases of building up these systems. But, this way is suitable for the case that accumulated information is small comparatively.

If a company is big and the accumulated information is so huge, building a knowledge management system may not an appropriate solution because the accuracy of the search result may be poor. Eventually, the most valuable information is who help me and how to get the information.

Companies can easily deal with the problems through the power of SNS service. As well known, like Twitter[8], the ripple effect of SNS is so powerful. Personal human networks based on SNS communities always help to find the latest information distributed among employees of a company. Also it enables to enhance the utilization of the existing company-inside systems.

2.1   Features of Business SNS

The introduction of a business SNS platform has the following benefits.

More Accurate information. It supports the more accurate and faster information search using SNS's human networks. The results from searching may be more accurate than those of the existing information search systems of companies.

Fast Communication. It enables to support fast communications through a variety of devices such as PC and smartphone because of no limitations of time and place.

Easy and Simple Communication. SNS's microblogging serivce enables free discussions of employees. SNS helps every employee to participate free conversations in an equal position without the responsibility of the existing reports and formal conversations.

Companies which are considering the introduction of SNS, first of all, should prepare for the negative effect of SNS such as illegal trespassing and using[5,6].

Protection of company information. Because only employees of the company have to be the members of their own business SNS system, it can exchange safely the internal information and opinions, and protect the leakage of companies' information.

Lost of mobile devices. Because the introduction of business SNS causes the activation of using mobile devices such as smartphone, the security policy of personal mobile devices is also required. If any person losts his/her own mobile device, the company's information can be leaked to an external society and malicious attackers may access illegally to the company-inside systems using the lost mobile device. Therefore, if lost, security functions like a remote deletion function of all data and programs in the lost device should be embedded into personal devices.

Protection from illegal trespasses. For improving SNS's utilization, it should recommend using mobile devices. But, mobile devices which require access to company-inside network from external network can be used illegally as a trespass channel. To this end, a SNS Server should be deployed in DMZ to isolate company-inside systems from external illegal attacks.

Personal information access policy. Even if it could not prevent from any external attack due to the weakeness of a firewall system, the company's other systems should not be attacked. In order to protect other company's systems, the SNS system need to be built based on secureOS like selinux[10].

Encrypting communication channels. During access to their own company's SNS service, employees might be directed to enter at phishing sites for identify theft. It is possible to exchange the sharing information in SNS system with information containing malicious code. So, all communication channels between external and internal network have to be encrypted to prevent illegal access.

Monitoring use of SNS. For blocking the employees' inappropriate leakage of a company's information, ways of motoring continuously the company's network need to be arranged.

2.2 Case Study of Building Business SNS Platforms

Companies can consider the two following ways to build their own SNS services. The first way is to utilize external organizations' company SNS solutions. The latter way is to develop and deploy their own SNS platforms for themselves. The two ways have advantages and disadvantages in the aspect of cost, time, and utilization.

Utilization of SNS provider's solutions. It is a way to utilize SNS solutions of which IT providers developed for an unspecified number of companies. Some companies are trying to build their own business SNS service based on the existing SNS solutions such as Twitter[8] and Yammer[9]. For example, SK merged SNS service of Twitter at their group portal service. Also, several departments of LG electronics are using Yammer as communication channels among employees.

This way has a lot of benefits in the aspect of cost and time due to using SNS solutions. But, because all information is accumulated out of the companies, it is difficult to reflect for the companies' management through processing and refining the information. Also, it is difficult to develop a variety of services integrated into company-inside systems. The biggest problem is that it is possible to lose the long time accumulated information if a SNS solution provider supports no longer the service.

Fig. 1. The snapshot of Yammer' Social Network Service

Development of Company-inside SNS platform. This is a way to develop their own SNS platform reflecting the requirements of the company. Typical IT Company, Motorola, developed a SNS platform called as MotMot[4] based on open-source StatusNet[7]. Although this way requires much cost and time, there are many benefits in security, improvement of SNS's utilization, and practical use because the company information is accumulated inside the company.

Fig. 2. Motorola's Social Network Service, MotMot

3   Features of Open-Source Statusnet

StatusNet is an open-source platform for micro-blogging service developed by Identi.ca [7]. The supporting functions of StatusNet are similar to those of Twitter[8] and Yammer[9]. The source code of StatusNet is freely available for download and install under GPL. Using StatusNet has a few of the following benefits.

A low level of entrance barrier. Recently, many of companies are seeking business opportunities in Social community. But they, from now, are not sure that Social community is very useful for their business area. These make many companies hesitate to introduce social community to their own business area. If any company wants to process a pilot project to understand gains which they get from SNS, StatusNet can be a good candidate for low-cost and easy-use.

Functionality. The current version of StatusNet provides supporting full sets of functions for Social community. There are many of open-sources across the world, but the most of open-source are not enough to apply to business areas due to the lack of functionality, stability, and supporting. One of open group projects, StatusNet, provides a variety of documents and source code of extensive functions at their own web-site.

Integration with company-inside systems. StatusNet can be easily integrated into internal company systems because they support concepts of plug-in and Event for functional extension and integration with company-inside systems. Many companies have, for past a decade, integrated company-inside systems under IT-governance policy to collect data in their own system and refine the data for employees. The integration supporting of StatusNet makes many companies to recognize that StatusNet is useful for them.

Cost-effective. Using an open-source solution such as StatusNet is highly cost-effective. Companies can utilize StatusNet under GPL policy free of charge.

4 KepriSNS Components

KepriSNS is a prototype version before the introduction of our company's official SNS platform. Like Twitter [8] and Yammer [9], KepriSNS is supporting the access of personal PCs and smartphones for heterogeneous environment and fast communication. To support diverse environments like this, web service overcoming heterogeneous environment problems is so efficient.

Also, the deployment of KepriSNS needs to be inside a firewall system for the consideration of security incidents. Web service has a feature that occupies a network port (80) which all firewall system permits. Thus, it is relatively easy to build within a firewall system.

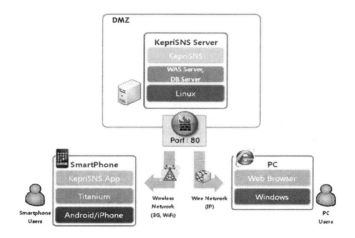

Fig. 3. KepriSNS's Network Environment for service access

TimeLine. This is a very important concept in social network service and is distinguished from the existing cafe and blog services. Until now, cafe and blog services have focused on information not human. So, cafe and blog services have categorized all articles into bulletin boards based on information. But, the users' focus is moving from information into human. The important thing is not the articles on bulletin boards, but what my favorite's people are doing. Timeline is a new conceptual bulletin board collecting the articles of my favorite's people.

KepriSNS are, currently, supporting three types of timelines; *Public timeline, Personal timeline, Group timeline. Public timeline* is a free-talking place for unspecified employees. All messages on *Public timeline* can be read and replied by all employees in a company. Various topics's messages in *Public timeline* might make employees turn their face away *Public timeline*. To this end, KepriSNS provide *Personal timeline*. *Personal timeline* shows only messages of people related to the human network of each employee.

Both of *Public* and *Personal timeline* have a principal that all messages are freely open to everyone. On the other hand, *Group timeline* has a principal that all messages in a group are open to the group's members. All access to information of a company has to be controlled by each person's authorization. And people with the same authorization need free-talking places only for them. Or it needs a support of company-inside clubs meeting sharing same specific issues. In order to support such needs, KepriSNS provide *Group timeline*.

All messages on timelines are categorized into two types; public message, group message. Public message can be registered to public and personal timeline, whereas group message can be to group timeline. And KepriSNS's users can send directly private messages to other users.

Table 1. Relationship between timelines and messages

	Public timeline	Personal timeline	Group timeline	
Public Msg.	O	O	X	
Group Msg.	X	O	O	
Private Msg.	X	X	X	Directive delivery

Groupping. Compared to the existing SNS systems for internet users, business SNS systems have a unique function called as Groupping. Groupping is necessary to support the cooperations of departments with various cultures. Also, it can use to back up operations of company-inside meeting such as departments, habit, and projects. KepriSNS's Groupping provides unprecedented functions like human connection, invitation, talking places, privacy setting in the existing café and blog services.

Heterogeneous Service Access. KepriSNS aimming to anywhere and anytime supports access of PCs and Smartphones for fast communication and diverse environments. To support diverse environments like this, web service overcoming heterogeneous environment problems is so efficient. The technology of Web feed service, RSS and ATOM, are applied to KepriSNS for mobile devices such as smartphones. In order to support fast communication, the well-defined XML data formats of RSS and ATOM have features that it is easy to process and refine the data.

It does not overload KepriSNS server due to a character there is only data delivery among clients and KepriSNS server.

Notification Service. Employees in companies have their own email address provided by companies and the utilization of Email is very high. Therefore, in order to make social community active, KepriSNS service has to be linked to company email service. When an employee registers their own message, KepriSNS send the new registered message to followers using company email.

6   Prototype Version of KepriSNS

In order to develop KepriSNS, we developed a protype version based on open-source StatusNet and are currently evaluating the version against employees of our company. The development consisted of the improvement of SNS server and the enhancement of smartphone app of android and iPhone. The server-side development was archived using PHP and MySQL and the smartphone-side app development was based on Appcelerator' Titanium Mobile library.

Supporting Korean language. While the architecture of StatusNet is supporting multi-language, it could increase more confuses of many users because the Korean translation of messages is incomplete. In order to solve the issue, we corrected all inacceptable messages of StatusNet's menu and notice.

Grouping. The existing groupping functions of StatusNet provide a simple filtering feature that only messages of a group show. Because the messages in a group are open to everyone, KepriSNS can not use the function without modification. Therefore, we develped group timeline which the only members of a group can read the group's messages.

Fig. 4. The snapshot of a KepriSNS group

Notification Service. Employees in companies have their own email address provided by companies and the utilization of Email is very high. Therefore, in order to make social community active, KepriSNS service has to be linked to company email service. When an employee registers their own message, KepriSNS send the new registered message to followers using company email.

Fig. 5. Setting email environment on KepriSNS

Smartphone app. StatusNet has used Appcelerator' Titanium Mobile library for the development of smartphone app. Titinaium Mobile library has many problems in the aspect of functionality and stability. For example, if a user subscribes to follow another user, it is aways fail to do it. We changed logical processing ways within the source codes to avoid the limitation. The current KepriSNS smartphone app operates ordinarily basical functions such as Following and user blocking.

Fig. 6. KepriSNS's smartphone app

7 Conclusion

In order to develop our company's SNS platform, we developed a prototype version based on open-source StatusNet. The development consisted of the improvement of SNS server and the enhancement of smartphone app of android and iPhone. We are currently evaluating the version with the participation of employees of our company and collecting requirements for building our official SNS platform through the evaluation. The current collected requirements are mainly related to the improvement of security rather than functionality. Based on the requirements, we will make a plan to introduce our official SNS platform.

References

1. Nimetz, J.: Jody Nimetz on Emerging Trends in B2B Social Networking. Marketing Jive (November 18, 2007)
2. Tynan, D.: As Applications Blossom, Facebook Is Open for Business. Wired (July 30, 2007)
3. Heejong, K.: The evolution of SNS. Newspaper Digitaltime (February 27, 2011), http://www.dt.co.kr/contents.html?article_no=201102280201015 1780001
4. Phillips, J.: statusnet revolutionizes Motorola's Internal communications, http://status.net/2010/07/08/statusnet-revolutionizes-motorolas-internal-communications
5. McDowell, M., Morda, D.: Socializing Securely: using social networking services. Carnegie Mellon University. Produced for US-CERT, a government organization (2011), http://www.us-cert.gov/reading_room/safe_social_networking.pdf
6. Writer, S.: Hacker Exposes Private Twitter Documents. Bits (blog of The New York Times) (July 15, 2009), http://bits.blogs.nytimes.com/2009/07/15/hacker-exposes-private-twitter-documents/?hpw
7. StatusNet's main web page, http://status.net/wiki/Main_Page
8. Twitter's main web page, http://twitter.com/
9. Yammer's about web page, https://www.yammer.com/about/product
10. selinux's main web page, http://selinuxproject.org/page/Main_Page

A View of Future Technologies and Challenges for the Automation of Localisation Processes: Visions and Scenarios

Lamine Aouad, Ian R. O'Keeffe, J.J. Collins,
Asanka Wasala, Naoto Nishio, Aram Morera,
Lucia Morado, Lorcan Ryan,
Rajat Gupta, and Reinhard Schaler

Localisation Research Centre
Centre for Next Generation Localisation (CNGL)
Dept. of Computer Science and Information Systems
University of Limerick, Limerick, Ireland
{firstname.lastname}@ul.ie

Abstract. Scenarios are often used to cope with uncertainty in areas with long planning horizons and research. They provide a vehicle that can play an important role in the development of shared visions of the future: creating powerful expectations of the potential of emerging technologies and mobilising resources necessary for their realisation. This paper describes a Web-based problem solving environment providing such a scenario by tracing, sharing, executing, and monitoring examples through a number of current localisation research topics, prototypes, and imagined as well as real technologies. For each of its components, it will consider related opportunities, drivers, barriers and challenges, and describe the key technological building blocks required.

1 Introduction

Current batch localisation processes are widely manual, rigid, and provide limited scalability. Language Service Provides (LSPs) often connect their in-house services with external resources and technologies, in natural language processing, using various workflow technologies, different metadata, and by 'cutting and pasting' data between these components. This makes them error-prone, hard to retrace, to retrieve, and to re-use as well.

Alternatives? Current approaches to the automation of localisation processes are largely based on complex, proprietary, and closed systems, which assume large scale, relatively static and well prepared source material in one single source language.

These assumptions are no longer valid and current systems are no longer suitable for the type of multilingual, user produced, multi-modal and dynamic content localisers are increasingly dealing with today. In this scenario, we address the need for an open, highly configurable, and loosely coupled aggregation

G. Lee, D. Howard, and D. Ślęzak (Eds.): ICHIT 2011, CCIS 206, pp. 371–382, 2011.

of heterogeneous services that can meet the varying demands of a wide range of localisation needs in the non-profit, the enterprise and the SME sectors. It facilitates the leverage of the provided infrastructure encapsulated in the platform and tailors it to specific needs through further component development. Indeed, it embraces the Service Oriented Architecture (SOA) philosophy, which enables the development of a component marketplace. It also promotes and supports an open metadata standard allowing a high degree of flexibility and interoperability.

The proposed framework, called SOLAS, for Service-Oriented Localisation Architecture Solution, is a best-in-class concept integrating novel components in workflow and translation recommendation, business process management, services mapping, in addition to novel linguistic assets integration and adaptation. The platform facilitates the training and learning of user-specific and domain-specific patterns through its recommendation component. It also supports the dynamic reconfiguration and selection of distributed services.

On the other hand, an increasing number of web-based collaborative translation and localisation platforms have recently appeared providing easy access and collaboration facilities to many companies and NGOs [4], [5]. While this is a very appealing paradigm, there are still many challenges, especially in terms of designing efficient and adaptive workflows; the huge volume of often user-generated source content, not just in one but possibly in a number of source languages; the dynamic and ever evolving nature of the source; multi-author approaches; and other language and domain-specific issues. The proposed framework can be seen as a problem-solving environment which in addition of providing the possibility of learning and sharing workflows and data, intra- and inter- instances, will also provide milestones in the field of intelligent and collaborative design of localisation processes.

This work presents then a set of contributions that addresses some of the most considerable challenges that translation and localisation are nowadays being faced with. We argue that current systems are not properly evaluated, neither evolved, with the specific needs of localisation processes, in terms of usability, flexibility, re-use, and support to collaboration in its broader sense, i.e. including the design of the process itself. The paper also presents the current components of SOLAS, considering an end-to-end localisation process, from the quality assessment to the delivery of the localised content.

2   Requirements Analysis

Before presenting the SOLAS system, this section identifies some of the most important requirements for the development of efficient automated localisation systems. These include some of the requirements, identified in [9], for next generation workflow systems, which represents a core component of any process management system, including in localisation.

Interoperability. Interoperability and the ability to link different localisation tools and services are of foremost necessity in today's localisation processes.

Linking distributed services requires efficient standardised metadata and infrastructural support, which should facilitate the discovery and integration of components, services, the metadata, and the data flowing between components and processes.

Supporting Collaboration. Localisation is becoming an inherently collaborative process, in which translators, localisors, and experts from many other domains participate; such as software engineers and data analysts. There is a need to manage, leverage, and bridge the differing expertise and contributions of collaborating members.

Flexibility. Given the large variety of localisation processes, and services involved, a localisation system must be flexible, both in terms of supporting different analyses in defining workflows, and in terms of the combinations of services. The possibility for the user to interact with the workflow at predefined events in the workflow would also be of great interest.

Service Discovery. Service discovery is a very important step in constructing a localisation workflow. Typically this would be in finding an applicable service that meets the needs of a specific task and matches it to the available exposed services.

Processes Sharing. Research in other areas has previously identified the necessity to share workflows with peers, in a process similar to publishing experiments [10]. However, in localisation, companies generally do not share their process data. When attempting to straighten collaboration, collaborative design, and to learn from previous processes, this is of foremost interest.

3   The SOLAS System

In this section, we present the SOLAS system and its components, which have been specifically designed to address the requirements of today's localisation processes, as discussed in the previous section.

3.1   LocConnect

LocConnect is a core component in SOLAS, supporting the interoperability requirement, and providing workflow management of the end-to-end localisation processes, through a novel framework based on a service oriented architecture and a full support of the XLIFF standard.

As already mentioned, localisation process management can be a highly complex task, involving a large number of files, many language pairs, and translation and engineering teams from different locations. Different projects, languages, and teams, require tools and technologies that have been chosen specifically for their particular needs. These tools and technologies need then to be interoperable and to work with the same data packets sent by localisers (or their corresponding electronic 'agents') along the localisation process represented essentially by connected component technologies. Standards are vital to achieve interoperability in this case. Indeed, one of the major problems is the non-standard interfaces (i.e. mechanism by which software interact with each other) of processes that prevent distributed applications, components, and services, to be successfully integrated.

Although the adoption of localisation standards would give benefits such as reusability, accessibility, interoperability, and reduced cost to the consumers, software publishers either refrain from the full implementation of standards or do not pay attention to conformance testing of their products with standards, especially due to lack of evidence for improved outcomes and associated high costs. One of the biggest problems existing today with regards to the software is the pair-wise product drift [6], i.e. the output of one software needs to be transformed in order to compensate for other software's non-conforming behaviour. This problem is commonly seen in localisation. This is especially due to the lack of adoption of localisation standards as well as incomplete implementation of standards by the vendors.

Indeed, conformance testing is an area typically not given much attention to by localisation software vendors. To exacerbate this situation, most of the localisation standards investigated as part of this research only have loosely worded conformance clauses. In the absence of a clearly defined conformance clause, vendors are unable to develop proper test suites for testing their products for conformance with standards. Some tools have been developed for validating XLIFF, TBX and TMX content. However, no tools have been reported to assess the actual level of conformance with these standards, for a given file. The XLIFF Technical Committee is currently working on a conformance clause for the XLIFF standard, while no conformance criteria other than XML compliance have been specified in other localisation standards discussed earlier. The absence of proper conformance clauses for these standards has contributed to the improper implementation of the standard as well as a lack of tool support for these standards.

Interoperability is becoming hugely important due to the distributed nature of today's applications. However, this is one of the areas that have not been paid much attention to in the localisation domain. It is one of the primary aims of LocConnect and SOLAS to fill this gap in the area of interoperability in localisation processes. While interoperability testing suites have been reported about in relation to several XML technologies [7], no such work has been reported about in relation to localisation standards. The research around LocConnect will lead to the development of a test suites, tools, and guidelines, for conformance testing for key localisation standards.

3.2   Localisation Knowledge Repository

The Localisation Knowledge Repository (LKR) is a pre-translation quality assurance system designed to facilitate the development of usable, readable and translatable source language digital content. The first objective of the LKR is to check that source language text adheres to a set of user-defined guidelines. Any text string contradicting these guidelines is flagged by the system, and may be edited by the user to rectify the issue. The second objective of the LKR is to generate and store metadata relevant to the localisation process during the content development phase. Authors of source language content are often subject matter experts and can add metadata to provide useful context for translation and localisation professionals. Metadata such as author name, email address, domain and comment are added manually by the author, while other types of metadata such as job ID, word count and source language are added automatically by the system.

The LKR was developed as part of research into improving the quality of source language digital content. The motivation for this research is that digital content that is developed with localisation in mind is easier and cheaper to translate [20]. The LKR facilitates the process of âĂĲwriting for translationâĂİ by enabling authors to automatically check documentation for violations of guidelines, such writing in the active rather than the passive voice. The system enables authors to select pre-defined guidelines such as those published in the Microsoft Manual of Style [21], or to create their own customised guideline checks, such as checked for âĂĲbannedâĂİ words. The LKR is unique in that no existing technology enables authors to both check for violations of content development guidelines, and to add and store metadata relevant to localisation in an XLIFF project file.

3.3   Workflow Recommender

One of the most important strategies that have been used to reduce the cost and increase the speed and efficiency of localisation processes is automation [11]. Automated management in localisation has indeed become unavoidable and a foremost necessity. To respond to this need, many systems have been proposed to facilitate the automation, access and execution of localisation workflows. These include GlobalSight [12], Idiom WorldServer and Across [13], among others. However, these Translation Management Systems (TMSs) allow limited workflow configuration and flexibility. In addition, none of the tools currently available are able to automatically generate a workflow according to the characteristics of the project at hand. Choosing and creating the most suitable workflow still fall in the hands of the project manager. As part of the SOLAS system, a workflow recommender is proposed to automate the creation of a suitable workflow for a given project.

One of the precursors to this recommender system is called Transrouter [14]. Transrouter is a desktop-based decision support system that checks projects characteristics, some of which are obtained through the analysis of the source files, while others have to be manually entered. Transrouter then produces a number of recommended routes (sequences of tasks including the agents and resources that would enact them), ranked according to their suitability for the project. The project manager could then use them to carry out the project.

These routes, however, did not provide any automation for the management of the process.

Another recommendation system, called IGNITE, was also developed by the same team [15]. IGNITE is a server-based application that aims at 'demonstrating an up-to-date automated workflow model'. A description of its process is as follows: a client uploads a file to the system; the system turns then the file into an XLIFF format. The client would then select a suitable LSP (Language Service Provider) from a list, the LSP would download, translate and upload the XLIFF file back, and the client would then have access to a translated file leaving the LSP to manage their internal workflow as usual, i.e. IGNITE basically acts as a front-end to different systems and companies.

The team has acquired a valuable experience with these projects over the years, which has helped in developing a novel recommender system. Amongst the innovations of the new system is the inclusion of a wider range of components-specific rules to decide when the use of these technologies is appropriate. It also integrates community management and quality review in the workflow generation logic. The current rules are continually evolving and we are looking for feedback from industry and our own processes to adjust them. The proposed system is also web-based and fully XLIFF-compliant.

3.4   LMC and XLIFF Phoenix

The LMC and XLIFF Phoenix are a set of TM (Translation Memory) leveraging tools which aim at answering questions on what data and metadata can effectively help in improving the end-to-end localisation process; making it better, faster, and cheaper, by using efficient metadata.

The design of a data container, called LMC (Localisation Memory Container) was the first step towards encapsulating localisation data and metadata. The second step was the development of a localisation tool that will actually automate the process of recovering these data and metadata. The purpose of the two steps strategy is to organise the data (in an LMC file), its capture (with the extraction and merging tool, XLIFF Phoenix), and its usage (with a traditional Computer Assisted Translation, CAT tool, supporting the XLIFF format). The resulting files will be then: a) Enriched XLIFF files that will be used in CAT tools, and b) LMC files where XLIFF files could be stored for a later reuse. An important part of this work is to test whether this new localisation metadata, introduced in the enriched files, will be helpful for the translators, and subsequently improve their work and productivity.

Localisation Metadata and the XLIFF Standard. Metadata usually refers to 'data about data'. In localisation, metadata also refers to data that allows to smoothly and efficiently connect the data flowing throughout the end-to-end localisation process. The usage of open, rich, and flexible metadata is an important step towards the effective aggregation and sharing of data between localisation processes. To this end, the XLIFF standard was adopted by the SOLAS system. A complete analysis of the current XLIFF specification has been carried out in order to identify attributes that will encapsulate relevant localisation metadata. The interested reader can find a detailed description in [1].

Localisation Memory Cntainer. The Localisation Memory Container is an XML-based vocabulary that was developed as a data descriptor allowing an efficient storage of previous XLIFF documents in a single file. Instead of using extracted translated data, as it is done in the standard TMX, we use full XLIFF files as localisation memory components that could be reused in future processes. While it is possible to encapsulate several XLIFF files into one XLIFF file (XLIFF TC 2008), this possibility did not allow to have our desirable self description document, i.e. the information about what we have in the container in the same document.

The novel approach here was to develop a new XML-based language able to store previous XLIFF files. The language is also a self descriptor, i.e. includes information about the files it contains, the author, its creation and last modification date, and so on. The structure and technical details of the LMC are out of the scope of this paper. We refer the interested reader to [1] and [2].

XLIFF Phoenix. The XLIFF Phoenix is a tool that actually extracts localisation data and metadata from an LMC file and introduces it into an untranslated XLIFF document. XLIFF Phoenix allows the reusage of previously localised XLIFF documents. This is achieved by filtering the information included in the files and matching it with a new document introduced to the system. The resulting document is an enriched file that contains translation recommendations and embedded rich metadata.

3.5 Services Mapper

Most of the current localisation processes are manually intensive operations that require frequent human intervention [17], [18]. We have already mentioned the automated generation of a suitable abstract workflow in a previous section. This section presents the actual mapping of the abstract workflow to the set of suitable and available service instances [16]. Our previous experience of selecting and integrating Machine Translation (MT) systems illustrated the problem of selecting appropriate services as part of a localisation workflow. Indeed, since currently available services are not described in a standard manner, the selection is dependent on the analysis and intervention of skilled personnel. Moreover, service descriptions may not be available beyond corporate boundaries. This makes the selection and integration of services very difficult, as a unified description is not available.

In the current version of SOLAS, the MT mapper illustrates ongoing work towards an effective solution to support the full automation of localisation processes mapping. It demonstrates the optimum selection of web services, including MT, for dynamic workflow execution by descriptors supporting functional and non-functional service attributes. The MT mapper will implement a localisation service descriptor supporting both functional and non-functional attributes of a service, and it will demonstrate the plausibility of realising automated optimum service selection, using MT service selection as an example.

The selected localisation workflow will be based on the discovery of the requirements of this process. For instance, when the workflow requests an MT service, it specifies an abstract view of the service, such as functionality, performance, quality of the service, or constraints pertaining to that service. The MT mapper will then look up service descriptors supporting functional and non-functional attributes provided by each Machine Translation (MT) service provider, and then select the optimum MT service that fulfils these requirements. The same process applies to other language technologies and localisation services.

The development of the service mapper is carried out in two phases. The first phase is focused on the specification of a template that enumerates all of the functional and non-functional attributes of a localisation service. The second phase concentrates on the design of a methodology that allows the attributes of a service to be captured and documented in the template, and is dependent on the use of benchmarks and metrics that facilitate cross comparison.

3.6   Community Translation and Rating

Traditional localisation models are still dominating today's localisation approaches, mostly based on work with professional translators. On the other hand, new technologies such as Web 2.0 have rendered geographical boundaries irrelevant as evidenced by the rapid evolution and uptake of the social networking paradigm. Another example is the crowdsourcing paradigm in which organisations outsource their Research and Development work to a community of possibly unknown contributors [22]. It provides a means of harnessing the global intellect in order to expand the boundaries of the solution space, and is a rich source of diverse submissions from the community that can capture local social and cultural preferences. Crowdsourcing allows companies to involve a wide client base in the design of next general solutions. It is a paradigm that is being frequently used to evaluate and rate proposed solutions prior to construction. Crowdsourcing is seen as a key component in bridging the digital divide through the localisation of digital content for locales that are not sufficiently catered for. The demand for content is particularly acute in emerging economies where mobile telephony has enabled large sections of the population to connect to the web, and where access to information is a key driver for multidimensional impact on the upside [24]. Crowdsourcing can and is being incorporated into traditional localisation enterprise-oriented platforms as a means of harnessing the community. Motivational drivers for contributors can include monetary rewards as well as the traditional altruistic returns.

Three issues need to be addressed when leveraging the crowdsourcing paradigm, these being

– Quality: how can quality be maximised?
– Motivation: what are the design guidelines for the construction of a crowd-sourcing infrastructure that motivates participation and longevity?
– Control: what protocols will maximise quality and contributions?

This research focuses on the issue of quality. Given a crowdsourcing paradigm for localisation in which multiple translations are allowed for each translation unit, and where translations have been rated by the community using a declarative rating mechanism such as thumbs up or thumbs down, can the system facilitate the project manager identify the higher quality solutions? This research is focused on the specification of an algorithm that generates an overall rating for a translation based upon the application of procedural rating mechanisms to declarative ratings. Examples of procedural rating mechanisms are Friend Of A Friend (FOAF) [23], Pootle Translate Toolkit Metric, Bayesian Rating, etc. The authors argue that a strategy based on selecting translations with the highest average declarative rating is insufficient to guarantee better than average quality because this approach fails to take cognizance of biases that raters may have towards certain translators, the competency of the rater, the previous track record of the translator, and other such considerations. These factors require a much richer fully automated aggregate rating mechanism. Each component in the aggregate is weighted, and computation of the set of weights will be domain specific for each application area. The SOLAS rating component is still in the research phase focused on the capture of sufficient data to enable computation of weights through machine learning paradigms [8].

4   Discussion and Future Work

In this paper, we highlighted some of the most important challenges in localisation, which will need a substantial effort in the near future in order to address the unimaginably vast amount of content that need to be localised. Indeed, a leading market research firm, namely Common Sense Advisory, claims that only 0.5% of what needs to be localised is actually being localised today. This content is getting ever vaster more rapidly than it never had; as an example, enterprise data is predicted to grow by 650% in the coming 3 years.

In this work, we argued that flexible localisation web services can be an efficient solution provided that we address issues in interoperability, in the discovery, mapping, and services intercommunication, in providing intelligence to the system to assess and deal with a wide range of projects and content, and in rich and effective metadata allowing a smooth flow of data and tasks.

As already mentioned, current tools, components, and services are developed and deployed by various sources, and use different localisation / translation standards, and various notations. Open standardised-based localisation web services constitute an important step towards our aim. Also, in this aspect, introducing semantic constraints or objects will help the user in validating the components, such as templates or components represented in ontologies for instance, which will include information about the metadata.

Augmenting the abstract level of the recommended workflow with semantic descriptions will also support the search of previous instances or templates in the case of similar processes, and reduce the cost of designing and conducting new workflows. Introducing the user experience in the workflow design would also constitute a valuable input to the workflow recommendation engine. For the mapping system, the possibility to allow more than one instance per service type (similar to parallel batching) would also be an interesting feature. This would allow a decrease in the execution time for very large or time critical tasks.

Prospective future components could include a dedicated media localisation engine focusing on the automated cultural adaptation of any embedded media to match the target locale of the project file, as while the primary focus of localisation today is on translation, and the transformation of text, there has been a large growth in media-rich content online in recent times. Indeed, music localisation itself has already been proposed as a relatively unexplored research area [19]. For example consider music clips, or other such media which may not be suitable for the target culture, being automatically replaced with their local equivalents as part of the workflow. This would require a database of possible media, stored by locale, clip length, intended domain/meaning, file type and so on, and a tag definition to support the location and replacement of such media, as while XLIFF supports image tagging, it does not as of yet support such tagging of audio clips, MIDI files, or video clips.

5   Conclusion

Automated natural language processing components and services are becoming increasingly viable and readily available. These advances in technologies, along with more efficient and flexible computational and infrastructural support, can play a critical role in the future of localisation. Indeed, this will allow to tackle the huge volume of content with high flexibility, efficient interoperability, and acceptable quality, for a variety of digital content.

In this paper, we have discussed the use of a service-oriented approach in the development of future generation localisation workflows, which integrate novel components in process recommendation, business process management, services mapping into actual instances, and other novel linguistic assets integration and reuse. The maturity of these techniques and standards are key aspects to the support of efficient localisation architectures. We are addressing the lack of standard interfaces and terminologies in providing the necessary support to large-scale, future generation service-oriented localisation.

References

[1] Morado Vazquez, L., Torres del Rey, J.: The relevance of metadata during the localisation process: An experiment. In: Internacional T3L Conference: Tradumatica, Translation Technologies & Localization, June 21-22, Universitat Autonoma de Barcelona, Spain

[2] Morado Vazquez, L., Mooney, S.: XLIFF Phoenix and LMC Builder: Organising, capturing and using localisation data and metadata. In: The Annual Conference Proceedings of the Localisation Research Centre, Brave New World, University of Limerick, LRC XV (2010)

[3] XLIFF TC (2008), XLIFF Version 1.2 Specification,
http://docs.oasis-open.org/xliff/xliff-core/xliff-core.pdf

[4] http://www.therosettafoundation.org/

[5] Losse, K.: Facebook - Achieving Quality in a Crowd-sourced Translation Environment. In: XIII Localisation4All, Limerick, Ireland (2008)

[6] Kindrick, J.D., Sauter, J.A., Matthews, R.S.: Improving conformance and interoperability testing. Standard View 4(1), 61–68 (1996)

[7] Shah, R., Kesan, J.: Interoperability challenges for open standards: ODF and OOXML as examples. In: Proceedings of the 10th Annual International Conference on Digital Government Research: Social Networks: Making Connections between Citizens, Data and Government (Digital Government Society of North America, 2009), pp. 56–62 (2009)

[8] Gupta, R., Collins, J.J., O'Keeffe Ian, R.: Collaborative Localisation Platform: How could one investigate and evaluate perceived quality of the translated text using various procedural and declarative rating mechanisms in a collaborative platform? Technical report, University of Limerick (2010)

[9] Gil, Y., et al.: Examining the Challenges of Scientific Workflows. IEEE Computer 40(12), 24–32 (2007)

[10] De Roure, D., Goble, C., Stevens, R.: The design and realisation of the ^{my}Experiment Virtual Research Environment for social sharing of workflows. Future Generation Computer Systems 25(5), 561–567 (2009)

[11] Cadieux, P.: Automated Localization Workflow A Reference Model: How CMS and GMS collaborate in producing multilingual enterprise content

[12] Bergmann, F.: The Race for Open Source. Multilingual Computing 19, 52–57 (2008)

[13] Sargent, B., DePalma, D.: Translation Management Systems: Assessment of Commercial and LSP specific TMS Offerings. Common Sense Advisory (2008)

[14] Hammwohner, R.: TransRouter revisited-Decision support in the routing of translation projects. In: Informationskompetenz Basiskompetenz in der Informationsgesellschaft, Internationalen Symposiums fur Informationswissenschaft, Darmstadt, UVK Verlagsgesellschaft mbH (2000)

[15] Reinhard Schaler, K.B., Bourke, M.: Linguistic Infrastructure for Localisation: Language Data, Tools and Standards. LRC - XI The Localisation Factory (2006)

[16] Aouad, L.M.: Mapping Localisation Workflows on the Cloud. In: Translating and the Computer, ASLIB 2009, vol. 31 (2009)

[17] Van Der Meer, J.: Impact of Translation Web Services. Localisation Focus 1(2), 9–11 (2002)

[18] Lewis, D., et al.: Web service integration for next generation localisation. In: Proceedings of the Workshop on Software Engineering Testing, and Quality Assurance For Natural Language Processing, ACL Workshops, June 05, pp. 47–55. Association for Computational Linguistics, Morristown (2009)

[19] OKeeffe, I.: Music Localisation: Active Music Content for Web Pages. Localisation Focus - The International Journal of Localisation 8(1), 67–81 (2009)

[20] Cobbold, G.L., Pontes, R.: Five steps from local to global. MultiLingual Computing & Technology (October/November 2007)

[21] Microsoft Manual of Style for Technical Publications. 3rd edn., Microsoft Press, Washington (2004)

[22] Howe, J.: Crowdsourcing: Why The Power Of The Crowd Is Driving The Future Of Business. Crowne Publishing (2008)

[23] Marmolowski, M.: Real-life Rating Algorithm. DERI Deliverable D1.6.3 (2008), http://sw.deri.ie/fileadmin/documents/DERI-TR-2008-05-22.pdf (accessed June 2, 2011)

[24] The World Summit on the Information Society (WSIS), http://groups.itu.int/wsis-forum2011/ (accessed June 2, 2011)

Design of Reference Model for Improvement
of Security Process

Eun-Ser Lee

Andong National University Computer Engineering 388 Songcheon-dong, Andong-city,
Gyeongsangbuk-do 760-749, South Korea
eslee@andong.ac.kr

Abstract. This paper is intended to proposal maturity model of security
process. There are many risk items that cause the security requirement problems
during software development. This paper evaluates the efficiency of security
lifecycle that detection of new risk items and remove ratio at the security
requirement lifecycle. For the similar domain projects, we can remove security
risk items and manage to progress them by using security lifecycle, which can
greatly improve the software process.

Keywords: Security process, Process improvement, Maturity model.

1 Introduction

Web Application security is a complex subject tha must be fully understood before
effective security testing can be accomplished[1][8].

Security tests are designed to probe vulnerabilities of the client-side environment,
the network communications that occur as data are passed from client to server and
back again, and the server-side environment. Each of these domains can be attacked,
and it is the job of the security tester to uncover weakness that can be exploited by
those with the intent to do so[5].

In this paper provides the level for the activity of security process. Also we can
identify maturity of security process to progress them by using concept of
improvement, which can greatly improve the all around system.

2 Related Works

2.1 ISO/IEC 15480

The Common Criteria (CC), is meant to be used as the basis for evaluation of security
properties of IT products and systems. By establishing such a common criteria base,
the results of an IT security evaluation will be meaningful to a wider audience[2][3].

The CC is useful as a guide for the development of products or systems with IT
security functions and for the procurement of commercial products and systems with
such functions. During evaluation, such an IT product or system is known as a Target
of Evaluation (TOE). Such TOEs include, for example, operating systems, computer
networks, distributed systems, and applications.

G. Lee, D. Howard, and D. Ślęzak (Eds.): ICHIT 2011, CCIS 206, pp. 383–390, 2011.
© Springer-Verlag Berlin Heidelberg 2011

The CC defines three types of requirement constructs: package, PP(Protect Profile) and ST(Security Target). The CC further defines a set of IT security criteria that can address the needs of many communities and thus serve as a major expert input to the production of these constructs. The CC has been developed around the central notion of using wherever possible the security requirements components defined in the CC, which represent a well-known and understood domain. Figure 1 shows the relationship between these different constructs[1][2].

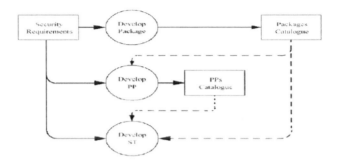

Fig. 1. Use of security requirements

2.2 Protection Profile

Evaluation criteria are most useful in the context of the engineering processes and regulatory frameworks that are supportive of secure TOE development and evaluation. This sub-clause is provided for illustration and guidance purposes only and is not intended to constrain the analysis processes, development approaches, or evaluation schemes within which the CC might be employed[2][3][4].

The CC requires that certain levels of representation contain a rationale for the representation of the TOE at that level. That is, such a level must contain reasoned and convincing argument that shows that it is in conformance with the higher level, and is itself complete, correct and internally consistent. Rationale directly demonstrating compliance with security objectives supports the case that the TOE is effective in countering the threats and enforcing the organizational security policy.

The CC layers the different levels of representation as described in Figure 2, which illustrates the means by which the security requirements and specifications might be derived when developing a PP or ST. All TOE security requirements ultimately arise from consideration of the purpose and context of the TOE. This chart is not intended to constrain the means by which PPs and STs are developed, but illustrates how the results of some analytic approaches relate to the content of PPs and STs.

2.3 ISO/IEC 15504

ISO/IEC 15504 defines the basis for process assessment. It is primarily addressed to the competent assessor and other stakeholders, such as the sponsor of the assessment, who need to be assured that the requirements of this International Standard have been met. It will also be of value to developers of assessment methods and of tools to support an assessment[7][8].

ISO/IEC 15504-2 sets out the minimum requirements for performing an assessment that ensure consistency and repeatability of the ratings. The requirements help to ensure that the assessment output is self-consistent and provides evidence to substantiate the ratings and to verify compliance with the requirements. ISO/IEC 15504-1 provides a general introduction to the concepts of process assessment and a glossary for assessment related terms. ISO/IEC 15504-3 provides guidance for interpreting the requirements for performing an assessment. ISO/IEC 15504 identifies the measurement framework for process capability and the requirements for:

a) Performing an assessment;
b) Process Reference Models;
c) Process Assessment Models;
d) Verifying conformity of process assessment.

Process assessment, as defined in this International Standard, is based on a two dimensional model containing a process dimension and a capability dimension.

The process dimension is provided by an external Process Reference Model, which defines a set of processes characterized by statements of process purpose and process outcomes.

3 Theory and Case Study

This chapter will be proposed a level to identify capability of security factor that needed during actual systems. Therefore, the structure and contents to improve the items and analyze the result are presented in this chapter.

In this paper is using published paper data. We are already published other paper about the security lifecycle. Section 3.1 is published data. And, the content is like the following.

3.1 Security Profile Lifecycle

First of all, we are making the lifecycle by the protection profile for the extraction of the security requirement. Protection profile progress is like following. And next steps are removes the risk items by the redundancy[6][9][10].

Protection profile identification

a) Protection profile developer must be provided the explanation of TOE by the part of profile.

b) TOE explanation must be described that product type of TOE and character of the general IT.

c) Evaluator must be confirmed that information was satisfied the all of the evidence requirement.

d) Evaluator must be confirmed that TOE explanation of quality and consistency.

e) Evaluator must be confirmed that TOE explanation and protection profile consistency of the relationship.

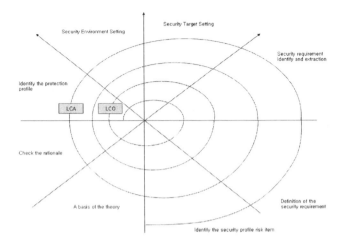

Fig. 2. Lifecycle milestone of the protection profile

In this figure is like the providing the activity of the risk item remove for the security requirement extraction.

Each of the factors is use of the repeatable milestone (LCO, LCA) for the progress management.

Each of stage is like the following.

Table 1. Description of the protection profile

Stage	Contents
Identify the protection profile	Identify and analysis the protection profile for the extraction of the security requirement.
Security Environment Setting	H/W and S/W, personal element setting for the security.
Security Target Setting	Security requirement Target Setting
Security requirement identify and extraction	Identify and extraction of the Security requirement get use of the UML (or other method)
Definition of the security requirement	Definition of the security requirement for the system building
Identify the security profile risk item	Identify the security profile risk item at the domain
A basis of the theory	Build the repeatable lifecycle for the Milestone (LCO, LCA)
Check the rationale	Check the rationale of the repeatable milestone and risk analysis

In this figure is like the providing the activity of the security requirement extraction for the security target.

Each of the factors is use of the repeatable milestone (LCO, LCA) for the progress management. Each of stage is like the following.

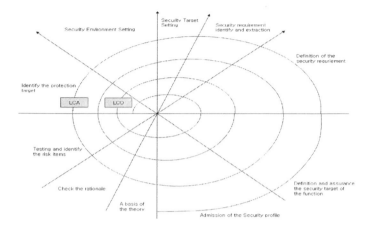

Fig. 3. Lifecycle milestone of the Security Target

Table 2. Description of the security target

Stage	Contents
Identify the protection target	Identify and analysis the protection target for the extraction of the security requirement
Security Environment Setting	H/W and S/W, personal element setting for the security.
Security Target Setting	Security requirement Target Setting
Security requirement identify and extraction	Identify and extraction of the Security requirement ge t use of the UML(or other method)
Definition of the security requirement	Definition of the security requirement for the system building
Identify the security profile risk item	Identify the security profile risk item at the domain
A basis of the theory	Build the repeatable lifecycle for the milestone (LCO, LCA)
Check the rationale	Check the rationale of the repeatable milestone and risk analysis
Testing and identify the risk items	Testing and identify for the extraction of new risk items.

3.2   Maturity Level of Security Process

In this paragraph is proposal that maturity level of security process. Therefore, we are needed classify the level. It is level consist of five levels. Each of level is like the following.

Table 3. Maturity level of security process

Level	Definition
Level 0	Not available.
Level 1	Being the activity and work product for security quality.
Level 2	Performance of Security activity according to planning.
Level 3	Definition of the standard of procedure and methodology for security activity.
Level 4	Management of quantified of security quality.
Level 5	Provide the consistency of security quality.

Table 4. is represent of each of levels for security process and management. Also, each of levels means and estimate of decision that is the following.

Table 4. Explain of security process level

Level	Explain
Level 0	This level is not being any of security process and activity.
Level 1	In this level must be input and output about security activity. Also security process attribute consists of activities and security process performance, security target.
Level 2	In this level must be performance policy and plan work product of security data. Also security process attribute consists of activities and security process performance, performance management, work product.
Level 3	In this level must be standard of security model and rule of management, policy of tailoring. Also security process attribute consists of activities and security process performance, performance management, work product, security process definition, security process tailoring, security process resource, security process customizing.
Level 4	In this level must be standard and management of measurement and alternative, control of security model. Also security process attribute consists of activities and security process performance, performance management, work product, security process definition, security process tailoring, security process resource, security process customizing, quantified of security data, classification of control.

Table 4. (*continued*)

Level 5	In this level must be provide changeable security process. Also security process attribute consists of activities and security process performance, performance management, work product, security process definition, security process tailoring, security process resource, security process customizing, quantified of security data, classification of control, quality of security data, quality of security process.

3.3 Case Study

We are evaluates the efficiency level of Security Process. Therefore, we are use of the numerical value for the level of Security Process. Also, we are compares before application with applying to the level of Security Process. In this result is like the following.

Table 5. Defection of risk number

Phase	Before application	Application of security process
Inception	33	25
Elaboration	35	30
Construction	40	30
Transition	50	28

Table.5 is shown that efficiency of the security process and numeric means detection of risk number. "Before application" is not applying to security process. Therefore, we are gained the result that security process is efficient rather than waterfall process.

4 Conclusion

In this paper we are proposed the maturity level of security process applicable to the extraction of security requirement. Also, in this paper provide the criteria for judgment in the development of security area. This security process is able to the various area. Therefore, various systems and types is require the architecture for the flexibility and security and usability of the support.

For the future studies will be provide the development tool for the applicable to extract of security requirement and evaluate the architecture of the personal security. And we are supplements architecture after applying of the real world. Also, we are analysis of the architecture efficiency against security of the defense and flexibility of rate. Therefore, we are put to use the reliable method of the evaluation.

References

1. Garfinkel, S., Spafford, G.: Web security, Privacy and commerce. O'Reilly & Associates, Sebastopol (2002)
2. ISO. ISO/IEC 15408-2:1999 Information technology - Security techniques - Evaluation criteria for IT security - Part 2: Security functional requirements
3. ISO. ISO/IEC 15408-3:1999 Information technology - Security techniques - Evaluation criteria for IT security - Part 3: Security assurance requirements
4. ISO/IEC Guide 65—General Requirements for Bodies Operating Product Certification Systems (1996)
5. Pressman, R.S.: A practice's approach, 6th edn (2005)
6. Lee, E.-s., Lee, K.W., Kim, T.-h., Jung, I.-H.: Introduction and evaluation of development system security process of ISO/IEC TR 15504. In: Laganá, A., Gavrilova, M.L., Kumar, V., Mun, Y., Tan, C.J.K., Gervasi, O. (eds.) ICCSA 2004. LNCS, vol. 3043, pp. 451–460. Springer, Heidelberg (2004)
7. Software Process Improvement Forum, KASPA SPI-7 (December 2002)
8. Dunn, R.H.: Software defect removal. McGraw-Hill, New York (1984)
9. Fenton, N., Ohlsson, N.: Quantitative analysis of faults and failures in a complex software system. IEEE Trans. Software Eng. 26, 797–814 (2000)
10. Lee, E., Lee, K.W., Lee, K.: Development Design Defect Trigger for Software Process Improvement. In: Ramamoorthy, C.V., Lee, R., Lee, K.W. (eds.) SERA 2003. LNCS, vol. 3026, pp. 185–208. Springer, Heidelberg (2004)

A Formal Framework for Aspect-Oriented Specification of Cyber Physical Systems

Lichen Zhang and Jifeng He

Software Engineering Institute
East China Normal University
Shanghai 200062, China
Zhanglichen1962@163.com

Abstract. Cyber physical systems consist of three parts: the dynamics and control (DC) parts, the communication part and computation part. In this paper, we propose an aspect-oriented specification framework for cyber physical systems. The proposed aspect–oriented formal framework is such a formwork. On the one hand, it can deal with continuous-time systems based on sets of ordinary differential equations. On the other hand, it can deal with discrete-event systems, without continuous variables or differential equations. We present a combination of the formal methods Timed-CSP, ZimOO and differential (algebraic) equations or differential logic. Each method can describe certain aspects of a cyber physical system: CSP can describe communication, concurrent and real-time requirements; ZimOO expresses complex data operations; differential (algebraic) equations model the dynamics and control (DC) parts. This aspect oriented formal specification framework simplifies the requirement analysis process of cyber physical systems. A case study of train control system illustrates the specification process of aspect-oriented formal specification for cyber physical systems.

Keywords: Aspect-oriented, Cyber Physical Systems, ZimOO, Timed-CSP, Differential Logic.

1 Introduction

Cyber-Physical Systems (CPS)[1] have the particularity to combine a discrete behavior, specified with traditional test and assignment operations, with a continuous behavior, specified by the mean of differential equations or inclusions. They primarily allow us to model a physical environment ruled by physical laws, which may be either purely continuous, or mixing discrete and continuous aspects.

Cyber physical systems consist of three parts: the dynamics and control (DC) parts, the communication part and computation part. The DC part is that of a predominantly continuous-time system, which is modeled by means of differential (algebraic) equations, or by means of a set of trajectories. The evolution of a hybrid system in the continuous-time domain is considered as a set of piecewise continuous functions of time. The computation part is that of a predominantly discrete-event system. A well-known model is a (hybrid) automaton, but modeling of discrete-event systems is also based on, among others, Z,VDM, process algebras, Petri nets, and data flow

G. Lee, D. Howard, and D. Ślęzak (Eds.): ICHIT 2011, CCIS 206, pp. 391–398, 2011.

languages. Clearly, cyber physical systems represent a domain where the DC, communication and computation aspects must be met, and we believe that a formalism that integrates the DC, communication and computation aspects is a valuable contribution towards integration of the DC, communication and computation methods, techniques, and tools.

Aspect oriented software development (AOSD)[2] is a relatively new type of development that simplifies maintenance and increases usability of software. Cyber physical systems could be separated into different crosscutting concerns and designed independently by using AOP techniques. When designing the system, aspects are analyzed and designed separately from the system's core functionality, such as real-time, security, error and exception handling, log, synchronization, scheduling, optimization, communication, resource sharing and distribution. After the aspects are implemented, they can be woven into the system automatically.

In this paper, we provide some ideas for the aspect–oriented formal specification of cyber physical systems and one well known case study to validate aspect-oriented formal specification.

2 Aspect-Oriented Formal Specification

A formal software specification is a statement expressed in a language whose vocabulary, syntax, and semantics are formally defined. Researchers have developed a large number of formal specification languages [3]. Many of them are suitable for expressing and analyzing particular characteristics of a system. For example, Z [4]is a formal language used to define data types and to show the effect of operations on these types. It lacks, however, features to express the order in which the operations are executed. Process algebras[5], like CSP, on other hand, are suitable for showing the order of the occurrence of events but lack the ability to handle complex abstract data types and operations. Finally, formalisms like temporal logics and its derivatives concentrate on time aspects. The increasing complexity of real-time systems has made the use of formal specification more frequent in this area, and new languages have proliferated. Timed CSP is an extension of CSP that includes new operators like *Wait* and _ (timeout). It has a new semantic model, derived from the untimed models of CSP, to represent dense time information, and a proof system. Timed CCS has an operational semantics; a communication on a described by $a(t)e$, e is restricted to happen in the closed time interaval , and, after the communication, the variable t holds the time at which it occurred. Proof rules for timed CCS have been elaborated and shown to be independent of the time domain used. Logic approaches to specification benefit from clear notations and automated validation using existing theorem provers. Modal logic adds time reasoning in logical formulae. Temporal logic is a modal logic in which new operators for quantification either in the future or in the past are included. To overcome difficulties with modularity, temporal logic is often used in combination with other techniques. An example is the work of Duke and Smith , in which temporal logic is used in the invariants of Z specifications to define liveness properties. Duration calculus (DC) is concerned with intervals instead of time instances. Real-time logic (RTL) extends predicate logic by relating events with the time in which they occur. In contrast to other logics, RTL allows specification of the absolute timing of events, and not only their relative ordering; it also provides a uniform way of incorporating different scheduling disciplines. Timed automata extends state-transition systems with finitely many real-valued clock variables that are used in annotations. Analysis is based on a finite quotient of the infinite space of clock valuations. Work has been carried out on verification algorithms, including heuristics, and tools. Petri Nets allow mathematical modeling of discrete event systems in terms of conditions and events, and the relationship between them. Time information has been added to Petri Nets in a number of forms. The most common approach is to add time delays to transitions. A similar approach assigns delays to places instead of transitions, and creates a delay between the time the token arrives in a place and the time it enables a transition to fire. A more flexible approach assigns intervals to the transitions and time stamps to the tokens. Specifications of complex systems normally involve a mixture of data types, operations, and time constraints; current research has focused on more comprehensive languages. Circus combines CSP, Z, specification statements and guarded commands to provide a notation for both specification and programming, and for verification by refinement. Time is continuous by nature, but a discrete representation of time is also satisfactory in most cases. In specification languages, time is often represented by real numbers, but in programming languages, time is represented by integers. In principle, the continuous time model is more appropriate because it can express time in both forms, and time in the real world is continuous. A continuous time model, however, cannot be implemented by a software system[3].Object-Z [6]is an object-oriented extension of the formal specification language Z. It adds, to Z, notions of classes and objects, and inheritance and polymorphism. TCOZ[7] is a blending of Object-Z and Timed CSP.

ZimOO[8] is based on Object-Z, an object-oriented extension of Z, A system specification in Z consists of a description of the global system state and a set of operations describing how this state can be changed. Object-Z provides additional means for describing systems in an object-oriented style, an Object-Z specification consists of a set of classes, each containing a state description and a set of operations Z and Object_Z support only specifications of discrete systems. ZimOO is an extended subset of Object-Z allowing descriptions of discrete and continuous features of a system in a common formalism ZimOO supports three different kinds of classes: discrete as in Object-Z, continuous and hybrid classes. Thus, the system can be structured better and the well-known suitable formalisms can be applied to describe, analyze, and refine the different parts of the system. The bridge between the continuous and the discrete world is built by hybrid classes.

The differential dynamic logic (dL) [9][0] is a logic for specifying and verifying hybrid systems [17][15]. The logic dL can be used to specify correctness properties for hybrid systems given operationally as hybrid programs [9][10]. The basic idea for dL formulas is to have formulas of the form $[\alpha]\varphi$ to specify that the hybrid system α always remains within region φ, i.e., all states reachable by following the transitions of hybrid system α statisfy the formula φ. Dually, the dL formula $<\alpha>\varphi$ expresses that the hybrid system α is able to reach region φ, i.e., there is a state reachable by following the transitions of hybrid system α that statisfies the formula φ.

For instance, the following formula expresses that for the state of a train controller train, the property $y \leq m$ always holds true when starting in a state where $v^2 \leq 2b(m-y)$ is true: $v^2 \leq 2b(m-y) \rightarrow [train]y \leq m$[9][10].

Aspect-oriented approaches use a separation of concern strategy, in which a set of simpler models, each built for a specific aspect of the system, are defined and analyzed.

Each aspect model can be constructed and evolved relatively independently from other aspect models. Aspect-oriented specification is made by extending TCOZ and ZIMOO notation with aspect notations. The schema for aspect specification in has the general form as shown in Fig.1, Fig.2 and Fig.3.

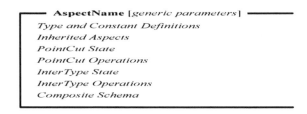

Fig. 1. Aspects of Model Structure

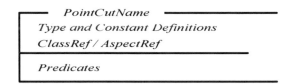

Fig. 2. PointCut Operation Schema of Structure

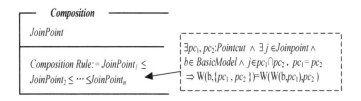

Fig. 3. Composition Schema of Structure

3 Case Study: Aspect-Oriented Formal Specification of a Train Control Systems

The Problem that must be addressed in operating a railway are numerous in quantity, complex in nature, and highly inter-related . For example, collision and derailment, rear-end, head-on and side-on collisions are very dangers and may occur between trains. Trains collide at level crossing. Derailment is caused by excess speed, wrong switch position and so on. The purpose of train control is to carry the passengers and goods to their destination, while preventing them from encountering these dangers. Because of the timeliness constraints, safety and availability of train systems, the design principles and implementation techniques adopted must ensure to a reasonable extent avoidance of design errors both in hardware and software. Thus a formal technique relevant to design should be applied for train systems development. The purpose of our exercise is to apply aspect-oriented formal methods to develop a controller for train systems In order to keep the description focused, we concentrate on some particular points in train control systems rather than the detailed descriptions of all development process. The specification is made by integrating Object-Z, Timed CSP, ZimOO and DL[9][10].

Assuming the train starts in a controllable state, the following global and unbounded-horizon safety formula about the system y≤m holds. As system invariant we choose[9][10]:

$$inv \equiv v^2 \leq 2b(m-y) \wedge \varepsilon > 0 \wedge v \geq 0$$

Which expresses that it is possible to completely stop the train within the distance left to the end of the movement authority. This constraint describes a controllable state of the train and therefore we choose inv as initial configuration of the system. When an movement authorities(MA) has been granted up to the track position m and the train is currently located at position y then dL can analyse, for example, the following safety statement about the (simplified) acceleration system[9][10]:

$$\Psi \rightarrow [((m-y < s?; a :=-b) [(m-y \geq 2s?; a := 0.1)); \ddot{y}= a] \ y < m$$

It expresses that, under a condition Ψ about parameters, trains always remain within their MA m. Further, it specifies that the train decelerates using engine brakes of force b if the safety envelope is underrun (m−y < s). It slowly accelerates if there is sufficient distance (m − y≥2s).

A train controller [11][12]limits the speed of the train, decides when it is time to switch points and secure crossings, and makes sure that the train does not enter them too early as shown in Fig.4.

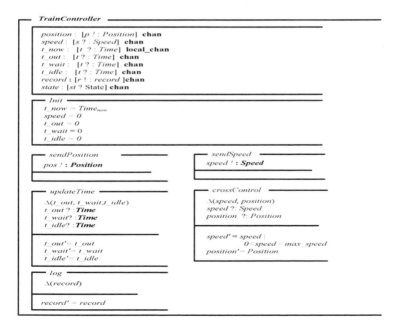

Fig. 4. Control Model of Train Dispatching System

In automatic railway crossing system, The sensors detect the presence of the train near rail crossing and barrier shuts down when train is approaching to the railway crossing[11][12]. Once train crosses the rail crossing barriers opens by itself as shown in Fig.5 and Fig.6.

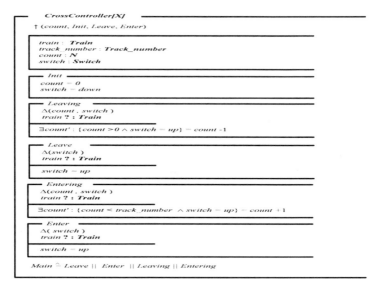

Fig. 5. Trains through the intersection mode

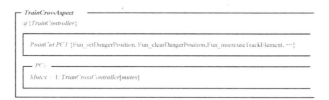

Fig. 6. Cross roads aspects model

We integrates CSP and Object-Z with aspect-oriented technique to specify real-time aspect as shown in Fig.7 and Fig.8.

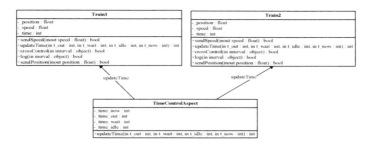

Fig. 7. Aspects of Time Model

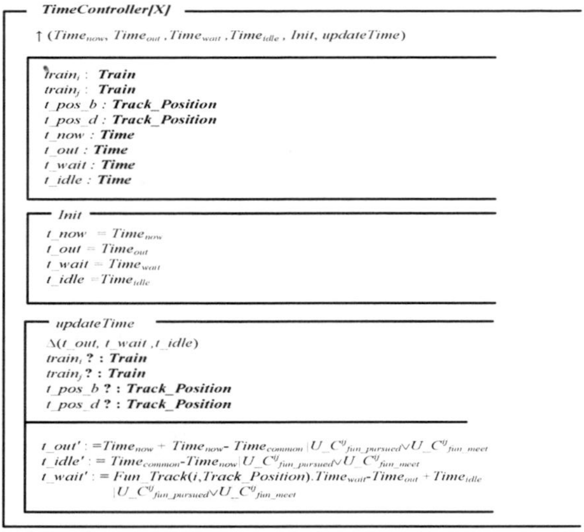

┌─ *TimeController[X]* ───────────────────────────────
│ ↑ (*Time*$_{now}$, *Time*$_{out}$, *Time*$_{wait}$, *Time*$_{idle}$, *Init*, *updateTime*)
│
│ ┌─────────────────────────────────
│ │ *train*$_i$: **Train**
│ │ *train*$_j$: **Train**
│ │ *t_pos_b* : **Track_Position**
│ │ *t_pos_d* : **Track_Position**
│ │ *t_now* : **Time**
│ │ *t_out* : **Time**
│ │ *t_wait* : **Time**
│ │ *t_idle* : **Time**
│ │
│ │ ┌─ *Init* ───────────────
│ │ │ *t_now* = *Time*$_{now}$
│ │ │ *t_out* = *Time*$_{out}$
│ │ │ *t_wait* = *Time*$_{wait}$
│ │ │ *t_idle* = *Time*$_{idle}$
│ │
│ │ ┌─ *updateTime* ─────────────────
│ │ │ Δ(*t_out*, *t_wait*, *t_idle*)
│ │ │ *train*$_i$? : **Train**
│ │ │ *train*$_j$? : **Train**
│ │ │ *t_pos_b* ? : **Track_Position**
│ │ │ *t_pos_d* ? : **Track_Position**
│ │ │
│ │ │ *t_out'* := *Time*$_{now}$ + *Time*$_{now}$- *Time*$_{common}$ [$U\_C^{ij}_{fun\_pursued}$∨$U\_C^{ij}_{fun\_meet}$]
│ │ │ *t_idle'* := *Time*$_{common}$-*Time*$_{now}$ [$U\_C^{ij}_{fun\_pursued}$∨$U\_C^{ij}_{fun\_meet}$]
│ │ │ *t_wait'* := *Fun_Track*(*i*,*Track_Position*).*Time*$_{wait}$-*Time*$_{out}$ + *Time*$_{idle}$
│ │ │ [$U\_C^{ij}_{fun\_pursued}$∨$U\_C^{ij}_{fun\_meet}$]

Fig. 8. Time Aspect Formalization Method

Logs monitor system behavior as shown in Fig.9.

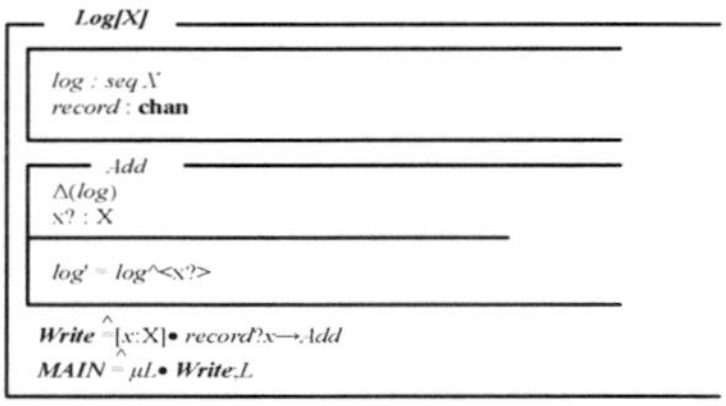

┌─ *Log[X]* ───────────────────
│ *log* : *seq X*
│ *record* : **chan**
│
│ ┌─ *Add* ─────────
│ │ Δ(*log*)
│ │ *x*? : *X*
│ │
│ │ *log'* = *log*⌢⟨*x*?⟩
│
│ *Write* ≙ [*x*:X]• *record*?*x*→*Add*
│ *MAIN* = μL• *Write*;L

Fig. 9. Log Class Model of Train

Finally, woven model is shown as Fig.10

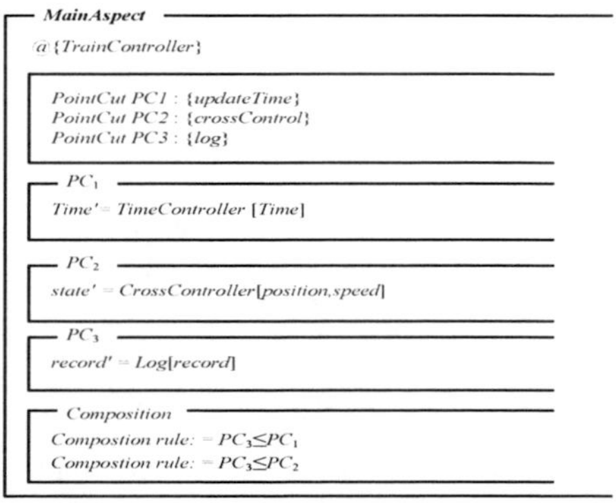

┌─ *MainAspect* ───────────────────────────
│ @ {*TrainController*}
│
│ *PointCut PC1* : {*updateTime*}
│ *PointCut PC2* : {*crossControl*}
│ *PointCut PC3* : {*log*}
│
│ ┌─ *PC*$_1$ ───────────────
│ │ *Time'* = *TimeController* [*Time*]
│
│ ┌─ *PC*$_2$ ───────────────
│ │ *state'* = *CrossController*[*position*,*speed*]
│
│ ┌─ *PC*$_3$ ───────────────
│ │ *record'* = *Log*[*record*]
│
│ ┌─ *Composition* ───────────────
│ │ *Compostion rule:* = *PC*$_3$≤*PC*$_1$
│ │ *Compostion rule:* = *PC*$_3$≤*PC*$_2$

Fig. 10. Woven Aspects of Diagram

4 Conclusion

In this paper we proposed to use aspect-oriented formal specification for cyber physical systems based on the combination of the formal methods Timed-CSP, ZimOO and differential (algebraic) equations or differential logic. A case study of train control system was used to illustrate the specification process of aspect-oriented formal specification for cyber physical systems.

The further work is devoted to integrated aspect-oriented formal specification with UML further.

Acknowledgments. This work is supported by National Natural Science Foundation of China under Grant.No.90818008 and No. 61021004), National HighTechnology Research and Development Program ofChina (No. 2011AA010101), National Basic Research Program of China (No. 2011CB302904), Docto-ral Program Foundation of Institutions ofHigher Education of China (No. 200802690018). The authors alsogratefully acknowledge support from the Danish National Research Foundation and The National Natural Science Foundation of China(Grant No. 61061130541) for the Danish-Chinese Center for Cyber Physical Systems.

References

1. Sha, L., Gopalakrishnan, S., Liu, X., Wang, Q.: Cyber-Physical Systems: A New Frontier. Springer, Heidelberg (2009); ISBN 978-0-387-88734-0
2. Kiczales, G., et al.: Aspect-Oriented Programming. In: Aksit, M., Auletta, V. (eds.) ECOOP 1997. LNCS, vol. 1241, Springer, Heidelberg (1997)
3. Sherif, A., Cavalcanti, A., He, J., Sampaio, A.: A process algebraic framework for specification and validation of real-time systems. Formal Asp. Comput. 22(2), 153–191 (2010)
4. Spivey, J.: The Z Notation: A Refernce Manual, 2nd edn. Prentice Hall, UK (1992)
5. Davies, J., Schneider, S.: A Brief History of Timed CSP. Theoretical Computer Science 138(1), 243–271 (1995)
6. Smith, G.: The Object-Z Specification Language. Software Verification Research Centre University of Queensland (2000)
7. Mahony, B.P., Dong, J.S.: Blending Object-Z and Timed CSP: An introduction to TCOZ. In: ICSE 1998 (April 1998)
8. Friesen, V.: An Exercise in Hybrid System Specification Using an Extension of Z, http://citeseerx.ist.psu.edu/viewdoc/download?doi=10.1.1.30.2010&rep
9. Platzer, A.: Differential dynamic logic for hybrid systems. Journal of Automated Reasoning 41(2), 143–189 (2008)
10. Platzer, A.: Differential dynamic logic for verifying parametric hybrid systems. In: Olivetti, N. (ed.) TABLEAUX 2007. LNCS (LNAI), vol. 4548, pp. 216–232. Springer, Heidelberg (2007)
11. Hoenicke, J.: Specification of Radio Based Railway Crossings with the Combination of CSP, OZ, and DC, http://citeseerx.ist.psu.edu/viewdoc/summary?doi=10.1.1.21.4394
12. Faber, J., Jacobs, S.: Viorica Sofronie-Stokkermans Verifying CSP-OZ-DC Specifications with Complex Data Typesand Timing Parameters. Integrated Formal Methods (July 3, 2007)

Aspect-Oriented QoS Specification for Cyber-Physical Systems

Lichen Zhang and Jifeng He

Software Engineering Institute East China Normal University
Shanghai 200062, China
Zhanglichen1962@163.com

Abstract. Cyber-physical systems having quality-of-service (QoS) requirements driven by the dynamics of the physical environment in which they operate. The development of cyber physical is challenging due to conflicting quality-of-service (QoS) constraints that must be explored as trade-offs among a series of alternative design decisions. The ability to model a set of possible design alternatives and to analyze and simulate the execution of the representative model helps derive the correct set of QoS parameters needed to satisfy cyber physical system requirements. This paper proposes an aspect-oriented QoS modeling method based on UML and formal methods. We use an aspect-oriented profile by the UML meta-model extension, and model the crosscutting concerns by this profile. In this paper, we build the aspect-oriented model for specifying Quality of Service (QoS) based on the combination of UML, RTL and ERTL. An examples depicts how aspect-oriented methods can be used.

Keywords: QoS, Cyber Physical Systems, Aspect-Oriented, ERTL.

1 Introduction

A cyber-physical system (CPS)[1] is a system featuring a tight combination of, and coordination between, the system's computational and physical elements. Today, a pre-cursor generation of cyber-physical systems can be found in areas as diverse as aerospace, automotive, chemical processes, civil infrastructure, energy, healthcare, manufacturing, transportation, entertainment, and consumer appliances. This generation is often referred to as embedded systems. In embedded systems the emphasis tends to be more on the computational elements, and less on an intense link between the computational and physical elements.The dependability of the software [1]has become an international issue of universal concern, the impact of the recent software fault and failure is growing, such as the paralysis of the Beijing Olympics ticketing system and the recent plane crash of the President of Poland. Therefore, the importance and urgency of the digital computing system's dependability began arousing more and more attention. A digital computing system's dependability refers to the integrative competence of the system that can provide the comprehensive capacity services, mainly related to the reliability, availability, testability, maintainability and safety. With the increasing of the importance and urgency of the

G. Lee, D. Howard, and D. Ślęzak (Eds.): ICHIT 2011, CCIS 206, pp. 399–406, 2011.

software in any domain, the dependability of the distributed real-time system should arouse more attention.[2] Fundamental limitations for Cyber-Physical Systems (CPS) include:

• Lack of good formal representations and tools capable of expressing and integrating multiple viewpoints and multiple aspects. This includes lack of robust formal models of multiple abstraction layers from physical processes through various layers of the information processing hierarchy; and their cross-layer analyses.

• Lack of strategies to cleanly separate safety-critical and non-safety-critical functionality, as well as for safe composition of their functionality during humanin-the-loop operation.

• Ability to reason about, and tradeoff between physical constraints and QoS of the CPS.

Aspect-oriented programming (AOP) [3] is a new software development technique, which is based on the separation of concerns. Systems could be separated into different crosscutting concerns and designed independently by using AOP techniques. Every concern is called an "aspect". Before AOP, as applications became more sophisticated, important program design decisions were difficult to capture in actual code. The implementation of the design decisions were scattered throughout, resulting in tangled code that was hard to develop and maintain. But AOP techniques can solve the problem above well, and increase comprehensibility, adaptability, and reusability of the system. AOSD model separates systems into tow parts: the core component and aspects.

With the deepening of the dependable computing research, the system's dependability has becoming a important direction of the distributed real-time system, the modeling and design of dependable and distributed real-time system has become a new field. The dependable real-time system has a high requirement of reliability, safety and timing, these non-functional properties dispersed in the various functional components of system, so the Object-Oriented design has lost its superiority very obviously. The QoS of dependable real-time system [4]is very complex, currently the QoS research still hasn't a completely and technical system, and there isn't any solution meeting all the QoS requirements. We design the QoS of dependable real-time system as a separate Aspect using AOP, and proposed the classification of complex QoS, divided into the timing, reliability and safety and other sub-aspects. These sub-aspects inherit t members and operations from the abstract QoSaspect. We design each sub-aspects through aspect-Oriented modeling, to ensure the Quality of dependable real-time system meeting the requirements of the dependability.

This paper introduce the aspect-oriented modeling method based on UML, mainly form an aspect-oriented profile by the UML meta-model extension[5], and model the crosscutting concerns by this profile. Then we make a in-depth study of UML extension of QoS, including general model, framework, metamodel, profile and the QoS catalog, and highlights the dependability characteristic of QoS, the QoS aspect inherits the properties of the QoS model. Finally, we analyze the QoS aspect- oriented modeling via an example of fire real-time system, dividing the complex QoS into some sub-aspects, and the time, capacity and level aspect inherit the members and opera problems, enhancing the system's modular degree, lowering coupling between modules.

2 Aspect-Oriented Specification of QoS

The integration of physical systems and processes with networked computing has led to the emergence of a new generation of engineered systems: Cyber-Physical System s(CPS). Such systems use computations and communication deeply embedded in and interacting with physical processes to add new capabilities to physical systems. These cyber-physical systems range from miniscule (pace makers) to large-scale (the national power-grid). Because computer-augmented devices are everywhere, they are a huge source of economic leverage. A CPS is a system in which computation/information processing and physical processes are so tightly integrated that it is not possible to identify whether behavioral attributes are the result of computations (computer programs), physical laws, or both working together; Real/Continuous time models take physical durations into account. These are important for doing various time-related analyses (e.g., deadline matches) and, in particular, for real-time scheduling as in RMA approaches [8]. They are also used for modeling the temporal characteristics of the physical environment or system with which the embedded system is interacting (usually before discretization).A real-time and embedded modeling language for cyber physical systems needs concepts for dealing with different models of time. In order to meet the challenge of cyber-physical system design, We need to realign abstraction layers in design flows and develop semantic foundations for composing heterogeneous models and modeling languages describing different physics and logics.

One of the fundamental challenges in research related to CPSs is accurate modeling and representation of these systems. The main difficulty lies in developing an integrated model that represents both cyber and physical aspects with high fidelity.Among existing techniques, aspect-oriented modeling is a suitable choice, as it can encapsulate diverse attributes of cyber physical systems.

The control-centric nature of the programmatic QoS adaptation extends beyond software concepts, e.g., issues such as stability and convergence become paramount. In cyber physical system, QoS is specified in software parameters, which have a significant impact on the dynamics of the overall physical system. Due to complex and non-linear dynamics, it is hard to tune the QoS parameters in an *ad hoc* manner without compromising the stability of the underlying physical system. The QoS adaptation software is, in effect, equivalent to a controller for a discrete, non-linear system. Sophisticated tools are therefore needed to design, simulate, and analyze the QoS adaptation softwarefrom a control system perspective.

Programmatic QoS adaptation approaches offer a lower level of abstraction, i.e., textual code based. For example, implementing a small change to CDL-based adaptation policies requires manual changes that are scattered across large portions of the cyber physical system, which complicates ensuring that all changes are applied consistently. Moreover, even small changes can have far-reaching effects on the dynamic behavior due to the nature of emergent crosscutting properties, such as modifying the policy for adjusting communication bandwidth across a distributed surveillance system.

QoS provisioning also depends on the performance and characteristics of specific algorithms that are fixed and cannot be modified, such as a particular scheduling algorithm or a specific communication protocol. These implementations offer fixed

QoS and offer little flexibility in terms of tuning the QoS. Consequently, any QoS adaptation along this dimension involves structural adaptation in terms of switching implementations at run-time, which is highly complex, and in some cases infeasible without shutting down and restarting applications and nodes. Moreover, issues of state management and propagation, and transient mitigation, gain prominence amid such structural adaptations. Programmatic QoS adaptation approaches often offer little or no support for specifying such complex adaptations.

In Object-Oriented Programming, crosscutting concerns are elements of software that can not be cleanly captured in a method or class. Accordingly, crosscutting concerns has to be scattered across many classes and methods. OOP fails to provide a robust and extensible solution to handle these crosscutting concerns. AOP is a new modularity technique that aims to cleanly separate the implementation of crosscutting concerns. It builds on Object-Orientation, and addresses some of the points that are not addressed by OO. AOP provides mechanisms for decomposing a problem into functional components and aspectual components called aspects[4]. An aspect is a modular unit of crosscutting the functional components, which is designed to encapsulate state and behavior that affect multiple classes into reusable modules. Distribution, logging, fault tolerance, real-time and synchronization are examples of aspects. The AOP approach proposes a solution to the crosscutting concerns problem by encapsulating these into an aspect, and uses the weaving mechanism to combine them with the main components of the software system and produces the final system. We think that the phenomenon of handling multiple orthogonal design requirements is in the category of crosscutting concerns, which are well addressed by aspect oriented techniques. Hence, we believe that system architecture is one of the ideal places where we can apply aspect oriented programming (AOP) methods to obtain a modularity level that is unattainable via traditional programming techniques. To follow that theoretical conjecture, it is necessary to identify and to analyze these crosscutting phenomena in existing system implementations. Furthermore, by using aspect oriented languages, we should be able to resolve the concern crosscutting and to yield a system architecture that is more logically coherent. It is then possible to quantify and to closely approximate the benefit of applying AOP to the system architecture.

UML is acquainted to be the industry-standard modeling language for the software engineering community, and it is a general purpose modeling language to be usable in a wide range of application domains. So it is very significant to research aspect-oriented real-time system modeling method based on UML[6]. However they didn't make out how to model real-time systems, and express real-time feature as an aspect. In this section, we extend the UML, and present an aspect-oriented method that model the real-time system based on UML and Real-Time Logic (RTL). Real Time Logic is a first order predicate logic invented primarily for reasoning about timing properties of real-time systems. It provides a uniform way for the specification of both relative and absolute timing of events. Real Time Logic (RTL) was introduced as a formalism for reasoning about the relative and absolute timing properties of computational tasks of discrete real-time systems. Extended Real Time Logic (ERTL)[13] is a formalism for the modeling and analysis of relative and absolute timing properties of cyber physical systems (systems that combine continuous variables and discrete event dynamics). The extensions provided by ERTL enable the modelling of system behaviour ranging from activities of the physical entities that form part of the environment of a computing system, to the temporal ordering of the computational tasks of the computing system itself, thus providing a formal notation that can be used in all stages of software development. The specification of the Object Constraint Language (OCL) [7] is a part of the UML specification, and it is not intended to replace existing formal languages, but to supplement the need to describe the additional constraints about the objects that cannot be easily represented in graphical diagrams, like the interactions between the components and the constraints between the components' communication[9].

As the QoS concern needs to be considered inmost parts of the system, it is a cross-cutting concern.[10] Cross-cutting concerns are concerns that span multiple objects or components. Cross-cutting concerns need to be separated and modularized to enable the components to work in different configurations without having to rewrite the code. If the code for handling such a concern is included in a component, it can make the component tied to a specific configuration. This code will typically be scattered all over the component implementation and tangled with other code in the component. Modularizing it will make it more robust for change, and separating it totally from the component implementation will save the component programmers from having to implement it. Aspect oriented programming is a new method formodularizing cross-cutting concerns. By using AOP, concerns can be modularized in an aspect and later weaved into the code.

3   Aspect Oriented Model of QoS of Fire Alarm System

An automatic fire alarm system is designed to detect the unwanted presence of fire by monitoring environmental changes associated with combustion. In general, a fire alarm system is either classified as automatically actuated, manually actuated, or both. Automatic fire alarm systems can be used to notify people to evacuate in the event of a fire or other emergency, to summon emergency services, and to prepare the structure and associated systems to control the spread of fire and smoke.

QoS Constraint Q1,Q2, Q5 of the fire alarm system is expressed as follows with formal technique RTL and ERTL.[11] [13]:

[Q1]:
$\forall i \exists j @ (\uparrow data.collect, j) @ (\downarrow stop, i) \geq COLLECT\_MIN\_TIME \wedge @ (\downarrow data.open, j) - @ (\downarrow stop, i) \leq COLLECT\_MAX\_TIME$

[Q2]: $\forall i \exists j @ (\uparrow data.process, j) - @ (\downarrow stop, i) \geq DATA\_PROCESS\_MIN\_TIME \wedge @ (\downarrow data.process, j) - @ (\downarrow stop, i) \leq COLLECT\_MAX\_TIME$

[Q5]: $\forall i \exists j @ (\uparrow alarm.process., j) - @ (\downarrow command.send, i) \geq ALARM\_PROCESS\_MINTI ME \wedge @ (\uparrow alarm.process, j) - @ (\downarrow command.send, i) \leq ALARM\_PROCESS\_MAXTIME$

QoSConstraint Q3 and Q4 of the fire alarm system is expressed as follows with XML[12]:
[Q3]: <QoS *type*="Level">

<Firelevel val = "FIRE_MAX_LEVEL"/>
</QoS>
[Q4]: <QoS *type*="Contraint">
    <frame_rate val = "FRAME_RATE_CONSTRAINT"/>
    <audio_sample_rate val = "FRAME_RATE_CONSTRAINT"/>
    </QoS>

We separate QoS from real-time fire system as an aspect, the aspect-oriented model of QoS of real-time fire system is shown as Fig.1.

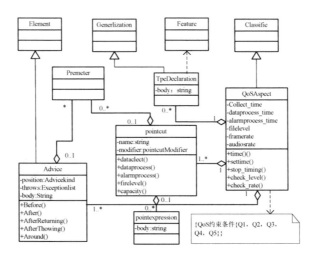

Fig. 1. QoS aspect-oriented model of Fire real-time system

We use QoSAspect to express QoS of real-time fire alarm system. The class diagram of Fire Real-time System with the aspect-oriented extension is shown as Fig.2.

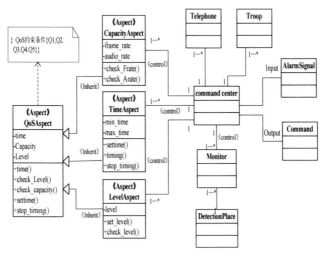

Fig. 2. Class Diagram of Fire Real-time System

The QoS Aspect Weaving Diagram of real-time fire alarm system is shown as Fig.3.

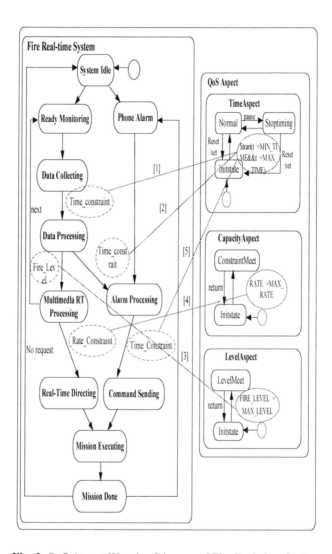

Fig. 3. QoS Aspect Weaving Diagram of Fire Real-time System

4   Conclusion

In this paper, we presented an aspect-oriented model for specifying Quality of Service (QoS) based on the combination of UML, RTL and ERTL. Two types of notation, graphical (semi-formal) and, respectively, formal, can efficiently complement each other and provide the basis for an aspect-oriented specification approach that can be both rigorous and practical for QoS modeling. An examples depicted how aspect-oriented methods can be used during QoS analysis and design process.

Future works will focus an automatic weaver for aspect oriented model of QoS.

Acknowledgments. This work is supported by National Natural Science Foundation of China under Grant. No. 90818008 and No. 61021004), National High Technology Research and Development Program of China (No. 2011AA010101), National Basic Research Program of China (No. 2011CB302904), Docto-ral Program Foundation of Institutions of Higher Education of China (No. 200802690018). The authors also gratefully acknowledge support from the Danish National Research Foundation and The National Natural Science Foundation of China (Grant No. 61061130541) for the Danish-Chinese Center for Cyber Physical Systems.

References

1. Lee, E.A.: Cyber physical systems:Design challenges. In: Proc. of the 11th IEEE Int. Symp. on Object Oriented Real-Time Distributed Computing, pp. 363–369. IEEE, NJ (2008)
2. Svizienis, A., Laprie, J.C., Randell, B.: Dependability of computer systems: Fundamental concepts, terminology, and examples. Technical report, LAAS-CNRS (October 2000)
3. Kiczales, G., et al.: Aspect-Oriented Programming. In: Proceedings of the 11th European Conference on Object-Oriented Programming (June 1997)
4. Frolund, S., Koistinen, J.: Quality of Service Specification in Distributed Object Systems. IEE/BCS Distributed Systems Engineering Journal 5, 179–202 (1998)
5. Aldawud, Elrad, T., Bader, A.: A UML Profile for Aspect Oriented Modeling. In: Workshop on AOP (2001)
6. Clemente, P.J., Sánchez, F., Perez, M.A.: Modelling with UML Component-based and Aspect Oriented Programming Systems. In: Seventh International Workshop on Component-Oriented Programming at European Conference on Object Oriented Programming (ECOOP), Málaga, Spain, pp. 1–7 (2002)
7. Lavazza, L., Quaroni, G., Venturelli, M.: Combining UML and formal notations for modeling real-time systems. ACM SIGSOFT Software Engineering Notes 26, 196–206 (2001)
8. Object Management Group. UML Profile for Modeling Quality of Service and Fault Tolerance Characteristics and Mechanisms Joint Revised Submission, OMG Document realtime/03-05-02 edition (May 2003)
9. OMG UMLTM Profile for Schedulability, Performance, and Time Specification (2005), http://www.omg.org/cgi-bin/doc?formal/2005-01-02
10. Wehrmeister, M.A., Freitas, E.P., Pereira, C.E., et al.: An Aspect-Oriented Approach for Dealing with Non-Functional Requirements in a Model-Driven Development of Distributed Embedded Real-Time Systems. In: 10th IEEE International Symposium on Object and Component-Oriented Real-Time Distributed Computing, May7-9, pp. 428–432. IEEE Computer Society, Greece (2007)
11. Jahanian, F., Mok, A.K.: Safety Analysis of Timing Properties in Real-Time Systems. IEEE Trans. Software Eng. 12(9), 890–904 (1986)
12. Frolund, S., Koistinen, J.: QML: A Language for Quality of Service Specification, Technical Report HPL-98-10 (Februry 1998)
13. de Lemos, R., Hall, J.: ERTL: An extension to RTL for requirements analysis for hybrid systems. Technical report, Department of Computing Science, University of Newcastle upon Tyne, UK (1995)

V & V Practices of SP Quality Certification Model

Sun Myung Hwang

Computer Engineering Department, Daejeon University
sunhwang@dju.kr

Abstract. In order to make high quality software and high reliability software, systematic development management and organizational support are essential needed. Since 1990's Software process models such as ISO/IEC 15504 and CMMI have been used to improve organization capability. Korean Software Process Quality Certification is established by law in 2008 year. The way of eastern Asian's thinking and western thinking is different This model contains the difference. Especially, verification and validation practice implementation of eastern Asian is different from western.

Keywords: ISO/IEC 15504, CMMI, Quality Certification, Validation and Verification.

1 Introduction

In this paper, it will be introduced how Korean Software Process Model contains verification and validation process, and the example of implementation for verification and validation process.

Korean Software Process Quality Certification Model does not have any process area for verification and validation but, has detail practices in many processes.

An organization having higher development maturity has better software process in the overall areas of their organization and they can implement software more consistently. Software process capability refers to the ability of the organization to produce these products predictably and consistently. A capability level is a set of process attributes that work together to provide a major enhancement in the capability to perform a process. Each level provides a major enhancement of capability in the performance of a process [1].

2 Software Process Model

2.1 ISO/IEC 15504

ISO/IEC 15504 also assesses process maturity by identifying present process condition to support organization's process enhancement activities.

ISO/IEC 15504 in Table 1 consists of 3 categories, 10 process groups and 48 processes.

G. Lee, D. Howard, and D. Ślęzak (Eds.): ICHIT 2011, CCIS 206, pp. 407–414, 2011.

Table 1. ISO/IEC 15504 process

Category	Process	

PRIMARY
Life Cycle Processes
	1. Acquisition Group
	  ACQ.1 Acquisition Preparation
	  ACQ.2 Supplier selection
	  ACQ.3 Supplier monitoring
	  ACQ.4 Customer acceptance
	2. Supply Group
	  SPL.1 Supplier tendering
	  SPL.2 Contract agreement
	  SPL.3 Software release
	  SPL.4 Software acceptance
	3. Operation group
	  OPE.1 Operational use
	  OPE.2 Customer support
		4. Engineering Group
		  ENG.1 Requirement elicitation
		  ENG.2 System requirement analysis
		  ENG.3 System architectural design
		  ENG.4 Software requirement analysis
		  ENG.5 Software design
		  ENG.6 Software construction
		  ENG.7 Software integration
		  ENG.8 Software testing
		  ENG.9 Software installation
		  ENG.10 System integration
		  ENG.11 System testing
		  ENG.12 System & software maintenance

ORGANIZATIONAL
Life Cycle Processes
	1. Management Group
	  MAN.1 Organizational alignment
	  MAN.2 Organization management
	  MAN.3 Project management
	  MAN.4 Quality Management
	  MAN.5 Risk Management
	  MAN.6 Measurement
	2. Process Improvement Group
	  PIM.1 Process establishment
	  PIM.2 Process assessment
	  PIM.3 Process improvement
		3. Resource & Infrastructure Group
		  RIN.1 Human resource management
		  RIN.2 Training
		  RIN.3 Knowledge management
		  RIN.4 Infrastructure
		4. Reuse Group
		  REU.1 Asset management
		  REU.2 Reuse program management
		  REU.3 Domain engineering

SUPPORTING
Life Cycle Processes
	1. Configuration control Group
	  CFG.1 Documentation Management
	  CFG.2 Configuration Management
	  CFG.3 Problem Management
	  CFG.4 Change Request Management
		2. Quality Assurance Group
		  QUA.1 Quality assurance
		  QUA.2 Verification
		  QUA.3 Validation
		  QUA.4 Joint review
		  QUA.5 Audit
		  QUA.6 Product Evaluation

2.2 CMMI

CMMI that consists of 5-levels, 4 categories and 21 processes, provides a framework for introducing new disciplines about systems engineering and software engineering as needs arise[2]. CMMI process can be simplified as a Table 2.

Table 2. CMMI Process

Category	Process Area	Level
Project Management	1. Project Planning(PP2) 2. Project Monitoring and Control (PMC) 3. Supplier Agreement Management (SAM)	2
	4. Integrated Project Management(IPM) 5. Risk Management(RSKM)	3
	6. Quantitatively Project Management (QPM)	4
Support	1. Configuration Management(CM) 2. Product and Process Quality Assurance(PPQA) 3. Measurement and Analysis(MA)	2
	4. Decision Analysis and Resolution (DAR)	3
	5. Causal Analysis and Resolution (CAR)	5

Table 2. *(continued)*

Engineering	1. Requirement Management(REQM)	2
	2. Requirement Development(RD) 3. Technical Solution(TS) 4. Product Integration(PI) 5. Verification(VER) 6. Validation(VAL)	3
Process Management	1. Organizational Process Focus(OPF) 2. Organizational Process Definition (OPD) 3. Organizational Training(OT)	3
	4. Organizational Process Performance (OPP)	4
	5. Organizational Innovation and Deployment(OID)	5

2.3 Korean Software Process Quality Certification Model

This model is korea standard software process certification model for korea small medium sized companies NIPA performs certification system using this model, and supports process improvement for raising software product quality in korea

This model consists of 3 Levels, 5 categories 17 assessment items and 76 detailed items as a figure 1.The certification level is differently applied to valuation factors by each certification degree as an indicator representing the degree of activity capability level related with software development project performance and its meaning is also different.

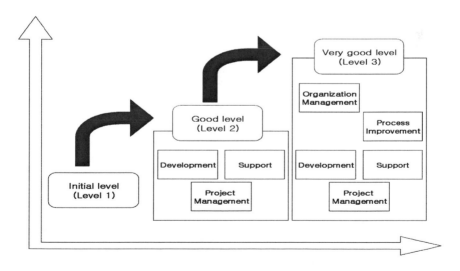

Fig. 1. steps of Korean Software Process Quality Certification model

Table 3 shows the characteristics of each level

Table 3. Characteristics of 3 Level in Korean Software Process Quality Certification model

Certification Degree	Characteristic
Level 1	- to perform project according to circumstances - the level to make and use process for oneself to perform individual tasks - Not to share similar process for each use to make and use - repeatedly happening the trial and error in person and system not sharing the outcome of trial error
Level 2	- successful performance of individual project - level interested in project performance efficiency focusing on individual projects in project level - to perform project by projected process in project, to share and manage the outcomes only in team unit - not repeatedly happening the trial and error in project team but repeatedly happening it in system
Level 3	- to perform project securely and consistently - level interested in consistently performing without environment change by using experience or cases during performing each - to develop task performance method as system guideline process in system level, to regulate and apply the process in various ways according to various traits of each project, to share the outcome in whole system - prevention of repeatedly happening the trial and error in system

The process group, assessment process and practice of level 2 can be shown from Table 4.

3   V & V in Korean Software Process Quality Certification Model

The verification and validation practices are in each requirement management, analysis, design, implementation, test process of development domain. And peer review process is contained in the practice of quality assurance process.

The verification practices are in requirement management, analysis, design, implementation, test process, quality assurance processes. The validation practices are in requirement management, analysis, test process.

3.1 Process Relationship to Related with Verification and Validation

Development practices are a series of sequence from requirement to product. Work products that are documented in development procedure are reviewed by various methods. The quality assurance process contains the verification practice.

The requirement process has 3 practices, customer requirement definition, requirement change management, requirement traceability management.

Customer requirement definition practice makes requirement specification by using project objective and data derived from customer. The requirement change management practice is the activity for application requirement change request of customer during the project. The requirement traceability management practice is maintaining bidirectional traceability among the requirements and project plan and work product. The bidirectional traceability is not required horizontal traceability but vertical traceability necessarily.

The design process has 3 practices, performing high level design, performing detail design, establishing test plan. In high level design, system architecture and structure is made by using the software requirement analysis result Decision of architecture selection can be in a part of this practice.

The analysis process has 3 practices, software requirement definition, software requirement analysis, software requirement review. The software requirement definition practice is activity documenting a function specification by using customer requirement specification. Next are making to-be process and review the documentation.

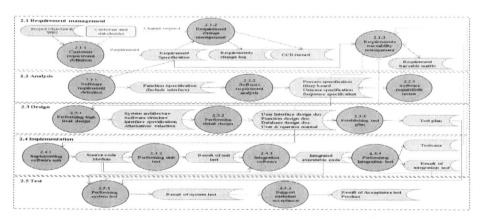

Fig. 2. Development practices relationship

The implementation process has 4 practices, implementing software unit, performing unit test, software integration, performing integration test. The software unit modules are developed and tested by the developer. And integration software practice is to assemble all software modules according to the integration sequence.

Next practice is to perform the integration test for the interface among software modules and to meet requirements.

The test process has 2 practices, performing system test, supporting customer acceptance. The system test practice is to validate whether the product meets the requirements, and can be delivered to customer. The last practice is supporting customer to accept and approve the product.

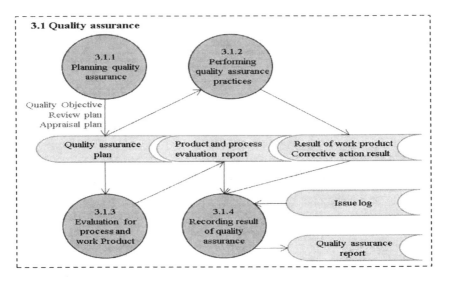

Fig. 3. The quality assurance practice relationship

The quality assurance process has 4 practices, planning quality assurance, performing quality assurance practices, evaluation for process and work product, recording result of quality assurance.

The performing quality assurance practices contains the peer review process to verify the work product to follow the standard process and to meet the quality parameter.

3.2 Verification and Validation Practices

The verification practices are the software requirement review and performing unit test, performing integration test in development process domain and are performing quality assurance practices of quality assurance process. The validation practice is a supporting the customer acceptance.

When they are shown the practices for just name, the practices for the verification and validation have 4 practices. But, more detail practices are related with verification and validation practice shown by table 4 Korean Software Process Quality Certification model practices and verification and validation relationship matrix.

The customer requirement definition practice contains to verify requirements and to review with customers and stakeholders. The software requirement definition practice contains to review with stakeholders.

The performing high level design of design process contains to review requirement traceability and whether design specifications follow the standard process and meet the quality parameter.

Table 4. Korean Software Process Quality Certification model practices and verification and validation relationship matrix

Process	Practice	Detail practice	Relation
2.1 Requirement management	2.1.1 Customer requirement definition	To Verify the requirements.	VER
		To review with customers and stakeholders.	VAL
2.2 Analysis	2.2.1 Software requirement definition	To review with stakeholders.	VER
	2.2.3 Software requirement review	All	VAL
2.3 Design	2.3.1 Performing high level design	To review traceability and whether to follow the standard process to meet the quality parameter.	VER
	2.3.2 Performing detail design	To review traceability and whether to follow the standard process and to meet the quality parameter.	VER
2.4 Implementa-tion	2.4.1 Implementing software unit	To review whether to follow the standard process and to meet the quality parameter with inspection method.	VER
	2.4.2 Performing unit test	All	VER
	2.4.4 Performing Integration test	All	VER
2.5 Test	2.5.1 Performing system test	All	VER/VAL
	2.5.2 Support customer acceptance	All	VAL
3.1 Quality assurance	3.1.2 Performing quality assurance practices	To review whether to follow the standard process and to meet the quality parameter.	VER

The implementation software unit practice contains to review whether source code follows the standard process and meets the quality parameter with inspection method.

The performing quality assurance practices contains to review whether work products follow the standard process and meet the quality parameter.

Figure 4 shows development activity flow and verification validation activity like review for work product, unit test, integration test, QA test.

The requirement function definition review is verification and validation activity involved with stakeholder and higher manager and internal customer. In organization, internal customer is almost real customer and the real customer can't be in any project. The requirement analysis document review, the system structure and interface review, detail design document review, test plan review are verification activity with the project team member. Of course, a customer can be participated in the meeting but it is not necessary.

The code review activity is defined using inspection method necessarily and verification activity.

The unit test and integration test activity are verification practices by project team member. And integration test contains system test practice like function test, performance test, volume test, endurance test, stress test in real operational environment.

The QA test activity is performed with validation practice by third party tester of QA team.

4 Conclusion

This implementation way of the organizational standard software process for the verification and validation practice is a series of the sequence and focused on activity flow. This is not the implementation way of process component unit.

So this way is easy to understand for organization's members. but it is difficult to implement to the organizational standard software process and it can be missing a practice to implement. It needs to observe continuously the organization that Korean Software Process Quality Certification model is applied whether to achieve the organizational process performance.

Acknowledgement. This work was supported by the Security Engineering Research Center granted by the Korea Ministry of Knowledge Economy.

References

[1] OMG.: Business Process Model and Notation (BPMN). version 2.0-Beta2 (June 2010)
[2] ARC. Assessment Requirements for CMMI, Version 1.0 CMU/SEI-2000-TR-011. Software Engineering Institute, Carnegie Mellon University, Pittsburgh (2000)

P2P Based Collision Solving Technique for Effective Concurrency Control in a Collaborative Development Environment

Hyun-Soo Park[1], Dae-Yeob Kim[2], and Cheong Youn[3]

[1] Baekseok Culture University, Computer Science Dept.,
393 Anseo-dong, Dongnam-gu Cheonan 330-705, Korea
hspark@bscu.ac.kr
[2,3] Chungnam National University, Computer Engineering Dept.,
79 Daehangno, Yuseong-gu Daejeon 305-764, Korea
kdymn2@cnu.ac.kr,
cyoun@cnu.ac.kr

Abstract. This paper provides a way to overcome the limitations of general collaborative software development tools that completely restrict co-ownership of resources among individuals in a team oriented developmental environment. It also provides a solution for users to co-own resources and at the same time manage version control and solve collision problems that may occur due to the co-ownership of resources. The collaborative software development tool presented in this paper is made up of the classical client/server structure with the P2P (peer to peer) method which supports information exchange among individuals. This tool is developed based on open source software CVS (Concurrent Version System). Functional efficiency was confirmed by comparing it to the utility of prior existing collaborative software development tools.

Keywords: Collaborative Development, Concurrency Control, Peer-to-Peer (P2P), Software Configuration Management (SCM), Concurrent Version System (CVS).

1 Introduction

The software development organization is becoming larger and the distribution of software developmental environment is expanding as well. Due to these environmental changes, the number of software collaborative development support tools that improve the quality of software and support the mutual information exchange and communication by internet is also increasing [1]. The importance of collaborative skills required for operations done in a team unit is becoming more significant.

Information exchange and systematic communication skills between members of the team are required in order to effectively carry out software collaborative development in a team unit. However, when more than two members approach and

G. Lee, D. Howard, and D. Ślęzak (Eds.): ICHIT 2011, CCIS 206, pp. 415–422, 2011.
© Springer-Verlag Berlin Heidelberg 2011

adjust a specific resource at the same time, this may cause a collision problem due to the modification of the same part of the resource. These situations can lead to waste of human resources as well as serious confusion when overlapping operation on the same source is done, and can cause significant financial expense waste when the problem is found out too late. Thus, "concurrent access control" is extremely important when trying to synchronize the members' work in real-time operations done in a collaborative developmental environment.

"Concurrent access control" can be largely divided into optimistic and pessimistic techniques [2]. Although the pessimistic technique is used at the point of execution of each transaction in order to maintain the consistency of the data, the optimistic technique is used at points where each transaction is done. Despite the fact that most of the commercial database system use pessimistic techniques, which have a foundation on the locking protocol, many show interest in the optimistic techniques because optimistic techniques have the capacity to produce a much more improved transaction management performance if enough hardware resources can be used [3, 4].

We can see that the "concurrent access control" used to keep the consistency of the data has the same purpose as to solve problems caused by overlapping operation on shared resources in a collaborative developmental environment[5-7]. Traditionally, many collaborative development support tools chose to use the pessimistic technique, which has its foundation on the exclusive locking that uses reserved checkout. This technique only allows one member to modify the resource and can cause unnecessary delay in the team working environment.

On the contrary, the optimistic technique uses unreserved checkout. This technique allows numerous members to modify their own local copy. This environment allows many members to work on shared resources at the same time [8]. This prevents any unnecessary delay in the development process. However, in certain situations, the possible conflicts are realized too late, and this can lead the waste in financial expense as well as unnecessarily overlapped work.

Our method provides an opportunity for all users to freely modify the copies of the provided data instead of restricting it to one selected user. This shortens the time required for an inquiry to be authorized to use the shared data. However, the original data that is used to make copies is managed through the server. This is based on the traditional optimistic technique but is hoping to make positive adjustments on conflicts that may be caused by overlapping operations on the same data.

This paper is composed as followed. After the introduction in the first section, the second section describes the techniques that prevent conflicts based on P2P (peer to peer) and the system architecture are introduced. Also, the software system implementation and execution is described. The third section shows the effectiveness of our system through the comparison of the existing collaborative software development systems. The conclusion is in the fourth section.

2   P2P Based Conflict Prevention Technique

2.1   Introduction of Conflict Prevention Technique

The P2P based conflict prevention technique that is introduced in this paper is fundamentally based on the optimistic technique. The optimistic technique allows a free modification process of the local copy that was downloaded from the source code server. The server distinguishes the possible conflicts when uploading occurs at the server, and notifies the user. This process is shown in Fig.1.

As seen in Fig.1, the optimistic technique's downside is that when overlapping operation occurs among many users, each user has to finish his or her work and incorporate it into the server in order to find out whether a conflict has occurred or not. If the conflict is recognized, other versions that were uploaded by other users need to be downloaded from the server and be compared in order to find out the difference and find out how to incorporate them together. Due to this time consuming process, this causes more delay in the resolving process as well.

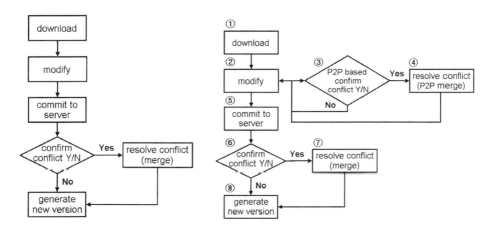

Fig. 1. Optimistic Technique   **Fig. 2.** P2P based conflict prevention technique

The core process of the P2P based conflict prevention technique is shown in Fig.2, where when the source is being modified, the possibilities of conflicts can be detected and solved. The time to detect conflicts and solve them is shortened and this functionality makes collaboration easier and more effective. To make this possible, this system consists of four major functions.

- File System Change Monitor function
- P2P Status Observer function
- P2P Conflict Notifier function
- P2P Conflict Resolver function

With these functions as the foundation, let us assume as the following in order to explain the process described in Fig.2. User A and B download Source P from the server (process ①) and decide to modify it as they would need. If User A modifies source P and produces P', the file system change monitor detects this and notifies User B through the P2P status observer function. If User B uses the same Source P to produce P'' (process ②), P2P Status observer function detects that a conflict has occurred and that specific file that has been overlapped is notified to User B through the P2P conflict notifier (process ③). When User B receives this information, User B uses P2P conflict resolver function to receive P' that was produced by User A in order to visually compare the differences and to incorporate these different information (process ④). If User B needs to communicate with User A in the process of incorporating the information, they can use the messenger function to communicate.

When the conflict is resolved through the ④ processes, the user reflects the integrated result into the server (process ⑤). At a point of integration, the possible conflicts with the server's recent version are examined (process ⑥) and if conflict is not detected, the integration is completed to make a new version (process ⑧). If another version was produced by another user before integration, the modified information needs to be incorporated together like the optimistic technique (process ⑦). This occurs when the conflict between the users is not detected very well or due to the time difference of updating the new version to the server, and this is an exceptional case of conflict prevention technique.

2.2   System Architecture and Embodiment

For the P2P based conflict prevention's four major functions, Fig.3's system architecture is provided. Each user's local system has the four components and the integrated workspace environment is installed. With this, P2P communications between the users are provided. Through the client/server architecture, necessary resources can be downloaded or modified results can be incorporated.

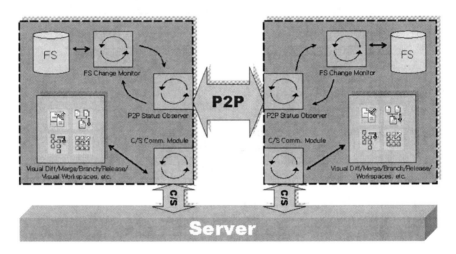

Fig. 3. System Architecture

The P2P collaborative development supporting tools dealt in this research usually do not use exclusive locking like the traditional collaborative development supporting tools, but helps many users to receive checkouts from the same source. The checkout in this case means the module of basic unit that is managed in the server being downloaded into the user's own local area (process □ of Fig.2). When the module is saved in each user's local area, this creates a workspace.

The concept of checkout is based on the preexisting CVS, which is an open system that provides the foundation for the version management, comparing differences, and functions relating to amalgamation [10]. First, the module that was checked out by many users becomes the shared resource, and the users can be identified through the system's user list. A module that has been checked out has many files and each of these files becomes the subject for modification. Also, the user can check the list of users who are using the same module that he or she has checked out.

The system observes the possible changes, such as modification or deletion, of the module that is being shared by many users. If certain files are modified within the module, the system recognizes the information and notifies other P2P users. Fig.4 shows the workspace application which include checked out modules and how the notification looks like when a specific file is modified within specific module that is being shared by other users.

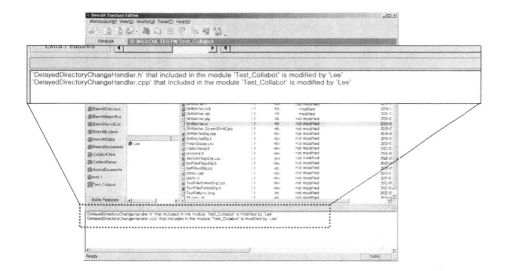

Fig. 4. Receive the Notification Message that File Modified

3   Evaluation of Effectiveness

P2P based Intelligence Agent is a tool used to help provide effective developmental environment in cases of conflicts in the process of software development.

In this section, the technical effectiveness of the system introduced in this paper will be discussed through the comparison of numerous well-known products that support software collaborative software development.

The criteria for evaluation are focused on whether parallel development and distributed environment is supported or not and the availability of the tools' implementations that are core in the collaborative developmental environment [11, 12]. Table 1 shows the criteria for evaluation.

Table 2 shows the evaluative results on functional effectiveness with other well-known collaborative software development tools based on the features in Table 1.

Other tools have a limitation to support branching except ClearCase. CVS, Subversion and VSS are uses single repository model. If all users use the same repository, and one user make a change in the main tree, all users are affected by this change. ClearCase has typical centralized system architecture. This system allows accessibility to clients in a local area only where the server is installed. This feature has a limitation to support accessibility in distributed environments. To overcome this problem, users have to purchase the high-priced product package separately and it can be a burden to users. Furthermore, this policy to support distributed environments does not provide complete functionality. The idea applied in ClearCase to supports distributed environments is assigning branches in each site. However, the branching of ClearCase is different from dynamic branching that is proposed in this paper. The branch of ClearCase is modified by one site only. That is, the branch in the other sites is possible to read only. This is an example of the pessimistic technique[14].

The Peer-to-Peer(P2P) architecture is a communication architecture for the sharing of resources that are expected to be operated on. Unlike the other tools that were previously mentioned, this architecture allows all users to modify the copy of master resource that is managed by the server and also supports the conflict prevention and solving algorithm. Through this architecture, diverse forms of parallel development will be enabled and the limitation of distributed environments can be overcome.

Dynamic branching, P2P architecture, True parallel development, or multi-site development can be interpreted in the same perspective. 'Post-event trigger' enables the work flow performed by developers, in a conflict, this function allows perform to notify the event and compare the differences. 'Accurate recording of all history' is a function for more complete parallel development, this record and notify the working history of all users. In the P2P environment, each client can detect the result of works on specific resources made by all users, and the server allows the users to refer to who worked on which operations at what time. In the case of VSS, because not enough background or objectives are recorded in the new version of the same resource, it is limited in parallel development since it is hard to understand the intentions of the changes of the other users.

Table 1. Qualitative Evaluation Criteria for Effectiveness Comparison

Features	Explanation
Graphical checkin tool	Graphical tool for file and changeset checkins which promotes more useful comments to speed up development process
Dynamic branching	Any workspace can be turned into a branch. Advanced planning for branching is not needed.
Peer-to-Peer architecture	Supports any workflow for enhanced development process. Supports the rapid open source style of development.
True parallel development	Enhanced productivity. Faster time to market.
Multi-site development	Provides functionality and productivity at all distributed sites.
Post-event triggers	Supports notification of events for provides easier process management.
Accurate recording of all history	Accountability: Easy to find who did what when. Provides a complete picture of parallel development.

Table 2. Comparison Functionalities with Other Well-Known Collaborative Software Development Tools

	CVS	Subversion	ClearCase	VSS	Proposed system
Graphical checkin tool	No	No	Yes	Limited	Yes
Dynamic branching	No	No	No	No	Yes
Peer-to-Peer architecture	No	No	No	No	Yes
True parallel development	No	No	Yes(complex)	No	Yes
Multi-site development	No	No	Yes(add on)	No	Yes
Post-event triggers	Weak	Yes	Yes	No	Yes
Accurate recording of all history	No	No	No	No	Yes

4   Conclusion

As the software types and scale is increasing as time passes by, the need for effective collaborative development support tools to lessen the trials and errors to control the high quality of the products is increasing, as well. However, the reality is that the domestic software developmental companies are unable to construct an effective collaborative environment.

This research suggests a P2P based software collaborative development support tools that will enable the system to work more in harmony in the software development process. CVS is relatively popularized version management system and with its functional elements, the goal is to make a high quality software collaborative development environment construction. By using the version management function that uses the branch/merge model, it is expected to increase the flexibility of the software development.

Furthermore, this research is hoping to construct the combined environment with configuration management system that supports project management, process management, and change management functions. This will increase the maturity of the software development organization, and will contribute to the improvement of the software quality.

References

1. DeFranco-Tommarello, J., Deek, F.P.: Collaborative Software Development: A Discussion of Problem Solving Models and Groupware Technologies. In: Proceedings of the 35th Hawaii International Conference on System Sciences (2002)
2. Makni, A., Bouaziz, R., Gargouri, F.: Formal Verification of an Optimistic Concurrency Control Algorithm using SPIN. In: Proceedings of the 13th International Symposium on Temporal Representation and Reasoning, pp. 160–167 (2006)

3. Choi, H.Y., Hwang, B.H.: A Concurrency Control Technique Using Optimistic Atomic Broadcast in Replicated Database Systems. Korea Information Processing Society Transactions, Part D 8-D(5) (2001)
4. Yang, G.L.B., Chen, J.: Efficient Optimistic Concurrency Control for Mobile Real-Time Transaction in a Wireless Data Broadcast Environment. In: Proceedings of the 11th IEEE International Conference on Embedded and Real-Time Computing Systems and Applications, pp. 443–446 (2005)
5. Strom, R., Banavar, G., Miller, K., Prakash, A., Ward, M.: Concurrency control and view notification algorithms for collaborative replicated objects. IEEE Transactions on Computers 47(4), 458–471 (1998)
6. Gao, M.L.S., Fuh, J.Y.H., Zhang, Y.: A Fine Granular Concurrency Control Mechanism for a Peer-to-Peer Cooperative Design Environment. In: 11th International Conference on Computer Supported Cooperative Work in Design, pp. 180–185 (2007)
7. Tang, M., Chou, S.C., Dong, J.X.: Concurrency Conflict Solving for Collaborative Feature Modeling. In: Proceedings of the 9th International Conference on Computer Supported Cooperative Work in Design, vol. 1, pp. 50–55 (2005)
8. Shen, H., Zhou, S., Sun, C.: Flexible concurrency control for collaborative office systems. In: 3rd International Conference on Information Technology and Applications, vol. 2, pp. 45–50 (2005)
9. Webster, M.: An End-User View of the Collaborative Software Development Market. In: IDC (2003)
10. Hunt, A., Thomas, D.: Pragmatic version control with CVS, Insight (2004)
11. http://www.bitkeeper.com/Comparisons.html
12. http://better-scm.berlios.de/comparison/comparision.html

Optical Properties of Photoluminescence of Carbon Doped Silicon Oxide Films Annealed by Rapid Temperature for Passivation

Teresa Oh

School of Electronics and Information Engineering, Cheongju University, 36 Naedukdong Sangdangku, Cheongju 360-764, Korea
teresa @cju.ac.kr

Abstract. To research the chemical and optical properties, SiOC films made by the inductively coupled plasma chemical vapor deposition were analyzed the chemical shift by the Fourier transform infrared spectra and photoluminescence spectra. The chemical shift obtained in Fourier transform infrared spectra resulted from the increase of right shoulder Si-O bond in the main bond due to the delocalization and lowering polarization of carbon in Si-CH$_3$. Increasing of the right shoulder bond of Si-O in side of higher wave number caused to change the degree of amorphous bonding structure depending on the reduction of polarization after annealing. The phenomenon of the delocalization in the carbon related bond of Si-CH$_3$ was also confirmed by the chemical shift in the photoluminescence spectra. The PL intensity increased after annealing at samples 9 and 13 with low polarity. Low polarization induced to decrease the thickness of SiOC film, which could contribute to improve the stability due to the low surface energy and the effect of electron deficient of carbon. It was observed the blue shift by FTIR (Fourier transform infrared spectra) and red shift by PL (photoluminescence) spectra in LP (low polarization) SiOC films.

Keywords: Photoluminance, Low polarization, Chemical shift, Thickness.

1 Introduction

The progress of the ultra large scale integrated (ULSI) circuits for the reduction in size involves the increasing of the capacitance between wiring, and results in a signal propagation delay. To solve this problem, there is a need for new interlayer dielectric materials with low dielectric constants (k < 2.5) instead of conventional SiO$_2$ film [1-2]. The promising candidate for low-k (low dielectric constant) materials is SiOC film, which is produced by the chemical vapor deposition (CVD) and spin on coating deposition (SOD). The nano pore formed in the film by the steric hindrance effect of the alkyl group and low polarization is the reason of lowering the dielectric constant in SiOC film. The chemical shift SiOC film observed in FTIR spectra has been many attention by many researcher, because the chemical shift observed in FTIR of carbon based organic thin films gave us much information to understand the bonding structure about Si-O-C and its related sites [3-5] and the chemical shift in SiOC film

G. Lee, D. Howard, and D. Ślęzak (Eds.): ICHIT 2011, CCIS 206, pp. 423–428, 2011.
© Springer-Verlag Berlin Heidelberg 2011

is also important factors to define the low dielectric constant. Recently, many researcher has been referred the correlation between the thickness and the lowering of the dielectric constant in SiOC film. The low dielectric constant of SiOC film deposited by CVD originates from the reduction of polarization and annealing process. The annealing process progresses the mechanical-electrical properties of SiOC film, so the hardness becomes higher than that of as deposited film and the dielectric constant also decreases because of the lowering the polarization and the reduction of the thickness [6-8].

In this study, SiOC film was produced by the ICP-CVD (inductively coupled plasma chemic vapor deposition) with MTMS (methyltrimethoxysilane) and oxygen mixed precursor. Optical properties of SiOC film were measured by the Photo Luminescence and also researched the correlation between the polarization and optical properties.

2   Experiments

The low-k SiOC films were obtained using the mixed gases of oxygen and methyltrimethoxysilane (MTMS) by inductive coupled plasma chemical vapor deposition (ICP-CVD). The films were processed at annealing temperature 500℃ for 10 minutes by using the rapid temperature annealing. The MTMS was vaporized and carried by argon gas at 35 ℃ with a thermostatic bubbler and the MTMS flow rate with Ar gas was varied for 3 sccm ~ 13 sccm. The base pressure of the mixture was kept at 3 Torr and the rf power was 600 W in each experiment. The chemical-optical properties of SiOC film was analyzed by Photo Luminescence (SPEX, SPEX) and Fourier Transform Infrared spectrometer (FTIR, Galaxy 7020A). The thickness and refractive index were measured by the Ellipsometer (uvsel/fpd-12, Horiba Jobin Yvon) with the source of 632.8 nm. PL spectra were researched by the UV-visible spectrometer at Korea Basic Science Institute, Gwang Ju , South Korea.

3   Results and Discussion

SiOC films were deposited by inductively coupled plasma chemical vapor deposition (ICP-CVD) and it was researched the relationship of polarity and optical properties by using the photo luminance spectra. Sample numbers were the MTMS (Ar) flow rate.

Figure 1(a) shows the FTIR spectra in the narrow range of 950~1250 cm^{-1}. SiOC film consists of absorption bands due to the three bonds of Si-O-C of 950~1250 cm^{-1}, Si-CH$_3$ at 1270 cm^{-1} and Si-O-C under 950 cm^{-1} in FTIR spectra. Intensity of the main bond of 950~1250 cm^{-1} decreased after annealing. For the main Si-O-C bond of 950~1250 cm^{-1}, the main bond was united two bonds of C-O and Si-O. Figure 2(a) displays the difference between the as deposited and annealed films. The right shoulder of Si-O in the main bond increased after annealing. It means that the increasing of Si-O bond and shifting to higher wave number of the main bond became more density cross-link in a bonding structure.

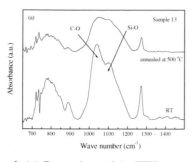
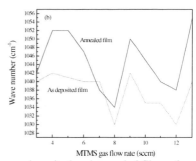

Fig. 1. (a) Comparison of the FTIR spectra between as deposited and annealed films of sample 13 in the main bond range of 950~1250 cm^{-1}, (b) FTIR peak position of the main bond range of 950~1250 cm^{-1} in SiOC films with various flow rater ratios.

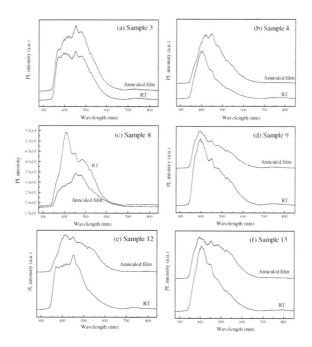

Fig. 2. Comparison PL spectra between as deposition and after annealing of SiOC films with various flow rate ratios; (a) sample 3, (b) sample 4, (c) sample 8, (d) sample 9, (e) sample 12, (f) sample 13

Figure 1(b) indicates the peak of Si-O-C in the main bond of 950~1250 cm^{-1} in FTIR spectra. The peaks of samples 4, 9 and 13 were higher wave number than that of any others, because the right Si-O bond of main bond increased at samples 4, 9 and 13 as shown in Fig. 2(a). Most peaks moved to higher wave number after annealing process. Increment of Si-O and decreasing of the relative carbon content made the cross link structure owing to the low polarity, which causes to decrease the thickness of SiOC film and increase the film density.

Figure 2 is the comparison of the PL spectra of 350~650 nm between the as-deposited films and annealed films of samples 3, 4, 8, 9, 12 and 13. There were two kinds of samples such as samples 3, 8 and 12 with high wavelength and samples 4, 9 and 13 with low wavelength in as deposited films as next mentioned in Fig. 3. In view of the peak position in the range of 350~650 nm, samples 4, 9 and 13 were almost same after annealing process, in spite of the chemical shift of samples 3, 8 and 12 in PL spectra. Lowering the polarization and the reduction of the thickness gives the stability of final materials due to the reduction of π-π* band gap originated from the electron donation effect of electron deficient group.

Figure 3 indicates the peak position of PL spectra of as deposited and annealed SiOC film with various flow rate ratios. The band peaks at samples 4, 9 and 13 were almost similar after the deposition and annealing, but that of samples 3, 8 and 12 showed a large difference. Samples of 4, 9 and 13 with low polarity were observed very little variation in the optical properties. Moreover, the peaks of PL spectra of samples 4, 9 and 13 are lower wavelength than those of other films. The PL peak position of samples 4, 9 and 13 changed very little in spite of annealing at 500℃, because of lowering the polarization due to the dispersion of proton charge. In other words, the other samples 3, 8 and 12 had polar properties, so the effects of after annealing occurs much variation in the PL analysis as an optical results generated from the chemical reaction of oxidation.

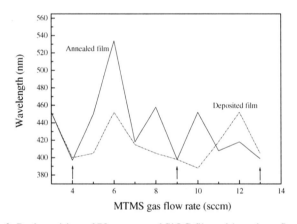

Fig. 3. Peak position of PL spectra of SiOC film with various flow rate ratios

Figure 4 is the thickness and refractive index of annealed SiOC films with various flow rate ratios. The thickness decreased at samples 4, 9 and 13 with high wave number in the Si-O-C main bond, and also showed the trend of the reduction in the refractive indexes at those samples without sample 13. The thickness abruptly decreased at the sample 4, 9 and 13. The diminution of thickness was owing to the increment of Si-O cross link and the reduction of polarization, which is caused by the weak boundary condition due to the elongation effect of bonding length. The effect induced the stability owing to the low surface energy, so the density of the film increases by the lowing of the thickness depend on the increment of the Si-O cross link bond. But sample 8 produces the steric hindrance in the film by the aloof force between polar sites and then increases the thickness.

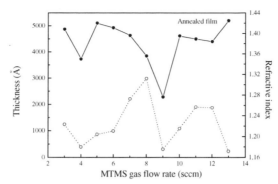

Fig. 4. Thickness and refractive index of annealed SiOC films with various flow rate ratios

Generally, the refractive index is in proportion to the dielectric constant, so we know that the dielectric constant of samples 4, 9 and 13 could be decreased. Low polarity in SiOC film was due to the chemical reaction between alkyl and hydroxyl group, and the PL spectra at SiOC film of low polarization did not change by the annealing process, but that of film with polarity had much variation after annealing.

4   Conclusion

SiOC films were prepared at various flow rate ratios by the inductively coupled plasma chemical vapor deposition and annealed at 500 ℃. The chemical shift strongly influenced the lowering of the polarization between the carbon bond in Si-CH_3 and hydroxyl sites. The reduction of the polarity in FTIR spectra increased the right shoulder bond of Si-O in the main bond range of 950~1250 cm^{-1}, so the moving to higher wave number of the main peak induced the blue shift. Lowering the polarization depends on the electron deficient group as the contribution of carbon increased in accordance with the carbonation. In the case of PL spectra, the peak intensity was found to be decreased with the reduction of the polarity. The chemical shift observed in FTIR and PL spectra of SiOC film with low polarity gives an insight into reduced thickness of SiOC films and then increased density.

Acknowledgments. This research was supported by Basic Science Research Program through the National Research Foundation Korea (NRF) funded by the Ministry of Education, Science and Technology (20110005954).

References

1. Oh, T., Choi, C.K.: Journal of the Korean Physical Society 56, 1150–1155 (2010)
2. Widodo, J., Lu, W., Mhaisalkar, S.G., Hsia, L.C., Tan, P.Y., Shen, L., Zeng, K.Y.: Thin Solid Films 462-463, 213–218 (2004)
3. Grill, A., Neumayer, D.A.: J. Appl. Phys 94, 6697–6707 (2003)
4. Yu, L.D., Lei, S., Dong, Z.S., Yi, W., Yan, L.X., Qi, H.R.: Chin. Phys. Soc. 16, 240 (2007)

5. Park, H.S., Ryu, H.R., Rhee, C.K.: Talanta 70, 481–484 (2006)
6. Oh, T.: Jpn. J. Appl. Phys. 44, 4103–4107 (2005)
7. Damayanti, M., Widodo, J., Sritharan, T., Mhaisalkar, S.G., Lu, W., Gan, Z.H., Zeng, K.Y., Hsia, L.C.: Materials Science and Engineering B 121, 193–198 (2005)
8. Kim, M.S., Kim, J.H., Kim, E.K.: Journal of the Korean Physical Society 48, 1552–1555 (2006)

Design and Implementation of an Ultra-Low Power Energy Harvesting Sensor Network

Jung Kyu Yang, Jun-Seok Park, Yeong Rak Seong, and Ha-Ryoung Oh

Department of Electrical Engineering, Kookmin University,
861-1 Jeongneung-Dong Seongbuk-Gu Seoul, Korea
{poohyjk3,jspark,yeong,hroh}@kookmin.ac.kr

Abstract. This paper proposes an ultra-low power sensor network system for energy harvesting environments. The system is composed of two types of sensor nodes: master nodes, which can acquire relatively sufficient energy from batteries or mains, form the backbone of the entire system; and slave nodes, which are powered by using energy harvesting technology, are scattered around the master nodes. To overcome the limitation of harvestable energy, a slave node is usually turned off and is activated when a master node sends a request signal which contains the ID of the slave node. To verify the energy effectiveness of the proposed architecture and techniques, a simple pilot system is implemented and tested. The experimental results show that the proposed techniques drastically reduce energy consumption of the sensor nodes.

Keywords: energy harvesting, ultra-low power, sensor network.

1 Introduction

A lot of sensor networks have been studied during the last decades. Most of sensor nodes use batteries as their power source. A sensor node can be operated for a long time, since it consumes very low power. However, it may eventually stop when the power of its battery is completely exhausted. As the number of stopped sensor nodes increases, operation of the entire sensor network may be severely restricted as well. This issue can be moderated by making low-priced sensor nodes and replacing stopped sensor nodes with new ones. Unfortunately, however, environmental pollution due to the used sensor nodes and batteries will cause a new problem.

Recently, energy harvesting has been focused as a solution to the battery problem of the sensor nodes. Energy harvesting is a renewable energy technology which acquires electrical power from surrounding natural or artificial energy sources. Energy harvesting is environment-friendly and we can acquire energy permanently as long as the energy source exists.

Heliomote [1], [2], [3] is the most famous sensor network node using the energy harvesting technology. It acquires relatively sufficient energy by using solar cells in outdoor environments and is able to perform the functions of general sensor nodes. However, the amount of energy that can be harvested from most of other energy sources is very small as shown in Table 1. Many well-known traditional low-power

G. Lee, D. Howard, and D. Ślęzak (Eds.): ICHIT 2011, CCIS 206, pp. 429–436, 2011.
© Springer-Verlag Berlin Heidelberg 2011

techniques, e.g. DVS [4] and duty cycle control [5], [6], [7], are hardly useful for this degree of energy deficiency. Therefore, a novel approach should be devised to run a sensor network in such ultra-low power environments.

Table 1. Common energy harvesting sources [3], [8], [9]

Power Source	Power Density
Solar(outside)	15(mW/cm^2)
Solar(inside)	10(μW/cm^2)
Temperature(5°C Gradient)	40(μW/cm^2)
Human Power	330(μW/cm^3)
Airflow(5m/s)	380(μW/cm^3)
Vibrations	200(μW/cm^3)
Acoustic Noise(100dB)	960(nW/cm^3)

This paper proposes an ultra-low power energy harvesting sensor network system. The system is composed of two types of sensor nodes: *master nodes*, which can acquire relatively sufficient energy from batteries or mains, form the backbone of the entire system; and *slave nodes*, which are powered by using energy harvesting technology, are placed around the master nodes. To reduce energy consumption of slave nodes having a very limited energy budget, a slave node is usually turned off; and it is turned on only when a master node sends it a request signal. At this time, the request signal contains an ID of a specific slave node to be operated. Thus, only one slave node is activated for a request signal, and no communication collision occurs unless multiple master nodes in the same region send request signals simultaneously. Moreover, for less power consumption during an active period, the slave node is designed so that each module can be individually turned on and off by CPU.

2   System Design

This section proposes an ultra-low power sensor network system for energy harvesting environments. First, the architecture of sensor network is expanded for the environment. Then, two techniques for sensor nodes which are powered by using energy harvesting technology are proposed. The operation procedure of the entire system is explained at the end of this section.

2.1   Network Architecture

We assume that the energy harvesting environment discussed in this paper cannot supply sufficient energy to energy harvesting sensor nodes to directly execute all functions of a general sensor network node. Nonetheless, the sensor network system which contains such energy harvesting sensor nodes should be still able to support those functions. For that, this paper proposes novel sensor network architecture for ultra-low power energy harvesting environments by expanding traditional sensor network architecture. Fig. 1 illustrates the traditional sensor network architecture as well as the expanded sensor network architecture. The traditional sensor network in Fig 1(a) has a mesh topology. The sensor nodes in the architecture can execute various functions, such as sensing, message routing, etc. As shown in Fig. 1(b), however, the sensor network proposed in this paper has an expanded star topology based on Fig. 1(a). In the proposed architecture, a new layer which is composed of a set of slave nodes is added. The slave nodes are implemented by using energy harvesting technology and act as remote sensors of the upper layer sensor nodes, i.e. master nodes. Due to the energy limitation, a slave node has restricted capability. It gathers information of attached sensors and delivers the information to the master node. On the other hand, a master node has relatively sufficient energy. Thus, it can communicate with neighboring slave nodes and can execute energy consuming functions including message routing.

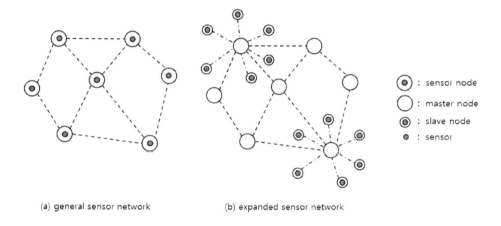

(a) general sensor network   (b) expanded sensor network

Fig. 1. The general and expanded sensor network architecture

The expanded sensor network structure can be easily applied to various environments. For example, when the sensor network is established at home, a slave node can be installed at any places without considering electric power supply, and a master node can be installed at the places where the master node can steadily receive power, e.g. a ceiling.

2.2 Slave Node

The slave node is constructed as shown in Fig. 2. To reduce power consumption of slave nodes, two techniques are devised: the *two-stage activation technique* and the *selective power control technique*.

The two-stage activation technique energizes only one slave node, although many slave nodes receive the same request signal from a master node. This technique is implemented by two modules of each slave node: a *pre-wakeup module* and a *main wakeup module*.

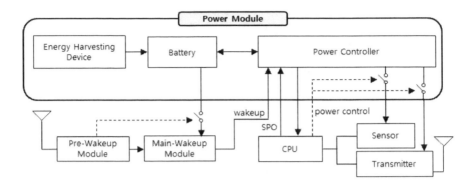

Fig. 2. The block diagram of a slave node

The pre-wakeup module turns on the main wakeup module without consuming the energy stored in the slave node. A slave node is usually in idle state. During the state, a slave node is turned off [1]. Therefore, the pre-wakeup module and the main-wakeup module are also turned off in the idle state. To reduce consumption of the energy stored in slave nodes, the pre-wakeup module does not use the energy; instead it employs the battery-less startup technology used in general passive RFID tags. When a master node needs to communicate with a specific slave node, it sends a request signal with an amplitude-modulated continuous wave. Then each pre-wakeup module receiving the signal extracts energy from the wave to use power for its own sake, and turns on the main-wakeup module.

Although many main-wakeup modules are activated by a single request signal, only one main-wakeup module turns on the rest part of the slave node. For this, each slave node stores a unique ID in its own main-wakeup module, and a request signal sent from a master node contains the ID of a specific slave node. When a main-wakeup module is activated, it demodulates the request signal which activates itself, and compares the ID extracted from the signal with its own ID. The main-wakeup module whose ID is equal to the ID contained in the request signal sends a *wakeup signal* to its own power module. Then the power module supplies energy to its CPU by using the selective power control technique, which will be described later. Note that the demodulation operation in the main-wakeup module also does not consume the energy stored in the slave node; instead the demodulation operation use the energy extracted from the request signal as the pre-wakeup module does. Thus, the main-wakeup module consumes a very small amount of the stored energy only while comparing IDs. Also, note that, since only one slave node is activated for a request signal by the benefit of the main-wakeup module, no communication collision occurs unless multiple master nodes in the same region send request signals simultaneously.

Selective power control technique enables that the sensor and the transmitter in a slave node can be individually turned on or off by the CPU of the slave node. This technique is implemented in the *power controller* of a slave node. Operation of the

---

[1] However, a miniscule amount of energy is consumed by the power module of a slave node even in the idle state.

power controller is initiated by a wakeup signal of the main-wakeup module as explained earlier. Then the power controller supplies energy to the CPU. After the CPU executes its startup routine, it controls the power controller to turn on or off its sensor and transmitter. Since the CPU can control energy supply to the devices individually, unnecessary energy consumption can be eliminated. When the CPU eventually finishes processing the request received from the master node, it sends a *self-power-off (SPO) signal* to the power controller to turn off the CPU itself. Thus, the slave node is completely turned off until it receives a request signal from a master node again.

2.3 Operation Procedure

Fig. 3 illustrates the operation of the proposed sensor network system. The system operates as follows.

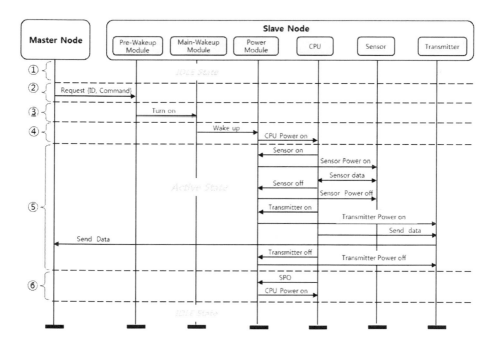

Fig. 3. The operation procedure of the entire system

① All slave nodes are in the idle state. The pre-wakeup modules and the main-wakeup modules in slave nodes are completely turned off.
② A master node M sends a request signal containing a command and the ID of a specific slave node S.
③ The pre-wakeup modules of all the slave nodes which receive the request signal turn on their own main-wakeup modules.
④ The main-wakeup modules demodulate the received request signal and extract the ID information. The main-wakeup module of the slave node S detects that

the extracted ID equals to its own ID and then sends a wakeup signal to the power module to supply energy to its CPU. Now, the slave node S is activated; but other slave nodes are still in the idle state. The master node M ceases to send the request signal and waits until the slave node S sends a response. Therefore, all main-wakeup modules and pre-wakeup modules are turned off.

⑤ The CPU of the slave node S gathers sensor data and broadcasts the data using a transmitter. At this time, the sensor and the transmitter acquire energy from the power module only while they operate. The master node M receives the broadcasted data.

⑥ After the CPU of the slave node S finishes the broadcasting, it sends a SPO signal to its own power module to turn off itself. Then the slave node S is to be in the idle state until it receives the next request signal.

3   Implementation

The ultra-low power energy harvesting sensor network system proposed in this paper is implemented as shown in Fig. 4. Especially, the main-wakeup module of a slave node is implemented as an IC with the use of the TMC 0.18μm process in order to reduce consumption of the energy obtained from an energy harvesting module. The energy harvesting module is implemented by using a 150mm x 38mm thin film solar cell.

(a) master node   (b) slave node

Fig. 4. The implementation of the master node and the slave node

The implemented system is tested in an indoor environment. With a light intensity of 600 lux, the energy harvesting modules of slave nodes acquire $17.3\mu W/cm^2$. Operation of slave nodes is tested with varying output power of master nodes and the distance between master nodes and slave nodes. The experimental result shows that the operable distance can be extended up to 10m when the output power is 24dBm. At this time, it was measured that a slave node consumes $0.0231\mu W$ in idle state and $167.53\mu W$ in active state. The power consumption in the idle state is caused by the standby current of the power module. Also, it is measured that the main-wakeup module operates for about 300usec to process a single request signal. Thus, the main-wakeup module consumes energy of just $0.503\mu J$ per a request signal. Note that the standby current of the power module is independent of characteristics of other devices in the slave node. Thus, the amount of the current is not considerably changed in other hardware configurations.

4 Discussion and Conclusion

Now we will discuss how much the proposed techniques can reduce energy consumption. Usually, the total energy consumption of a sensor node strongly relies on the electrical characteristics of the used IC chips. To guarantee fairness, the energy reduction performance is discussed with the Mica2 mote, the most well-known sensor node, in this paper. Table 2 presents detail energy consumption of the Mica2 mote. Mica2 mote consumes 21.33mW if the CPU and the radio device for an idle state operate at a power-save mode and an Rx mode respectively. Even the radio device is turned off during the idle state, Mica2 consumes 0.33mW. In comparison, the proposed slave node consumes just 0.0231μW caused by the standby current of the power module in the state with the same operating voltage.

Table 2. Mica2 platform current draw measured with a 3V power supply [10]

Device/Mode	Current	Device/Mode	Current
CPU		*Radio*	
Active	8.0mA	Rx	7.0mA
Idle	3.2mA	Tx(-20dBm)	3.7mA
ADC Acquire	1.0mA	Tx(-8dBm)	6.5mA
Extended Standby	0.223mA	Tx(0dBm)	8.5mA
Standby	0.216mA	Tx(10dBm)	21.5mA
Power-save	0.110mA	*Sensors*	
Power-down	0.103mA	Typical Board	0.7mA

This paper proposes noble expanded sensor network architecture to overcome energy limitation of the ultra-low power energy harvesting environment in which the amount of harvestable energy is extremely small. Also two-stage activation technique and selective power control technique are proposed to effectively implement the architecture. To verify the energy effectiveness of the proposed architecture and techniques, a simple pilot system is implemented and tested. From the experimental results, we can conclude that the proposed techniques drastically reduce energy consumption of the sensor nodes. The techniques proposed in this paper can be applied not only to the energy harvesting environment but to other energy-critical environments. For example, the lifetime of a traditional battery-powered sensor node can be lengthened by using the techniques.

Acknowledgments. This work was supported by research program 2011 of Kookmin University in Korea.

References

1. Kansal, A., Srivastava, M.B.: An Environmental Energy Harvesting Framework for Sensor Networks. In: International Symposium on Low Power Electronics and Design, ISLPED 2003, Seoul, pp. 481–486 (2003)
2. Kansal, A., Hsu, J., Zahedi, S., Srivastava, M.: Power Management in Energy Harvesting Sensor Networks. ACM Transactions on Embedded Computing Systems, TECS 6(4), article 32 (2007)

3. Raghunathan, V., Kansal, A., Hsu, J., Friedman, J., Srivastava, M.: Design Considerations for Solar Energy Harvesting Wireless Embedded Systems. In: 4th International Symposium on Information Processing in Sensor Networks, IPSN 2005, Los Angeles, pp. 457–462 (2005)
4. Min, R., Bhardwaj, M., Cho, S.-H., Sinha, A., Shih, E., Wang, A., Chandrakasan, A.: Low-Power Wireless Sensor Networks. In: International Conference on VLSI Design 2001, Bangalore, pp. 205–210 (2001)
5. Hsu, J., Zahedi, S., Kansal, A., Srivastava, M.: Adaptive Duty Cycling for Energy Harvesting Systems. In: International Symposium on Low Power Electronics and Design, ISLPED 2006, Tegernsee, pp. 180–185 (2006)
6. Moser, C., Thiele, L., Brunelli, D., Benini, L.: Adaptive Power Management in Energy Harvesting Systems. In: Design, Automation and Test in Europe, DATE 2007, France, pp. 773–778 (2007)
7. Vigorito, C.M., Ganesan, D., Barto, A.G.: Adaptive Control of Duty Cycling in Energy-Harvesting Wireless Sensor Networks. In: 4th Annual IEEE Communications Society Conference on Sensor, Mesh and Ad Hoc Communications and Networks, SECON 2007, San Diego, pp. 21–30 (2007)
8. Roundy, S., Steingart, D., Frechette, L., Wright, P., Rabaey, J.: Power Sources for Wireless Sensor Networks. In: Karl, H., Wolisz, A., Willig, A. (eds.) EWSN 2004. LNCS, vol. 2920, pp. 1–17. Springer, Heidelberg (2004)
9. Roundy, S.: Energy Scavenging for Wireless Sensor Nodes with a Focus on Vibration to Electricity Conversion. Ph.D. Dissertation, Dept. of EECS, UC Berkeley (May 2003)
10. Hempstead, M., Tripathi, N., Mauro, P., Wei, G.Y., Brooks, D.: An Ultra Low Power System Architecture for Sensor Network Applications. ACM SIGARCH Computer Architecture News 33(2), 208–219 (2005)

Implementation of Location Awareness UWB PHY

Byoung-Sup Shim[1] and Hyoung-Keun Park[2]

[1] Research & Development Center, Leotek Co. Ltd., Korea
[2] Dept. of Electronic Eng. Namseoul Univ., Korea
bshim@leotek.co.kr, phk315@nsu.ac.kr

Abstract. This study searched the IEEE 802.15.4a technology which is the standard of the location awareness WPAN system which provides the communication and location. The UWB technology can provide the function of the location awareness as it is adopted as the standard technology for the communication and location which requires the degree of precision within several dozens of centimeters in a relatively narrow area. Relevant common products will be released from now on because the research and development have actively been proceed at home and abroad as the IEEE 802.15.4a standard was completed.

Keywords: UWB, Ranging, Location awareness, MAC, WPAN.

1   Introduction

The UWB technology which comes into the spotlight as one of the new generation wireless system technology is the wireless technology which takes more than 20% or 500MHz of the occupied band-width in the center frequency. This UWB technology has received attention as the PHY of the location awareness WPAN system because it can offer functions of the minute location awareness and tracking by using the characteristic of the very wide frequency band occupation [1, 2].

The location awareness using the UWB technology can enable the location awareness indoors or in the shaded area and the service for the location awareness which requires the degree of precision within several dozens of centimeters in a relatively narrow area can create the new application technology and market. The IEEE 802.15 Alternate Task Group (TG4a) launched for the low-power PHY standard based on the location as the new networking technology to provide the minute location awareness, the basis of the ubiquitous environment which will become the main issue from now on has come to the fore. The technical specification required for the standardization of IEEE 802.15.4a was presented. The work for the integration by the adjustment of each proposal based on the UWB and CSS method by receiving the total 26 proposals was done. As a result, the final standard baseline was decided. This study realized the UWB PHY technology adopted for the use of the communication and localization as hardware and measures its characteristics.

G. Lee, D. Howard, and D. Ślęzak (Eds.): ICHIT 2011, CCIS 206, pp. 437–445, 2011.
© Springer-Verlag Berlin Heidelberg 2011

2   UWB Technology for the Location Awareness

2. 1   UWB Frequency Band and Channel

In case of IEEE 802.15.4a, a 3.1~10.6GHz frequency brand allowed as the communication in the FCC and frequency brand under 1GHz were divided into 3 types of brands such as a sub-band, low-band, and high-band as in Table 1 and all the 16 channels were allotted. Channel number 0, 3, and 9 are the mandatory channel each brand and one of them should surely be realized.

It can flexibly be responded to the case that the further interference problem happened through the management of these channels. For example, even if some specific brand can use the UWB without any problem in any country, the interference with other communications can be in a serious situation in other country. In this case, it is necessary to make some specific brand not to be used in some region to avoid the interference. This problem can be responded as the brand which contains that brand is not used. The radio techniques which is a basics does not need to be revised because the frequency bandwidth of the rest of channels including the mandatory channel are the same   except channel number 4, 7, 11, and 15 in Table 1.

Table 1. Channel frequency of IEEE 802.15.4a(MHz)

Ch.#	Sub-Band	4	3994.6	10	6486.4
0	499.2	Ch.#	high-band	11	7987.2
Ch.#	low-band	6	6988.8	12	8985.6
1	3494.4	7	6489.6	13	9484.8
2	3993.6	8	7488.0	14	9984.0
3	4492.8	9	7987.2	15	9484.8

That is, the same technology as the existing frequency band can be applied. The frequency band of 3.1~4.8GHz and 7.2~10.2GHz was allowed to be used as the UWB in Korea [3]. However, the interference avoidance technology (DAA) should be applied in a 3.1~4.8GHz band. The application of the DAA was decided to be postponed in a 4.2~4.8GHz band [3,4]. Therefore, the current status of the frequency band should certainly be identified when the WPAN system of location awareness corresponding to the IEEE 802.15. 4a standard is developed.

2.2   UWB PHY System Technologies

The UWB PHY is based on an impulse system and can compose two pans by using its own acquisition code each frequency band. Two pans from one band among the frequency band number 0, 3, and 9 in Table 1 is supposed to be realized. The chip-rate is the same with 499.2MHz regarding three mandatory frequency bands.

The modulation system for the UWB PHY assumes the form of combining the BPM and BPSK to support all the synchronization and asynchronous reception, and each symbol consists of the burst which several the UWB pulse. Various data service is provided by adjusting the length of a burst.

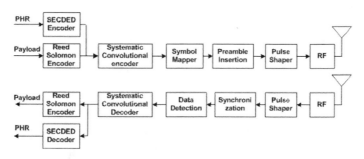

Fig. 1. The transmission and reception of the UWB PHY

The sending end is supposed to go out into the channel through RS encoder, convolutional code, symbol mapper, preamble inserter, pulse shaper, and antenna. The receiving end restores the signal entered as a transmitting signal through RF through pulse shaper, synchronization, data demodulation, convolutional decoder, and RS decoder.

2.3   Ranging Technology

The degree of precision in the location awareness at the level of several dozen cm is required between devices in IEEE 802.15.4a. There is TOA (time of arrival) which seeks the location by measuring the wave transfer time between two nodes as the method of the location measurement for this. In addition, there are AOA (angle of arrival) and TDOA(time difference of arrival).

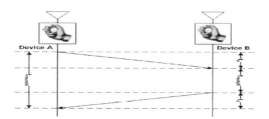

Fig. 2. Two-way ranging based on TOA

The device A operates its own counter by sending a ranging message as in Fig. 2. The device B starts to operate the counter and sends ACK message to the device A after the receiving process time (t_{replyB}) if it receives the corresponding ranging message from the device A. The Device A stops its counter and obtains the value of RTT(round-trip time, t_{roundA}) as soon as it receives a message from device B. The value t_p can be shown as in the formula (1) corresponding to TOA as it can be known in the device A by a method such as loading a message.

$$t_p = \frac{t_{roundA} - t_{replyB}}{2} \qquad (1)$$

There may be many errors because of the leading edge detection of a signal, the precision of a counter, crystal offset, internal delay of a transmitter-receiver, when calculating t_p based on Two-way ranging. Therefore, the algorithm which can do the precise ranging should be drawn considering these errors.

To ensure that the reproduction of your illustrations is of a reasonable quality, we advise against the use of shading. The contrast should be as pronounced as possible.

If screenshots are necessary, please make sure that you are happy with the print quality before you send the files.

2.4   MAC Technology

IEEE 802.15.4a MAC follows IEEE 802.15.4 MAC basically. 16 slots were placed during a super-frame, a characteristic of IEEE 802.15.4 MAC. It was made by transmitting most of data and adopting the method of operating as a sleep mode in the inert state. The data operates based on CSMA/CA for the CAP period. A super frame was designated based on a beacon. Accordingly, it increased the data transfer rate and liquidity of the applied distance as it supported the mode of messaging a data in a state without a beacon as well as the mode of transmitting data according to this.

The characteristic of IEEE 802.15.4a MAC is the addition of MLME-DPS and MLME-DITHER, a service access point (SAP) by the UWB PHY method regarding a ranging. MLME-DPS is a primitive regarding the DPS setting which can prevent "Spoof Attack" as it uses a preamble different form a normal mode in the UWB PHY on ranging. The MLMEDITHER is a primitive to give the "Dither Time" on the ACK message response time so that other device cannot be involved in ranging on the TWR ranging in the UWB PHY [4].

3   Realization of IEEE 802.15.4a PHY

3.1   The Transmission Structure of UWB PHY

The form of UWB frame realized was composed SHR preamble, PHY header, and payload. SHR, a preamble signal in Fig. 3 is used for the algorithm at the sending end such as AGC, Signal acquisition, estimation of frequency offset, and ranging. It was divided into SYNC, a section in which a preamble symbol is repeated as in Fig. 3 and SFD section in which the preamble was ended.

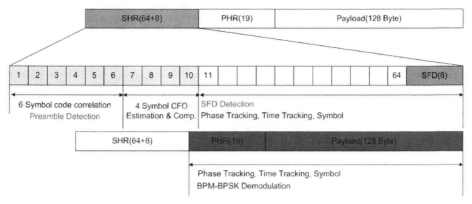

Fig. 3. The form of UWB frame

Preamble symbol suggested in IEEE 802.15.4a is a ternary code with the length of 31 or 127. Two ternary codes were allotted each channel. The ternary code had the peak value when the characteristic of a periodic correlation was multiplied by its own code, and is used as a code of separating different PAN because the rest part is 0.

Table 2. The ternary code with a 31 length

Index	Ternary Code	Ch. #
1	-0000+0-0+++0+-000+-+++00-+0-00	0,1,8,12
2	+0+0-0+0+000-++0-+---00+00++000	0,1,8,12
3	-+0++000-+-++00++0+00-0000-0+0-	2,5,9,13
4	0000+-00-00-++++0+-+000+0-0++0-	2,5,9,13
5	-0+-00+++-+000-+0+++0-0+0000-00	3,6,10,14
6	++00+00---+-0++-000+0+0-+0+0000	3,6,10,14

The PHR in Fig. 3 is composed of the data transfer rate, section length of preamble, field 8bit which has the flag information telling ranging, and length field 8bit which has the length information on a pay-road which must be transmitted. It is altered by the BPM-BPSK modulation method.

The transmission structure needed to generate this UWB frame was the same as in Fig. 4.

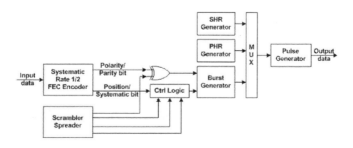

Fig. 4. FEC encoder

A FEC encoder in Fig. 4 uses by connecting RS(Reed-Solomon) encoder and Convolutional encoder in series as shown in Fig. 5.

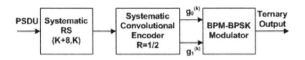

Fig. 5. FEC and Modulation Process

Above all, 378 bit (b0,...,b329, p0,...,p47) is generated in the by attaching 48 parity to the data of 330bit(b0,...,b329) in the RS encoder. The code rate R=1/2, Constraint Length K=3, and generation polynomial generates systematic-bit $g_0^{(k)}$ and parity-bit $g_1^{(k)}$ by using $(2,6)_8$. The Modulation of UWB PHY is altered by produced $g_0^{(k)}$ and $g_1^{(k)}$ through a FEC encoder and the BPM-BPSK modulation method. The kth transmit signal of the UWB system changed according the modulation method of the BPM-BPSK can shown as in formula (2).

$$x^{(k)}(t) = [1 - 2g_1^{(k)}] \sum_{n-1}^{N_{cpb}} [1 - 2s_{n+kN_{cpb}}] p(t - g_0^{(k)} T_{BPM} - h^{(k)} T_{burst} - nT_c) \qquad (2)$$

where, $g_0^{(k)}$ is altered as the location information of a burst, $g_1^{(k)}$ is falsified as the polarity information of a burst. The sequence $s_{n+kN_{cpb}}$ is the spreading code for the kth transmission and $h^{(k)}$ decides the hopping location of kth burst. $p(t)$ is transmission pulse type of antenna input.

3.2   The Reception Structure o f the UWB PHY

As mentioned earlier, the transmitter and receiver were altered as BPM-BPSK as that it could support all the coherent and non-coherent reception.

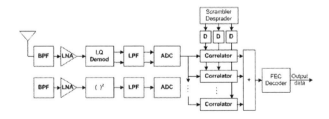

Fig. 6. The structure of the synchronous and asynchronous reception

The coherent reception is a structure of finding the location and polarity information through the back-diffusion and correlator after it separates I, and Q, and takes the ADC each. On the other hand, the non-coherent structure stores the location information by judging whether the energy of the burst signal is located in front or at the back end. At this time, the location information stored went through only the RS decoder without convolutional decoder.

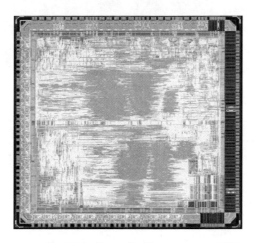

Fig. 7. The result of P&R planned as the process of TSMC 0.13

4   Chip Test and Verification

The saw tooth wave and pulse data packet were generated to investigate functions and performance of the ADC and DAC in a modem chip. In addition, the verification was possible through the test mode setting in a board design as an ADC-to-DAC look-back path was reflected into a plan.

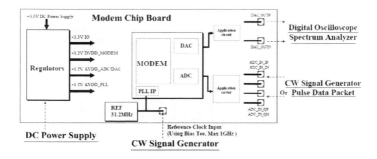

Fig. 8. Test board setup

A ramp wave, a pattern signal known the inside of a modem chip was directly generated for a chip test, permitted into the input of the DAC. As a result of checking the output signal as a digital oscilloscope, the normal ramp wave was output in Fig. 9 and Fig. 10.

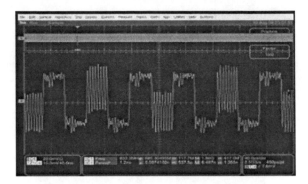

Fig. 9. Internal PLL Mode[@1.2V]

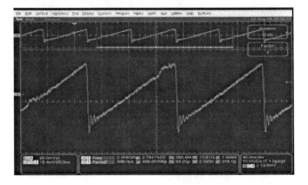

Fig. 10. External Bypass Mode[@1.2V]-998.4MHz

FD_Flag was continuously measured as in Fig. 11 and Fig. 12 after the first packet received when the packet data transmitted from a modem platform was succeeded in receiving.

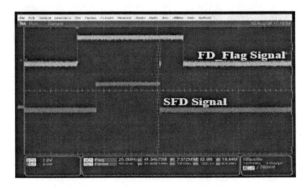

Fig. 11. FD_Flag and SFD when the reception of the first packet was a success

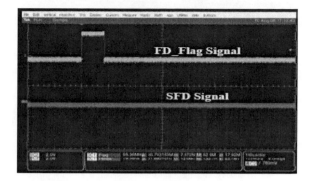

Fig. 12. FD_Flag and SFD when the packet receives after the first packet receives

4   Conclusion and Directions for Further Research

This study searched the IEEE 802.15.4a technology which is the standard of the location awareness WPAN system which provides the communication and location. The UWB technology can provide the function of the location awareness as it is adopted as the standard technology for the communication and location which requires the degree of precision within several dozens of centimeters in a relatively narrow area. Relevant common products will be released from now on because the research and development have actively been proceed at home and abroad as the IEEE 802.15.4a standard was completed. It is also expected that this study will play an important role in realizing the ubiquitous sensor network. In addition, it is considered that the research and development for the location awareness WPAN system should be continued because it aims at the low price and low consumed electric power.

Acknowledgments. Funding for this paper was provided by Namseoul university.

References

1. Foerster, J., Green, E., Somayazulu, S., Leeper, D.: Ultra-Wideband Technology for Short or Medium Range Wireless Communications. Intel. Technology Journal Q2 (2001)
2. Gezici, S., Tian, Z., Giannakis, G.B., Kobayashi, H., Molisch, A.F., Poor, H.V., Sahinoglu, Z.: Localization via Ultra-Wideband. IEEE Signal Processing Magazine, 70–84 (July 2005)
3. Radio Regulation Law of Korea (Aticle 9 of Frequency distribution) (2006)
4. Oh, M.K., Kim, M.J., Kim, J.Y.: Low data-rate location awareness UWB technology for ubiquitous home. Electronics and Telecommunications Trends, ETRI 21(5), 30–39 (2006)

Hardware Implementation and Performance Analysis of NLM-128 Stream Cipher

Soh Yee Lee and HoonJae Lee*

Dept. Information Communication Eng.,
Dongseo University,
Sasang-gu, Busan, 617-716, Korea
eyestarslee@gmail.com, hjlee@dongseo.ac.kr

Abstract. A NLM generator is an alteration of LM generator with the addition of NFSR. NLM generator is high speed with low power consumption and has high security level. NLM-128, an instance from NLM family, is implemented and analyzed. This paper examines the hardware performance of NLM-128. An ALTERA FPGA Cyclone II device (EP2C35F672C8) is chosen to analyze the performance of NLM-128. The results show an output performance of 83.33Mbps with low power consumption. In addition, it is easy to be implemented in hardware because NLM-128 requires little logic gates only.

Keywords: NLM generator, stream cipher, simulation, FPGA, Galois LFSR, keystream generator.

1 Introduction

The cryptographic ciphers are generally divided into two types, which are public key encryption (also known as asymmetric encryption) and private key encryption (also known as symmetric encryption). We can further divide private key ciphers into two categories – stream ciphers and block ciphers. Stream ciphers have a speed advantage against block ciphers. In addition, it is more efficient than block ciphers and much simpler to be implemented. RC4 is one of the commercial applications that use stream cipher. It is almost two times faster than nearest block cipher and can be written in 30 lines of code as compare to a typical block cipher which needs several hundred lines of code [1, 2].

In year 1985, a summation generator was purposed [3]. It has excellent cryptographically properties such as the ability to produce random binary sequences with maximum period. Besides that, it has high linear complexity which makes the analysis of the sequences harder. However, the summation generator is suffered from correlation attack [4, 5], fast correlation attack [6] and divide and conquer attack [5, 7]. Then, a LM generator had been proposed in year 2000 with an addition of a memory bit to the summation generator [8]. Since the construction of summation generator and LM[1] generator is analogous, similar attacks can be used to attack both

---

* Corresponding author.

G. Lee, D. Howard, and D. Ślęzak (Eds.): ICHIT 2011, CCIS 206, pp. 446–453, 2011.
© Springer-Verlag Berlin Heidelberg 2011

generators [9]. A NLM generator, based upon summation generator was proposed in year 2009 [10]. This scheme was proposed with an attempt to defeat the known attacks against the summation-type generators. NLM generator is an alteration of LM generator [8] with the replacement of one of the Linear Feedback Shift Register (LFSR) with Nonlinear Feedback Shift Register (NFSR). The purpose of the additional NFSR in the design is to increase the nonlinearity of the generated output keystream [10]. By doing so, the security level can be enhanced.

This paper examines the hardware performance of NLM-128 in FPGA.

2 Summation-Like Generator

This section will briefly describe summation generator, LM generator, and NLM generator.

2.1 Summation Generator

In summation generator, there are r regularly clocked binary LFSRs and $\log_2 n$ bits of carry [3]. Two LFSRs are denoted as L_a and L_b respectively. c represents carry bit. a_j is the output of L_a at time j. Similarly, b_j is the output of L_b at time j. The output c_j represents the carry bit of the generator at time j, and z_j represents the generated keystream bit at time j. c_j and z_j are defined as:

$$c_j = f_c = a_j b_j \oplus (a_j \oplus b_j) c_{j-1} \tag{1}$$

$$z_j = f_z = a_j \oplus b_j \oplus c_{j-1}. \tag{2}$$

Initial state (at time $j = 0$) of carry bit in the generator, $c_{-1} = 0$.

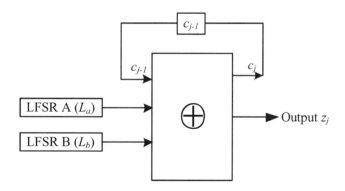

Fig. 1. Summation generator ($r = 2$)

2.2 LM Generator

Fig. 2 shows a LM generator [8] with $r = 2$. LM generator (as shown above) and summation generator (fig. 1) are very similar. A LM generator is an improved-summation generator with an additional memory bit, d. Notations L_a, L_b and c in LM generator are defined exactly the same as in summation generator. The carry bit of LM generator at time j, c_j is defined by equation (1), which is totally identical to summation generator. At time j, the additional memory bit of the generator is denoted as d_j. d_j is calculated by function f_d as shown in equation (3). The addition of extra memory bits d affects the calculation of function z (f_z). The f_z is changed to include d in the calculation, as shown in equation (4).

$$d_j = f_d = b_j \oplus (a_j \oplus b_j) d_{j-1}.$$ (3)

$$z_j = f_z = a_j \oplus b_j \oplus c_{j-1} \oplus d_{j-1}.$$ (4)

Initial state (at time $j = 0$) of carry bit (c_{-1}) and memory bit (d_{-1}) is defined to be 0.

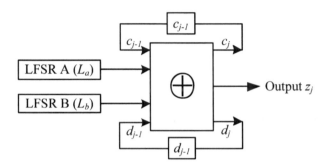

Fig. 2. LM generator ($r = 2$)

2.3 NLM Generator

NLM generator [10] as shown in fig. 3, substitutes one of the LFSR in LM generator with NFSR. It is believed that the substitution of NFSR can produce more unpredictable outputs. All the notations used in NLM generator are defined exactly the same as in LM generator, except b_j. b_j is defined as the output of NFSR (instead of LFSR's output in LM generator) at time j. At time $j = 0$, $c_{-1} = 0$, $d_{-1} = 0$. The function f_c, f_d and f_z are identical to LM generator, except the definition of b_j in the equation is different.

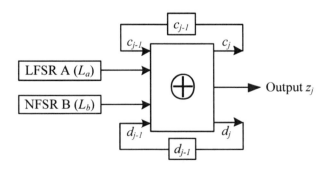

Fig. 3. NLM family generator

NLM-128. An instance from NLM family generator is NLM-128. It has 127 bits of LFSR and 129 bits of NFSR. The two polynomials of NLM-128, $p_a(x)$ and $p_b(x)$ are defined as follwing:

$$p_a(x) = x^{127} \oplus x^{109} \oplus x^{91} \oplus x^{84} \oplus x^{73} \oplus x^{67} \oplus x^{66} \oplus x^{63} \oplus x^{56} \oplus x^{55} \oplus$$
$$x^{52} \oplus x^{48} \oplus x^{45} \oplus x^{42} \oplus x^{41} \oplus x^{37} \oplus x^{34} \oplus x^{30} \oplus x^{27} \oplus x^{23} \oplus \tag{5}$$
$$x^{21} \oplus x^{20} \oplus x^{19} \oplus x^{16} \oplus x^{13} \oplus x^{12} \oplus x^{7} \oplus x^{6} \oplus x^{2} \oplus x^{1} \oplus 1.$$

$$p_b(x) = x^{129} \oplus x^{125} \oplus x^{121} \oplus x^{117} \oplus x^{113} \oplus x^{109} \oplus x^{105} \oplus x^{101} \oplus x^{97} \oplus x^{93} \oplus x^{89} \oplus$$
$$x^{85} \oplus x^{81} \oplus x^{77} \oplus x^{73} \oplus x^{69} \oplus x^{65} \oplus x^{61} \oplus x^{57} \oplus x^{53} \oplus x^{49} \oplus x^{45} \oplus \tag{6}$$
$$x^{41} \oplus x^{37} \oplus x^{33} \oplus x^{29} \oplus x^{25} \oplus x^{21} \oplus x^{17} \oplus x^{13} \oplus x^{9} \oplus x^{5} \oplus (\prod_{i=1}^{129} x^{i}).$$

$p_a(x)$ is a primitive polynomial, and $p_b(x)$ is a de Bruijn sequence irreducible polynomial [10]. NLM-128 uses 128 bits of key, k and initialization vector, iv in initialization process.

The initialization process runs the generator twice. In first iteration, starting state of 127-bit $L_a = (k \oplus iv) \bmod 2^{127}$. Modulus of 2^{127} after the XOR operation is essential to obtain 127 bits from the 128-bit result. Starting state of 129-bit $L_b = (k << 1) \oplus (0|iv)$. This means the key (128 bits) embedded in 129-bit word is first shifted 1 bit to the left. Then, it is XORed with iv with a leading zero. Now, the cipher is run to produce an output string of length 256 bits, R_1. In second iteration, the first 127 bits from 256-bit R_1 are used as the initial state of L_a, and the remaining 129 bits form the initial state of L_b. The cipher is run once again to produce a 256-bit output, R_2. As previously, the first 127 bits of R_2 form initial state of L_a, and the remaining 129 bits of R_2 are used as the initial state of L_b. At this stage, the initialization process is accomplished. The LM generator initial state is now set and it is ready to generate keystream. The generated keystream is then XORed with the plaintext bit to produce ciphertext.

3   Performance Analysis of NLM-128

3.1   FPGA Implementation

Galois implementation is used for LFSR and NFSR in NLM-128 rather than Fibonacci implementation. A FPGA design was done to assess the performance of NLM-128. FPGA schematic approach was used in Quartus II v9.1 to target device EP2C35F672C8 (Cyclone II family). An assumption was done for the assessment. The input for LFSR and NFSR in the design is assumed as the final output of initialization process, R_2. That is, the two iterations in initialization process are assumed done. The generator is ready to be run after the input is fed to LFSR and NFSR. Thus, the implementation of initialization process is not included in this design. The focus of the design is mainly on the key generation. Table 1 shows the design environment for the assessment.

Table 1. Design environment

Item	Design Environment
Platform	PC, Windows 7 32-bit
Design Tool	ALTERA Quartus II v9.1
Device family	FPGA Cyclone II Family
Device type	EP2C35F672C8

Karnaugh map was used to simplify the equation (3). The simplified equation is:

$$d_j = b_j \overline{d_{j-1}} + a_j d_{j-1}. \tag{7}$$

Simplification was not done on equation (2) because an adder's megafunction provided in Quartus II tool could be used for equation (2). The Altera megafunctions are ready-made, pre-tested blocks of intellectual property that are optimized to make efficient use of the architecture of the targeted programmable device [11].

Fig. 4 is the total block diagram for FPGA design of NLM-128. Only a small part of LFSR and NFSR implementation is shown here due to the design diagram of LFSR and NFSR are too huge. The output of LFSR and NFSR at time j is denoted as a_j and b_j respectively in the diagram. a_j, b_j and d_{j-1} are then go through some combinational logic gates which are designed according to equation (7) to produce d_j. On the other hand, the outputs of LFSR and NFSR together with c_{j-1} are passed to the megafunction *mega_add* at the same time to produce c_j and *partial result of* z_j (a_j *xor* b_j *xor* c_{j-1}). The *partial result of* z_j is then XORed with d_{j-1} to produce z_j. After that, a plaintext bit can perform XOR operation with z_j to produce ciphertext bit.

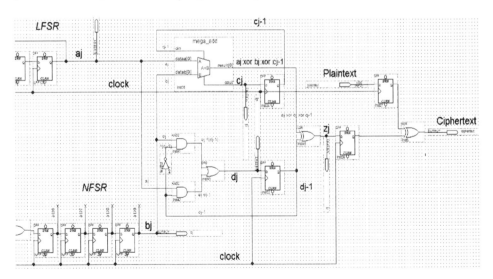

Fig. 4. Design view of nonlinear function block in FPGA for NLM-128.

Fig. 5 shows the implementation of LFSR and NFSR. LFSR and NFSR are formed by numbers of D flip-flop. Be noted that the XOR gates are appeared between D flip-flops since Galois implementation is used. Design of NFSR is similar as LFSR, except the feedback function was implemented differently. A clock is used to control both LFSR and NSFR and is assigned as global clock to minimized clock skew.

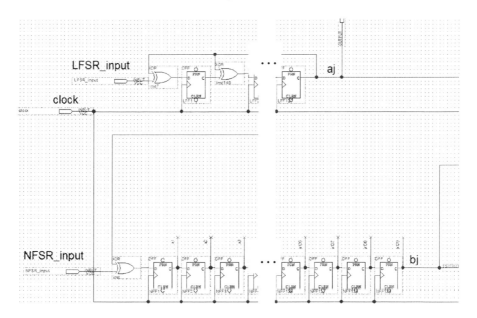

Fig. 5. Design view of LFSR and NFSR block in FPGA for NLM-128

Fig. 6 is part of the diagram showing how the feedback function of NFSR is done. AND operation is performed to all the 129 D flip-flops' output at certain time. Again, megafunction AND was employed to reduce the number of gates used in the design. AND megafunction can support 128 input ports only. Thus, an additional AND gate was essential to get the final result of feedback function.

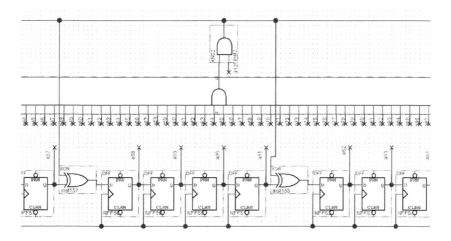

Fig. 6. Design view of NFSR feedback function block in FPGA for NLM-128

The design meets all the timing requirements in timing analysis. This reflects the setup and hold relationship in the design is not violated. *set_input_delay -min* is set to -1ns, and *set_input_delay -max* is set to 3ns. It means the input signal can take anywhere from -1ns to 3ns relative to the rising clock edge to reach the FPGA. Similarly, *set_output_delay -min* is set to -1ns, and *set_output_delay -max* is set to 3ns. The goal for timing analysis on all paths is to have a positive slack. A slack is used to indicate how well a design is meeting or missing its requirements.

3.2 Performance Analysis

In fig. 7, the *LFSR_input* and *NFSR_input* are loaded to LFSR and NFSR respectively. As mentioned earlier, the two iterations in initialization process is assumed done and we are at the stage where the output of the initialization process, R_2 is ready to be loaded to L_a and L_b. *LFSR_input* and *NFSR_input* represent substring of R_2, where *LFSR_input* represents the first 127 bits of R_2 and *NFSR_input* is the remaining 129 bits of R_2. In fig. 7, after 129 clock cycles, all registers in L_a and L_b is loaded with bit 1. This indicates all bits of R_2 are 1 in this case. The initial state of the LM generator is now set and it is ready to generate keystream starting from 130th clock cycles.

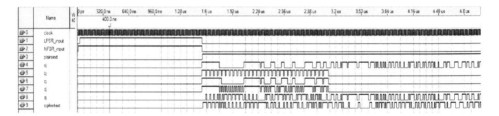

Fig. 7. Timing simulation result for NLM-128

Table 2. Summarization of result

Performance Category	Result
Throughput of data rate	83.33 Mbps
Elements used	257 cells
Total power dissipation	126 mW

With this design, since the longest delay is 9.761ns, a clock with period equal to 12ns is able to produce a throughput of 83.33 Mbps with a positive slack of 2.494ns. This proves NLM-128 can generate keystream rapidly.

Total logic elements used in this design are 257 cells out of 33216, which mean the usage of logic elements is less than 1% over the total availability. This indicates NLM-128 requires little logic gates and it is easy to be implemented in hardware.

The estimation of total power dissipation of the design is 126mW. It reflects just a little power is enough to run the generator. The value is an estimation value because there are external factors like toggle rate which affect the power consumption.

4   Conclusion

NLM-128 is a stream cipher with good properties. It has high linear complexity and it is easy to be implemented in software and hardware. Apart from that, as shown in this paper, NLM generator can generate keystream in high speed and consumes little power only. The implementation of NLM-128 is easy and it is adequate to be used in various applications.

Acknowledgments. This research was supported by Basic Science Research Program through the National Research Foundation of Korea (NRF) funded by the Ministry of Education, Science and Technology.

References

1. Schneier, B.: Applied Cryptography, 2nd edn. John Wiley & Sons, Inc., Chichester (1996)
2. Wedbush Morgan Securities – Industrial Report, http://www.vikasqupta.com
3. Rueppel, R.A.: Correlation Immunity and the Summation Generator. In: Williams, H.C. (ed.) CRYPTO 1985. LNCS, vol. 218, pp. 260–272. Springer, Heidelberg (1986)
4. Meier, W., Staffelbach, O.: Correlation Properties of Combiners with Memory in Stream Ciphers. In: Damgård, I.B. (ed.) EUROCRYPT 1990. LNCS, vol. 473, pp. 204–213. Springer, Heidelberg (1991)
5. Dawson, E.: Cryptanalysis of Summation Generator. In: Zheng, Y., Seberry, J. (eds.) AUSCRYPT 1992. LNCS, vol. 718, pp. 209–215. Springer, Heidelberg (1993)
6. Golic, J., Salmasizadeh, M., Dawson, E.: Fast Correlation Attacks on the Summation Generator. Journal of Cryptology 13(2), 245–262 (2000)
7. Siegenthaler, T.: Design of Combiners to Prevent Divide and Conquer Attacks. In: Williams, H.C. (ed.) CRYPTO 1985. LNCS, vol. 218, pp. 273–279. Springer, Heidelberg (1985)
8. Lee, H., Moon, S.: On An Improved Summation Generator with 2-Bit Memory. Signal Processing 80(1), 211–217 (2000)
9. Chen, K., Dawson, E.: Security Analysis of the LM Generator. Report (2004)
10. Lee, H., Sung, S., Kim, H.: NLM-128, An Improved LM-type Summation Generator with 2-bit memories. In: Proceedings of 4th International Conference on Computer Sciences and Convergence Information Technology, pp. 577–582 (2009)
11. Altera Megafunctions, http://www.altera.com/products/ip/altera/mega.html

Optical Clock Signal Extraction of Nonreturn-to-Zero Signal by Suppressing the Carrier Component

Jaemyoung Lee

Korea Polytechnic University
2121 Jungwang Shihung Kyunggi, Republic of Korea
lee@kpu.ac.kr

Abstract. We propose a simple optical clock signal extraction scheme of a nonreturn-to-zero signal and experimentally demonstrate the clock signal extraction using a single FBG filter. The proposed optical signal processing shows the 10Gb/s PRZ pattern which can be used for optical clock recovery.

Keywords: optical signal processing, clock recovery.

1 Introduction

In order to cope with the recent explosive increase in IP traffic, many research groups have focused on all optical clock recovery for real time data processing which is expected to be realized in the near future for all optical transparent networks [1,8] . For real time optical signal processing, a number of methods for all optical clock recovery have been proposed and demonstrated [4,5,7,6]. To apply the reported methods, however, these schemes require specially designed a DFB laser[4,6], a Fabry-Perot laser diode[5], or semiconductor optical amplifier Mach-Zehnder interferometric switches[7], of which systems are complex to implement. On the other hand, an optical clock recovery system employing passive components has been reported for 40 Gb/s nonreturn-to-zero (NRZ) signal with two fiber Bragg grating (FBG) filters [2,3]. The scheme of Ref.[3] can not only generate the pseudoreturn-to-zero (PRZ) but eliminate noise components between the carrier and the clock component.

In this paper, we further simplify the previous scheme of Ref.[3] to convert the NRZ signal to pseudoreturn-to-zero (PRZ) for clock recovery using a single FBG filter. The experimental result shows that the proposed scheme employing a FBG filter converts the NRZ signal to PRZ.

2 Proposed Scheme for Clock Extraction of NRZ Signal

Figure 1 show a random digital sequence of an NRZ signal in which the bit of "1" holds the value of "1" throughout the bit period. In other words, an NRZ

G. Lee, D. Howard, and D. Ślęzak (Eds.): ICHIT 2011, CCIS 206, pp. 454–458, 2011.

Fig. 1. Waveform of an NRZ signal

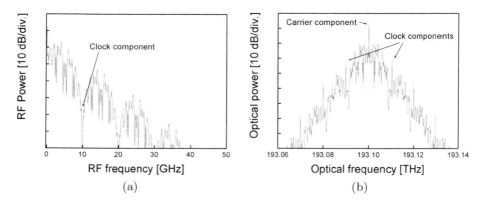

Fig. 2. Simulation results for RF and optical spectra of input 10Gb/s NRZ signal (a) and(b), respectively

waveform does not return to zero for the period of the bit interval for a signal value of "1". The power spectral density of an NRZ signal is

$$\Psi_{NRZ}(f) = \frac{A^2 T_b}{4} \left[sinc(fT_b) \right]^2 \left[1 + \frac{1}{T_b} \delta(f) \right] \tag{1}$$

where A and T_b are the amplitude of the bit of "1" and the bit period of the signal, respectively. Since the sinc function of the power spectral density of an NRZ signal in eq.(1) falls on the value of "0" at the multiples of the clock frequency of the NRZ signal, $1/T_b$, an NRZ signal does not have the clock components in its spectrum, as shown in the simulation result of Fig. 2(a). The real pulse shape of an NRZ signal in an optical transmission system, however, is not rectangular due to limited response times of devices used in systems. The rise/fall time of devices in a system transforms the ideal NRZ rectangular pulse to the shape with an exponential increase/decrease. Nonlinear properties of optical devices in terms of response time account for the clock components in the optical spectrum in Fig. 2(b). However, the amplitude of the clock components in the optical spectrum are not large enough to be used for clock extraction compared with the carrier frequency component. Spectral components $\Omega(\omega)$ of the NRZ signal in eq.(1) can be described as follows:

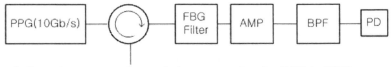

Fig. 3. Experiment setup for optical preprocessing for NRZ-to-PRZ conversion

$$\Omega(\omega) = \sum_i C_i \delta(\omega - i\omega_0) \tag{2}$$

Since the clock signal is generated through the beating process between two frequency components which are spaced the clock frequency from each other, the amplitude of the NRZ clock signal can be expressed as

$$C_{clk} = \sum_{-\infty}^{\infty} C_i C_{i-N} \tag{3}$$

where N is given from $\omega_0 = 2\pi/NT$. Note that the ideal NRZ signal does not have any clock components in its spectral domain as mentioned in this section and shown in the simulation result of Fig. 2(a). That is, the amplitude of the clock component, C_{clk} in eq.(3) in spectral domain is theoretically "0". However, by suppressing or enhancing amplitudes of some frequency components in eq.(3), we can have the clock component, C_{clk}, in an NRZ signal.

For the clock extraction for an NRZ signal, we propose a scheme in which the clock component is generated by suppressing the carrier component, which disturbs the balance between the positive and the negative terms in the eq.(3). Figure 3 shows the proposed scheme in which the amplitude of the carrier frequency component is reduced through the reflection of the FBG filter.

We conducted numerical simulation by reducing the carrier component by 99.9% as shown in Fig. 4(b), generating the clock component even in the rf

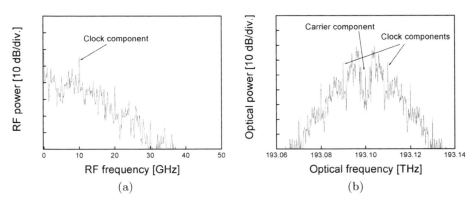

Fig. 4. Simulation results for RF and optical spectra of the output signal through the proposed scheme (a) and (b), respectively

spectrum, Fig. 4(a). In the numerical simulation, the amplitude of the clock component, Fig. 4(a), is enhanced and larger than 7 dB compared other frequency components adjacent to the clock component. The optical amplifier after the FBG filter in Fig. 3, amplifies the optical power of the PRZ signal to be received by the photodetector.

3 Experiment

We conducted an optical experiment to convert a 10 Gb/s NRZ signal to a PRZ signal using the proposed scheme, Fig. 3. The bandwidth and reflectivity of the FBG filter used in the experiment are 0.17 nm and 90%, respectively. The modulated optical NRZ signal of which RF spectrum and eye diagram are shown in Fig. 5(a) and (b), respectively, and propagates to the FBG filter through a circulator. The FBG filter reflects frequency components around the carrier frequency component in the optical frequency domain, reducing the carrier component amplitude and reshaping the optical spectrum of the NRZ signal. By passing the NRZ signal through the FBG filter, the proposed scheme enhances the amplitude of the clock component in the RF spectrum, Fig. 5(c), and converted the

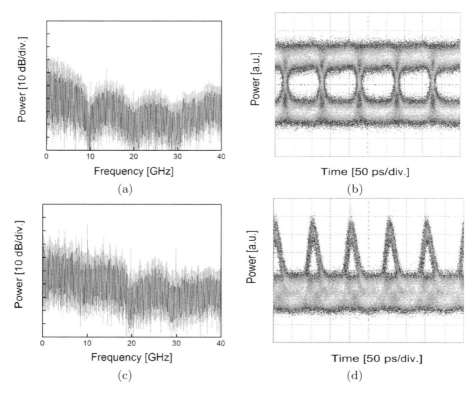

Fig. 5. RF spectra and eye diagrams of the input 10Gb/s NRZ signal (a), (b), and of the output signal through the proposed scheme (c), (d)

NRZ signal to the PRZ signal, Fig. 5(d). Figure 5(c) shows that the amplitude of the clock component in the rf spectrum is 10 dB larger than other frequency components adjacent to the clock component. The PRZ signal is amplified by the optical amplifier and other noise components which exist beyond the clock frequency components are eliminated by the bandpass filter.

4   Conclusion

We proposed a simple NRZ-to-PRZ conversion scheme and experimentally demonstrated the NRZ-to-PRZ conversion using a single FBG filter. Through the scheme, the amplitude of clock component increases and is 10 dB larger than other frequency components adjacent to the clock components. The proposed optical signal processing scheme shows the PRZ pattern which can be used for optical clock recovery.

References

1. Hanik, N., Ehrhardt, A., Gladisch, A., Christophe Peucheret, P.J., Molle, L., Freund, R., Caspar, C.: Extension of all-optical network-transparent domains based on normalized transmission sections. J. of Lightwave Technol. 22, 1439–1453 (2004)
2. Lee, J., Cho, H., Ko, J.S.: Enhancement of clock component in a nonreturn-to-zero signal through beating process. Opt. Fiber Technol. 12, 59–70 (2006)
3. Lee, J., Cho, H., Lim, S.K., Lee, S.S., Ko, J.S.: Optical clock recovery scheme for a high bit rate nonreturn-to-zero signal using fiber bragg grating filters. Optical Engineering 44(2), 020502-1–2 (2005)
4. Mao, W., Li, Y., Al-Mumim, M., Li, G.: All-optical clock recovery for both RZ and NRZ data. IEEE Photon. Technol. Lett. 14, 873–875 (2002)
5. Parra-Cetina, J., Latkowski, S., Maldonado-Basilio, R., Landais, P.: Wavelength tunability of all-optical clock-recovery based on quantum-dash mode-locked laser diode under injection of a 40-Gb/s NRZ data stream. IEEE Photon. Technol. Lett. 23(9), 531–533 (2011)
6. Sartorius, B., Bornholdt, C., Brox, O., Ehrke, H., Hoffmann, D., Ludwig, R., Möhrle, M.: All-optical clock recovery module based on self-pulsating DFB laser. Electron. Lett. 34, 1664–1665 (1998)
7. Spyropoulou, M., Pleros, N., Miliou, A.: SOA-MZI-based nonlinear optical signal processing: A frequency domain transfer function for wavelength conversion, clock recovery, and packet envelope detection. IEEE J. Quantum Electron. 47(1), 40–49 (2011)
8. Urban, P.J., Huiszoon, B., Roy, R., de Laat, M.M., Huijskens, F.M., Klein, E.J., Khoe, G.D., Koonen, A.M.J., de Waardt, H.: High-bit-rate dynamically reconfigurable WDM-TDM access network. J. Opt. Commun. Netw. 159, A143–A159 (2009)

Dynamic Message Server for Personal Health Data Transmission in u-Health Service Environment

Eun Jeong Choi[1] and Hee Joung Hwang[2,*]

[1] Department of Computer Science and Engineering,
University of Incheon, Incheon, Korea
sosim0735@naver.com
[2] Department of Computer Science,
Gachon University of Medicine and Science, Incheon, Korea
hwanghj@gachon.ac.kr

Abstract. As the social environment rapidly changes and health concerns increase, medical expenses have also been increasing. Also, the source of value in the medical service industry is changing from single service competitiveness to a total solution philosophy. In addition, any attempt to manage personal health data and U-Health continue to grow by innovations and are processed through IT convergence services such as Google Health or Touch Doctor. Basic U-health services are created by sending and receiving data between personal health devices, gateways, service providers and medical institutions. It is incomplete to implement U-Health, although a guideline based on standard focusing Continua is currently suggested. Thus in this paper, we designed a message server based on relevant requirements and standards. The proposed server can implement dynamic modules. It also allows many different standards to be accepted and expanded, so it is expected to help U-Health service provider systems to be implemented.

Keywords: OSGi, HL7, Personal Health Data, Message Gateway Server, U-Health.

1 Introduction

The interest in health has been growing socially as the quality of life has improved with the increase of personal income. Also, information technology develops while in the medical industry, a technical and institutional environment regarding U-Health has been implemented via Touch Doctor or Google Health. These services are changing the concept of the traditional medical services. So consumer demand is also increasing.

Previously, U-Health was independent, so health care services were available only in the hospital and depended on doctors and medical devices. The market of medical devices has expanded with the Internet and the changing paradigm of the medical

---

* Corresponding author.

G. Lee, D. Howard, and D. Ślęzak (Eds.): ICHIT 2011, CCIS 206, pp. 459–466, 2011.
© Springer-Verlag Berlin Heidelberg 2011

service industry [1][2][3]. Furthermore, as the number of chronic patients and elderly people has rapidly risen, and consumers become more aware of how to prevent diseases, an increasing number of people manage their health with Personal Health Devices (PHD) at home [4].

Basic U-Health services transmit biometric data to a gateway after measuring blood pressure, pulse, ECG, etc and transfer personal health information to medical institutions via Service Providers [5]. So it allows people to manage their health data and take proper medical treatment for their medical conditions. U-Health services currently have guidelines based on standard focusing Continua, but these guidelines are not yet complete enough to implement a u-Health service.

Thus, in this paper, we designed a message server based on relevant requirements and standards. The proposed server can be implemented as a dynamic module. It is also able to accept and extend diverse standards. It is expected to help the U-Health service provider system to be implemented.

2 Relation Study

2.1 Continua

Continua attempts to solve these problems by suggesting a Continua Design Guideline for Interoperability gaps which occur when adopting them, using as many current standards as possible instead of setting up new standards to provide connections between personal health devices.

In order to implement U-Health services, end-to-end architecture should be defined for personal health devices measuring biometric data to a medical information server which analyzes it. Continua End-to-End Architecture categorizes medical information devices required for u-Health services by interface according to their roles and purposes as shown in Fig. 1, and adopts standard techniques appropriate to each interface. Continua End-to-End Architecture solves end-to-end interoperability problems by suggesting a Continua Design Guideline [6][7].

Fig. 1. Definitions and graphical notation

Two interfaces – a Personal Area Network (PAN) and a Heath Record Network (HRN) - were defined in the Continua Design Guideline version 1.0. Basic End-to-End Architecture for U-Health services was completed after two more interfaces - Wide Area Network (WAN) and Local Area Network (LAN) - were additionally defined in the v1.5 specification published in 2010. The range of guidelines is expected to gradually expand after new standard techniques such as Bluetooth Low Energy (LE) by interface is adopted in the v 2.0 in 2011 as shown in Table 1 [8].

Table 1. Standard technique classified by interface adopted in Continua

Interface	Standard Technique	Area	Adopt Guideline
PAN	ISO/IEEE 11073-Device Specialization	Data	v1.0
	Bluetooth LE Thermometer		v2.0
	ISO/IEEE 11073-20601-2008	Message	v1.0
	ISO/IEEE 11073-20601A-2010		v1.5
	Bluetooth LE Protocol		v2.0
	USB PHDC	Transport	v1.0
	Bluetooth HDP		v1.0
	Bluetooth LE		v2.0
LAN	ISO/IEEE 11073-Device Specialization	Data	v1.5
	ISO/IEEE 11073-20601-2008	Message	v1.5
	ISO/IEEE 11073-20601A-2010		v1.5
	ZigBee HCP	Transport	v1.5
	ZigBee many-to-many connectivity		v2.0
WAN	IHE PCD-o1 (HL7 V2.6 Message)	Data	v1.5
	W3C WS-1 BP,BSP, RSP	Message	v1.5
	SOAP 1.2, HTTP v1.1	Transport	v1.5
HRN	HL7 PHMR(CDA R2)	Data	v1.0
	IHE PIX	Patient ID	v2.0
	IHE XDS	Message	v1.0
	IHE XDR	Transport	v1.0
	IHE XDM		v1.5

2.2 ISO/IEEE11073

ISO/IEEE 11073 has been developed and adopted by almost every country to provide complete connections between medical devices, interoperability, plug and play, transparency, ease of use and configurations. ISO/IEEE 11073 contains commonly-used standards in different levels which provide connections to relevant devices and complete solutions from low to high levels (abstract expression of data and services).

Those standards are divided into device data, general application services (event or polling), transmission (cable or radio), network communication (few network connecteds by a party telephone network) and gateway standards (DICOM or HL7 messages) [9]. ISO/IEEE 11073 enables communication between medical devices and external medical systems. It also automatically saves information regarding signal data of patients and equipment movement.

ISO/IEEE 11073 aims to provide real-time plug and play interoperability for medical devices connected to patients and make it easy to efficiently exchange life signals obtained from point-of-care with medical device data in every health care environment. Fig. 2 is a reference model for medical device communications [10].

ISO/OSI		CEN/IEEE (beginning)	ISO/IEEE (middle)	ISO/IEEE (current)
7	Application	13734-Vital	1073.1.x 11073-10x	11073-1x
6	Presentation	10735-Intermed	1073.2.x 11073-20x	11073-2x
5	Session			
4	Transport			
3	Network	1073.3.x	1073.3.x 11073-30x	11073-3x
2	Data Link			
1	Physical	1073.4.x		

Fig. 2. Reference model for medical device communications

3   Dynamic Message Server for Personal Health Devices

We designed a message gateway server which collected diverse PHD data and connected it to a hospital healthcare system.The message server based on OSGi collects data from diverse devices and PHD by communication and improves productivity by embedding a framework usable in the server.

Data collected by the server is provided to users in diverse forms such as HL7 and CDA. The proposed server can implement devices and services in the form of bundling and expansion. The framework based on the server can support applications regarding U-Health services without any expert knowledge for medical information standards and OSGi. Fig. 3 shows the architecture of the entire system.

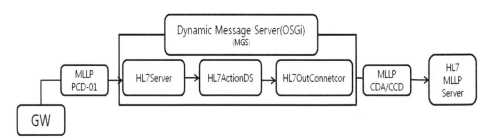

Fig 3. System architecture

As shown in Fig. 4, users provide data to the gateway after measuring it with a device. In the gateway, the user receives measured data and sends processed data to message the broadcast server if requested. The message gateway server transforms personal health data such as ECG and blood pressure into the forms of HL7 and PCD-01 suited to medical institutions. Moreover, the message gateway server searches for diverse bundle types of device protocols and services, and provides them to medical institutions through service providers.

Medical institutions provide users, hospital EMR, and other health care institutions with data in the form of graphs after transforming numerical data from a message gateway server into HL7, CDA, and PCD-01. Personal health information transferred to EMR enables applications in diverse services other than in medical systems.

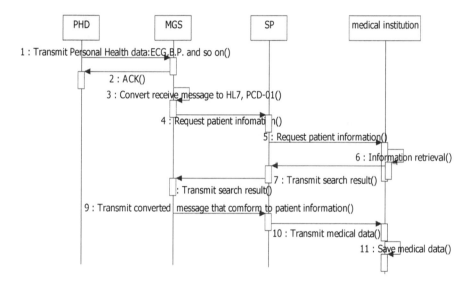

Fig. 4. Sequence diagram of the message gateway server system

4 Implementation and Experiment

4.1 Implementation Environment

Java 6.x and the embedded platform gateway based on an Intel Atom Process were used in this experiment as shown in Table 2. The message gateway server was implemented using Eclipse Equinox, an integrated development platform and library for Java and JPA in a Linux-based and Java v1.6 environment.

4.2 Experiment

We use a 1-channel personal ECG device (EP202, Parama-Tech, japan) in order to validate the proposed message gateway server. Fig. 5 shows the results connected to the gateway.

Table 2. Experiment environment

Item	Description	Remarks
H/W Spec	Linux Set-top based on Intel Atom Process	Embedded Platform
	EP 202	1 channel ECG Device
S/W Spec	JDK 6.x or higher OSGi -R4 Eclipse Equinox JPA(Database)	Message Transmission Server
Data Format	HL7, PCD-01, CDA	Data Format

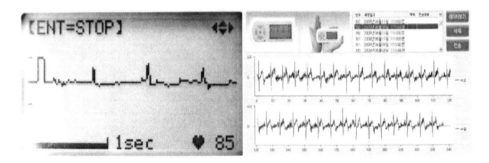

Fig. 5. ECG measurement (result of previous research)

This was transformed into a PCD-01 message as shown in Fig. 6 by the message gateway server, conversion module, and outbound module.

```
MSHI^~\&IE018IIII20110404090030IIORU^R01^ORU_R01IMSGID1234IPI2.6IIINE
IALIIIIBarcodeIIHE PCD ORU-R01 2006^HL7^2.16.840.1.113883.9.n.m^HL7
PIDIII12072507^^^GMC^PIIIKang^Dae^Seoung^^^^L^A
PV1IIEI
OBRI1IGilHospitalIIIII20110404090032
OBXI1ISTIProtieinI1.0.0.1I4.8I%IIIIIR
OBXI2ISTIAlbuminI1.0.0.2I21.7Ig/dlIIIIIR
```

Fig. 6. PCD-01 message

The transformed data can be sent to medical institutions through a message gateway server as shown in Fig. 7. As a result of the experiment, measured data can be confirmed in the medical institution, and the messages are normally sent to the external HL7 gateway system through the HL7 outbound module.

Fig. 7. Experimental results

5 Conclusion

In this paper, we designed and implemented a message server for broadcasting personal health data. The proposed message server supported dynamic component architecture based on OSGi, so it easily expanded and improved diverse modules. Also, medical information standards such as HL7 and IHE PCD-01 were observed by the proposed message server.

We succeeded in connecting data measured from PHD with Gil Hospital's EMR system through our experiment. Also, receiving messages could be transformed into CDA, the medical information exchange standard, and transmitted to other HL7 servers. The created message was verified by IHE and proved that the proposed message server was available.

References

1. Choi, E.J., Hwang, H.J.: Multiple User and Service Management Architecture for Medical Gateway. In: 19th Korean Society for Internet Information General Meeting and Spring Conference, pp. 315–319. Korea Society for Internet Information, Asan (2009)
2. EPCglobal, http://www.epcglobalinc.com

3. Kilner, T.: Triage decisions of prehospital emergency health care providers, using a multiple casualty scenario paper exercise. Emerg. Med. J., 348–353 (2002)
4. Kim, J.T., Choi, E.J., Hwang, H.J.: Development of u-Health system prototype based on dynamic management framework. Korean Institute of Information Technology 8, 87–95 (2010)
5. Park, C.Y., Lim, J.H., Park, S.J., Kim, S.H.: Technical Trend of U-Healthcare Standardization. Telectronics and Telecommunications Trends (August 2010)
6. Shepherd, M., Zitner, D., Watters, C.: Medical Portals: Web-Based Access to Medical Information. In: Proceedings of the 33rd Hawaii International Conference on System Sciences, HICSS 2000, Maui, p. 5003 (2000)
7. Continua Alliance, http://www.continuaalliance.org
8. Telecommunications Technology Association, http://www.tta.or.kr
9. Martínez, I., Fernández, J., Galarraga, M., Serrano, L., de Toledo, P., Jiménez-Fernández, S., Led, S., Martínez-Espronceda, M., García, J.: Implementation of an end-to-end standard-based patient monitoring solution. IET Commun. 2, 181–191 (2008)
10. Galarraga, M., Martinez, I., de Toledo, P.: Review of the ISO/IEEE 11073. Telemedicine Network (2005)

Customized Healthcare Infrastructure Using Privacy Weight Level Based on Smart Device

Namje Park

Department of Computer Education, Teachers College, Jeju National University,
61 Iljudong-ro, Jeju-si, Jeju-do, 690-781, Korea
namjepark@jejunu.ac.kr

Abstract. Personalized radio-frequency identification (RFID) tags can be exploited to infringe on privacy even when not directly carrying private information, as the unique tag data can be read and aggregated to identify individuals, analyze their preferences, and track their location. This is a particularly serious problem because such data collection is not limited to large enterprise and government, but within reach of individuals. In this paper, we describe the security analysis and implementation leveraging globally networked mobile RFID service. We propose a secure mobile RFID service framework leveraging mobile networking. Here we describe the proposed framework and show that it is secure against known attacks. The framework provides a means for safe use of mobile phone-based RFID services by providing security to personalized RFID tags.

Keywords: Mobile RFID, Privacy, Security, Hospital, Healthcare.

1 Introduction

The recent medical security guidelines and the development of information technology make hospitals reduce the expense in surrounding environment and it requires improving the quality of medical security of the hospital. That is, with the new guidelines and technology, hospital business escapes simple fee calculation and insurance claim center. Moreover, MIS (Medical Information System), PACS (Picture Archiving and Communications System) are also developing. Medical Information System is evolved toward integration of medical IT and situation si changing with increasing high speed in the ICT convergence. These changes and development of ubiquitous environment require fundamental change of medical information system. Mobile medical information system refers to construct wireless system of hospital which has constructed in existing environment. Through mobile RFID development in existing system, anyone can log on easily to Internet whenever and wherever.

RFID technology is widely used in supply chain management and inventory control, and is recognized as a strong potential vehicle for ubiquitous computing. However, continued development and global adoption has also raised fears of the potential for exploiting such tags for privacy infringement in 'Big Brother' type scenarios. We propose a secure framework for mobile-phone based RFID services

G. Lee, D. Howard, and D. Ślęzak (Eds.): ICHIT 2011, CCIS 206, pp. 467–474, 2011.

using personal privacy-policy-based access control for personalized ultra-high frequency (UHF) tags employing the Electronic Product Code (EPC). The framework, called mobile RPS, has dynamic capabilities that extend upon extent trust-building service mechanisms for RFID systems. This new technology aims to provide absolute confidentiality with only basic tags.

2   Security Framework Architecture

2.1   Privacy Protection Framework for Mobile RFID Services

The objective of personal privacy in mobile RFID services is to allow individuals to control their personal information related to RFID services. In other words, unauthorized distribution of personal information carried on the tag shall be prevented and a privacy protection mechanism shall be applied to the information collection process through the use of terminals. This paper aims to provide privacy protection services by adopting a privacy protection system (RPS) in the mobile RFID service network. Figure 1 shows the structure of mobile RFID service including RPS.

Privacy protection in mobile RFID services refers to technological measures against unauthorized access of personal information. Access to platform resources can be controlled based on each user's privacy protection level. Privacy protection in mobile RFID services is based on the following concepts.

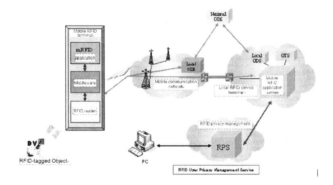

Fig. 1. Mobile RFID service

1) For privacy-secure mobile RFID services, the privacy protection system guarantees confidentiality and integrity of privacy information on the network and ensures authorization of entities.
2) Mobile RFID application and contents provides detailed access control mechanisms that can manage object information, log data, and personal information by user group.
3) Mobile RFID application and contents provision systems communicate with RPS systems through secure communication paths.

4) Mobile RFID application and contents provision systems provide auditing functions with stronger privacy based on the privacy protection policy that each individual user defined in the RPS system.

5) The Mobile RFID application and contents provision system manages personal privacy information based on the rules that individual users defined in the RPS system. The system operators are obliged to protect personal privacy information in earnest.

6) Mobile RFID application and contents provision systems have a mechanism to negotiate privacy policies with mobile RFID terminals to prevent them from gathering personal information.

2.2 Application and Contents Information Server of the Service Provider

The application and contents information server of the service provider provides an extended access control for greater stability. Depending on the tag owner's policy transmitted from RPS, the application and contents information server manages privacy contents, checks who is accessing information, and controls access based on the privacy protection level that shall be set by the object holder.

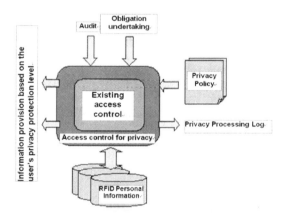

Fig. 2. Application and contents information system of service provider

2.3 Procedure for Secure Mobile RFID Services

There are three privacy protection scenarios for mobile RFD services.

2.3.1 Privacy Policy-Setting Stage
1) Subscribing to RPS and Setting the Privacy Policy

Figure 3 shows the procedure for subscribing to RPS. To use the privacy protection service, the user shall subscribe to RPS and define his/her privacy protection policy. In the same way, the service providers that intend to provide privacy-secured services shall also subscribe to RPS and comply with the default privacy policy of the corresponding service or the privacy protection policy set by the user.

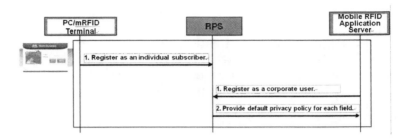

Fig. 3. Subscribing to privacy protection service in a mobile RFID environment

2) Personalization of Tag-attached Object (Privacy Information Combining Phase)
In Figure 4 above, the privacy policy is applied when the tag-attached object is personalized or when privacy information is combined. The following describes the procedure in more detail.

①The mobile RFID terminal reads the tag. Depending on the user's decision, the RFID terminal starts the application program to personalize the tag-attached object (such as purchase.)

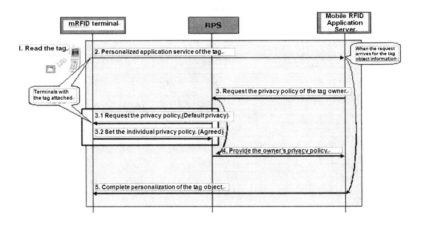

Fig. 4. Personalization of tag-attached object

②The RFID terminal finds the mobile RFID application server in the ODS server and sends the request. At this time, the mobile RFID application server receives information requests and checks whether there is any privacy policy for the owner's tag-attached objects.

③In case the application server does not have a privacy policy for the tag-attached object owner, the RFID terminal requests the policy from the RPS server. Then, RPS checks whether there is a privacy policy for the tag-attached object.

3) Provision of Privacy-protected Information

In the above procedure, privacy-protection information is provided in three ways as shown in Figure 5. The following describes the procedure.

①The mobile terminal reads the tag's information. At this time, the privacy policy stored in the tag is also sent to the mobile RFID terminal.

②The mobile RFID terminal is coupled with the user's terminal application service and finds the mobile RFID application server in the ODS server. At this time, the mobile RFID application server receives the request and checks whether it has the privacy policy of the tag-attached object owner.

③In case the mobile RFID application server does not have the privacy policy of the tag-attached object owner, the mobile RFID application server will request the RPS server to send the policy. Then, RPS checks whether it has the privacy policy of the tag-attached object owner.

④RPS stores personal privacy policy when the object is personalized and sends the privacy policy to the application server. In case RPS does not have a policy, the owner will send the privacy information request in a short text message and will inform who is requesting the privacy information. To avoid delays, the default privacy protection policy determined based on the privacy impact assessment result is sent to the application server.

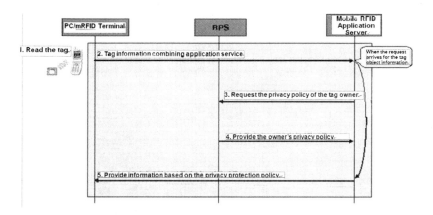

Fig. 5. Mobile RFID privacy protection scenario

⑤The mobile RFID application server sends information of which privacy protection level is lower than the one defined by the user. In other words, only privacy-protected information is sent to the one who is requesting the information. Appendix 1 shows an example.

2.4 Classification of Privacy Levels

The following table is an example of how the privacy levels are classified and how each level is applied. The privacy level is from 0 to 10. In Level 0, virtually no privacy protection is provided, and in Level 10, tags are killed or the levels are not in use.

As shown in Table 1, levels actually used for privacy protection are from 1 to 9. These levels are again classified into Low Level (1 ~ 3), Medium level (4 ~ 6), and High Level (7 ~9.) Each level is for the privacy protection in each application service. However, it is recommended that the privacy protection system should support the following levels to ensure compatibility with the RPS system. In other words, privacy platforms have three groups of privacy protection levels that are from 1 to 10. Three groups of privacy protection levels include Low Level (where most information is disclosed), Medium Level (where object information and history are disclosed) and High Level (where only part of the object information and object category are disclosed.) The default privacy level is applied to the tag and the RPS system.

1) Low Level (Open)

Low levels refer to levels where privacy is least protected among all privacy protection levels. When the privacy level is a low level, most mobile RFID terminals can access the system and related information including parts of personal information. Low levels are allowed only for those who are reliable.

2) Medium Level (Object Information and History)

When mobile RFID accessing individuals are reliable or information carried on the tag does not infringe on a users' privacy, Medium levels are applied. In Medium levels, parts of information are not protected because some security keys are disclosed and disclosed information does not affect security.

3) High Level (Part of Object Information and Object Category)

Access to High-level information is not reliable, and all access is controlled. Only limited parts of information such as object names or object categories are allowed in High levels. For example, in a High level, mobile RFID service is sensitive to privacy and the object owner allows the least information to be exposed to third parties.

Table 1. Default privacy protection level

Privacy Level Object Information	Low Level (1~3)			Medium Level (4~6)			High Level (7~9)		
	1	2	3	4	5	6	7	8	9
Object Category	O	O	O	O	O	O	O	O	X
Object Name	O	O	O	O	O	O	O	O	X
Object Code	O	O	O	O	O	O	O	X	X
Object History	O	O	O	O	O	O	X	X	X
Price	O	O	O	O	O	X	X	X	X
Distribution Information	O	O	O	O	X	X	X	X	X
Object Description	O	O	O	X	X	X	X	X	X
Owner ID	O	O	X	X	X	X	X	X	X
Owner Account	O	X	X	X	X	X	X	X	X
Owner Personal	O	X	X	X	X	X	X	X	X

X: Not to be disclosed/ O: To be disclosed.

3 Implementation of Application Solutions

3.1 Implementation of Customized Healthcare Service

In the proposed hospital data management system, RFID-tagged medical card are given to patients on registration. Patients with sensitive conditions, for example, heart disease or cerebral hemorrhage, can use the medical card to rapidly provide medical history that can used for fast application of first aid. Further, biosensors can be incorporated to provide real-time data to the doctor for each specific patient. The RFID patient tags also can be used to verify patient identity to ensure the correct treatment is administered. Thus, the system allows chartless service.

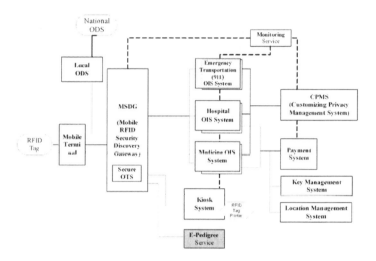

Fig. 6. Proposed customized ubiquitous hospital model

3.2 Implementation

The hospital generated an initial set of control data, which included the patient code, medical ID, and related information. The default privacy level was used and the patient was not allowed to control security policy. In order to provide authentication and privacy interface to patient as a agent in medical discovery gateway and hospital's information server system. Essentially, each bit of sensitive data was initially classified by the default privacy weight, which was then modified by the end user's detailed policy. The user-controllable privacy policy in this system evaluation is considered a basic part of RFID privacy management. The compatibility and scalability may be limited, which will hamper system migration, but the mechanism is suitable for policy based privacy control. The proposed privacy management mechanism was implemented in an actual medical emergency room, including a networked medical information RFID kiosk, RFID networked emergency rescue system, and medical examination service. There is some approach applying the RFID to medicine and hospital. From above, proposed privacy scheme has advantages in custom centric approach aspect for constructing a privacy aware ubiquitous medical system.

4 Conclusion

RFID technology will evolve to become ubiquitous, allowing automatic detection and delivery of information on the surrounding environment, and interconnecting them through the network. This will require RFID implementation of security measures as the technology is vulnerable to privacy infringement via counterfeiting, falsification, camouflage, tapping, and tracking. Therefore, it is necessary to enact laws and regulations that meet the expectations of consumer protection organizations that are sensitive to individual privacy, and develop and apply secure technologies that can follow such laws and regulations.

Mobile RFID readers are being actively researched and developed throughout the world, and more efforts are underway for the development of related service technologies. Though legal and institutional systems endeavor to protect privacy and encourage data protection, the science and engineering world must also provide suitable technologies. Seemingly, there are and will be no perfect security/privacy protection methods. The technologies proposed in this paper, however, would contribute to the development of secure and reliable RFID systems.

Acknowledgments. This paper is extended from a conference paper presented at The 3rd International Conference on Computational Collective Intelligence. The author is deeply grateful to the anonymous reviewers for their valuable suggestions and comments on the first version of this paper.

References

1. Mobile RFID Forum of Korea: Mobile RFID Privacy Protection Framework (Framework for Privacy Protection of Mobile RFID Services). MRFS-4-08. Standard Paper (2006)
2. Park, N., Song, Y., Won, D., Kim, H.: Multilateral Approaches to the Mobile RFID Security Problem Using Web Service. In: Zhang, Y., Yu, G., Hwang, J., Xu, G. (eds.) APWeb 2008. LNCS, vol. 4976, pp. 331–341. Springer, Heidelberg (2008)
3. Park, W., Lee, B.: Proposal for participating in the Correspondence Group on RFID in ITU-T. Information Paper. ASTAP Forum (2004)
4. Park, N., Kwak, J., Kim, S., Won, D., Kim, H.: WIPI Mobile Platform with Secure Service for Mobile RFID Network Environment. In: Shen, H.T., Li, J., Li, M., Ni, J., Wang, W. (eds.) APWeb Workshops 2006. LNCS, vol. 3842, pp. 741–748. Springer, Heidelberg (2006)
5. Park, N., Kim, H.W., Kim, S., Won, D.H.: Open Location-Based Service Using Secure Middleware Infrastructure in Web Services. In: Gervasi, O., Gavrilova, M.L., Kumar, V., Laganá, A., Lee, H.P., Mun, Y., Taniar, D., Tan, C.J.K. (eds.) ICCSA 2005. LNCS, vol. 3481, pp. 1146–1155. Springer, Heidelberg (2005)
6. Park, N.: Security scheme for managing a large quantity of individual information in RFID environment. CCIS, vol. 106, pp. 72–79. Springer, Heidelberg (2010)
7. Park, N.: Secure UHF/HF Dual-band RFID: Strategic Framework Approaches and Application Solutions. In: ICCCI 2011. LNCS, Springer, Heidelberg (2011)

Implementation of Mobile Healthcare Monitoring System with Portable Base Station

Jiunn Huei Yap, Yun-Hong Noh, and Do-Un Jeong

Divsion of Computer & Information Engineering, Dongseo University, South Korea
jhyap85@gmail.com, noh108@nate.com, dujeong@dongseo.ac.kr

Abstract. We propose mobile healthcare monitoring system embedded using a portable base station. The portable base station is integrated with ECG sensor node and responsible to act as a gateway to interface two wireless technologies: Zigbee and Bluetooth. We show the working principle of portable base station and also demonstrating it with a few practical approaches. Wearable biosensors, mobile application for vital signal monitoring, PC program for remote vital signals monitoring are also demonstrated.

Keywords: Portable base station, body area network, healthcare monitoring.

1 Introduction

In the past 10 years, wearable and mobile healthcare monitoring had been marked up as a convenience solution for real time and non-invasive healthcare monitoring. User trial experience on mobile healthcare concluded that users are demanding for good telemedicine service [1]. Jonese and Valerie Gay suggest wearable physiological signal device to have Zigbee linked with user's mobile system [2], but a setback is to have a Zigbee compatible interface board connecting to user's mobile system. Lubrin and Lawrance work out another body area network prototype by communicating several wireless motes to user's mobile system using Zigbee solution [3], but yet, a bulky Zigbee compatible interface board is required to plug on user's mobile system. Aleksandar, Chris, and Emil Jonanov discuss about the basic working principle of body senor network on personal healthcare monitoring but yet still promoting the same idea of using Zigbee compatible board to interface Zigbee bio sensor node with user's mobile system [4]. In this paper, we propose an idea of using a portable base station to replace the existing Zigee interface board solution. Portable base station receives Zigbee transmitted data and sends to user's mobile system using Bluetooth connection. Thus, user's mobile system need no to connect with an extra interface board and hence promoting better flexibility and reserving battery power for user's mobile system.

G. Lee, D. Howard, and D. Ślęzak (Eds.): ICHIT 2011, CCIS 206, pp. 475–480, 2011.

2   Overview of System Architecture for Mobile Healthcare Monitoring

We propose a portable base station to integrate with ECG sensor node within a body area network. ECG sensor is attached to user's chest area to acquit electrocardiography (ECG) signal, where ear PPG sensor is clipped to user's ear lope and finger PPG is clipped to user's thumb to obtain photoplethysmograph (PPG) signal. Ear PPG and finger PPG sensors are Kmote integrated. Therefore both of the PPG signals are sampled and transmitted to portable base station via Zigbee transmission. Then, sampled ECG data, ear PPG data, and finger PPG data are sent to user's mobile system (Smart Phone) in real time and simultaneously display on screen for monitoring purpose. At the same time, Smart Phone uploads the vital signal data to Server PC via Bluetooth connection. At server PC, a remote PC monitoring program is developed to display vital signal graph for healthcare monitoring purpose. Vital signal graphs are then uploaded to Wide Area Network (WAN) or Local Area Network (LAN) via Internet. Third party users like doctors, medical service provider, and patient's family are able to access the very first hand vital signal information by tracing the corresponding IP address of the Server PC within WAN or LAN.

ECG signal is band limited at the range of 1.59Hz~40Hz and amplification factor is set to be closed to 1000. Ear PPG signal is band limited at 0.1~3HJz and Finger PPG signal is band limited at 0.35Hz~10.5Hz . Amplification factor for both PPG amplifiers is set to be closed to 20. Both PPG sensors are integrated with Kmote, where Kmote is an ultra low power wireless node in which used for Zigbee communication [5].

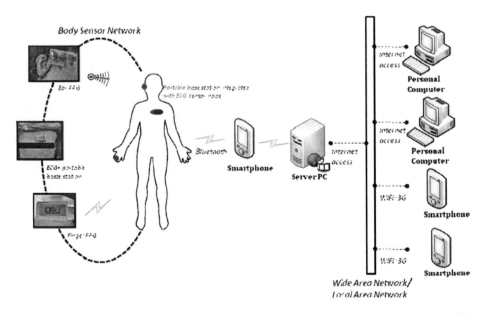

Fig. 1. Overview of system architecture for mobile healthcare monitoring system using portable base station

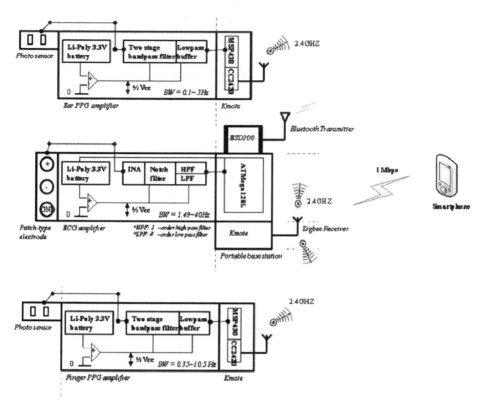

Fig. 2. Portable base station receive ear PPG and finger PPG data using Kmote (Zigbee communication) and send ECG, ear PPG and finger PPG using ESD200 (Bluetooth communication)

Portable base station consists of one microcontroller integrating with ECG senor node, Kmote and ESD200 Bluetooth transmitter [6]. ECG signal is sampled using the ADC port from the microcontroller (ATMega128) [7] and transmitted to Smart Phone using Bluetooth transmitter. On the other hand, Kmote as part of the control unit of the microcontroller act as a Zigbee receiver to receive Zigbee transmitted data from both ear PPG sensor and finger PPG sensor. All the vital signals data are handled with appropriate time synchronizing scheme and then send to Smart Phone via Bluetooth transmission.

Since the portable base station is integrated with ECG sensor node and embedded control on three bio sensor node for wireless transmission to user's Smart Phone, an extra Zigbee interface board which in used in [2], [3], [4] is no longer a must. This promotes mobile flexibility and power saving for Smart Phone battery. Another main idea of portable base station is to demonstrate the possibility of interfacing two or more wireless technologies within a body area network. Portable base station as a gate way to integrate with difference wireless technologies is foreseen to be a potential market in mobile healthcare.

3   Demonstration of Mobile Healthcare Monitoring System Architecture

In this section, we illustrate the proposed body area network architecture with three kinds of demonstrations.

3.1   Demonstration of Mobile Healthcare Monitoring System Using Portable Base Station

Patch-type ECG sensor node integrated with portable base station is attached at user's chest [8]. Ear ring-type PPG sensor is attached to user ear lope [9]. Finger-clip type PPG sensor node is clipped at user's thumb. Both PPG signals are digitize sampled at a rate of 300Hz using TI MSP430 microcontroller in Kmote. Sample data are transmitted to portable base station using CC2420 radio chip. PPG signals are sent and receive simultaneously at portable base station using Zigbee transmission method. ECG data, ear PPG data, and finger PPG data are then sent to user's Smart Phone via Bluetooth using ESD200 as a Bluetooth transmitter. Three vital signal graphs are displayed simultaneously on user's Smart Phone as shown in Fig. 3.

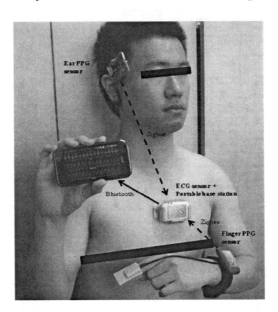

Fig. 3. Demonstration of mobile healthcare monitoring system using portable base station

3.2   Demonstration of Displaying Vital Signals on Smart Phone

Nevertheless, the core idea of this body area network is to receive vital signal data by Smart Phone via Bluetooth communication and displaying it on the screen for convenience healthcare monitoring purpose. Fig 4 shows the android application for displaying three simultaneous vital signals: ECG, Finger PPG, and Ear PPG signals on the Smart Phone screen.

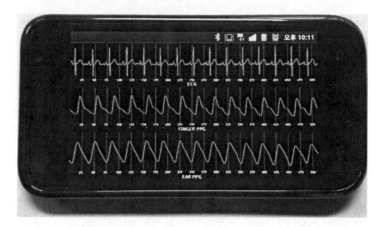

Fig. 4. Screen shot on android phone while monitoring three simultaneous vital signals

3.3 Displaying Vital Signals on a Remote Personal Computer for Third Party Monitoring Purpose

Remote monitoring program for personal computer is developed to monitor three simultaneous vital signals data. With this program, third party user like doctor, physician or any other medical service providers are able to monitor a user's real time vital signal information. Real time and continuous monitoring of vital signal serve crucial purpose for healthcare monitoring. Any abnormal occurrence in vital signal monitoring gives us a pre-alert on our healthy level in daily life.

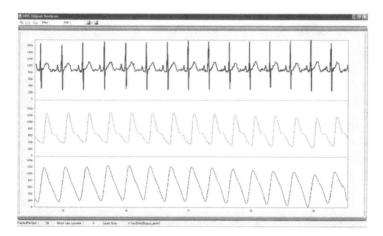

Fig. 5. Screen shot on a remote personal computer for simultaneous vital signal monitoring

4 Conclusions

In this paper, mobile healthcare monitoring system is presented. From system architecture expect, we suggest portable base station to be integrated with ECG senor node to have optimum system design architecture within the body area network. Portable base station as a central control unit for body area network, serve as a gateway to handle Zigbee transmission from sensor node and Bluetooth transmission to user's Smart Phone. From demonstration expect, we show several practical demonstration of the proposed mobile healthcare monitoring system. Several programming effort and hardware design effort are covered in section 3.1, 3.2 and 3.3.

Body area network for mobile healthcare monitoring are made possible with the embedded control of three bio sensor node using portable base station. Mobile healthcare monitoring system is designed to serve the purpose of convenience, real time and continuous healthcare monitoring. Patch-type ECG sensor integrated with small size portable base station serve the convenience purpose of mobile healthcare monitoring. Ear-ring type PPG sensor design is as user friendly as Bluetooth ear phone available in the market. Finger-clip type PPG system is expected to be improved to glove-type in order to have better user flexibility. Vital signals can be continuously monitoring using Smart Phone and even on a remote personal computer, thus, meet the objective of mobile healthcare monitoring.

Acknowledgments. This research was supported by Basic Science Research Program through the National Research Foundation of Korea (NRF) funded by the Ministry of Education, Science and Technology (No. 2011-0004910).

References

1. Jonese, V., Gay, G., Leijekkers, P.: Body Sensor Networks for Mobile Health Monitoring: experience in Europe and Australia. University of Twente Enshede, The Netherlands
2. Jung, J., Ha, K., Lee, J.: Wireless Body Area Network in a Ubiquitous Healthcare System for Physiological Signal Monitoring and Health Consulting, International Journal of Signal Processing, Image Processing, and Pattern Recognition
3. Lubrin, E., Lawrence, E., Navarro, K.F.: Wireless Remote Healthcare Monitoring with Motes. University of Technology, Sydney
4. Milenkovic, A., Otto, M., Jonanov, E.: Wireless Sensor Networks for Personal Health Monitoring: Issues and an Implementation, University of Alabama, Hustville
5. Kmote datasheet, TinyOS Mall
6. ESD200/210 datasheet, Parani-ESD Series, http://www.sena.com
7. ATmega128 datasheet, ©, Atmel Corporation (2010)
8. Puurtinen, M., Hyttinen, J., Malmivuo, J.: Optimizing bipolar electrode location for wireless ECG measurement – analysis of ECG signal strength and deviation between individuals. IJBEM 7(1) (2005)
9. Poh, M.-Z., Swenson, N.C., Picard, R.W.: Motion Tolerant Magnetic Earring Sensor and Wireless Earpiece for Wearable Photoplethymography. IEEE Transactions on Information Technology in Biomedicine (TITB-00242-2009.RI) (2009)

Detection of P300 in a BCI Speller

Monica Fira

Institute of Computer Science, Romanian Academy, Bl. Carol I, No. 8, Iaşi, Romania
mfira@etti.tuiasi.ro

Abstract. This paper has presented the algorithm for P300 Speller Paradigm. The dataset used has been dataset II of the BCI competition III 2005. The novelty in the approach is that for each epoch the dataset has been split, organized and then preprocessed by average function of number of raw / column that was intensified. For each character classification is based on the Euclidean distance between a pattern P300 calculated on the training set and patterns of six rows and six columns calculated on the testing set.

Keywords: EEG, BCI, P300, speller paradigm.

1 Introduction

The usage of electroencephalographic signals EEG as a vector of the communication between humans and machine is one of the new challenges existing in the theory of signals. The main element of such a communication system, known as the "Brain Computer Interface-BCI", is represented by the interpretation of the EEG signals interpretation corresponding to the characteristic parameters and of the electric cerebral activity [1] [2]. A definition of a BCI, subsequently accepted by most of the researchers in the field is that *a BCI is a communication system in which the messages or the commands sent to the exterior environment by an individual do not pass through the normal output paths of the brain, paths constituted by the peripheral nerves and muscles.*

The purpose of BCI is to establish a communication system which translates the human intentions – represented by appropriate signals – into control signals for an output device, for example a computer or a neural prosthesis.

A BCI system can offer to the paralyzed persons (for example from lateral amyotrophic sclerosis, stroke or severe polyneuropathy) or lacking the muscle control, the possibility of giving quick responses to simple questions, of controlling the environment, to process slowly certain words or even to control a neural prosthesis [2]. At the same time, the performance of this new technology, measured in terms of speed and precision, or by an inclusive measure, the transfer speed of the information, it is modest [3]. The current systems do not have a transfer speed of the information of more than 25 biti/min. The exceptions are the system based on the visual evoked potentials. These systems do not depend directly on the muscular control, but they require direct control on the gaze. Therefore, they cannot be used for the persons with complete paralysis. [6]

G. Lee, D. Howard, and D. Ślęzak (Eds.): ICHIT 2011, CCIS 206, pp. 481–487, 2011.

As any communication system, a BCI has inputs (electrophysiological signals resulting from the monitoring of the brain's activity), outputs (actions executed by an active device), elements which transform the inputs into outputs and a protocol based on which the system is working [6], [7].

This paper presents the algorithm for classification on the dataset produced by a P300 speller matrix during the BCI III competition. The P300 Speller is a paradigm where a user sits in front of a 6x6 matrix of letters and numbers on a computer monitor and using his/her thoughts is able to spell words on the computer screen.

2 The Data Set - Description

The P300 speller paradigm [4] [5] use visual stimulation enabling to write by spelling. The method consists in displaying a 6x6 matrix composed by the figures and letters. Lines and columns of the matrix are successively highlighted. When the line or the column contains the chosen letter, a P300 ERP appears. A classifier is then used to determine if this signal correspond to a positive response or not.[1]

For BCI III competition the dataset has been recorded from two different subjects in five sessions each and signals was been bandpass filtered from 0.1 - 60Hz and digitized at 240Hz. Each session is composed of runs, and for each run, a subject is asked to spell a word. For a given acquisition session, all EEG signals of a 64-channel scalp have been continuously collected. [2]

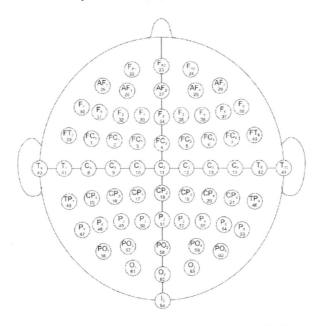

Fig. 1. Placing the 64 electrodes to acquire the EEG signals in BCI competition III

Row/column intensifications were the block is randomized in blocks of 12. The sets of 12 intensifications were repeated 15 times for each character epoch (i.e., any specific row/column was intensified 15 times and thus there were 180 total intensifications for each character epoch). Each character epoch was followed by a 2.5sec period, and during this time the matrix was blank. The train set contains 85 characters and the test set is on 100 characters for each of the two subjects A and B. A more detailed description of the dataset can be found in the BCI competition paper [1].

A	B	C	D	E	F
G	H	I	J	K	L
M	N	O	P	Q	R
S	T	U	V	W	X
Y	Z	1	2	3	4
5	6	7	8	9	_

Fig. 2. Example of a 6 × 6 user display in P300 Speller

3 Methods

We present the methods we followed for building our classifier in this section. Our method has two major stages, namely:

1. The preprocessing stage: a preprocessing based on the average of several channels and the 15 repetitions in order to calculate the P300 pattern from all 85 epochs and for the calculation of 12 the average signals for each epoch (corresponding to the six lines and six columns)
2. The classification stage: a classification stage for signals obtained from the previous stage, based on the Euclidean distance from P300 pattern.

1 Preprocessing Stage
At first, for some predefined channels of all 64 of channels, we extracted and normalized all data samples between 0 and 1 sec posterior to the beginning of an intensification. We obtained such, from a training set of 85 characters spelling a

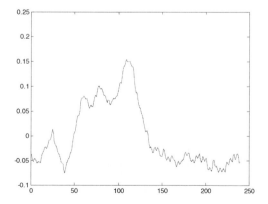

Fig. 3. The P300 patterns

database of 12*15*85 = 15300 signals of 240 samples for each analyzed channel. These 15 300 signals (for every channel analysis) are then divided into signals with P300 and nonP300 signals, resulting two subsets, the subset of 2550 signal with P300 and the subset of 12750 signals without P300. From all subsets P300 by averaging a pattern P300 we obtained.

For classification we constructed 12 test signals (corresponding to the six lines and six columns) by averages of the all 15 repetitions on the testing set, for each character (epoch). In other words, mediating all 15 intensifications corresponding to each row or column will get a pattern corresponding to the respective line or column. Patterns thus obtained are subjected to median filtering with a window of 10 samples.

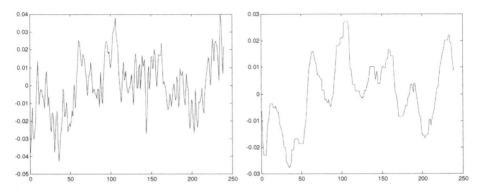

Fig. 4. Example of pattern and filtered pattern (with P300)

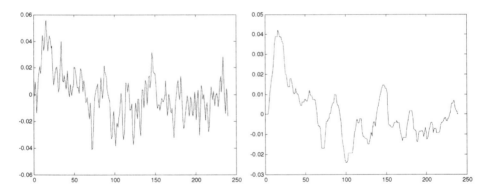

Fig. 5. Example of pattern and filtered pattern (without P300)

2 Classification Stage

The classification of the patterns obtained on the previous stage is actually a binary classification and the results are converted into a decision for the 36-class by crossing the line and column corresponding binary classification.

The problem is to find only a single line of six lines containing P300 and the like to find a single column of six columns containing P300. The decision which the line (or column) contains the P300 is based on the Euclidean distance of the six patterns corresponding to the six lines (or the 6 columns) and P300 pattern (calculated on the preprocessing stage). In other words, all 6 Euclidean distances are calculated and the row (column respectively) which corresponds to the smallest distance is categorized as having P300.

For each character (time) once marked the row and column containing the P300, the final classification decision is at the intersection line with the column containing P300. In other words, for a correct classification the both row and column containing P300 should be identified. A wrong classification of the corresponding line or column of P300 leads to an invalid final decision. If a character string followed a decision, the correct classification probability is 1 / 36, i.e. 2.77%.

4 Experimental Results

In literature there is no universal recipe of channels to be used to ensure success in BCI P300 Speller, relevant channels for the same methods ranging from one topic to another. Therefore, most algorithms that have shown outstanding performance involving a training stage in which to determine which channels are optimal in terms of classification performance, following that once this channels are determined the classification will be based only on signals from these channels.

In the proposed method, a first stage was the channel sensitivity analysis, i.e. to find which channels are best in terms of classification based on the proposed method. Thus, using one single channel and using the proposed algorithm it the classification performances of these channels were found. Table 1 summarizes the performance for several channels.

Table 1. The classification performances for a single channel

channel	Classification %	channel	Classification %
F3 - 32	31	Cz - 11	51
F1 - 33	47	C2 - 12	39
F2 - 35	44	CP3 - 16	17
FC3 - 2	42	CP1 - 17	38
FC1 - 3	54	CPz - 18	44
FCz - 4	35	Pz - 51	47
FC2 - 5	53	P1 - 50	30
C3 - 9	23	POz - 58	20
C1 - 10	40	PO7 - 56	17

Starting from the idea that using EEG signals from multiple channels will increase the classification performance, and taking into account the sensitivity of the channel for (presented in table 1) the proposed algorithm, we have tested various configurations of channels used and classification performance is presented in Table 2.

Table 2. The classification performances using various configurations of channels

Channels used	Classification performances
3, 5, 11	**64.7**
3,5	60
3,5,18	58.82
3,5,11,51	58.8
3,3,5,10,11,12,17,18,19	58.5

5 Discussion and Conclusions

This paper has presented the algorithm for P300 Speller Paradigm. The dataset used has been dataset II of the BCI competition III 2005. The novelty in the approach is that for each epoch the dataset has been split, organized and then preprocessed by average function of number of raw / column that was intensified. Thus, for each epoch, from the signals corresponding the 12 intensification repeated on 15 times by preprocessing presented has obtained only 12 signals (one for each row or column) on which we decide the character follow. The decision is taken on the basis of comparison of 12 signals with a P300 pattern (this pattern is constructed from training dataset).

The best classification performance is obtained when using three EEG channels, i.e. channels 3, 5, 11 ie, FC1, FC2 and Cz, when we obtain a classification of 64.7%, on a testing set of 100 characters.

The proposed algorithm is sensitive to the channels used and therefore it is necessary to analyze the sensitivity channels. After finding the sensitivity of channels to obtain improved results can be used more channels (choose channels with higher sensitivity).

Acknowledgments. This work has been jointly supported by CNCSIS –UEFISCSU, project PNII – RU - PD 347/2010 and CNMP –PC – 12115/2008.

References

1. Blankertz, B.: BCI competition III webpage,
   http://ida.first.fraunhofer.de/projects/bci/competition
2. Blankertz, B., Mueller, K.-R., Curio, G., Vaughan, T., Schalk, G., Wolpaw, J., Schloegl, A., Neuper, C., Pfurtscheller, G., Hinterberger, T., Schroeder, M., Birbaumer, N.: The BCI competition 2003: Progress and perspectives in detection and discrimination of EEG single trials. IEEE Trans. Biomed. Eng. 51(6), 1044–1051 (2004)
3. Blankertz, B., Mueller, K.-R., Krusienski, D., Schalk, G., Wolpaw, J., Schloegl, A., Pfurtscheller, G., Millan, J., del, R., Schroeder, M., Birbaumer, N.: The BCI competition III: Validating alternative approaches to actual BCI problems. IEEE Transactions on Neural Systems and Rehabilitation Engineering 14(2), 153–159 (2006)

4. Donchin, E., Spence, K., Wijeshinge, R.: The mental prosthesis: assessing the speed of P300-based brain-computer interface. IEEE Transactions on Rehabilitation Engineering 8(2), 174–179 (2000)
5. Farwell, L., Donchin, E.: Talking off the top of your head: toward a mental prosthesis utilizing event-related brain potentials. Electroencephalography and Clinical Neurophysiology 70(6), 510–523 (1988)
6. Schalk, G., McFarland, D., Hinterberger, T., Birbaumer, N., Wolpaw, J.: BCI2000: a general-purpose brain-computer interface (BCI) system. IEEE Transactions on Biomedical Engineering 51(6), 1034–1043 (2004)
7. Serby, H., Yom-Tov, E., Inbar, G.F.: An improved p300-based brain-computer interface. IEEE Trans. Neural Syst Rehabil Eng. 13(1), 89–98 (2005)

Analysis for Characteristics of Electroencephalogram (EEG) and Influence of Environmental Factors According to Emotional Changes

Jeong-Hoon Shin and Dae-Hyeon Park

Dept. of Computer & Information Communication Eng.
Catholic University of Dae-Gu, Korea
{only4you,ttnsoo}@cu.ac.kr

Abstract. Neurofeedback is a kind of bio feedback that induces the brain wave pattern or the regional cerebral blood flow (rCBF). For the last thirty years, this technique was clinically applied in order to increase the effect of the psychotherapy. However, most of the studies which were carried out until now were just simple ones that analyze the patterns of the brain waves according to specific stimulations. It seems that the studies related to the environmental factors for the specific time have not been carried out yet. There was a problem caused by the difference according to the surrounding environment when a specific type of stimulation is applied to a specific patient. Such a problem was an obstacle in the clinical utilization of this technique. In order to solve such a problem, this study has analyzed the characteristics of the brain waves along with the changes of the surrounding environment for the time when the stimulation is applied to the patient. Based on the results, the trend for inducing emotional changes in the environmental factors was analyzed. Also, a way to utilize brain waves for stable neurofeedback treatments, and related training techniques and health-care tools, which were not influenced by the environmental changes, was suggested.

Keywords: brain waves, changes of emotional status, changes of temperature, environmental factors, influence.

1 Introduction

The most basic and sensitive condition required in the u-health industry is the ability to measure the private health-related information in the u-health environment. The level of sensitivity, which is related to the user, is very high for the medical data. However, regarding the acquisition of the data for biological measurements, since the level of change is too severe due to the status of the user and the surrounding environment, the precision and credibility of the obtained data will influence the usage of the system. [1], [2]

The technology of measuring biological information, which is expected to be widely used in the ubiquitous environment, measures the signals of the current or

G. Lee, D. Howard, and D. Ślęzak (Eds.): ICHIT 2011, CCIS 206, pp. 488–500, 2011.
© Springer-Verlag Berlin Heidelberg 2011

voltage created by the neurons or muscle cells in the brain. The source of such signals is believed to be the membrane potential.

Electroencephalogram is one of the biological signals that were widely used in the fields of the u-healthcare tools and the neurofeedback treatment. The study about such a signal has received positive attention and feedback. By utilizing various methods of handling the signals, the field related to such a signal was continuously developed. Several methods of using the signals to diagnose and treat various diseases were suggested. [3],[4]

However, regarding many EEG-related studies, which were carried out until now, such fields as the exact measurement of the signals related to the brain waves and the elimination of the artifacts related to such signals were analyzed. Also, the field of pattern recognition was analyzed. Such fields of studies, which were analyzed more than others, have caused many unexpected results to occur, due to the different ways of using the study results. As a result, the potential problem caused by such a situation has become an obstacle for the utilization as tools by the medical and the health-care industries, which focus on the level of credibility.

By considering the environmental factors, which were regarded as diverse variables when measuring biological signals; this study analyzes the influence of the environmental factors on the biological signals. The results of the analysis for the changes of the characteristics related to the biological signals influenced by the environmental factors make it possible to provide the results of the analysis for the stable biological signals which were not ultimately influenced by the environmental factors. Based on such results, it is expected that the findings will be utilized in various fields including the stable neurofeedback treatment and training, and various health-care tools. [1-7]

This study is composed of the following parts. In the second chapter, the usage of the brain waves for the medical tools and the study trend related to the clinical diagnosis were introduced. In the third chapter, the experiment related to "the characteristics of the brain waves when exposed to the emotional changes by considering the changes of various environmental factors," which was carried out in the study, is introduced. The environmental structure and the content of such an experiment were discussed. Also, in the fourth chapter, the results provided by the experiment which was executed in the study were suggested. In the fifth chapter, the conclusion and the study trend in the future were discussed.

2 Related Works

In order to utilize the biological signals in the fields of the health-care technology and neurofeedback, some major studies were globally carried out. The themes included in such studies were classified into four categories, including the sensor technology for measuring the biological signals, the technology for transmitting the biological signals and the related monitoring technology, the technology related to the tools for measuring the biological signals, and the technology for standardization.

The study related to the brain waves among the biological signals was actively carried out. According to the study, the potential difference of the weak brain waves occurring in the physiological activity of the brain provides electrodes on the skin of the head. Such electrodes were used to analyze the frequency elements of the brain waves. Also, various preceding studies and various types of utilization were carried out to find out the functional status of the central nervous system accompanied with the brain tumor, cerebrovascular injuries, and external head injuries.

However, most of the studies which were carried out until today, regarding the brain waves, were independently limited to the four fields mentioned above. The composite study considering major environmental factors has not been carried out yet.

Various techniques were suggested, regarding such things as the utilization of the biological information for the field of the health-care technology and the technique for treating various psychological diseases by using the neurofeedback. At this time, in order to determine which clinical utilization is most credible, it is necessary to carry out composite studies related to the environmental factors.

3   Configuration of Experimental Environment

3.1   Environmental Factors

In order to analyze the changes of the emotional status according to the changes of the surrounding temperature, this study was carried out in an experimental environment composed of three independent rooms, as shown in Figure 1, in order to measure the brain waves after applying the changes of the room temperature and the emotional status.

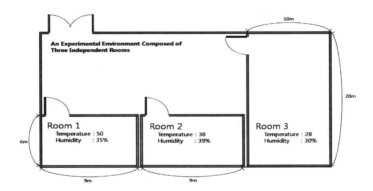

Fig. 1. Experimental Environment

During the experiment, the room temperature and humidity of each independent room were kept at 28℃ and 30%, 38℃ and 39%, and 50℃ and 35% respectively, in order to prevent the composite environmental changes and increase the level of credibility for the analysis of the influence according to the changes of the temperature.

3.2 Subjects

The subjects of the study included 20 men and 20 women, in good health and in their 20s. The range of ages varies between the early 20s to the late 20s. Also, in order to make the subjects feel comfortable, the experiment was carried out between three in the afternoon and eight in the evening. The changeability according to the time of measurement was eliminated.

3.3 Placement of Electrodes and Device for Measurement of Brain Waves

In order to carry out the analysis for the changes of the emotional status and the characteristics of the brain waves according to the changes in temperature, the same conditions were applied for all the elements that influenced the experiment other than the changes in temperature. The measurement device used in the study is the eight-channel brain-wave measurement tool (Laxtha Co., Ltd., Korea). Regarding the electrodes for the measurement of the brain waves, such locations as Fp1, Fp2, T3, T4, C3, C4, O1 and O2 were selected among the placement of electrodes, according to the 10-20 methods with the international standard for arranging the locations of electrodes. When measuring the brain waves, the data related to the brain waves of the subjects were digitalized by using the sampling rate of 256Hz.

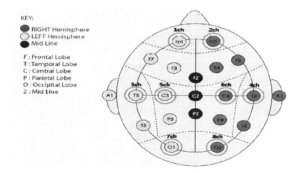

Fig. 2. International 10-20 system of Electrode Placement

Regarding the selection of the placement of electrodes, Fp1 and Fp2 were selected to analyze the changes of the status for the brain activity of the frontal lobe. Regarding such physiological factors as the body temperature and the blood pressure, T3, T4, C3 and C4 were selected to analyze the changes of the status for the brain activity of the hindbrain which is involved in the activity of the diencephalon and other activities related to digestion, circulation, respiratory control and reflex action. Also, O1 and O2 were selected to analyze the changes of the status for the brain activity of the occipital lobe, which deals with the time-related information.

3.4 Methods

☐☑ Step 1

Electrodes were attached to the scalp of the subject participating in the experiment. The subject enters the room where the measurement takes place according to the prearranged order and remains there, without moving, for ten minutes.

☐☑ Step 2

After the time for stabilization passes, the subject's pulse and the blood pressure were measured, and then their brain waves were measured for three minutes.

☐☑ Step 3

After measuring the brain waves, the subject was given one minute to rest. The subject was advised to remain in a natural state. During such time, the brain waves were not measured.

☐☑ Step 4

After the measurement of the brain waves was completed in the state of stabilization, the subject was directed to watch the moving images for five minutes. Such moving images could induce changes in the emotional status. While the subject was watching the moving images, it was necessary to measure his or her brain waves. The moving images which the subject watches in this step need to contain the contents that could influence the emotional status or the mind of the subject. In this study, images related to traffic accidents were shown, in order to make the subject become emotionally unstable and nervous, and even felt disgusted.

☐☑ Step 5

After completing the step 4, the subject goes out of the room where the measurement takes place and takes a 10-minute break. After the break, the subject needs to enter the room where the next measurement takes place. After he or she enters the room, the whole process from the step 1 to the step 4 is repeated. If the moving images shown in the step 4 were same as before, the subject would not show emotional changes as before. In order to prevent such a situation, new moving images with a similar story were shown. The study is repeated until the measurement is complete in three rooms with the room temperatures of 28℃, 38℃ and 50℃.

4 Experimental Results

In the study, the cross-correlation coefficients between the EEG signal channels were analyzed to trace the movement of the emotional information when the brain carries out its activities. Also, the cross-correlation coefficients were used as the typical standard to analyze the influence of the changing temperature on the movement and change of the emotional information. By activating the 'All Pair-Cross Pearson's Correlation Function' method, it is possible to generalize the process in the statistical way. Also, in order to analyze the influence of the element in the specific frequency contained in the brain waves, the band-path filter was used. Through such a process, the frequency of the EEG signal was classified into various fields. By activating the 'All Pair-Cross Pearson's Correlation Function' method to analyze the cross-correlation coefficients between the channels according to the individual frequency, it is possible to check the central field of the signals in the brain waves, which plays a significant role of delivering the emotional information.

Also, in order to prevent such a factor as the rapid noise from decreasing the credibility for the results of the analysis, it is necessary to analyze the 'Ensemble Averaging over Moving Window' method in the interval of measurement. Figure 3 shows the analysis of the signals in the brain waves in the block diagram.

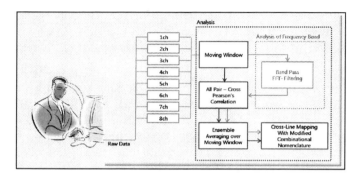

Fig. 3. Block diagram of analyzing process of the brainwave

4.1 Analysis of Characteristics for Brain Waves in Stabilized Intervals According to Changes of Temperature

In order to analyze the characteristics of the brain waves in the state of stabilization and the cross-correlation of the brain activities according to the changes of temperature, the eight electrodes in such locations as Fp1, Fp2, T3, T4, C3, C4, O1 and O2 provided the required data for the brain waves. Such data can be used to analyze the cross-correlation. The average value for the data of 40 subjects was measured and shown in Figure 4.

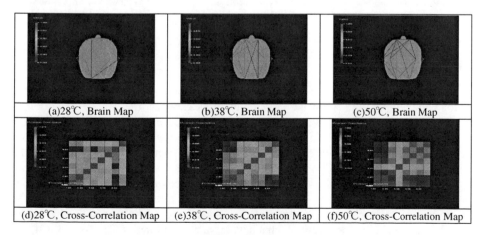

Fig. 4. Cross-Correlation between channels in the state of stabilization according to the changes of temperature

After analyzing the characteristics of the signals in the brain waves in the state of stabilization according to the changes of temperature, it seems that active exchanges of information occur in the brain at the low temperature rather than the high temperature. Also, as the room temperature increases, the Cross-Correlation between the left hemisphere of the brain and the hindbrain increases as well, activating the exchange of information. (In other words, the Cross-Correlation among T3, T4, C3, C4, O1 and O2 increases.)

4.2   Analysis of Characteristics for Brain Waves According to Changes of Temperature when the Emotional Status Changes

In order to analyze the characteristics of the brain waves and the correlation of the brain activities according to the changes of temperature when the emotional status changes, the eight electrodes in such locations as Fp1, Fp2, T3, T4, C3, C4, O1 and O2 provided the required data for the brain waves. Such data can be used to analyze the cross-correlation. The average value for the data of 40 subjects was measured and shown in Figure 5.

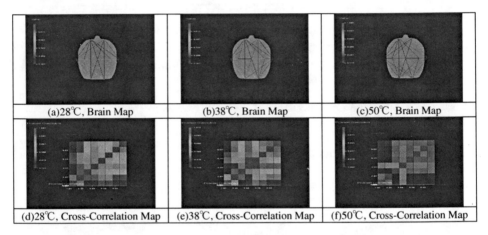

| (a)28℃, Brain Map | (b)38℃, Brain Map | (c)50℃, Brain Map |
| (d)28℃, Cross-Correlation Map | (e)38℃, Cross-Correlation Map | (f)50℃, Cross-Correlation Map |

Fig. 5. Cross-Correlation between channels according to changes of temperature when the emotional status changes

By analyzing the characteristics for the signals in the brain waves according to the changes of temperature when the emotional status changes, the same stimulation as shown in Figure 5 was applied to the subject in order to change the emotional status of the subject. As the room temperature increases, the intensity of the reaction becomes bigger. As a result, it seems that a lot of information is exchanged in the entire brain. Especially, when the emotional status changes at the high room temperature, the amount of information which is exchanged in the brain becomes bigger in the left hemisphere of the brain and the hindbrain than the time when the room temperature changes in the state of stabilization. As a result, the cross-correlation coefficient between the channels in such locations as T3, T4, C3, C4, O1 and O2 increases more than the case of the stabilized situation.

4.3 Analysis of Factors Influencing Changes of Emotional Status for Each Element of Brain Waves in Each Frequency Band

According to the analyzed results shown in 4.1 and 4.2, the changes of the room temperature influence the amount of information exchanged in the brain. Also, such changes influence the emotional status. In the study, in order to analyze the factors influencing the exchange of information in the brain and the emotional status, the measured data of the brain waves were classified into each frequency band. Then the cross-correlation between the channels was analyzed. Based on such results, the analysis for the characteristics was carried out.

4.3.1 Analysis of the Influence Provided by the Delta Wave (0.1Hz~4Hz) on the Change of the Emotional Status

When the emotional status changes in the band of the delta wave, as shown in Figure 6, the change of the cross-correlation coefficient between the channels is similar to the one of the entire signals in the brain waves shown in Figure 5. At this time, it is possible to measure the amount of exchanges information between the channels. Also, the intensity of the reaction in the band of the delta wave is analyzed by applying the same kind of stimulation to the subject as shown in Figure 5. Even if the emotional status changes, the intensity becomes stronger as the room temperature increases. It is evident that the amount of information exchanged becomes bigger in the entire brain. (The cross-correlation coefficient between the channels in the locations of T3, T4, C3, C4, O1 and O2 increases.) Such a result is similar to the one shown in the analysis for the state of stabilization in the band of the delta wave. Also, it is similar to the one shown in the analysis for the signals in the brain waves in the state of stabilization throughout the entire frequency band.

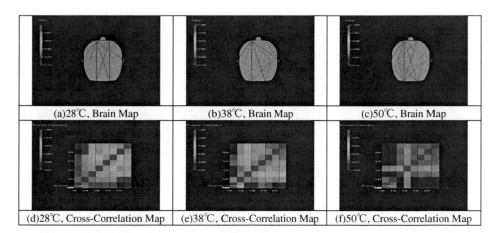

| (a)28℃, Brain Map | (b)38℃, Brain Map | (c)50℃, Brain Map |
| (d)28℃, Cross-Correlation Map | (e)38℃, Cross-Correlation Map | (f)50℃, Cross-Correlation Map |

Fig. 6. Cross-Correlation between the channels in the band of the delta wave according to the changing room temperature and the changing emotional status

4.3.2 Analysis of the Influence Provided by the Theta Wave (4Hz~8Hz) on the Change of the Emotional Status

As shown in Figure 7, there is a big difference between the changing cross-correlation coefficient between the channels, which can be used to measure the amount of information exchanged in the brain when the emotional status changes in the band of the theta wave, and the changing correlation for the entire signals in the brain waves, which is shown in Figure 4 and Figure 5.

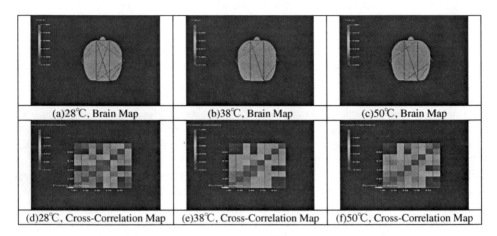

| (a)28℃, Brain Map | (b)38℃, Brain Map | (c)50℃, Brain Map |
| (d)28℃, Cross-Correlation Map | (e)38℃, Cross-Correlation Map | (f)50℃, Cross-Correlation Map |

Fig. 7. Cross-Correlation between the channels in the band of the theta wave according to the changing room temperature and the changing emotional status

If (a), (b) and (c) shown in Figure 7 were compared with (a), (b) and (c) shown between Figure 4 and Figure 5, It is evident that the amount of information exchanged in the brain is relatively low when the temperature and the emotional status change. Also, it is evident that the cross-correlation coefficients between the channels, which were shown from Figure 4 to Figure 6, were greatly influenced by a small change (such as the change from 28℃ to 38℃). However, the cross-correlation coefficient between the channels in the band of the theta wave was hardly changed. When the room temperature was rapidly changed, there seem to be small changes for the cross-correlation coefficients between the channels. Also, the amount of information exchanged as shown between Figure 4 and Figure 6 is shown as the changing cross-correlation coefficient between the channels of such locations as T3, T4, C3, C4, O1 and O2. When the cross-correlation coefficient between the channels in the band of the theta wave is influenced by the rapid change of the surrounding temperature, there seemed to be a small change among such locations as T3, C3, C4, O1 and O2. By combining all the technical contents mentioned above, the cross-correlation coefficients including the band of the theta wave among the values that were used to measure the amount of information exchanged in the brain, were not greatly influenced by the changing temperature and environment.

4.3.3 Analysis of the Influence Provided by the Alpha Wave (8Hz~13Hz) on the Change of the Emotional Status

It is evident that the change of the cross-correlation coefficient between the channels, which can be used to measure the amount of information exchanged in the entire brain when the emotional status changes in the band of the alpha wave as shown in Figure 8, is similar to the one in the theta wave shown in Figure 7.

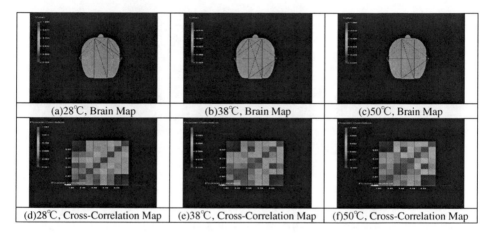

| (a)28℃, Brain Map | (b)38℃, Brain Map | (c)50℃, Brain Map |
| (d)28℃, Cross-Correlation Map | (e)38℃, Cross-Correlation Map | (f)50℃, Cross-Correlation Map |

Fig. 8. Cross-Correlation between the channels in the band of the alpha wave according to the changing room temperature and the changing emotional status

When (a), (b) and (c) of Figure 8 were compared with (a), (b) and (c) of Figure 7, it was possible to know that the cross-correlation coefficient that was used to measure the amount of information exchanged in the brain is relatively smaller than that of Figure 7. When the cross-correlation coefficients between the channels were compared in the room for the measurement where the temperature is kept at 50℃, it is evident that the amount of information exchanged in such locations as T3, T4, C3, C4, O1 and O2 is active in Figure 7. However, the amount of information exchanged (the cross-correlation coefficient between the channels) among all the electrodes excluding T3 and C3 is not increased. If all the technical contents shown above are summarized, it is evident that the cross-correlation coefficient between the channels including the band of the alpha wave, among all the correlation coefficients used to measure the amount of information exchanged in the brain, is not influenced by the changing temperature and environment. Especially, in the case of the theta wave, the cross-correlation coefficient between the channels is changed a little when the room temperature is changed. In the case of the alpha wave, the cross-correlation coefficient between the channels is hardly influenced by the rapidly changing temperature.

4.3.4 Analysis of the Influence Provided by the Beta Wave (13Hz~30Hz) on the Change of the Emotional Status

It is evident that the change of the cross-correlation coefficient between the channels, which can be used to measure the amount of information exchanged in the entire brain when the emotional status changes in the band of the beta wave as shown in Figure 9, is similar to the one in the alpha wave shown in Figure 8. However, the amount is relatively lower than that of the alpha wave.

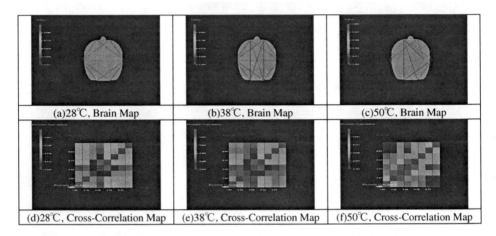

| (a)28℃, Brain Map | (b)38℃, Brain Map | (c)50℃, Brain Map |
| (d)28℃, Cross-Correlation Map | (e)38℃, Cross-Correlation Map | (f)50℃, Cross-Correlation Map |

Fig. 9. Cross-Correlation between the channels in the band of the beta wave according to the changing room temperature and the changing emotional status

When (a), (b) and (c) of Figure 9 were compared with (a), (b) and (c) of Figure 8, It is evident that the cross-correlation coefficient that was used to measure the amount of information exchanged in the brain is relatively smaller than that of Figure 8. The change of the cross-correlation coefficient between the channels in the room for measurement at the room temperature of 50℃ is not small. However, it seems that the amount of information exchanged in the locations of O1 and O2 is relatively active.

When all the technical contents mentioned above were summarized, it was evident that the cross-correlation coefficient between the channels including the band of the beta wave, among the cross-correlation coefficients that were used to measure the amount of information exchanged in the brain, was not influenced by the changing temperature.

5   Conclusions and Future Study

In the study, various types of analyses were carried out in order to investigate the characteristics of the brain waves according to the changes in these characteristics and the changes of the emotional status, by considering the environmental changes. Especially, the 'Cross-Correlation' method was carried out to measure the correlation coefficient for the amount of information exchanged in the brain, and the 'Ensemble

Average over Moving Window' method was carried out to minimize the errors of the analysis, which can be caused by the variation of the biological signals between the individuals and the absorption of the noise waves in the same environment for the subject.

Also, regarding the analysis for the cross-correlation between the channels, which was carried out in this study, the widely-used method of analyzing the signals in the brain waves for each frequency band was utilized. By executing such a method for the analysis of the cross-correlation between the channels, it has become widely known that it is possible to use such a method in the practical field.

By summarizing the experiments and the analysis techniques used in this study, it is possible to show the following results in Table 1, regarding the results of the analysis for the characteristics in each frequency band according to the changing temperature and emotional status.

Table 1. Analysis for Characteristics of Brain Waves according to Changing Temperature and Emotional Status

	Stabilized State	Emotionally Changing State (Visual Stimulation)
Low Temperature (28℃)	- Maintaining a relatively flat state of activation for the cross-correlation coefficients in the entire channel.	- Maintaining a relatively flat state of activation for the cross-correlation coefficients in the entire channel. - Increasing cross-correlation coefficients compared to the state of stabilization.
Medium Temperature (38℃)	- Increasing state of activation among the channels of T3, T4, C3, C4, O1 and O2. - The cross-correlation between the channels with the element of the theta wave is not influenced by the changing temperature.	- As the room temperature increases, the state of activation for the brain and the amount of information exchanged among the channels increase. - The cross-correlation between the channels with the element of the theta wave is not influenced by the changing temperature.
High Temperature (50℃)	- Increasing state of activation among the channels of T3, T4, C3, C4, O1 and O2. - The cross-correlation between the channels with the element of the theta wave is partially influenced by the changing temperature. - The cross-correlation between the channels with the elements of the alpha wave and the beta wave is hardly influenced by the changing temperature.	- As the room temperature increases, the state of activation for the brain and the amount of information exchanged among the channels increase according to the external stimulation of the same degree. - The cross-correlation between the channels with the element of the theta wave is partially influenced by the changing temperature. - The cross-correlation between the channels with the elements of the alpha wave and the beta wave is hardly influenced by the changing temperature.

Recently, the neurofeedback treatment method was widely used to treat the psychological and the mental diseases. Such a method was used to analyze the individual frequency band contained in the brain waves of the patient. As a result, it is possible to provide the external stimulation that helps to contain the elements with the level of a normal person. The results of the analysis in this study can be used immediately in the practical field.

The study for the movement of the information in the brain and the influence of the changing temperature, for the time when the emotional status is changed, will be used to add variable environmental factors other than temperature, in order to provide results that can be used in the practical field.

Acknowledgments. This work was supported by the Korea Science and Engineering Foundation(KOSEF) grant funded by the Korea government(MOST) (No. 2010-0017098).

References

1. Hsiu, H., Hsu, W.-C., Hsu, C.L., Huang, S.-M., Hsu, T.-L., Wang, Y.-Y.L.: Spectral analysis on the microcirculatory laser Doppler signal of the acupuncture effect. In: IEEE-EMBS 2008, 30th Annual International Conference of the 2008 Engineering in Medicine and Biology Society, pp. 2916–2919 (August 2008)
2. Li, N., Wang, J., Deng, B., Dong, F.: An analysis of EEG when acupuncture with Wavelet entropy. In: IEEE-EMBS 2008. 30th Annual International Conference of the 2008 Engineering in Medicine and Biology Society, pp. 1108–1111 (August 2008)
3. He, W.-X., Yan, X.-G., Chen, X.-P., Liu, H.: Nonlinear Feature Extraction of Sleeping EEG Signals. In: IEEE-EMBS 2005, 27th Annual International Conference of the 2005 Engineering in Medicine and Biology Society, pp. 4614–4617 (September 2005)
4. Murata, T., Akutagawa, M., Kaji, Y., Shichijou, F.: EEG Analysis Using Moving Average-type Neural Network. In: IEEE-EMBS 2008, 30th Annual International Conference of the 2008 Engineering in Medicine and Biology Society, pp. 169–172 (August 2008)
5. Kaji, Y., Akutagawa, M., Shichijo, F., Nagashino, H., Kinouchi, Y., Nagahiro, S.: EEG analysis using neural networks to detect change of brain conditions during operations. In: IFMBE Proceedings, pp. 1079–1082 (April 2006)
6. Sun, Y., Ye, N., Xu, X.: EEG Analysis of Alcoholics and Controls Based on Feature Extraction. In: The 8th International Conference on Signal Processing (2006)
7. Zhang, S.Z., Kawabata, H., Liu, Z.-Q.: EEG Analysis using Fast Wavelet Transform. In: IEEE International Conference on Systems, Man, and Cybernetics, pp. 2959–2964 (October 2000)

Development of Real-Time Learning Components Using Expanded SCORM

Junghyun Kim[1], Doohong Hwang[2], Kangseok Kim[3],
Changduk Jung[4], and Wonil Kim[1,*]

[1] College of Electronics & Information Engineering at Sejong University, Seoul, Korea
+82-2-3408-3795
junghyun64@sju.ac.kr, wikim@sejong.ac.kr
[2] Department of Information & communication at Hanyang University, Seoul, Korea
michelmk77@paran.com
[3] Department of Knowledge Information Security at Ajou University, Suwon, Korea
kangskim@ajou.ac.kr
[4] Department of Computer and Information Science at Korea University, Korea
jcd1234@korea.ac.kr

Abstract. e-Learning 2.0, which is based on Web 2.0 technology characterized by users' participation, sharing, and social networking, is being changed to a new e-Learning paradigm. Existing LMS/LCMS have limitations in providing various types of interactive components of video contents. It lacks real-timeness and interactivity between teachers-learners and learners-learners in video contents operation. Thus, in order to overcome such limitations of e-Learning and to maximize learning effects, this study proposes a system emphasizing the characteristic (interactivity) of e-Learning 2.0 by implementing real-time interactive video contents (bidirectional learning components) usable in online video lectures. For this purpose, we design a standard by expanding the SCORM standard used currently in building LMS/LCMS, and implement e-Learning 2.0 environment based on the design.

Keywords: Video-contents, SCORM, LMS/LCMS, interactive, real-time.

1 Introduction

Learning methods like e-Learning, which is being used throughout the world, applicable in various education environments have advantages such as low expenses of education and availability of education to everybody at any time and in any place. The current e-Learning can be defined as fusion technology that integrates various information technologies such as CG (computer graphics), VR (virtual reality), network, game, vision and mobility with learning systems [1, 2]. e-Learning consists of three components, which are learners, contents and platforms, and is used as means of self-development, human resource development (HRD), lifelong education, etc.

---

* Corresponding author.

G. Lee, D. Howard, and D. Ślęzak (Eds.): ICHIT 2011, CCIS 206, pp. 501–512, 2011.
© Springer-Verlag Berlin Heidelberg 2011

The introduction of such a new concept as Web 2.0 has brought many changes to the environment of e-Learning [3, 4]. According to O'Reilly's definition of 7 principles on Web 2.0, technological trends in the age of new e-Learning 2.0 that adopted the concept of Web 2.0 are the expansion of SNS (Social Networking Service) and the use of blogs or podcasting in the field of education [1]. These trends require e-Learning environment based on user-centered LMS (Learning Management System) / LCMS (Learning Content Management System), which is more cooperative than existing e-Learning and allows learners to direct their learning. In this way, e-Learning 2.0 equipped with the function of SNS is spreading steadily in the process of solving problems in existing e-Learning environment using Web 2.0 technology, and this makes it possible to personalize contents and is diversifying the contents of e-Learning.

Some of trends expected in the development of e-Learning in 2011 is "Social learning", "Rapid Learning", "Mobile Learning" and "Cloud Computing for Learning" [1]. Analyzed through such changes and forecast of the e-Learning paradigm, existing learning environment based on videos and flashes is still inadequate for supporting real-time interaction. Changes in the e-Learning paradigm suggest the necessity to develop technologies for real-timeness and high interactivity that enhance learners' satisfaction and, at the same time, maximize the efficiency and outcome of learning. This requires the evolvement of existing video-based environment, which provides unidirectional education only, into real-time interactive learning environment.

With the emergence of new paradigm of education environment (e-Learning 2.0), e-Learning is attracting people's attention throughout the world, spotlighted as a new knowledge industry, and expected to make substantial achievements in human resource development. At present, a lot of investments and efforts are being made in the development of e-Learning technologies, and many studies are being conducted on new technologies for next-generation e-Learning.

Table 1. Representative e-Learning 2.0 services [1, 5]

Service	Contents
elgg.	http://elgg.org/ is an open source system that provides personal blogs, file storages, RSS readers, etc.
edublogs	http://edublogs.org/ is a blog-based e-Learning service that provides functions such as automatic storage, automatic spelling check, and convenient upload of files and images.
DIGIfcation	http://www.digication.com/ is an online classroom service through which students submit assignments online and teachers manage students.
Wikispaces	http://www.wikispaces.com/ is a Wiki-based e-Learning service that is easy to edit and update and allows free exchange of opinions.
ChinesePod	http://chinesepod.com/ is a Chinese language learning site using podcasting, RSS and blogs, through which learners can participate in discussions and have bidirectional learning and multimedia learning.
	http://www.dickinson.edu/ forms language learning communities so that language learners find native speaker partners in overseas and practice conversations through Skype.

In international e-Learning markets, new online services such as integrating Web 2.0 technology and e-Learning are emerging. Such services are spreading rapidly because they pursue sharing and openness rather than monopoly.

In order to overcome the existing passive environment of e-Learning 1.0 in response to trends in the age of e-Learning 2.0, this study proposes the base of learning environment suitable for e-Learning 2.0 that has emerged from Web 2.0. In this study, we develop real-time interactive video contents (bidirectional learning components) that maximize learning effects, and design an expanded standard based on SCORM, which is used as a current standard to construct LMS/LCMS [6, 7]. In this paper, Chapter 2 proposes an expand standard based on the SCORM standard. Next, we propose e-Learning environment implemented based on the results of this study, and draw conclusions.

2 Designing LMS/LCMS Using Expanded SCORM Standard

In consideration of the flow and trends of the e-Learning paradigm, we can analyze the shortcomings of LMS/LCMS, which manages existing e-Learning systems that are mostly video lectures, as follows. In existing video-based e-Learning systems, it is hard to measure learners' progress rate accurately, real-timeness is not sufficient in contents operation, and there is a limitation in providing interactive components in contents. Although there are problems also in the volume or availability of video contents, these shortcoming mentioned above impair the functions or effects of e-Learning. Moreover, the operation of video contents can impose restrictions on the operation of learning courses. In order to overcome these limitations in e-Learning, this study proposes an expand standard based on SCORM so that new bidirectional components can be added to existing video lecture systems. In addition, we implement LMS/LCMS that maximizes the effect of learning through supporting real-time interactive video contents.

2.1 System Flow Diagram

The Webpage and learning window (video contents) of LMS/LCMS are built on the Internet and flash video (FLEX), which are installable to all users (teachers, learners, administrators), for the efficient expression of e-Learning 2.0 functions. This system is implemented using the expanded SCORM standard so that a running video content may be interlocked with LMS/LCMS and execute required actions. [Figure 1].

Lastly, learning components with various functions are implemented for real-time interaction during a video lecture and, consequently, video contents can be organized as in [Figure 1]. As video contents such as real-time chatting and discussion, Q&A and bulletin board are implemented based on the expanded standard, users can gain new learning effects by clicking desired icons in the interactive learning environment.

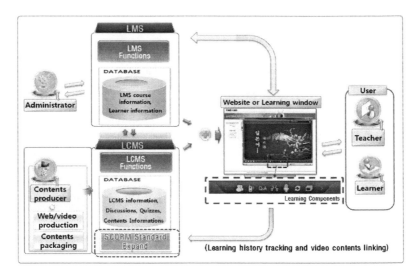

Fig. 1. Interactive e-Learning 2.0 environment in real time

2.2 Designing a Standard for Expanding SCORM

First, in order to connect LCMS developed based on the SCORM standard to real-time video contents, we design a new XML-type standard by expanding the existing SCORM standard. Using the designed expanded XML schema, contents (lectures) developers produce contents. The produced contents are packed, parsed, and stored in the database. In this study, we define these processes and then expand the standard so

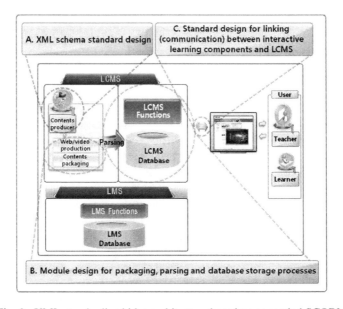

Fig. 2. XML standardized idea architecture based on expanded SCORM

that video contents can be connected to (communicate with) interactive components (chatting, discussion, quizzes, learning materials, etc.) through LCMS. [Figure 2].

2.2.1 (A) XML Schema Standard

Manifest XML defined in SCORM is described as four layers Metadata, Organizations, Resources and Sequencing. This study expands the Metadata and Organization tags into new structures in order to support the packaging of interactive video contents in real time.

2.2.2 (B) Contents Packaging Import Module Standard

When LCMS packages contents, it ports contents using the contents of database (contents information) based on import modules (Upload /Unzip module, Parsing module, and Data Process module). At that time, it builds contents packaging import modules through LCMS by defining three modules, which are packaging module, XML file parsing module, and database storage module.

2.2.3 (C) Standard for Linking Interactive Learning Components

Table 2 shows information defined for linking LCMS to interactive learning components (e.g. discussion, quizzes, supplementary learning materials, etc.) in video contents running online in real time.

Table 2. Information related to connected components

Information	Description
Discussion	· Information on discussions in packaged contents, connected to LCMS and presented and managed in a table for discussions. Used as standard information on discussions in LMS when a LMS courses are opened · LCMS course information - System serial number, course classification serial number, course serial number, and session serial number · Discussion information –serial number, name, object ID
Quiz	· When a quiz-related event is triggered while a learner is learning video contents, the following information is created in the learning history information table. - The total number of quizzes registered in contents - The number of quizzes to which the learner has answered correctly
Supplementary learning material	· When an event related to supplementary learning materials is triggered while a learner is learning video contents, the following information is created in the learning history information table. - The total number of supplementary learning materials - The number of supplementary learning materials in which the learner has participated

For example, when a real-time discussion function in video contents is implemented, the base model for nodes are included in the XML standard.

2.3 Standard for Linking Web Services

Second, because this study is for Web-based video learning, we design a standard for linking Web services. When a user logs in a LMS-based website and calls the learning window, information on the learner and the learner's learning history in LMS should be connected to and communicates with contents information in LCMS. Moreover, XML schema should be designed so that when the learning window (video lecture) is executed it is connected to tasks such as discussion, quiz and learning history tracking that appear between LMS/LCMS and videos. [Figure 3]

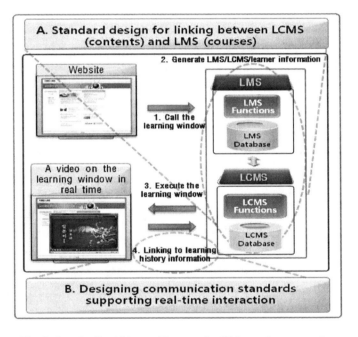

Fig. 3. Standardized idea architecture for Web service connection

Here we define standard information for linking between LCMS (contents) and LMS (courses) such as course information in LMS, and course and learner information in LCMS. Information necessary for the connection is as in Table 3.

Table 3 shows information necessary for linking related to LMS course, learner, bulletin board and discussion in order to support real-time interaction when the video player imports contents.

Table 3. Information necessary for linking

Information	Description
LCMS	· LCMS information on the contents of the course that the learner is taking · LCMS course information : System serial number, course classification serial number, course serial number, session serial number
LMS	· Information on the course that the learner is taking · LMS course information : System serial number, course classification serial number, course serial number, term serial number, session serial number
Learner	· Information on the learner · Learner information : Serial number, ID, Name
Bulletin board	· Information on the bulletin board of the course that the learner is taking · LMS course information: System serial number, course classification serial number, course serial number, term serial number, session serial number · Bulletin board information: Bulletin board serial number, post serial number, post title, post contents · Learner information : Serial number, ID, Name
Discussion	· Information on the course that the learner is taking · LMS course information : System serial number, course classification serial number, course serial number, term serial number, session serial number · Discussion information : Discussion serial number, contents ID, discussion name, discussion contents · Learner information : Serial number, ID, Name

2.4 Designing Learning History and Progress Rate Measuring Module

Third, we design standards for learning history and progress rate measuring modules. The existing learning management system (LMS) measures learners' progress rate using only learning hours. In order to overcome the shortcoming, we design logic that calculates learners' participation rate based on the designs proposed above and calculates the progress rate by tracking learning history using actual learning hours. The progress rate measuring process involves actual learning hours, participation in quizzes, participation in supplementary learning, etc. and the rate can be changed by contents producers. For this, we define a database structure that allows the efficient management of learning history, and design XML schema for communication between LCMS and learning history, which is learners' response to video contents. [Figure 4].

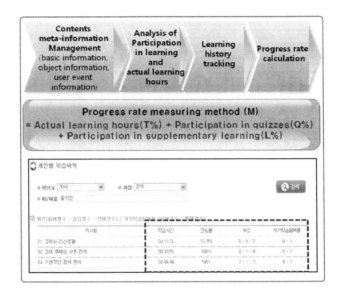

Fig. 4. Standardized idea architecture of learning record and progress measurement module

The standard database structure for managing learners' learning history is designed for data connecting between LCMS and LMS courses, between learner information in LMS and LCMS, and between learning history and quizzes/supplementary learning materials.

3   Real-Time Video-Contents Linking System (Implementation)

3.1   System Structure and Implementation

The system is designed for sharing information using video contents, real-time communication between users (teachers-learners, learners-learners) through chatting, replies or bidirectional information exchange, and viewing information on all interactive components available. In this part, we propose a platform for video contents operation by implementing a contents linking system, and implement an e-Learning system that can overcome the shortcomings and limitations of existing LMS/LCMS.

As in Figure 5, bidirectional learning components (video contents) implemented through linking between LMS and LCMS support learning environment personalized in the video learning window, and provide various functions including progress rate management and learning event support, and enable interactive learning through real-time edition and distribution. Such a system environment maximizes the effect of learning, and may become a new technology of e-Learning 2.0 paradigm that expresses technological functions based on Web 2.0 efficiently.

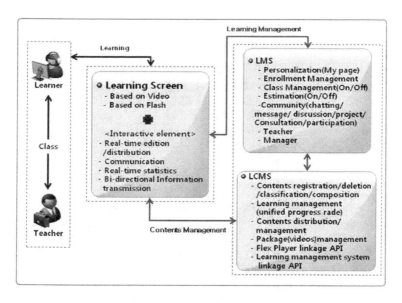

Fig. 5. Schematic diagram of system functions

The system was built by AJAX, which is Web 2.0 technology, based on AOP (aspect-oriented programming). The general architecture, which is independent from platform, was built on J2EE and implemented so that all services can be provided through a Web browser and therefore learners can access all services at any time and in any place. [Figure 6].

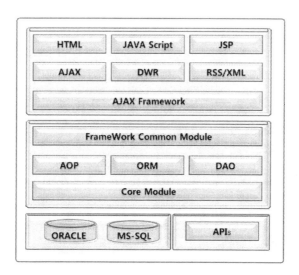

Fig. 6. Detailed structure of architecture

3.2 Implementing Various Real-Time Interactive Components

3.2.1 Real-Time Edition and Distribution

This component allows real-time edition of contents and immediate sharing among all users using the editing function. This function provides video learning contents services based on Web 2.0, which can be edited and executed on the Web [Figure 7]. Moreover, contents (texts, images, flashes, sounds, videos, etc.) can be edited, added, deleted or updated through online cooperation between teachers and learners, and contents can be tested and distributed freely [Figure 7].

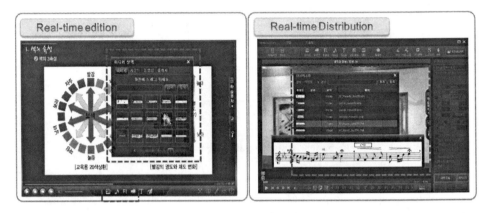

Fig. 7. Real-time edition and distribution functions

3.2.2 Communication

The communication component provides functions such as real-time chatting, replies, bidirectional information exchange (file transmission [Figure 8]), discussion, and Q&A.

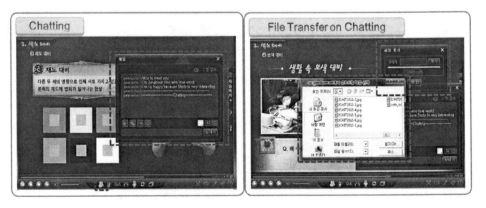

Fig. 8. Communication function (File transmission)

This function enables one-to-one communication with an online tutor. It provides a communication channel for conversation between a tutor and a learner who are connected, and they can use functions such as file transmission and walkie-talkie while talking through the screen. In addition, we can implement a real-time function as in Figure 8 by which a tutor can send a message or a material to multiple learners attending the class at once.

3.2.3 Real-Time Statistics

This component provides real-time view of useful statistics on videos such as view, edition, recommendation, and link registration. These functions also provide useful statistics to users by analyzing learners' participation in various interactive components included in learning contents and participation in quizzes, discussions, etc. Such statistics can be used to correct inadequacies in existing video contents and to manage learners' progress rate and actual learning hours. [Figure 9]

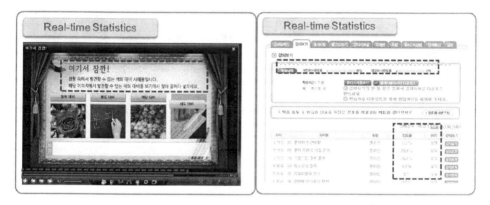

Fig. 9. Real-time statistics function

4 Conclusion

Changes in the e-Learning market triggered by the emergence of Web 2.0 have a very important meaning in that learners' participation began to be recognized as a new factor of competition [1]. In e-Learning 2.0 as well, learner-centered interactive components are being spotlighted.

With this background, this study expanded the existing SCORM standard model as well as the components of the packaging file (manifest.xml), and furthermore, expanded API used in real-time learning environment. Such a video contents linking system provides a platform for video contents operation, and solves problems in existing learning environment using videos.

Learning events of various functions designed based on the expanded SCORM standard and useful in watching videos allow users to monitor learners' state of learning and progress rate and provide various types of useful information. Furthermore, using real-time edition and distribution functions for editing contents and sharing them among all users in real time, users can create user-centered e-Learning 2.0

environment. In this way, collaborative real-time LMS/LCMS as a contents linking system can enhance the efficiency and effectiveness of learning and can be utilized as learner-centered customized learning environment.

Acknowledgment. This paper was conducted with the research fund for a research task (Task No. JP091004) under the Seoul R&BD Program.

References

1. Kumaran, K.S., Nair, V.M.: Future trends in E-Learning. In: 2010 4th International Conference on Distance Learning and Education (ICDLE), pp. 170–173 (2010)
2. Jee, H.K.: e-Learning Technology Trend. KIISE(Korean Institute of Information Scientists and Engineers) 26(12), 33–41 (2008)
3. O'Reilly, T.: What Is Web 2.0s: Design Patterns and Business Models for the Next Generation of Software (2005)
4. Bartolome, A.: Web 2.0 and New Learning Paradigms. eLearning Papers (April 2008) ISSN 1887-I542 , http://www.elearningpapers.eu
5. NIPA(National IT Industry Promotion Agency), The Appearance of e-Learning 2.0 and The Effect of Market , Software industry Trend, pp. 1–9 (2007)
6. KERIS(Korea Education Research Information Service), 2009 Adapting Education to the Information Age (2009)
7. Sharable Content Object Reference Model (SCORM) Version 2004 ,The SCORM Overview, Advanced Distributed Learning (2004) http://www.adlnet.org

CSP Based SCORM Interface for E-Learning System

HwaYoung Jeong[1] and BongHwa Hong[2,*]

[1] Humanitas College of Kyunghee University. Hoegi-dong, Seoul, 130-701, Korea
[2] Dept. of Information and Communication, Kyunghee Cyber University,
Hoegi-dong, Seoul, 130-701, Korea
hyjeong@khu.ac.kr,
bhhong@khcu.ac.kr

Abstract. Generally, CSP used to analyze and describe the system's process. And SCORM proposed the international standard to manage the learning contents for e-learning system. Therefore we developed e-learning system that is able to support learning materials include learning contents and topics according to SCORM interface. We also used CSP to the system to analyze the system's process and identified the logics efficiently.

Keywords: E-learning system, learning contents, SCORM API, CSP.

1 Introduction

Cyberphilosophy, according to Moor and Bynum [2], is a term which designates to the intersection of philosophy and computing. Similarly Floridi[3,4], implies that the [ethical]question of "what is the nature of right and wrong?" is one of the field questions of that [cyber]philosophy [of information and ICT](words in italics added for emphasis).Ethics is people 's desire to do good and the need to avoid doing harmful behavior [5, 1]. E-learning that is similar way is used for efficient learning progress. Numerous research works regarding to e-learning have been done to enhance teaching quality in e-learning environments. Among these studies, researchers have indicated that adaptive learning is a critical requirement for promoting the learning performance of students [6]. Actually, there are too many learning contents to use learning system to support user. Therefore, a developer has to make many learning contents according to learning courses or subjects. To develop and manage the contents on the web, teacher and developer of e-learning system used SCORM(Sharable Content Object Reference Model) that has reached great acceptance, since it brings together several standards of different standardization institutes in diverse fields of e-learning [7]. Additionally, it used CSP(Communicating Sequential Processes) to design and develop the system process exactly when we implement the system. CSP, originally developed by Hoare [8] and more recently by Roscoe [9], models concurrency via multiple CSP. However, CSP abstracts away

---

* Corresponding author.

G. Lee, D. Howard, and D. Ślęzak (Eds.): ICHIT 2011, CCIS 206, pp. 513–520, 2011.
© Springer-Verlag Berlin Heidelberg 2011

trueconcurrency through the nondeterministic sequential interleaving of simultaneously observed events by an Olympian. The notion of reasoning about a computation being equivalent to reasoning about its trace of observable events is central to the elegance – and utility – of CSP [10].

In this paper, we aimed SCORM based interface model for e-learning system. We also used CSP to design the process when we developed the system. The proposed system is under the environment as many learning contents used on the web. The rest of this paper is organized as follows: Section 2 reviews related works. Section 3 presents the proposed learning system mechanism include SCORM API interface with CSP. Finally, Section 4 provides a conclusion.

2 Related Works

With the development in communication and network technology in recent years, under the gradual improvement of network bandwidth and quality, the real-time transmission of high-quality video and audio becomes possible. Therefore, the transmission of multimedia and relative network application technologies have gradually been developed and become popular, such as the technology of Distance Education, Video Conference and Video on Demand. The advantage of e-Learning is that it can overcome the obstacle of geographical location; making students on remote sites feel that they are like being in the environment of attending classes in a classroom. Moreover, it can save cost and time of the students for their commuting to and fro the classroom. Currently, SCORM is the most popular standard for learning contents, and it is proposed by the U.S. Department of Defense's Advanced Distributed Learning (ADL) organization in 1997. The SCORM specifications are a composite of several specifications developed by international standards organizations [11]. SCORM references specifications, standards and guidelines developed by other organizations that are adapted and integrated with one another to form a more complete and easier-to-implement model. SCORM defines, in the SCORM Content Aggregation Model (CAM) book the components used to build a learning experience from learning resources and how they are aggregated and organized into higher-level units of instruction. It defines five different components as shown in Fig. 1. Assets are electronic representation of media that can be collected together to build other assets. If this collection represents a single launchable learning resource that utilizes SCORM RTE to communicate with an LMS, 1 it is referred to as an SCO (Sharable Content Object). An Activity is a meaningful unit of instruction that may provide a learning resource (SCO or asset) or be composed of several sub activities. To create a course, the activities compose a Content Organization, a map that represents the intended use of the content through structured units of instruction. Last but not least, a Content Aggregation is an entity used to deliver both the structure and the resources that belong to a course in a Content Package, which consists of a compressed file with the physical resources of educational content and at least one XML file - called manifest - that embodies a structured inventory of the content of the package [7].

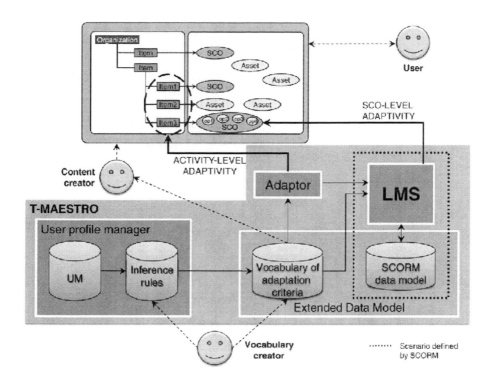

Fig. 1. Example of e-learning system with SCORM interface

The content packaging scheme defines a teaching materials package consisting of 4 parts, that is, (1) Metadata: describe the characteristic or attribute of this learning content, (2) Organizations: describe the structure of this teaching material, (3) Resources: denote the physical file linked by each learning object within the teaching material, and (4) (Sub) Manifest: describe this teaching material as consisting of itself and another teaching material [11].

CSP is a very useful programming model. Roscoe described [13] "CSP was designed to be a notation and theory for describing and analyzing systems where primary interest arises from the way in which different components interact". CSP came into the world of practical software development. In CSP related books, often (state) transition diagrams are used to illustrate CSP interaction patterns. Those are in fact Finite State Machine (FSM) kind of diagrams. Every node in such FSM represents a state of the process and every edge/transition is associated with some event. Figure 1 represents one CSP description and its associated visualization based on a FSM. In fact, the FSM in Fig. 2 is a typical UML-like visual representation of a state machine. In the picture, states 3 and 6 are named respectively Temp1 and Temp2, defining in such a way auxiliary processes needed as recursion entry points in the CSP description [12].

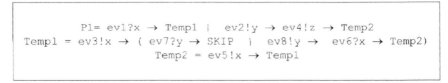

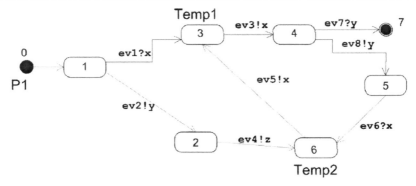

Fig. 2. Example of Classical FSM diagram with CSP

3   SCORM Interface Model for E-Learning System with CSP

This research focused interface model with CSP between SCOs or Items in SCORM. Proposed e-learning system considered activity and behavior of interface the SCORM API that designed by CSP model. The structure of the system was shown in Fig. 3.

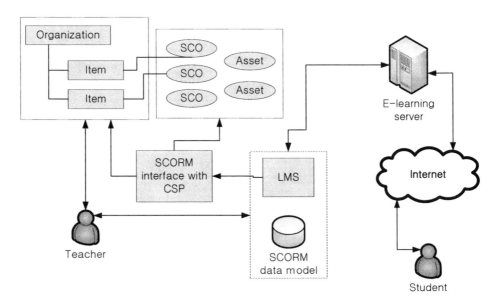

Fig. 3. The structure of proposed e-learning system

Organization constructed many items according to learning course that was made design by teacher or developer for e-learning system. Teacher is able to direct access LMS(learning management system) and compose the learning course to the organization. Item consist of SCO or Asset. In this environment, we can describe the behavior for interface to them as shown in **Definition 1**.

Definition 1 (behavior). *A process, P and Q, is an interface process to access Organization and SCO in SCORM. It is able to assemble as the law.*

$$P \parallel Q = Q \parallel P$$

Next, we define a search the learning contents by SCO or organization. Fig. 4 shows a simple structure of SCORM interface. The interface is able to connect SCOs or Items. Learning contents can make compose SCOs or Items according to learning course by teacher or developer.

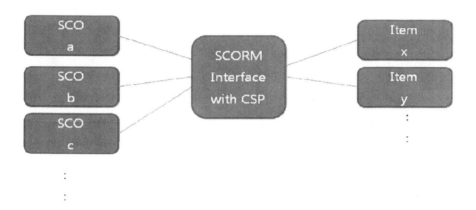

Fig. 4. A simple structure of SCORM interface with SCOs and Items

Definition 2 (connect). *when a, b and c are an alphabet to link SCO, and x and y are an alphabet to connect Items, the connect action to access SCORM API to interface learning materials (organization or SCO) should be*

$$(a : S_1 \rightarrow I(a)) \parallel (b : S_2 \rightarrow I(b)) \parallel (c : S_3 \rightarrow I(c)) = \text{CONNECT to SCO}$$
$$\text{if } a \neq b \neq c$$
$$(x : T_1 \rightarrow I(x)) \parallel (y : T_2 \rightarrow I(y)) = \text{CONNECT to items} \qquad \text{if } x \neq y$$

Definition 3 (compose). *If P and Q are learning process with the contents, then learning material of the course can be show the notation*

$$P \prod Q \quad (P \text{ or } Q)$$

That is to denote a process which behaves P or Q, without the knowledge of control of the external environment.

Example is : Learning material is consist of SCOs or Items, and can be made by composing with them.

Let LM_i is $i'th$ learning material according to the course, then
$$LM_i = (S_1 \rightarrow S_2 \rightarrow S_3 \rightarrow LM_i) \prod (T_1 \rightarrow T_2 \rightarrow LM_i)$$

Let P and Q are learning process with the contents as compose SCOs and Items, then

$LM_i = (P \parallel Q)$
$P = S_1 \rightarrow S_2 \rightarrow S_3 \rightarrow LM_i$
$Q = T_1 \rightarrow T_2 \rightarrow LM_i$

The SCOs or Items in learning process have interacted to *SCORM interface with CSP* between them.

$P = \mu I \cdot ((a \rightarrow \mu I) \prod (b \rightarrow \mu I) \prod (c \rightarrow \mu I))$
$Q = \mu I \cdot ((x \rightarrow \mu I) \prod (y \rightarrow \mu I))$

Let Sp be a finite nonempty set of learning process with SCOs and Sq be a set with Items.

$Sp = \{a, b, c\}, Sq = \{x, y\}$

Then we can define

$\prod_{\alpha:s} P(\alpha) = P(S_1) \prod P(S_2) \prod P(S_3)$
$\prod_{\beta:s} Q(\beta) = Q(T_1) \prod Q(T_2)$

Fig. 5 shows learning contents as SCO that is VOD and lecture note. Fig. 6 shows Item in this learning course.

(a)   (b)

Fig. 5. A sample contents as SCOs, a is VOD type of content and b is Text type.

```
[00:00] 9차시 1강 폼 뷰와 윈도우 공용 대화상자
[01:19] 01 폼 뷰(Form View)
[03:29] 학습정리
[03:34] 다음시간 안내
```

Fig. 6. Items of learning course. It shows the sequence of learning process

By composed method with CSP-SCORM interface, we constructed the contents and developed learning system as shown Fig. 7.

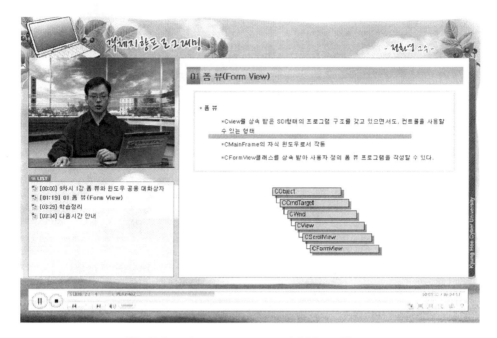

Fig. 7. Learning system composed SCOs and Items

4 Conclusion

In this article, we focused interface method between SCOs and Items to construct learning contents and to make the system. Actually, it normally used many learning contents using SCOs, and teacher or developer design the learning course using Items. Therefore, it needs efficient interface method between SCOs and Items, and to suggest making learning course using them. For this purpose, SCORM that is standards and guidelines to manage them was suggested. We proposed SCORM interface with CSP, formal method for interaction. Using the proposed interface, we constructed SCOs and Items to learning system.

References

1. Beycioglu, K.: A cyberphilosophical issue in education:Unethical computer using behavior –The case of prospective teachers. Computers &Education 53, 201–208 (2009)
2. Moor, J.H., Bynum, T.W.: Introduction to cyberphilosophy. Metaphilosophy 33(1-2), 4–10 (2002)
3. Floridi, L.: What is the philosophy of information? Metaphilosophy 35(1-2), 123–145 (2002)

4. Floridi, L.: Open problems the philosophy of information? Metaphilosophy 33(4), 554–582 (2004)
5. Macer, D.: Computing ethics:Intercultural comparisons.Ethical pluralism and social justice. In: Rooksby, E., Weckert, J. (eds.) Information Technology and Social Justice, pp. 1899–2204. Information Science Publishing, London (2007)
6. Chang, Y.-C., Kao, W.-Y., Chu, C.-P., Chiu, C.-H.: A learning style classi fication mechanism for e-learning. Computers &Education 53, 273–285 (2009)
7. Rey-López, M., Díaz-Redondo, R.P., Fernández-Vilas, A., Pazos-Arias, J.J., García-Duque, J., Hoare, C.A.R.: Communicating Sequential Processes. Prentice Hall International Series in Computer Science. Prentice-Hall International, UK (1985)
8. Roscoe, A.W.: The Theory and Practice of Concurrency. International Series in Computer Science. Prentice Hall Europe. Prentice Hall Europe (1998)
9. Smith, M.L.: A Unifying Theory of True Concurrency Based on CSP and Lazy Observation. In: Communicating Process Architectures 2005. IOS Press, Amsterdam (2005)
10. Shih, W.-C., Yang, C.-T., Tseng, S.-S.: Ontology-based content organization and retrieval for SCORM-compliant teaching materials in data grids. Future Generation Computer Systems 25, 687–694 (2009)
11. Orlic, B., Broenink, J.F.: SystemCSP – Visual Notation. In: Communicating Process Architectures 2006. IOS Press, Amsterdam (2006)
12. Roscoe, A.W.: The Theory and Practice of Concurrency. Prentice Hall, Englewood Cliffs (1997)

Implementation of Web Service Based U-Learning

HwaYoung Jeong[1] and BongHwa Hong[2]

[1] Humanitas College of Kyunghee University. Hoegi-dong, Seoul, 130-701, Korea
hyjeong@khu.ac.kr
[2] Dept. of Information and Communication, Kyunghee Cyber University. Hoegi-dong,
Seoul, 130-701, Korea
bhhong@khcu.ac.kr

Abstract. In this article, we aimed a web service based learning system for ubiquitous computing environment. In the environment, there used many devices and it can be different types according to company. To support the learning service with each different type, we implemented the web service considering meta data related the device types. The meta data include information of device characteristics such as screen size, memory capability, and type of operating system. The management service for the meta data register in UDDI. And the other web service is deal with control of learning contents.

Keywords: U-learning system, Ubiquitous computing, Learning system, Web service.

1 Introduction

The ultimate goal of ubiquitous computing is to provide services via a natural interface that can satisfy various user needs anytime, anywhere, with any device, running on any network. To reach this goal, services should be available dynamically in the ubiquitous network while considering users context or needs. Not only the dynamic nature of service delivery, but the diversity of user needs is also an important consideration. When user needs are highly diversified, new services are required to be dynamically composed [1].

With the development of e-learning area, it appears some new requirement [2]:

· The numbers of users increase fast, the load of the e-learning server make the system slower and the risk of crash increases. E-learning system needs a new method to balance the load.

· Users are more scattered. More users are cross-state and cross-province, the low-speed of internet results poor user experience. E-learning users hope to have the same user experience as they are in the same city.

G. Lee, D. Howard, and D. Ślęzak (Eds.): ICHIT 2011, CCIS 206, pp. 521–528, 2011.
© Springer-Verlag Berlin Heidelberg 2011

Cause of the requirement, development of e-learning system need to change technique in application environment to ubiquitous computing. Many students and teachers want that the system is able to use more learning contents and apply to any devices such as PC, mobile phones or any other equipment. The phenomenon that is often referred to as ubiquitous computing is one of the latest transformational educational paradigms that foster an anywhere, anytime learning environment [3]. Even more recently, there has been a change in the nature of ubiquitous computing. The proliferation of portable electronic devices and wireless networking is creating a change from e-learning (electronic) to m-learning (mobile) [4]. The emphasis on technology in education is not to imply that the technology is the goal of the educational process; however, a technological learning environment can alter the way students learn and the way professors teach [2].

It also used web services as business logics of e-learning process. Web services, which are developed for this purpose, represent a new technology that permits the exchange of information through the network, using standard protocols and allowing communication between heterogeneous architectures. Today Web services are essentially based on four standards: the eXtensible Markup Language (XML), the Simple Object Access Protocol (SOAP), the Web Services Description Language (WSDL), and the Universal Discovery, Description and Integration (UDDI) [5].

In this paper, we implemented learning system for ubiquitous computing environment using web service. For the development, we used the business logics as web service through UDDI on the web. And applied ubiquitous device used PDA phone.

2 Related works

2.1 Web Service

The world wide web (www)has become a universal platform where people can publish and receive all kinds of information through standardized protocols .In addition to simple content sharing, new Web services are emerging to enable heterogeneous application functionality with prescribed message exchange communications. People are able to rapidly design, implement, deploy and deliver various application functionalities using a standardized Web services model. To be more responsive and cost-effective in today's economy, many enterprises provide different Web services. Examples of this include Google SOAP Search API for information queries, Amazon Web services for various e-commerce solutions, Dealersphere's Web services based integration hub for promoting B2B process automation in the auto industry, and Galileo's GDS, which enable partners in the travel industry to build customized applications [6]. Service-oriented architecture (SOA) is essentially an architectural style to allow a collection of loosely coupled software agents interacting with each other .The most common way of implementing SOA is by the use of Web services. There are today many definitions for Web services. According to W3C [7], Web services are software systems designed to support machine-to-machine interaction over a network via well-defined interfaces. A Web service is specified in a standard way by a service descriptor using a service description language, Web Service Description Language (WSDL, [8]) for example. Each service descriptor must contain all the information needed to make the service interaction possible, including message format, transportation protocol and binding information. Fig.1 shows Web services architecture. Web services can interact with other systems, in the way described by the service descriptors, using Simple Object Access Protocol (SOAP) to receive and send information. SOAP exchanges XML-based messages over another application layer protocol, usually Hypertext Transfer Protocol (HTTP) or Multipurpose Internet Mail Extensions (MIME). Those messages can differ in type and style. The two most common messages types are Remote Procedure Call (RPC) and Document. The RPC messages wrap program methods into the message, allowing them to be remotely invoked. The body and all parameters are sub elements. By contrast, in the Document type, the message content is placed directly into the body element, making Document-based Web services loosely coupled and document driven [9].

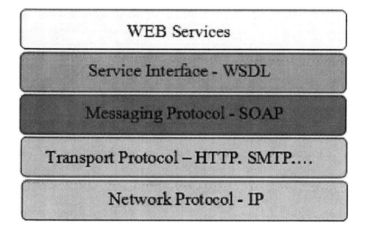

Fig. 1. Architecture of web service

Actually, developers and teachers tried to apply web service to e-learning system. In contrast to traditional e-learning platforms, LearnServe, being developed by Muenster University, makes e-learning offerings available though the emerging paradigm of Web services. Peter [10] proposed e-learning system using web service dividing in two parts: client software and Web services provided by several suppliers, as show Fig. 2. The client was the access point for users who could use the learning services. These services implemented on distributed servers and in particular include authoring, content, exercise, tracking, and discovery services as well as communication services such as email and message boards. The usage of learning services in their research was not limited to our clients because the implementation of the entire functionality as Web services enables an integration of the e-learning functionality directly into a business application to interact with applications, processes and information.

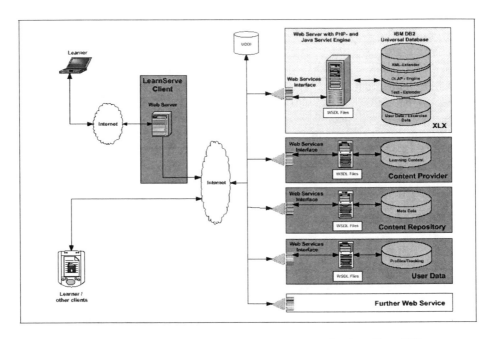

Fig. 2. Example of e-learning system using web service from Peter [10]

2.2 U-Learning System

Ubiquitous computing is a vision of invisible computing integrated in our everyday surroundings, introduced by M. Weiser and his group at Xerox PARC in 1988 [12]. Recently, the advance of wireless communication, sensor and mobile technologies has provided unprecedented opportunities to implement new learning strategies by integrating real-world learning environments and the resources of the digital world. With the help of these new technologies, individual students are able to learn in real situations with support or instructions from the computer system by using a mobile device to access the digital content via wireless communications [11].

Initially, the mobile learning researchers focused on applying distance learning techniques to mobile devices instead of desktops. Learners studied the sequence of instruction developed and experimented by transforming traditional distance education into a form more suitable for a mobile learning environment. The requirements for mobile learning environment include tailored contents, technologies, and suitable pedagogies [12]. Huang et. Al [12] proposed the enhanced student feedback model – the Interactive Service Module, which illustrates how the teacher and students interact with each other, as shown Fig.3. They said due to the restrictions of most mobile devices'keypad, especially cellular phones, keying in words is not easy. Therefore, a typed interactive solution, such as a chat room, is not suitable for mobile interaction. They have developed a friendly feedback process in which students can send their opinions to the instructor without keying in complex sentences or words. During the synchronous instruction process, the instructor can design a multimedia question/ questionnaire by using the web-based Questionnaire Design Tool.

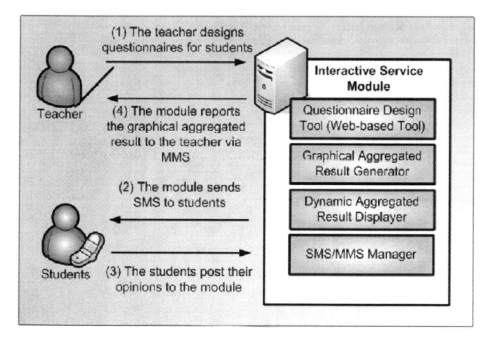

Fig. 3. The communication processes of Interactive Service Module by Huang et. Al [12]

3   U-Learning System Using Web Service

In this research, we focused development of u-learning system using web service. At the first time, we need business logic to perform the program and make/upload web service to UDDI. Then we are able to search the service for the system and composed the learning process according to learning progress or course. Fig. 4 shows proposed u-learning system's platform.

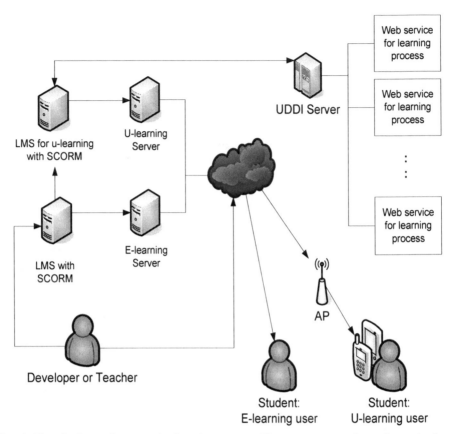

Fig. 4. The platform of proposed u-learning system. The system use web service as business logic for learning process

Fig. 5 shows the sequence diagram to learning process using web service. For this sequence, it consists of 6 objects: Search WS, UDDI Agent, Learning WS, U-learning meta handler, Comp u-learning, and Course ontology. Search WS is deal with the searching process to request web service's interface information. UDDI Agent control and manage handling process between search process and information interface in UDDI server. Learning WS has real business logic to perform the business of learning service. U-learning meta handler has meta data for various device's information, such as smart phones, mobile phones or any other devices using wireless network. Comp u-learning is to compose learning process and contents. It is main process to construct and make learning contents according to learning course. Finally, Course ontology has learning structure of the course. Therefore, it has each of learning unit's and section's information. Developer or teacher makes the course ontology for the purpose that student is able to get high performance to training the learning course.

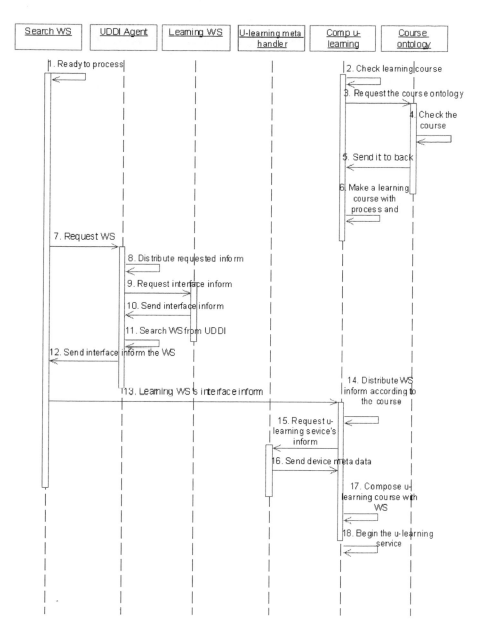

Fig. 5. The sequence diagram for u-learning with web service

4   Conclusion

In this article, we proposed u-learning system's platform using web service. To use web service as business logic for learning system, we have to search the web service to fit on our learning process or course. If there are not web service what we want the process in UDDI, we implemented the service and registered it on the web in UDDI server. However, after process finding web service for learning process, it is very easy to make and compose learning process. Additionally, this process support to user, developer and teacher, is able to control and manage the process for learning system.

References

[1] Lee, N.Y., Kwon, O.: A complementary ubiquitous service bundling method using service complementarity index. Expert Systems with Applications (2010)

[2] Wu, J., Zhang, S.: Broadband Multimedia e-learning system using Web Service. Fudan University (April 2, 2004)

[3] Wurst, C., Smarkola, C., Gaffney, M.A.: Ubiquitous laptop usage in higher education: Effects on student achievement, student satisfaction, and constructivist measures in honors and traditional classrooms. Computers &Education 51, 1766–1783 (2008)

[4] Dickson, G., Segars, A.: Redening the high-technology classroom. Journal of Education for Business 74(3), 152–156 (1999)

[5] Lee, M.J.W., Chan, A.: Exploring the potential of podcasting to deliver mobile ubiquitous learning in higher education. Journal of Computing in Higher Education 18(1), 94–115 (2005)

[6] Ardagna, C.A., Damiani, E., De Capitani di Vimercati, S., Samarati, P.: A Web Service Architecture for Enforcing Access Control Policies. Electronic Notes in Theoretical Computer Science 142, 47–62 (2006)

[7] Huang, A.F.M., Lan, C.-W., Yang, S.J.H.: An optimal QoS-based Web service selection scheme. Information Sciences 179, 3309–3322 (2009)

[8] W3C,Web services glossary (last access May 2011)

[9] W3C,Web services description language (WSDL)version 2 part 1:Core language, http://www.w3.org/TR/wsdl20/ (last access May 2011)

[10] de Melo, A.C.V., Silveira, P.: Improving data perturbation testing techniques for Web services. Information Sciences 181, 600–619 (2011)

[11] Westerkamp, P.: E-Learning as a Web Service. In: Proc. 7th International Conference on Database Engineering and Applications (IDEAS), Hong Kong (2003)

[12] Chu, H.-C., Hwang, G.-J., Tsai, C.-C.: A knowledge engineering approach to developing mindtools for context-aware ubiquitous learning. Computers &Education 54, 289–297 (2010)

[13] Wilhelm Bruns, F.: Ubiquitous computing and interaction. Annual Reviews in Control 30, 205–213 (2006)

[14] Huang, Y.-M., Kuo, Y.-H., Lin, Y.-T., Cheng, S.-C.: Toward interactive mobile synchronous learning environment with context-awareness service. Computers & Education 51, 1205–1226 (2008)

Intelligent System Modeling for Human Cognition

Youngdo Joo and Younghwa An

Computer Media Information Engineering, Kangnam University
111, Gugal-dong, Giheung-gu, Yongin-si, Gyounggi-do, 446-702, Korea
{ydjoo,yhan}@kangnam.ac.kr

Abstract. The development of computer-based modeling systems has allowed the operationalization of cognitive science issues. Human cognition has become one of the most interested areas in artificial intelligence to emulate high-level human capability. This paper introduces a methodology well-suited for designing of an intelligent system to evaluate human cognition. The research investigates how to elicit and represent cognitive knowledge from human being. Crucial to this research is to identify the collective perception to aggregate the congruence or sharing of individual psychological cognition. Unlike standard approaches based on the statistical techniques, the modeling presented employs a theory of fuzzy relation to deal with the required issues of similarity.

Keywords: Cognitive Structure, Collective Cognition, Fuzzy Relational Product.

1 Introduction

Human cognition has long been a problem for psychologists seeking to model the high-level mental capability of human beings. It is a major area in studies of intelligent systems where one tries to emulate the human behavior. In spite of the endeavoring efforts to model human cognitive perception, there still exist complexity and high degree of intricacies which traditional techniques have suffered to model human behavior properly. This paper describes a methodology well-suited for instantiation of an intelligent system to model human cognition in domain specific manner. The chosen domain of application is urban planning.

The goal of this research is to develop urban advisory system to evaluate the psychological cognition of the urban inhabitants and provide a decision-support mechanism. The environment is viewed as a series of *spatial* elements that are composed of the physical aspects of the city known as urban objects. Cognition is described in *aspatial* terms characterized by the socio-cultural needs, values, or emotions of urban residents. Research efforts start with establishing three main sets: *urban objects*, *constructs* and *respondents*. Urban objects are the physical or functional description of spatial entities existing in the urban environment. Constructs are verbal formulations and the aspatial aspects assigned to the urban objects. Finally, respondents are the set of individual urbanites that contribute relational information between the spatial and aspatial sets.

G. Lee, D. Howard, and D. Ślęzak (Eds.): ICHIT 2011, CCIS 206, pp. 529–536, 2011.
© Springer-Verlag Berlin Heidelberg 2011

The rest of this paper is organized as follows. We present the reasonable derivation of the cognitive structures as means of the cognition representation in section 2. A new approach for the assessment of the degree of congruence among individual cognitive structures is introduced in section 3. By the proposed scheme, we aim to seek collective cognition shared by like-minded groups of respondents to lead to the wisdom of crowds [1] or desired advice to solve the outstanding urban problems. Finally, our brief conclusions are given in section 4.

2 Representation of Cognition

We collected urban knowledge from the inhabitants using a knowledge elicitation technique based on the *Personal Construct Theory* [2]. A construct is a bipolar dimension, consisting of primary and contrasting poles. A construct is defined by two concepts acting as polar opposites such as "beautiful vs. ugly". An individual respondent is presented with a list of urban objects and labels of constructs where he can relate each construct to urban objects using a rating scale with a range of values 1 to 7. The points of 1, 2, and 3 on the rating scale indicating 'very', 'moderately', and 'slightly', respectively in verbal response can code a rating for the primary pole. Similarly, a rating for the contrasting pole can be coded by 5, 6, and 7.

Table 1. Constructs for Siena Group

No	Primary Pole	Contrasting Pole
1	Familiar to you (Familiar)	Unfamiliar to you (Unfamiliar)
2	Central	Peripheral
3	Beautiful	Ugly
4	You feel that it belongs to you (Yours)	You feel that it belongs to others (Others)
5	It makes you feel at ease (Myf at ease)	It makes you feel at unease (Myf uneasy)
6	Important to you (Important)	Dispensable to you (Dispensable)
7	Makes you feel proud (Myf proud)	Makes you feel ashamed (Myf ashamed)
8	You find it exhilarating (Exhilarating)	You find it depressing (Depressing)
9	It presents the spirit of Siena (Rep. Spirit)	It contradicts the spirit of Siena (Contrd. Spirit)
10	You go there often (Go often)	You avoid it (Avoid)
11	It promotes social contacts and encounters (Contactful)	It inhibits social contacts and encounters (Contactless)
12	Make you feel recognized (Myf recognized)	Make you feel anonymous (Myf anonymous)
13	Make you feel free (Myf free)	Make you feel controlled (Myf controlled)
14	You approve of the activities that take place there (Approve)	You disapprove of the activities that take place there (Disapprove)
15	It exerts a positive influence on the city (Pos. influence)	It exerts a negative influence on the city (Neg. influence)
16	You participate in the activities that go on there (Participate)	You hold aloof from the activities that go on there (Rejecdt)

The work in this paper focuses on "Siena data" gathered from 12 residents in local neighborhoods of Siena, Italy. 16 constructs are listed in Table 1 with abbreviated form used for the implementation of the purposed intelligent system. Urban objects consist of 18 buildings or areas such as city hall and hospital, etc. The application of constructs to urban objects forms a binary relation between both entities. If we convert the rating values in each cell of the matrix to fuzzy ones between 0 and 1, the matrix can be an ideal application area for Zadeh's *fuzzy set theory* [3]. The mapping for fuzzification is based on an even distribution with exception of the neutral point so that the crisp values; 1, 2, 3, 4, 5, 6, and 7 are converted to 1, 0.8, 0.6, 0.5, 0.4, 0.2, and 0, respectively. The output is a fuzzy binary relation between two sets, having constructs and objects as the rows and columns in a matrix form. Table 2 shows a binary relation fuzzified onto [0,1] from a specific rating of a respondent.

Table 2. Fuzzy Binary Relation Matrix

		O1	O2	O3	O4	O5	O6	O7	O8	O9	O10	O11	O12	O13	O14	O15	O16	O17	O18
	C1	0.8	1.0	1.0	0.8	0.8	1.0	1.0	1.0	1.0	1.0	1.0	1.0	0.8	0.8	1.0	n	0.8	0.6
	C2	1.0	0.0	0.8	0.6	0.2	1.0	1.0	1.0	1.0	1.0	1.0	1.0	1.0	1.0	1.0	n	0.6	0.2
C	C3	1.0	0.8	1.0	0.8	0.2	1.0	1.0	1.0	0.6	0.8	1.0	1.0	1.0	1.0	0.8	n	0.8	0.5
O	C4	1.0	0.8	0.6	0.5	0.2	1.0	1.0	1.0	0.8	0.8	1.0	1.0	0.6	0.8	0.8	n	0.8	0.2
N	C5	1.0	1.0	0.5	0.5	0.0	1.0	1.0	1.0	1.0	0.8	1.0	1.0	0.8	0.8	0.5	n	0.5	0.5
S	C6	1.0	1.0	0.5	0.5	0.0	1.0	1.0	0.8	0.8	0.8	0.8	1.0	0.8	0.8	0.5	n	0.6	0.5
T	C7	1.0	0.8	0.8	0.5	0.4	1.0	1.0	0.8	0.6	0.8	0.5	1.0	1.0	1.0	0.3	n	0.6	0.5
R	C8	1.0	0.6	0.6	0.5	0.0	1.0	1.0	1.0	0.6	0.6	0.5	1.0	0.8	0.8	0.5	n	0.5	0.5
U	C9	1.0	0.5	1.0	0.5	0.0	1.0	1.0	1.0	0.8	0.8	0.8	1.0	1.0	1.0	0.8	n	0.6	0.5
C	C10	1.0	0.8	0.6	0.4	0.0	0.8	0.6	0.8	0.8	0.6	0.6	0.8	0.8	0.6	0.5	n	0.6	0.0
T	C11	1.0	0.8	0.8	0.5	0.5	1.0	0.8	1.0	0.8	0.6	1.0	1.0	0.8	0.6	0.5	n	0.8	0.8
S	C12	1.0	0.8	0.6	0.5	0.5	1.0	0.8	1.0	0.8	0.6	0.8	0.8	0.8	0.6	0.5	n	0.8	0.5
	C13	1.0	0.8	0.8	0.5	0.5	1.0	0.8	1.0	1.0	0.8	0.8	0.8	0.8	0.6	0.5	n	0.8	0.5
	C14	1.0	0.8	0.8	0.5	0.8	1.0	0.5	0.8	0.8	0.5	0.6	1.0	0.6	1.0	0.5	n	0.8	0.0
	C15	1.0	0.5	0.8	0.5	0.2	1.0	1.0	1.0	0.8	0.5	0.5	1.0	1.0	1.0	0.5	n	0.8	0.5
	C16	0.8	0.8	0.6	0.5	0.5	0.8	0.5	0.8	0.8	0.5	0.6	0.8	0.6	0.8	0.5	n	0.6	0.0

Fuzzy relational theory [4] developed by Bandler and Kohout provides a proper mathematical implement to be used further analysis of the relational structure. They introduced *fuzzy relational products*, which are composed of two triangle products and one square product with two relations; R and S. The definition for each product is as follows. Here, $\rightarrow$ means the fuzzy implication operator.

$$\text{Triangle Subproduct, } \triangleleft : (R \triangleleft S)_{ik} = \min_j(R_{ij} \rightarrow S_{jk}) \tag{1}$$

$$\text{Triangle Subproduct, } \triangleright : (R \triangleright S)_{ik} = \min_j(R_{ij} \leftarrow S_{jk}) = (S^{-1} \triangleleft R^{-1})^{-1}{}_{ik} \tag{2}$$

$$\text{Square product, } \square : (R \square S)_{ik} = \min_j(R_{ij} \leftrightarrow S_{jk}) \tag{3}$$

Our major concern is of the triangle subproduct since we are searching for the degree to which whenever the respondent applies construct x_i, he also applies construct x_j. The formula for the subproduct translates as the degree to which x_i relating to y_j (R_{ij}) implies y_j relating to z_k(S_{jk}) for the two relations R and S where $x \in X$, $y \in Y$ and $z \in Z$. What we want to find is the degree to which construct x_i implies construct x_j based on how a respondent applied both constructs to the urban objects. Formally, it is the degree to which given x_i relates to y in R, x_j relates to y in R for all y. So, we want X and Z to be the same set and R and S to be the same relation. Using the inverse relation of R, we can have a formula of the subproduct on single relation.

$$(R \triangleleft R^{-1})_{ij} = \min_k(R_{ik} \to R_{kj}) \tag{4}$$

The degree may be interpreted as a certainty value in the statement, "whenever the respondent attributes construct x_i to an object, s/he also attributes construct x_j". The formula above is regarded as a harsh version since it takes the minimum of all those values. As the mean value has been more suitable to some applications [5], we adopt a mean version in which an equivalent definition is produced as follows.

$$\left(R \triangleleft R^{-1}\right)_{ij} = \frac{1}{N_k}\sum_k\left(R_k \to R_{kj}\right) \tag{5}$$

We first discover the degree of implication for the two constructs applied to the same object according to the implication operator used. Afterwards, we combine this result with the degree of implication for all other objects and take the mean of all the implications. The implication operators allow change to occur in the computational process of the triangle subproduct. Many definitions [6] exist for the fuzzy implication operator, $\to$. The operator used in this paper follows Lukasiewicz's definition, $\to_5$.

$$a \to_5 b = \min(1, 1-a+b) \tag{6}$$

As a preliminary stage for the representation of cognition, the triangle subproduct contains a cognitive view of the respondent implicitly. The result is a unary relation from constructs to constructs. The main concern is to transform the unary matrix into a graphical representation illustrating the order-like links among constructs. We achieve this transformation through the use of *a transitive closure* of the original relation and the *technique of α-cuts*. An order is a relation R from a set of elements X, to itself, that has the properties of reflexivity, transitivity, and anti-symmetry. These properties permit the partitioning of the elements of X into n exclusive subsets called *an equivalence class*. The resulting structure is a one-way hierarchical order. The relationship among constructs is expressed as *Hasse diagram*. The Hasse diagram is the directed graph of an order where each construct implies those which can be reached from it by paths upwards and is implied by those on paths downwards. Fig. 1 shows a Hasse diagram at a specific α-cut value which our system generates.

A construct hierarchy offers insight into cognition of an individual respondent's view about his environment. A Hasse diagram can be viewed as a cognitive structure through the dependencies of constructs specific to a respondent's thought and mentality about the particular set of the urban objects. The following can be spelled out as in Fig. 1. Whenever the respondent feels that a particular urban object belongs to him, he finds it is 'beautiful', and when he finds the urban object is beautiful, he finds it 'familiar'. By the transitivity, if the respondent feels it to 'belong to him' then he also feels it to be 'familiar'.

Urban inhabitants usually give different opinions on their environments or surroundings and consequently each respondent's cognitive structures are vastly different. Empirically, the majority of respondents in the Siena group is neither fully agreed nor disagreed, but very often overlapped in their arrangement of implications. Thus, *collective cognitive structures* by groups of individuals beyond individual structures exist. We try to identify the structures that characterize "like-mindedness" of groups by determining effective criteria in the following section.

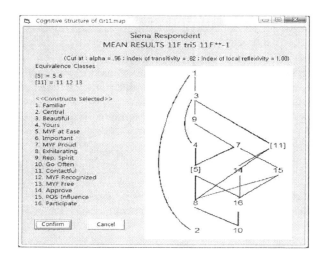

Fig. 1. Cognitive Structure of Individual Respondent

3   Elicitation of Collective Cognition

Two cognitive structures can be compared by examining how two individuals share meaning – the measurement of understanding and agreement between two individuals. Two respondents have an area of common knowledge or experience in their own matrix to reflect the application constructs to urban objects. The square product equation provides a valuable method to measure similarity between two matrices.

$$(R \ \square \ S)_{ik} = \frac{1}{N_j} \sum_j (R_{ij} \leftrightarrow S_{jk}) \qquad (7)$$

Using the inverse relation of R, instead of S, we derive the following equation.

$$(R \ \square \ R^{-1})_{ij} = \min\{(R \triangleleft R^{-1})_{ij}, (R \triangleleft R^{-1})^{-1}_{ij}\} \qquad (8)$$

The square product can be calculated by taking minimum values between the subproduct and inverse of the subproduct. $(R \ \square \ R^{-1})_{ij}$ gives the degree to which a respondent attributes construct x_i to exactly the same object to which s/he attributes construct x_j. In other words, it is the degree to which the respondent applies construct x_i and x_j, interchangeably, or mutually. Based on the square product of two relations extended easily from one on a single relation, the square product of relations, R_1 and R_2 from two different respondents is given by

$$(R_1 \sqcap R_2^{-1})_{ij} = \min\{(R_1 \lhd R_2^{-1})_{ij}, (R_2 \lhd R_1^{-1})^{-1}_{ij}\} \tag{9}$$

The formula gives the degree to which one respondent attributes construct x_i to the same object that the other respondent attributes construct x_j. Our interest is just $(R_1 \sqcap R_2^{-1})_{ii}$, occurring when $i = j$, as it may represent similar or different uses of each construct from two people. All the values of $(R_1 \sqcap R_2^{-1})_{ii}$ make up the main diagonal of the square product matrix. Accordingly, $(R_1 \sqcap R_2^{-1})_{ii}$ gives the degree to which two respondents attribute each construct, x_i to the objects mutually. This leads to a mutual implication value of x_i happening between two respondents. An average value of those values resulting from each construct is taken as a measure of similarity in that two respondents understand and empathize with each other in the context of each construct's mutual implication.

The mapping of pairs of matrices identifies subgroups of "like-minded" individuals, and may place these in the perspective of the entire group. Now every pair of matrices in the group is tried by the square product, thereby each pairing yields a measure of similarity regarded as mutual implication of corresponding two members of the group.

The goal of this research is to seek collective cognition to represent and reflect cognitive knowledge of all the individuals of a group. Each individual set of personal constructs may represent a person's thoughts and feelings. If some of his ideas are shared by other members of the group, as an expression of his construct system, they may benefit all the members. Our concern is to extract normative constructs which integrate individuals' ideas beneficially for the entire group. Remember that the square product for an individual and every other respondent produces mutual implications for every construct between each pairing. If we take an average of those construct mutual implications over the numbers of pairing, we can derive the degree to which the individual and others in the group apply each construct, mutually. In turn, every other person is considered to find this value.

Table 3 shows those values obtained for each member where respondents and constructs represent the columns and the rows of the matrix, respectively. Specifically, an entry at row i and column j indicates how similarly j^{th} respondent and every other one in the group use construct x_i with respect to objects. The respondent to show maximum degree marked in bold from Table 3 contributes his own construct and then constructs donated from some members in this manner complete 16 normative constructs. The respondent may be an influential [7] to have the greatest effect on the entire group in terms of the donating constructs.

The collection of these constructs make up *a normative matrix*. The normative matrix is not a consensus which discovers the mean of the individualities of the group, but it is believed that the matrix reflects the commonality or sharing within the group, adequately. Eventually, collective cognition can be instantiated in the form of *a collective cognitive structure* from the normative matrix in the usual way discussed previously. Fig. 2 shows a collective cognitive structure for Siena group. It may lead, hopefully, to ideals incorporating the overall viewpoints of respondents of the group to greatest extent. In other words, this representative cognition may involve overlapping and congruence among individual cognition when it is interpreted for investigation of group sharing. As noticed, collective cognition may be called normative in that it ought to be according to the ideals and values of the individual or one or more groups of individuals. Our technique to elicit collective cognition may be effectively applied to emerging web science to deal with issues of collective intelligence [8-9].

Table 3. Mutuality of Constructs among Siena Respondents

		RESPONDENTS											
		1F	2F	3F	4M	5F	6F	7F	8F	9M	10F	11F	12M
	C1	88.09	91.91	92.05	82.55	90.27	91.18	91.55	91.55	90.00	91.73	*92.09*	89.91
	C2	93.64	90.55	92.09	91.73	91.55	93.45	93.27	93.73	93.91	*94.09*	93.55	90.64
	C3	91.00	87.73	89.27	89.73	86.91	75.64	91.00	89.55	*91.36*	90.27	90.73	89.73
C	C4	81.91	83.64	85.00	83.36	79.82	83.00	84.91	*85.27*	84.27	84.36	83.82	78.45
O	C5	81.55	86.09	*86.36*	84.64	76.45	85.00	85.64	85.45	85.18	83.18	86.18	83.00
N	C6	80.18	83.18	84.55	83.82	81.27	86.64	86.18	84.45	84.36	81.55	*86.91*	84.91
S	C7	81.91	87.55	87.18	83.55	86.09	85.55	87.00	87.27	*89.00*	83.64	88.45	88.45
T	C8	85.55	87.82	84.36	86.91	78.36	87.36	87.55	85.64	86.82	84.45	*87.91*	85.45
R	C9	83.64	89.91	88.82	90.09	84.09	85.36	89.55	89.27	*90.45*	89.73	90.09	88.45
U	C10	82.45	86.55	*86.73*	85.09	83.00	86.09	85.82	86.09	82.18	82.09	85.55	84.73
C	C11	85.73	88.45	86.36	86.18	84.27	76.55	86.64	88.55	87.82	83.73	*89.09*	84.27
T	C12	83.27	84.55	79.45	83.36	81.09	74.73	82.73	*85.64*	79.64	76.36	85.00	76.18
S	C13	82.09	81.91	83.64	85.91	74.27	83.18	85.45	84.27	*86.45*	85.18	84.64	79.36
	C14	77.09	*86.82*	84.55	84.82	80.73	77.91	82.00	86.45	86.36	83.55	82.91	83.18
	C15	75.91	87.00	89.55	86.91	85.36	87.55	88.36	88.18	*89.64*	85.27	86.55	88.64
	C16	83.19	85.18	83.91	*87.73*	77.82	85.09	85.00	87.64	86.64	79.73	86.18	84.09

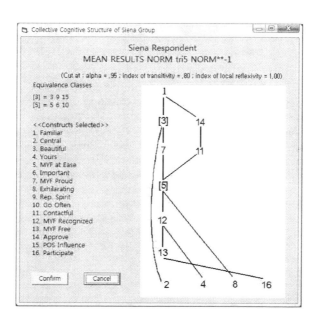

Fig. 2. Collective Cognitive Structure of Siena Group

536 Y. Joo and Y. An

4 Conclusions

The application of mental constructs to an urban environment in conjunction with the socio-cultural attributes of the urban inhabitants when expressed as fuzzy relations, represents a powerful knowledge presentation scheme for human cognition. Obviously, the fuzzy relational approach to the cognitive structure presented, provides a new view to evaluate cognition in a way that differ from the standard approach based on statistical techniques in that it may analyze personal construct systems directly by means of meaningful logic. Essentially, the square product from fuzzy relational theory yields good measure of similarity in individual knowledge by the mutual implication. In our work, the mutual implication algorithm yields desirable results as an effective criterion for similarity theory. It is appropriately applied to search for the collective cognition to represent individual perception of a group.

With the success of the model, the possibility exists to place inhabitants in the forefront of urban planning to lead a more human decision-making system or advisory system. Empirical results in the field work suggest that fuzzy relational approach is sound from an urban perspective. It is believed the applicability of these techniques from our research to other social science areas associated with human cognition is high.

References

1. Surowiecki, J.: The Wisdom of Crowds. Anchor (2005)
2. Bannister, D., Fransella, F.: Inquiring Man: The Psychology of Personal Constructs. Croom Helm, London (1986)
3. Zadeh, L.A.: Fuzzy Sets. Information and Control 8, 338–353 (1965)
4. Bandler, W., Kohout, L.J.: Semantics of Implication Operators and Fuzzy Relational Products. International Journal of Man-Machine Studies 12, 89–116 (1986)
5. Joo, Y., Noe, C.: Development of the Algorithm for the Selection of Clinical Investigations in Fuzzy Knowledge-Based System. Korea Telecom Journal 3(1), 22–33 (1998)
6. Bandler, W., Kohout, L.J.: Fuzzy Power Sets and Fuzzy Implication Operators. Fuzzy Sets and Systems 4(1), 13–30 (1980)
7. Watts, D., Dodds, P.: Influentials, Networks and Public Opinion Formation. Journal of Consumer Research 34(4), 441–458 (2007)
8. Berners-Lee, T., et al.: A Framework for Web Science. Foundation and Trends in Web Science 1(1), 263–275 (2006)
9. O'Reilly, T.: What is Web 2.0: Design Patterns and Business Models for the Next Generation of Software (2005), http://www.oreillynet.com/pub/a/oreilly/time/news/2005/09/30/what-is-web-20.html

Laser Pointer Based Constrained Input Scheme for Educational Game Software Framework

Jun Hee Cho[1] and Seong-Whan Kim[2,*]

[1] R&D Center of Famz Communication Inc. Seoul, Korea
netbelle@famz.co.kr
[2] School of Computer Science, University of Seoul, Seoul, Korea
simpo@uos.ac.kr, swkim7@uos.ac.kr

Abstract. In this paper, we present a new educational game interface using laser pointer based interaction with screen. Previous researches have focused on the display issue including stereoscopic and multimedia research, however they still use uneasy and un-natural interface based on keyboard/mouse or joystick along with PC for educational interaction. We implemented a laser based input device which composed of laser pointer and large active screen which tracks laser pointer. Laser based input device can give lots of flexibility on teaching staffs, and we propose an efficient character input configuration for the laser based large screen model for effective educational software framework. Experimental results show that we can easily input characters and numbers with more than 99 % of recognition rate.

Keywords: Laser based input, educational software, educational game, input configuration.

1 Introduction

Rapid development of multimedia research leads to new display device including stereoscopic and ultrasound display, however, there is little improvements on the input device for natural interaction. For example, first-person shooter (FPS) game is a video game that renders the game world from the visual perspective of the player character and tests the player's skill in aiming guns or other projectile weapons. All FPS feature the core game-play elements of movement and shooting, but many variations exist, with different titles emphasizing certain aspects of the game-play. Most modern FPS on the PC utilizes a combination of the WASD keys of the keyboard and mouse as a means of controlling the game (commonly referred to as "WASD/Mouse"). One hand uses the mouse, which is used for free look (also known as mouse look), aiming and turning the player's axis. On the keyboard, the arrow keys (or other keys arranged in the same manner, such as WASD, ESDF or IJKL) provide digital movement forwards, backwards, and sidestepping (often known as "strafing" among players) left and right. Usually these buttons make the player run, and a nearby button must be pressed in order to walk.

---

* Corresponding author.

G. Lee, D. Howard, and D. Ślęzak (Eds.): ICHIT 2011, CCIS 206, pp. 537–544, 2011.
© Springer-Verlag Berlin Heidelberg 2011

In this paper, we improve the interaction interface between player and game world by means of new control system: LaserTouch, which includes large screen, laser pointer, and directional device. LaserTouch creates a real life-like space for players. Because traditional display for FPS uses CRT, it cannot support large screen display due to limitation of CRT technology. To support more interaction including number, character, and gesture, we design a new typing method for laser touch. It is based on the constrained input configuration, whereby we can decrease the recognition errors. We review previous research issues in section 2, propose our input configuration in section 3, and show experimental results in section 4. We conclude in section 5.

2   Related Works

We review interaction scheme for FPS games, because FPS game is a good model for educational game framework. As shown in Figure 1, FPS is a video game that renders the game world from the visual perspective of the player character and tests the player's skill in aiming guns or other projectile weapons.

Fig. 1. Example of interaction device for First Person Shooter games

Most modern first-person shooters on the PC utilize a combination of the WASD keys of the keyboard and mouse as a means of controlling the game (commonly referred to as "WASD/Mouse") as shown in Figure 2 (a). WASD/Mouse interface is not natural way of FPS game interface because WASD/Mouse is initially designed for entering texts. Besides FPS game on PC, we can experience arcade FPS as shown in Figure 2(b). As shown in Figure 2(b), arcade FPS game uses CRT display, and the display cannot be larger than 30 inches due to the limitation of CRT technology. The primary advantage of CRT monitors is their color rendering. The contrast ratios and depths of colors displayed were much greater with CRT monitors than other type of monitors. The other advantage that CRT monitors is the ability to easily scale to various resolutions. This is referred to as multi-sync by the industry. By adjusting the electron beam in the tube, the screen can easily be adjusted downward to lower resolutions while keeping the picture clarity intact. While these two items may play an important role for CRT monitors, there are disadvantages. The biggest of these are the size and weight of the tubes. An equivalent sized LCD monitor for example is upwards of 80% smaller in size and weight compared to a CRT tube. The other major drawback deals with the power consumption. The energy needed for the electron beam means that the monitors consumer and generate a lot heat. Cons of large CRT display include (1) Very heavy and big, (2) large amounts of energy consumption, and (3) Generation of excess heat. Arcade FPS uses CRT for its main display and gets gunshot position (x, y) information using cathode ray position tracking.

Fig. 2. Traditional input device: (a) WASD/Mouse interface for PC FPS games, (b) CRT ray tracking for gunshot positioning

Recently a new gaming device and interaction method for FPS based on ChairIO has been developed by University of Hamburg [1, 2] as shown in Figure 3(a). ChairIO is based on a stool and is similar to a joystick, but controlled by the user's body motion.

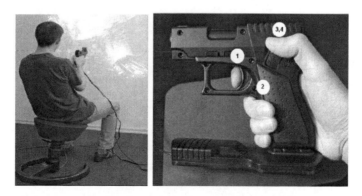

Fig. 3. Advanced input device for FPS: (a) ChairIO and InertiaCube2, (b) PistolMouse device

The ChairIO interaction is augmented by a game console gun to form a new interaction method for FPS. An initial evaluation compares several ChairIO and gun based interaction methods with traditional keyboard and joystick controls. They developed a gun in three degrees of freedom using InertiaCube2 a motion tracking product using Virtual Reality (VR) from InterSense Inc. This tracking device provides an absolute rotation in three axes with real-time update frequency and minimal drift. Another notable gaming interface is Pistol Mouse as shown in Figure 3(b). It has a joystick and gun interface. It is basically same as InterCube2 technology. Because traditional display for FPS uses CRT, it cannot support large screen display due to limitation of CRT technology. In our previous research works, we designed and implemented a new input device using laser recognizable display [3]. Our new control system is different from University of Hamburg's which has to deal with the problem of accuracy. Our approach comes from interplay between the large screen and a specially designed laser gun. The Laser beam coming from the gun draws a marker on the screen and its position is recognized and is inputted to the computer with other trigger information. As shown in Figure 4, there are three components in our new control system, laser gun, large screen and directional device. The Laser gun is installed away from the large screen, it can fire laser beam that will reach the screen and locate the position of target for the player. In addition, laser gun contains a set of buttons that deal with activities involved with weapon and ammunition such as switching weapon, firing. The Large screen is a special one made of optical sensors with capabilities of rendering image and receiving signals from laser gun. The Directional device in the form of a chair is developed similar to the ChairIO, for use as a joystick; and operation techniques allow the game characters to move around.

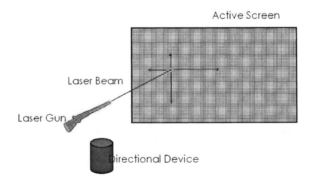

Fig. 4. LaserTouch: active laser recognizable screen, laser gun, and directional device for movement

3   Laser Based Character Input Scheme for Edu-Softwares

Most people are watching TV on the wall, sitting on the couch. Normally, when TV has bigger inches, distance between TV and couch gets more distant. The distance should be more than two miters. For the large audience lecture, the distance should be more than 6 meters. For this reason, laser based input device has strong advantage over touch based electronic board system for educational purpose. To support character recognition on IPTV or electronic board, we can use keyboard, mouse, and pen (enables curves recognition with pointing device). However, those devices have strong disadvantages that the audience should get closer to the display to touch or input characters [4, 5]. To solve this problem, we propose a laser pen input device which can recognize characters from far distance. To increase the character recognition rate without uneasiness, we provide a constrained input configuration as shown in Figure 5. The reason for design like above is people use straight lines and curved lines when they write the numbers and letters. To write the numbers and letters, there must be start and ending point of the writing. This method allows users to write characters on TV as like they write on the paper.

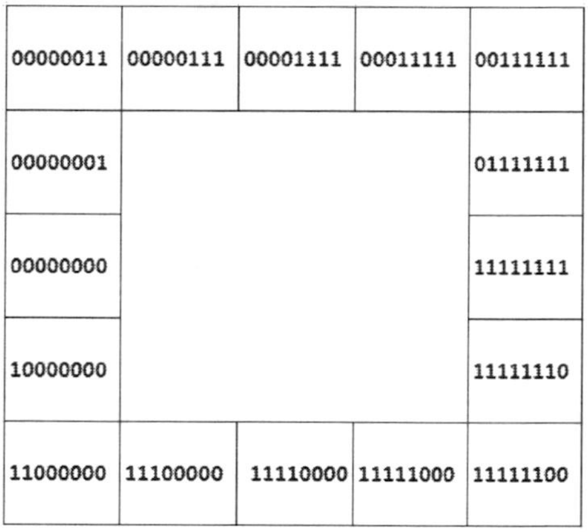

00000011	00000111	00001111	00011111	00111111
00000001				01111111
00000000				11111111
10000000				11111110
11000000	11100000	11110000	11111000	11111100

Fig. 5. Input zone was designed to constraint based input scheme for numbers and letters

The standard pattern offered to input numbers and letters are shown in Figure 6. They are standard pattern, and we don't need training pattern for recognition. Each target has 8digits binary, and we assign unique number on each zone.

Given unique number, we can compute the distance between each zone. For example, we can compute the distance between left-top and right-bottom zones as follows. The unique number of Left-top zone is 00000011. The unique number of right-bottom zone is 11111100. Now we compute XOR on those two unique numbers, which results in 00000011 XOR 11111100 = 11111111. The character recognition algorithm is summarized as follows.

1. Senses Laser beam input from the user's laser pointer.
2. Check what zone the laser beam input is touched on.
3. Recognize which process occurs when more than two inputs enters
4. Show character or number which has similar order and place on the center.
5. Find chosen character or number on the center with 3 to 5 zone chosen using the dynamic programming approach which is based on DTW (dynamic time warping) techniques [6, 7, 8] for our matching as follows.

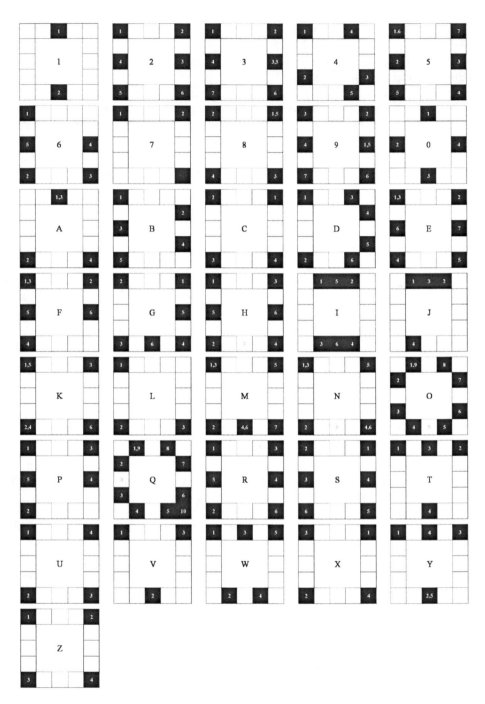

Fig. 6. Prototype for alphabet and number recognition with lasertouch pointer

```
Program DTWDistance (char s[1..n], char t[1..m], int d[1..n,1..m])
   declare int DTW[0..n,0..m]
   declare int i, j, cost
   for i := 1 to m
      DTW[0,i] := infinity
   for i := 1 to n
      DTW[i,0] := infinity
   DTW[0,0] := 0
   for i := 1 to n
      for j := 1 to m
         cost:= d[s[i],t[j]]
         DTW[i,j] := cost + minimum(DTW[i-1,j],
                     DTW[i  ,j-1], DTW[i-1,j-1])
   return DTW[n,m]
end.
```

4   Experimental Results

Comparing with previous input scheme used in PDA, cell phone, and tablet pens, our character recognition scheme is unique in the sense that we recognize characters via zone based constraint for efficient implementation of IPTV environments. In our experiments, we experimented with 100 recognition tests for each 36 alphabet letters and 10 digit numbers, which leads to total 4600 recognition tests. .Our experimental results show that we achieved 99% recognition rate for the alphabet and numeric input. We also experimented with 10 simple gestures including X, O, wipe-out, etc, and we achieved 90% recognition rate because of similarity of O and alphabet o. To decrease the miss-recognition, we used "mode" to discriminate alphabet input mode from gesture input mode.

Table 1. Experimental Results

Character	Recognition (%)	Character	Recognition (%)
0	100	I	100
1	100	J	100
2	100	K	100
3	100	L	100
4	100	M	100
5	100	N	100
6	100	O	80
7	100	P	100
8	100	Q	100
9	100	R	100
A	100	S	100
B	100	T	100
C	100	U	100
D	100	V	80
E	100	W	100
F	100	X	100
G	100	Y	80
H	100	Z	100

5   Conclusion

In this paper, we proposed a new character recognition scheme for laser based input device. Design goal includes the easy and natural interface with efficiently simple implementation and high recognition rate. To target for English and mathematical education for children, we focused on the number and character input recognition. Children do not like typing keyboard, rather they like to use joystick which offers more flexibility. Our experimental results show that children use our laser based input device as a kind of joystick. Our laser input device supports numbers [0-9], characters (English alphabet), and also simple gestures (10 major gestures). We designed and implemented a DTW based constrained recognition scheme, which achieves high recognition rate (average recognition rate is more than 99%) and also gives easiness of use.

Acknowledgments. This work was supported by Business for International Cooperative R&D between Industry, Academy, and Research Institute funded Korea Small and Medium Business Administration in 2010.

References

1. Beckhaus, S., Blom, K.J., Haringer, M.: A new gaming device and interaction method for a first-person-shooter. In: Computer Science and Magic 2005, GC Developer Science Track, Leipzig, Germany (2005)
2. Beckhaus, S., Blom, K.J., Haringer, M.: ChairIO - the chair-based interface. In: Magerkurth, C., Rötzler, C. (eds.) Concepts and Technologies for Pervasive Games, vol. 1, pp. 231–264. Shaker, Aachen (2007)
3. Shim, J.Y., Sung, H.S., Kim, S.W.: Cognitive laser: new gaming device for first person shooter games using laser shooter and laser-cognizable big screen. In: ACM SIGGRAPH , Asia (2010)
4. Microsoft Windows Hardware Developer Central,
   http://www.microsoft.com/whdc
5. Microsoft: Architecture of the Kernel-Mode Driver Framework. Microsoft Corporation (2006)
6. Myers, C., et al.: Performance tradeoffs in dynamic time warping algorithms for isolated word recognition. IEEE Trans. on Acoustics, Speech, and Signal Processing ASSP-28(6) (1980)
7. Sakoe, H., Chiba, S.: Dynamic programming algorithm optimization for spoken word recognition. IEEE Trans. Acoustics, Speech, and Signal Proc. ASSP-26, 43–49 (1978)
8. Berndt, D., Clifford, J.: Using dynamic time warping to find patterns in time series. In: AAAI 1994 Workshop on Knowledge Discovery in Databases, KDD 1994 (1994)

Public Transport Navigation System with Augmented Reality Interface

Jakub Królewski and Piotr Gawrysiak

Warsaw University of Technology, Institute of Computer Science,
Nowowiejska 15/19, Warsaw, Poland
{J.Krolewski, P.Gawrysiak}@ii.pw.edu.pl

Abstract. This paper provides an overview of an experimental mobile augmented reality system that combines the municipal public transport navigation capabilities with augmented reality visualization on a mobile device. The system is able to provide directions to tram and bus stops and monitors user journey in real time using geolocation data and schedule information. The system described herein has been implemented in a prototype form, that has been tested in a real case usage scenario in an urban environment.

Keywords: Augmented reality, municipal transport systems, navigation devices, mobile devices.

1 Introduction

Virtual Reality (VR) has become a term quite widespread both in popular culture and in science in recent years. However despite an initial enthusiasm it quickly turned out that contemporary computer systems are still incapable of producing "virtual worlds" that would be of high enough quality to be actually useful. Fortunately many technologies developed with VR in mind, proved to be adaptable to applications where the simulated elements of human visual stimuli are merely enhancing the real world perception, thus giving birth to Augmented Reality (AR) systems. This field of research is quite young and advanced dynamically, so it is not easy to provide a set of formal definitions of AR (these has been formulated – see e.g. [1] and [2], but not completely consistent), however for most practical purposes AR can be understood as a technique of overlying a realtime video feed (from a camera or via some kind of half-transparent screen) with computer generated imagery that is supposed to more or less seamlessly integrate with this feed.

Practical examples of such systems have been implemented in many different fields, such as military (helmet mounted displays modern combat aircraft), education, cinematography (virtual camera system used in production of the Avatar movie), or even advertising. Finally navigation and information systems (such as prototype campus information system created in the Columbia University [3]) have been created with this technique.

G. Lee, D. Howard, and D. Ślęzak (Eds.): ICHIT 2011, CCIS 206, pp. 545–551, 2011.

Augmented reality technology is especially attractive when combined with wearable computing and mobile devices. Wearable computing were – for quite a long time – a domain of costly and dedicated hardware platforms. Typical such AR system (including these mentioned above) would therefore consist of visual systems (cameras and displays), sensors and detectors required to "connect" the virtual world with the real one, and computers powerful enough to run the software. However, recent advances in mobile consumer electronics, such as creation and proliferation of powerful mobile devices such as *smartphones* allow to create such systems with COTS components, or – in some cases as demonstrated by our prototype system – even entirely in software.

Contemporary smartphones are not just expensive gadgets, but extremely powerful mobile computers. Newest models are equipped with a main processor with 1 GHz clock rate, which performance is greater than 10 MFLOPS. They also usually have a dedicated graphical chipset and hundreds of megabytes of RAM memory. What is more, nearly all smartphones have a camera, a GPS modules, accelerometers and magnetic field sensors [4].

Most existing mobile AR applications for mobile platforms, that could be classified as navigation and information systems, (like Wikitude World Browser [5] and Layar [6]) are however very simple. The main idea of such application can be summarized as follows: a user looks at his surrounding using his device camera and sees on the display not only real objects, but also some additional, virtual text boxes or markers, which provides information about what them. Using this kind of applications may be found much more convenient for finding nearby places than using traditional systems which are based on digital maps.

2 Augmented Reality for Personal Navigation

Mobile AR applications mentioned above have a real great potential in terms of providing information. However, it is not difficult to notice that their navigational functionality is very limited: only few of the currently available mobile AR applications offer a point-to-point (POI to POI) navigation; the majority of them can just track the location of the only one (final) point.

The only exception is AugSatNav created by the Phyora group [7], which is a car navigation system. What differs it from other similar software (e.g. Google Maps or NaviExpert) is the fact that it can draw the whole way to a destination point using the augmented reality technology. Another AR navigation application, called Wikitude Drive, is being prepared by Mobilizy. The project started in the middle of the year 2009 and was supposed to be ready before the end of 2010.

There is no reason why mobile AR navigation systems should be only targeted at users who travel by car. The other group of people that could benefit from such kind of systems are users of public transport. Although there are several AR applications which help travelling through a city (e.g. Nearest Tube for iPhone [8], which shows locations of underground stations in London), none of them has the ability of planning a journey and none of them offers a point-to-point navigation.

The aim of the project described in that paper was to investigate how augmented reality can enhance travelling speed and convenience in an urban environment. The market is already overcrowded with car navigation systems, so it was decided not to compete with them but rather focus only on alternative ways of travel, i.e. by means of a public transport and on foot. In order to be a fully-operative navigation system, the application should have the ability of finding the best way to the chosen destination. It should also use AR markers to guide a user point-to-point (stop to stop), so that he always knows where exactly he should go and which mean of transport is most appropriate.

The system, which received a codename 'NavAR', was implemented as an application for the Android platform. All routes calculations are performed by the jakdojade.pl service. Jakdojade.pl is a web site, which allows planning journey with public transport; currently, it supports eight major cities in Poland (available from: http://jakdojade.pl/ as per December 2010).

3 Prototype Application

Consider the following scenario. The user equipped with a smartphone with the NavAR application wants to travel to a place in the other part of the city. He only knows the address of that place, but does not know its exact location. He types the city name together with the street name and number into the NavAR search panel. While he is typing, the application connects to a Google location service in order to provide him with list of all known places that match the specified address. The user selects the place which he means and chooses what optimization technique the application should use to find the best routes, including selecting fastest and/or most convenient connections (e.g. minimizing walking distance).

The NavAR search module connects via Internet (GPRS, 3G connection or Wi-Fi) to the jakdojade.pl service and fetches information about the most appropriate routes. Each route is represented as the list of lines and each line consists of the list of stops. The last stop of every line has exactly the same coordinates as the first stop of the line that follows it. Walking parts of routes are treated as ordinary lines, which mean of transport is described as 'walk'. The user reviews the found connections and chooses the one that will be used for navigation.

The user holds the device in front of his eyes and looks at the camera image visible on the display. As he looks around, he can notice up to three virtual markers (presented on figure 1) that shows him the exact location of key points at the route:

- the closest stop, i.e. the line stop to which the user should travel directly;
- the last stop of the current line (right marker in Fig. 2), i.e. the place where the user should change the mean of transport;
- the route destination (checked black and white marker), i.e. the place the user wants to reach.

Fig. 1. The augmented reality view generated by the NavAR application. The right marker indicates the position of the last stop of the current line and the checked black and white marker indicates the position of the destination location.

The number of rendered markers may be less than three in a situation when two or three tracked key points are identical, e.g. when the closest stop is also the last stop of the current line. The user is also provided with information about distances to each of these locations and when he selects one of them (by touching an appropriate marker), he gains an access to the location details: stop name, information about the line which should be used to reach the location (name, direction and type), names and types of alternative lines which can be used instead of the main line, estimated time of arrival to the location and estimated time of departure from the location.

When the system detects approaching one of the tracked key points, it informs the user about that event and automatically starts showing guidelines to a new location. It also displays special messages, when an action is required from the user, e.g. when he should change the line (see figure 3) he is travelling with. When the user reaches the route destination, he is prompted if he wants to exit navigation or continue until he decides himself that he knows exactly where he is.

The main part of the NavAR application is the augmented reality engine. The purpose of that software module is to perform a real-time conversion of the geographical data about the positions of the user and a tracked location into screen coordinates of an AR marker. The engine uses four different input signals:

- coordinates of tracked object, received from an external data source;
- coordinates of the user, received from a GPS module;
- azimuth, i.e. the angle between the direction the device camera is pointing at and the geographical North, calculated using data from magnetic field sensors;
- roll, i.e. the angle of rotation of the device around the axis of the camera, calculated using data from accelerometers;
- inclination, i.e. the angle between the direction the device camera is pointing at and the plane of Earth, calculated using data from accelerometers.

The application uses the coordinates of the device and the tracked object to calculate the bearing value. The x coordinate of an AR marker is calculated according to the formula 1. The origin of the used coordinate system is placed in the top left corner of the screen. The X axis points right and the Y points down.

$$x = \frac{bearing - azimuth}{\frac{camera\ horizontal\ viewing\ anle}{2}} * \frac{horizontal\ screen\ size}{2} + \frac{horizontal\ screen\ size}{2} \qquad (1)$$

The y coordinate is calculated in a similar way, according to the formula 2.

$$y = \frac{inclination}{\frac{camera\ vertical\ viewing\ anle}{2}} * \frac{vertical\ screen\ size}{2} + \frac{vertical\ screen\ size}{2} \qquad (2)$$

The last step of the algorithm is multiplying the screen coordinates vector with a rotation matrix. The rotation value is equals to the device roll and the rotation pivot is in the screen centre, i.e. it is the point with coordinates

$$\left(\frac{horizontal\ screen\ size}{2}, \frac{vertical\ screen\ size}{2} \right). \qquad (3)$$

4 Evaluation and Conclusion

Field tests of the NavAR system were performed in Warsaw, Poland; it was used for travelling on foot, by bus, by tram and by underground. The device used for tests was HTC Desire smartphone (known also as T-Mobile G1), which was powered by the Android 1.6 platform.

Virtual markers generated by the NavAR augmented reality engine merge coherently with real-world objects captured by the device camera and they have only a limited tendency towards oscillations. When a device moves slowly, all AR markers change their positions smoothly and follow objects which are tracked by them. After a rapid turn of a device, it may be noticed that virtual markers need some time (up to 1.5 seconds) to fix their position on the screen. The delay is caused by the low-pass filter which is used to remove noise from signals generated by accelerometers and magnetic field sensors.

The augmented reality engine of another mobile application, Wikitude World Browser (version 9.04, the newest release available in October 2010) works in a complete different way on the test platform. AR markers rarely keep their position and have a strong tendency to oscillations. When a device is in move, all virtual objects change their position randomly what completely disrupt the coherence between them and real objects. When a device is held still again, it takes up to 2.5 seconds for virtual markers to fix their position.

The other top market application, Layar (version 4.0, the newest release available in October 2010), has performed much better on the test hardware. AR markers generated by it keep exactly their position when a device is still and fix immediately their position after a rapid move. The only noticed problem occurs when a device is moved slowly: virtual markers do not move smoothly but with visible breaks. The most probable reason of that behaviour is the fact that the Layar engine is not as lightweight as the one used by the NavAR system, so the test device has problems with the performance.

The NavAR system was used on many routes in different parts of the city and every time it allowed reaching chosen destinations in a fast and convenient way. Although it is very difficult to keep a constant eye contact with a mobile device display, this is not required to use the AR navigation system: it is enough to make a quick glance once every few seconds. Apart from that, using application with an augmented reality interface on a crowded pavement is much safer than navigating with a traditional map-based software, because it lets controlling the way in front of the user and reading navigation guides simultaneously.

The average accuracy of the information about the device position acquired from a GPS module in an urban environment is usually better than 16 m and worse than 8 m. Such values are enough for an effective navigation and automatically detection of approaching one of tracked stops.

However, the system does not work correctly in all conditions. The device determines its orientation by means of accelerometers, so the signals are disturbed when the smartphone is influenced by other accelerations type than gravitational, e.g. when a bus breaks or turns rapidly. When such situation happens, user may see that all AR markers "run out" from the screen and come back to their original position after a few seconds.

The application has also major problems when a user wants to travel by underground. First of all, the external data provider, jakdojade.pl, treats all stops as single geographical points. That approximation is good enough for bus and tram stops, which do not occupy large areas, but does not work at all for underground station, which (in Warsaw) have up to 300 m length. As a result of that data inaccuracy the system sometimes may suggest to its user that he should avoid the nearest entrance to a station and go to another one on the opposite side. Apart from that, there is no signal from GPS satellites in underground stations, so the system may rely only on location information estimated using the identifier of the nearest base transceiver station. The accuracy of such information is often worse than 1000 m, so it cannot be used for precise navigation.

The NavAR system in its current form remains a prototype which, while useful, is not yet ready for commercial deployment. Main issue that remains to be solved, and which will be worked upon by NavAR team, is positioning accuracy and determining user location in places where GPS signal and base station triangulation cannot be used (e.g. underground). Planned approaches include filtering sensor data and measuring device acceleration in order to improve virtual markers stability, especially while travelling on board of a bus or tram and implementing WiFi based localization together with position prediction system using detailed maps of underground station layouts.

References

1. Milgram, P., Kishino, F.: A Taxonomy of Mixed Reality Visual Displays. IEICE Transactions on Information Systems E77-D(12), 1321–1329 (1994)
2. Milgram, P., Takemura, H., et al.: Augmented Reality: A Class of Displays on the Reality-Virutality Continuum. In: Proceedings of SPIE: Telemanipulator and Telepresence Technologies, vol. 2351, pp. 282–292 (1994)

3. Feiner, S., Macintyre, B., Höllerer, T.: A touring machine: Prototyping 3d mobile augmented reality systems for exploring the urban environment. In: 1st IEEE International Symposium on Wearable Computers, pp. 74–81. IEEE Computer Society, Los Alamitos (1997)
4. The Smartphone Hub. Android 2.2 sets Android 2.1 on fire. From the friction. Because it's so fast (2010) http://thesmartphonehub.com/tag/mflops/ (May 16, 2010)
5. Mobilizy. Wikitude (2010), http://www.wikitude.org (April 21, 2010)
6. Layar. Augmented Reality Browser: Layar (2010), http://layar.com (April 2, 2010)
7. Phyora. Handcrafted Mobile Goodness (2009),
   http://geotact.appspot.com/augsatnav.html (October 9, 2010)
8. Acrossair. Nearest Tube Augmented Reality iPhone 3GS App,
   http://www.acrossair.com/acrossair_app_augmented_reality_
   nearesttube_london_for_iPhone_3GS.htm (October 10, 2010)

Best Recommendation Using Topic Map for On-Line Shopping System

HwaYoung Jeong and BongHwa Hong[*]

[1] Humanitas College of Kyunghee University. Hoegi-dong, Seoul, 130-701, Korea
hyjeong@khu.ac.kr
[2] Dept. of Information and Communication, Kyunghee Cyber University. Hoegi-dong,
Seoul, 130-701, Korea
bhhong@khcu.ac.kr

Abstract. Topic map is generally used to recommend shopping list as analyze a user preference in internet shopping mall. It is useful method to propose the preference that is related access rate and selecting time of production. In this research, we apply this method to analyze the user characteristics in on-line shopping system. To analyze user preference, we used the topic preference vector that is able to calculate the access rate with visiting time and production selection. And we designed and developed the path of on-line shopping site. It will help to navigate the shopping by recommendation of on-line shopping list.

Keywords: U-learning system, Ubiquitous computing, Learning system, Web service.

1 Introduction

Recent findings by industry analysts such that online consumers are impatient, easily dissatisfied and are likely to abandon their shopping carts and move to a different retailer if a site's features fail to meet their expectations [1]. When studying online shopping behavior, we should bear in mind that the Internet as a shopping channel is competing with the long established in-store channel and that the former has not yet reached the "mainstream "status that many had predicted. Despite steady continuous growth of online shopping, retailers are still concerned about the low conversion rates on their websites. But Moe and Fader [4] have highlighted that the low conversion rate from visits on retailers 'websites may be explained by the fact that the cost of browsing and searching for product information online is lower than doing these same activities of fline. Consequently, the consumer searches more frequently and intensely online than of fline. Furthermore, a retailer should not be concerned if a consumer decides to browse its website but does not buy online since the consumer may then decide to buy the product in-store [3]. By the Forrester Consulting report [2], however, also finds that attributes such as fast loading of pages, ease of navigation, efficient search and detailed product content are some of the features that online

---

[*] Corresponding author.

G. Lee, D. Howard, and D. Ślęzak (Eds.): ICHIT 2011, CCIS 206, pp. 552–559, 2011.
© Springer-Verlag Berlin Heidelberg 2011

consumers expect from retail sites and decrease the likelihood that consumers will leave sites without making purchases. In addition, investigations by scholars of online consumer behavior suggest that diversified selection, convenience, customization, availability of information, ability to interact with the site and efficiency of the shopping process are important to online consumers [1]. Brown et al.[8] find seven shopping styles [1]

(1)Personalizing: these shoppers look for retailers that provide personalized service.

(2)Recreational: shoppers who shop for the "pleasure of shopping "(Brown et al.,2003,p.1674).

(3)Economic: these consumers are interested in finding the lowest prices.

(4)Involved: consumers whose style includes multiple styles such as personalizing, recreational and economic.

(5)Convenience oriented recreational: these are shoppers who are primarily interested in the convenience of shopping but secondarily in the pleasure of shopping as well.

(6)Community oriented: shoppers who prefer to buy from the online sites of retailers that are local to where they live.

(7)Apathetic convenience oriented: these consumers buy online only for convenience but do not enjoy the shopping process.

By this the styles, the propensity to consume of user influence the propensity to purchase of production in on-line shopping mall. That is directly means success or failure whatever the shopping mall is. Therefore administrator or developer for the on-line shopping wants to know what the user wants or likes the production. So Knowledge management (KM) becomes a critical issue for the on-line shopping. Topic Map (TM) is a new standard, emerging as one of the structures for the semantic web. It is built on the ISO 13250 Topic Maps (TMs). TMs are used to organize information in a way that can be optimized for navigation [9].

In this paper, we proposed the recommendation using TM for on-line shopping mall. To analyze the user preference of the propensity to purchase, we make a block that is deal with process distribution of the user's purchase history.

2 Related Works

There is now ample evidence that the Internet has changed the way in which consumers purchase goods and services. Consumers are not interested in the technology per se but how the Internet can improve their shopping experiences and help them make better decisions. Some consumers have adopted the Internet for shopping whilst others have not. Many in the former group only use the Internet to search for product information and subsequently buy in-store [3]. Therefore, it is very important factor to know exactly what the user wants to purchase in on-line shopping mall. Fig. 1. shows an example of on-line shopping mall. The users define electronic shopping cart use as an online behavior in which a consumer places item(s) of interest in an online shopping cart.

Fig. 1. An example of on-line shopping mall

The motivations for these synergistic and at times opposing online behaviors (placing an item in a cart and purchasing during that session versus abandoning the purchase) may be different, as theories of motivation and online consumer behavior suggest [5].

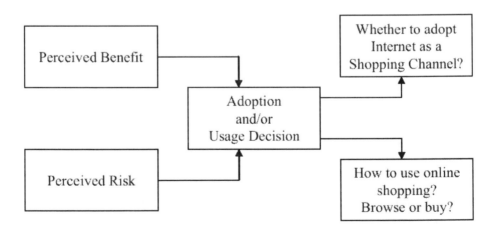

Fig. 2. Conceptual framework for adoption and usage behavior of online shopping

As the one hand of this approach, Didier and Alastair [3] developed the framework to apply in marketing, where behavioral attitude and beliefs are assumed to be predictors of behavior as shown Fig. 2.

The other method can be MBA(Market basket analysis). MBA refers to investigating the composition of the basket of products purchased by a household during a single shopping experience. Retailers have long been interested in learning about the cross-category purchase behavior of their customers, since such information makes it easier for the retailer to decide whether to group products by brand or by product type, for example. The market basket choice refers to the decision process in which a consumer selects items from several product categories during the same shopping experience. Therefore, basket analysis can provide the distribution of shoppers 'purchases based on different viewpoints, such as the product itself, the product category, the shopper's background, or the average purchases per shopper. Such distribution information will aid decisions on aspects such as planning and designing advertising, sales promotions, store layout, and product placement [6].

The other method will be Clickstream. Clickstream can be defined as the path a consumer takes trough one or more websites. It can include within-site information such as the pages visited, time spent on each page and between-site information such as the websites visited. Thus, researchers have investigated consumer behaviors across websites and within a particular website. In the latter category, some studies focused on single visits to a particular website, on multiple visits or on both types of visits. Within-website research has focused on clickstream or website related variables that help explain the goal pursued by consumers who visit a website, why consumers continue browsing on a website and which visitors are likely to make a purchase. Thus far, no study has investigated the different consumers' decision-making processes within one specific type of visit, namely directed-purchase visits [7].

To analyze the knowledge management of user's the propensity to purchase, TM was used. TM defines a standardized notation for interchangeably representing information about the structure of information resources used to define topics, and the relationships between topics [9]. Topic maps provide a bridge between the domains of knowledge representation and information management and link them to existing information resources. The basic concepts are shown as following.

- Topics correspond to the formal description of concepts or objects in the real world. An expressive topic name is helpful to understand an object or a concept. Topic types are naturally created to classify topics.
- Occurrences are the original information resource connected to the meaningful topic names. A topic is an abstract label, and the occurrences are substantial references.
- An association is formally a meaningful link that specifies a relationship among several topic names. These semantic association relationships help users understand the connections among topic names.

Topic map standardization can provide a clear structure in assisting an enterprise to organize knowledge from different information resources and to build a knowledge-sharing environment for users to gain knowledge. Topic maps are used to construct a navigable knowledge map of composite e-services [10].

3 Recommendation System for On-Line Shopping Mall Using Topic Map

This paper has focus to support recommendation to the user who purchase the some production using on-line shopping mall by internet. Existence on-line shopping mall system support purchase history list as shown in Fig 3.

Fig. 3. An example of user's purchase history list

We used the user's purchase history information as user's preference using topic map. The diagram explains the structure of the recommend system as shown in Fig. 4. The structure consists of four blocks; *TM, Analysis of purchase history, Analysis of user preference, and Make a recommendation.* Each block has database, *Analysis of purchase history* used *User's private information, User's history information, and Product information* to analyze user's propensity to purchase from previous to now a day. After process of *Analysis of user preference* finished, the system is able to make a recommendation for next purchase order with user preference of product information, that the process used *Product information* database.

Fig. 5. shows user's interesting product from user's the propensity to purchase with purchase history list. If the interesting product list was made by the process of *Analysis of user preference*, then the process of *Make a recommendation* make a decision for recommendation using user's the propensity to purchase, such as refer to product review, purchase count and so on.

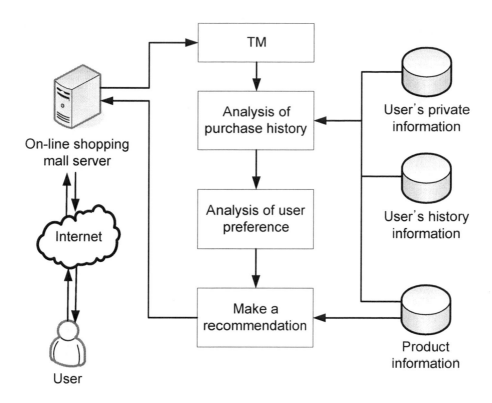

Fig. 4. The structure of proposed recommendation system for on-line shopping mall

Fig. 5. The structure of proposed recommendation system for on-line shopping mall

Fig. 6. shows the system suggests final recommendation among these product list.

· 상품만족도 : **100%**   · 후기/리뷰 : **82**

Fig. 6. The final recommendation for purchase product in on-line shopping mall system

4   Conclusion

In this paper, we focused recommendation system in on-line shopping mall. For this process, the system had four blocks; *TM, Analysis of purchase history, Analysis of user preference, and Make a recommendation. Analysis of purchase history* was used to analyze user preference information and user's history information. Application for this research, we could show the final recommendation from some product. The decision was refer to product review, purchase count and so on.

References

1. Papatla, P.: Do online shopping styles affect preferred site attributes? An empirical investigation and retailing implications. Journal of Retailing and Consumer Services 18, 362–369 (2011)
2. Forrester Consulting, eCommerce Web Site Performance Today: An Updated Look At Consumer Reaction To A Poor Online Shopping Experience (August 17, 2009)
3. Soopramanien, D.G.R., Robertson, A.: Adoption and usage of online shopping: An empirical analysis of the characteristics of "buyers" "browsers" and "non-internet shoppers. Journal of Retailing and Consumer Services 14, 73–82 (2007)

4. Moe, W., Fader, P.S.: Dynamic conversion behaviour at e-commerce sites. Management Science 50(3), 326–335 (2004)
5. Close, A.G., Kukar-Kinney, M.: Beyond buying: Motivations behind consumers' online shopping cart use. Journal of Business Research 63, 986–992 (2010)
6. Yang, T.-C., Lai, H.: Comparison of product bundling strategies on different online shopping behaviors. Electronic Commerce Research and Applications 5, 295–304 (2006)
7. Senecal, S., Kalczynski, P.J., Nantel, J.: Consumers' decision-making process and their online shopping behavior: a clickstream analysis. Journal of Business Research 58, 1599–1608 (2005)
8. Brown, M., Pope, N., Voges, K.: Buying or browsing? An exploration of shopping orientations and online purchase intention. European Journal of Marketing 37(11/12), 1666–1684 (2003)
9. Dong, Y., Li, M.: HyO-XTM: a set of hyper-graph operations on XML Topic Map toward knowledge management. Future Generation Computer Systems 20, 81–100 (2004)
10. Liu, D.-R., Ke, C.-K., Lee, J.-Y., Lee, C.-F.: Knowledge maps for composite e-services: A mining-based system platform coupling with recommendations. Expert Systems with Applications 34, 700–716 (2008)

Pattern and Event Based Logical UI Modeling for Multi-Device Embedded Applications

Saehwa Kim

Hankuk University of Foreign Studies, Yongin-si, Gyeonggi-do, 449-791 Korea
ksaehwa@hufs.ac.kr

Abstract. While there have been many research activities for model-based engineering for user interfaces (UIs) for multiple devices, they are still far from practical. We identify three major limitations for the conventional approaches. We propose a Pattern and Event based Logical UI Modeling framework (PELUM) to model UIs targeted for multiple embedded systems. PELUM encompasses (1) a pattern-based method for deriving a UI implementation from a UI model, (2) a meta-model for modeling both abstract UI and task model, whose name is Logical User Interface Model (LUM), and (3) its supporting tool. We incorporate events as a first-class modeling entity, essential to model UIs for embedded systems. PELUM does not employ the task model widely used in the conventional approach. Instead, we incorporate events as navigators to cover the task model in LUM. We also incorporate patterns as types of each LUM component. This makes UI modeling concise.

Keywords: User interface modeling, model-driven architecture (MDA), embedded systems, pattern-based transformational UI development, multi-device/plastic UI.

1 Introduction

The proliferation of various smart embedded devices, including smart phones, tablets, navigators, and smart TVs, has lead to the ever increasing needs for applications that are adaptable for multiple devices. One of the important bottlenecks for increasing such reusability is that the user interfaces (UIs) for multiple devices cannot be the same. Accordingly, there have been many research activities to provide a model-based user-interface development environment (MB-UIDE) or model-driven engineering (MDE) of user interfaces for multiple devices. They are listed in some survey work [3, 4].

While these approaches have some variances, most of them share a hierarchical model-driven architecture, as shown in the first column in Table 1, where there are five hierarchical models. The higher models are more concrete compared to the lower models. In the view point of OMG MDA (model-driven architecture) [5], the higher model is the platform specific model (PSM) and the lower model is the platform independent model (PSM). The lowest model in Table 1, the observed model, itself is not the part of UI models. However, it plays the role of base model for UI modeling.

G. Lee, D. Howard, and D. Ślęzak (Eds.): ICHIT 2011, CCIS 206, pp. 560–567, 2011.

The highest model, the final UI, is the executable UI implementation, and thus much previous research did not include this model in the framework. The task model is usually modeled with a concurrent task tree (CTT) [6]. Abstract UI models abstract UI components, such as inputs and outputs. Representative modeling languages for this are UMLi [7], UsiXML [2], and Canonical AUI [8]. The concrete UI maps the abstract UI components in the abstract UI to platform specific UI controls or widgets and specific location of some layout.

However, as [9-11] addressed, these previous research activities received criticisms about their practicality. While there were many different viewpoints for the inherent causes of the limitation of these approaches, we determined the following as dominant limitations.

- (L1) Modeling the abstract UI (AUI) models tends to be tedious and redundant work, since designers should model all the abstract UI components corresponding to all tasks in the task model.

- (L2) The abstract UI (AUI) model [1, 2, 8] is unsuitable for modeling embedded systems, since AUI cannot model events that are not associated with visible UI components.

- (L3) The task model [1, 2, 6] is instable [12] and unpredictable [9].

The first limitation (L1) was noticed by many pattern-based UI development approaches [13, 14] . While these research activities provide various reusable patterns and pattern-based languages, the difficult formality of these languages create a steep learning curve [9].

The second limitation (L2) is mainly due to the focus of conventional approaches on modeling UIs for web or PC (personal computer) together with mobile embedded systems [15, 16]. Events in embedded applications are much more diverse than that of web or PC-based applications. Events of web or PC-based applications are usually coupled with any kind of presentations or visible UI components. However, embedded systems, especially mobile embedded devices, have various kinds of sensors, such as gyroscope, accelerometer, compass, proximity sensor, ambient light sensor. The events generated by such sensors usually do not have corresponding visible UI components. However, these events play a very important role in constructing UIs especially in the control parts in the MVC (Model, View, and Controller) pattern.

Table 1. Comparison of model-driven architecture for UI modelling

Conventional Approaches (TERESA [1], CAMELEON [2], and etc.)	PELUM
Final UI (FUI)	Graphical Resource Model (GRM)
Concrete UI (CUI)	UI Controls and Layout Model (CLM)
Abstract UI (AUI)	Logical UI Model (LUM)
Task & Concept (Task Model)	
Observed Model	Programming Interface Model (PIM)

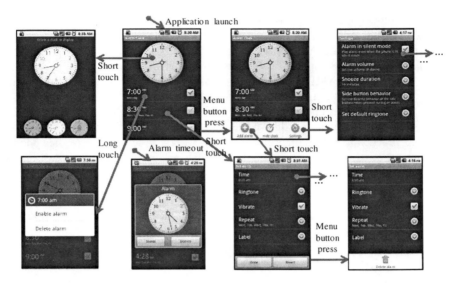

Fig. 1. A walk-through example: an Android alarm application targeted for a cell phone of HVGA with 160 ppi.

The third limitation (L3) is because tasks are anyhow related with input or output of data and if the content of data is changed, the task model should be changed accordingly [12]. In addition, the task model is not available at run-time, the links between the FUI and its original task model are lost [9]. With such observations, [9] focused on modeling or supporting flexibility at run-time, while [12] introduced the so-called interaction model similar to UML use-case models. However, the approach of [9] still leaves how to provide an appropriate initial UI to end-users an open problem, while the interaction model of [12] is too abstract to replace the task model.

This paper proposes a Pattern and Event based Logical User-interface Modeling framework (PELUM) for modeling UIs targeted for multiple embedded systems. PELUM encompasses (1) a pattern-based method for deriving a UI implementation from a UI model, (2) a meta-model for modeling both abstract UI and task model, whose name is Logical User Interface Model (LUM), and (3) its supporting tool.

First, our method for deriving UI implementations is based on the template-based UI transformation method in [17]. We associate each pattern in LUM with an implementation template. Here we also propose MDA-based hierarchical models as shown in the second column in Table 1, whose detail will be explained in Section 2. Second, we propose the Logical User Interface Model (LUM) that solves the limitations (L1)-(L3) mentioned above. Specifically, LUM adopts the following approaches, where each approach is to solve each limitation (L1)-(L3).

- (A1) LUM incorporates patterns, as types of each LUM component.
- (A2) LUM incorporates (abstract) events, as a first-class modeling entity.
- (A3) LUM incorporates events, as navigators.

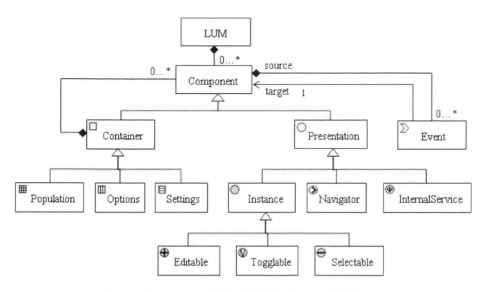

Fig. 2. Meta-model of PELUM LUM (Logical UI Model)

The approach (A1) enables LUM to be concise. (A2) enables modeling events that do not have visible UI components. (A3) enables LUM implicitly cover the task model. We will explain details of these approaches in Section 3.

Finally, we fully implemented the modeling tool for LUM as an Eclipse plug-in using the Eclipse Graphical Modeling Framework (GMF). We performed a case study with an alarm application in Android targeted for multiple embedded systems, such as a cell phone of HVGA with 160 ppi that is the same as that of Apple iPhone 3G, a 7 inch tablet device with 170 ppi that is the same as that of Samsung galaxy tab, and XGA with 130 ppi, the same as that of Apple iPad. We use an Android alarm application for HVGA with 160 ppi, as a walk-through example shown in Fig. 1.

The remainder of this paper is organized as follows. Section 2 explains our pattern-based method for deriving UI implementations from UI models. Section 3 presents the proposed logical user interface model. Section 4 concludes the paper.

2 Pattern-Based Method for Deriving UI Implementations from UI Models

The second column in Table 1 shows the proposed hierarchical models. As shown, we merged the AUI and task model into the Logical UI Model (LUM), while the names of other models were changed to explicitly emphasize the modeling entities in each specific modeling layer. Specifically, we renamed the observed model as the programming interface model (PIM), since this layer is usually given as API (Application Programming Interface) according to the specific platform, such as Android API level 7 or level 8. We also renamed concrete UI as UI Controls and Layout Model (CLM), since this layer is specifically to decide what kind of UI

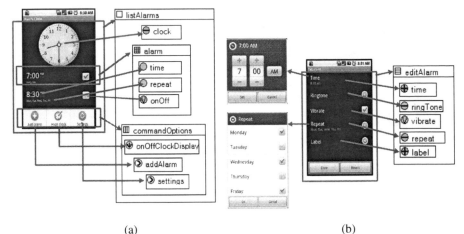

(a)   (b)

Fig. 3. LUM examples for an Android alarm application showing (a) containers of Container, Population and Options and (b) a container of type Settings

controls or widgets are used for each abstract UI element and how they are located with a specific layout. Finally, we renamed the final UI as the Graphic Resource Model (GRM), since this layer provides specific graphical resources for each UI element. In GRM, [18] can be integrated for automatic graphical rendering specialized in specific targeted devices.

Each UI component in LUM can be incorporated with a specific type that corresponds to a specific UI pattern [13]. We prepare GRM and CLM templates for each target device for each pattern in LUM. We can easily derive run-time flexible UI targeted for multi-devices, applying the template-based UI transformation method in [17] to this.

3   Logical User Interface Model (LUM)

In this section, we present the proposed logical user interface model (LUM). As explained in Section 1, LUM adopts three approaches (A1)-(A3) to solve the limitations (L1)-(L3). We explain each of them in the following subsections.

3.1   Incorporating Patterns as Types of LUM Components

Fig. 2 shows the proposed LUM meta-model. LUM is composed of two types of components, Container and Presentation. Container can contain any Component: that is, Container can also contain another Container. Presentation is for visible UI components, similar to the abstract UI components in AUI [2]. Along with Container and Presentation themselves, each child type of Container and Presentation corresponds to a pattern provided by LUM.

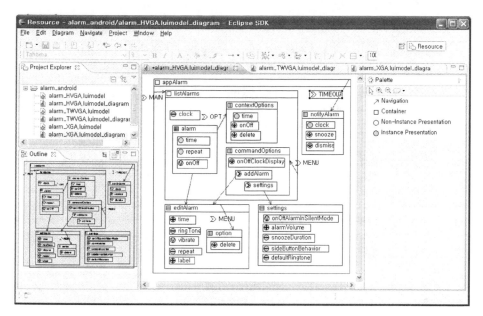

Fig. 4. LUM example for the walk-through example of Fig. 1 modeled in PELUM tool

The type names are self-explanatory. Fig. 3 shows an example usage of these types for the walk-through example in Fig. 1. Type `Population` was inspired by the population presentation pattern of JUST-UI [19]. In PELUM, a population represents a collection of instances of a class in PIM (programming interface model). As shown in Fig. 3 (a), we do not need to model repeated `alarm` items using `Population`. By also making the container `editAlarm` in Fig. 3 (b) be `Settings,` we do not need to model "Done" and "Revert" buttons, since the template in CLM (UI Control and Layout Model) will contain these items. Containers of `InternalService` are usually linked to an API of PIM, whose example is `onOffClockDisplay` in container `commandOption` in Fig. 3 (a). As shown in Fig. 3 (b), using `Editable` and `Selectable` types for `time`, `label`, `ringtone`, and `repeat,` we do not need to model all LUM components by specifying a specific type, since CLM will have templates for these required components. This makes LUM concise.

3.2   Incorporating Events as a First-Class Modeling Entity

LUM incorporates events as the first-class modeling entity. As shown in Fig. 2, the LUM meta-model explicitly contains `Event` as a modeling entity. Note that `Event` is not a child of `Component` but a compositional member of `Component`. This solves limitation (L2), since LUM can represent events that are not associated with visual UI components. As there are, and will be emerging, various kinds of sensors in embedded systems, events also drastically vary among different kinds of embedded devices. Thus, we do not model these events as device-specific but model as abstract events. Templates in CLM we presented in Section 2 map these abstract events to specific device-specific events.

3.3   Incorporating Events as Navigators

We do not employ the task-model usually adopted by conventional approaches. As mentioned in [20], the task model is not good at modeling the transition of dialogues. Our view is that the dominant reason for this is due to the task model being the lower model of AUI. We model events together with AUI components, by employing events as navigators with associations with LUM components. As the LUM meta-model in Fig. 2 shows, each `Event` has two associations with `Component`, whose role names are `source` and `target`. The key concept is that the `source` component of an event receives the event and the `target` component appears at the occurrence of the event.

Fig. 4 shows how LUM models the transitions of dialogues in the walk-through example in Fig. 1, modeled in our tool based on Eclipse GMF. Event `MAIN` from container `appAlarm` to container `listAlarms` represents that `listAlarms` appears when application `appAlarm` is launched. Event `TIMOUT` from container `appAlarm` to container `notifyAlarm` represents that `notifyAlarm` will appear when event `TIMOUT` is triggered at any time. Conversely, event `OPT` from `alarm` to `contextOptions` represents that `contextOptions` will appear only when event `OPT` is received by container `alarm`. Note that the event `OPT` is 'long touch' in Fig. 1. As such, LUM models abstract events and the templates in CLM will map these abstract events to specific events in the target device.

4   Conclusions

While there have been many research activities for model-based engineering for user interfaces (UIs) for multiple devices, they are still far from practical [9-11]. We identified three major limitations (L1)-(L3) for the conventional approaches. We proposed three approaches (A1)-(A3) to solve these limitations. We proposed a Pattern and Event based Logical UI Modeling framework (PELUM) to model UIs targeted for multiple embedded systems, by adopting these approaches. PELUM encompasses (1) a pattern-based method for deriving a UI implementation from a UI model, (2) a meta-model for modeling both abstract UI and task model, whose name is Logical User Interface Model (LUM), and (3) its supporting tool.

PELUM incorporated events as a first-class modeling entity, essential for modeling UIs for embedded systems. PELUM did not employ the task model widely used in the conventional approach. Instead, PELUM incorporated events as navigators to cover the task model in LUM. PELUM also incorporated patterns as types of each LUM component, which makes UI modeling concise.

Acknowledgments. This work was supported by the Hankuk University of Foreign Studies Research Fund of 2010.

References

1. Mori, G., Paterno, F., Santoro, C.: Design and Development of Multidevice User Interfaces through Multiple Logical Descriptions. IEEE Transactions on Software Engineering 30 (2004)
2. Calvary, G., Coutaz, J., Thevenin, D., Limbourg, Q., Bouil-lon, L., Vanderdonckt, J.: A Unifying Reference Framework for Multi-Target User Interfaces. Interacting with Computers 15, 289–308 (2003)

3. Pinheiro da Silva, P.: User interface declarative models and development environments: A survey. In: Paternó, F. (ed.) DSV-IS 2000. LNCS, vol. 1946, p. 207. Springer, Heidelberg (2001)
4. Pérez-Medina, J.-L., Dupuy-Chessa, S., Front, A.: A survey of model driven engineering tools for user interface design. In: Winckler, M., Johnson, H. (eds.) TAMODIA 2007. LNCS, vol. 4849, pp. 84–97. Springer, Heidelberg (2007)
5. Model Driven Architecture (MDA). Object Management Group
6. Mori, G., Paterno, F., Santoro, C.: CTTE: Support for Developing and Analyzing Task Models for Interactive System Design. IEEE Transactions on Software Engineering 28, 797–813 (2002)
7. de Silva, P.P., Paton, N.W.: User Interface Modeling in UMLi. IEEE Software 20, 62–69 (2003)
8. Constantine, L.L.: Canonical Abstract Prototypes for Abstract Visual and Interaction Design. In: Proceedings of International Workshop on Design, Specification, and Verification of Interactive Systems (2003)
9. Coutaz, J.: User interface plasticity: model driven engineering to the limit! In: Proceedings of ACM SIGCHI Symposium on Engineering Interactive Computing Systems (2010)
10. Vanderdonckt, J.: Model-driven engineering of user interfaces: Promises, successes, failures, and challenges. In: Proceedings of Annual Romanian Conference on Human-Computer Interaction, pp. 1–10 (2008)
11. Collignon, B., Vanderdonckt, J., Calvary, G.: Model-driven engineering of multi-target plastic user interfaces. In: Proceedings of International Conference on Autonomic and Autonomous Systems (2008)
12. Lu, X., Wan, J.: User Interface Design Model. In: Proceedings of the ACIS International Conference on Software Engineering, Artificial Intelligence, Networking and Parallel/Distributed Computing (2007)
13. Borchers, J.O.: A Pattern Approach to Interaction Design. In: Proceedings of Conference on Designing interactive systems: processes, practices, methods, and techniques (2001)
14. Pribeanu, C., Vanderdonckt, J.: A Transformational Approach for Pattern-based Design of User Interfaces. In: Proceedings of International Conference on Autonomic and Autonomous Systems (2008)
15. Paternò, F., Santoro, C., Spano, L.D.: Model-based design of multi-device interactive applications based on web services. In: Gross, T., Gulliksen, J., Kotzé, P., Oestreicher, L., Palanque, P., Prates, R.O., Winckler, M. (eds.) INTERACT 2009. LNCS, vol. 5726, pp. 892–905. Springer, Heidelberg (2009)
16. So, P.H.J.C.P.L., Shum, P., Li, X.J., Goyal, D.: Design and Implementation of User Interface for Mobile Devices. IEEE Transactions on Consumer Electronics 50 (2004)
17. Aquino, N., Vanderdonckt, J., Pastor, O.: Transformation templates: adding flexibility to model-driven engineering of user interfaces. In: Proceedings of ACM Symposium on Applied Computing (2010)
18. Meskens, J., Vermeulen, J., Coninx, K.L.K.: Gummy for Multi-Platform User Interface Designs: Shape me, Multiply me, Fix me, Use me. In: Proceedings of the Working Conference on Advanced Visual Interfaces (2008)
19. Molina, P.J., Melia, S., Pastor, O.: Just-UI: A User Interface Specification Model. In: Proceedings of Computer-Aided Design of User Interfaces (2002)
20. Caffiau, S., Scapin, D., Baron, P.G.M., Jambon, F.: Increasing the expressive power of task analysis: systematic comparison and empirical assessment of tool-supported task models. Interacting Computers (2010)

Smart Device Oriented Intelligent Context-Awareness System for U-SAFE Service Using Object Oriented Modeling Technique

Se-Hoon Jung[1] and Chun-Bo Sim[2]

[1] Master's Course, Dept. Of Multimedia Engineering, Sunchon National University,
Suncheon Jeonnam 540-742, Rep. of Korea
iam1710@hanmail.net
[2] Associate Professor, Dept. Of Multimedia Engineering, Sunchon National University,
Suncheon Jeonnam 540-742, Rep. of Korea
cbsim@sunchon.ac.kr

Abstract. Smart phone technology has the potential to improve crime figures by preventing certain crimes altogether or diminishing their frequency or severity. Elementary school sexual assault incidents as well as female serial murder cases have captured the public's attention, and when facing such social hazards, the user with a smart mobile device that is equipped with fusion technology may evade physical harm or evade the danger altogether. This paper discusses the components of context awareness system for this purpose and describes a putative implementation on a phone. The u-SAFE service aims to fight crime when it inputs phone sensing information (audio, video, image) and performs context acquisition and analysis, and pattern analysis of data tempered by knowledge of the user's activity history.

Keywords: Context-Awareness, Ontology, Data-mining, u-SAFE, UML.

1 Introduction

What two cruel murder cases, *'Jo Du Soon'* and *'Kim Gil Tae'*, that most shocked us recently, have in common is that they targeted female victims especially for children. Sex-related crime as sexual attack in elementary schools or serial killers targeting women is creating serious social problems. Consider that a mobile device empowered by software-hardware and sensors that fuse complex IT decision making into the wireless communication, now in the hands of a potential victim, has the potential to deter, stop or diminish the danger or damage that results from such crimes.

In other words, an environment recognition technology taking advantage of environment values (facial information, brightness, sounds, images, time, location and activity status and others) means a technology which enables users to receive a distinctive and proactive service by recognizing individual situations before and after incidents automatically when in an emergency[1]. Furthermore, the incredible market

G. Lee, D. Howard, and D. Ślęzak (Eds.): ICHIT 2011, CCIS 206, pp. 568–575, 2011.
© Springer-Verlag Berlin Heidelberg 2011

growth in the smart phone market offers the required wide spread availability of environment recognition capabilities and the high potential of deployment of all environmental sensing and sensor fusion technologies.

For these purposes, this paper intends to design a system by which users are able to escape from any dangerous situation. This system is a Context Aware Service that enables users to solve any emergency. It comprises a Context Sensing phase that is based on historical information as sensed by the phone.

We use UML to produce an object oriented specification for such a system and consider issues such as future maintenance after deployment.

2   Previous Work

Study [2] proposed a paradigm addressing issues of re-usability and scalability for software by securing re-usability and independence of component modules. In other words, it suggested that requirement analysis, re-usable component model, re-using existing model and strategy for purchasing components via a component distribution market all of these became highlighted through the life cycle in overall development.

Study [3] designed and proposed the so called Context awareness-based u-Silver care service encompassing a Context aware Agent with multi-agents and a modeling toolkit for context awareness. For these purposes, the study proposed a Context aware Agent phase which manages and analyzes context information acquired from surrounding sensors and agents, a modeling toolkit phase which defines various context awareness including users from context information defined at ontology and context awareness modeling information application phase.

Study [4] proposed an ontology-based context aware system for the ubiquitous environment. This study creates a context ontology using an ontology standard language, OWL to generate context information and infer context and share knowledge. In addition this study uses SWRL rule language to create a device context information employing rule inferring-based, Jess to provide the created device context with diagnostics and repairing services, and it proposed the use of Jess Tab API for inter-connectivity with the OWL-based context ontology.

3   Proposed System

3.1   Overview for the System

This paper proposes our so-called "u-SAFE Service system" for context aware applications which is designed based on object-oriented UML diagrams using data mining and ontology techniques to process context awareness.

Fig. 1 shows a construction plan of mobile context-awareness. And Fig. 2 shows a schematic overview for the system to be proposed in this paper.

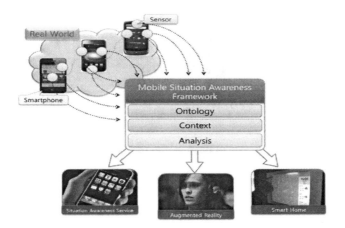

Fig. 1. Construction plan of mobile context-awareness

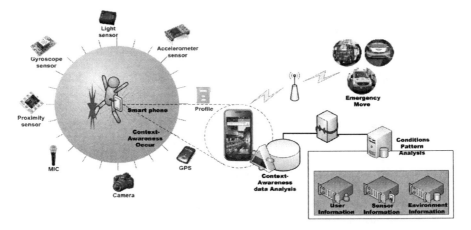

Fig. 2. System configuration for context aware system to be proposed

3.2   Design of Context-Awareness System for U-SAFE

3.2.1   Design of Context-Awareness

This system comprises: A) Sensors (location, images, sounds, temperature, humidity, brightness, contact, recognizing approaching objects and acceleration to surrounding information for user's emergent situation), B) smart phone records (receiving and sending records, history of SMS and LMS and application usage history), C) user created data and personal information (images, video clips, schedule, phone book).

A context aware information are converted into high level of meaning information although they could conceivably be directly used. For women who are weak at smart phone usage and when they are in an emergency situation, the system defines the user's activities and corresponding data which includes objects within the user's environments in the device.

Fig. 3 shows categories for technological structure to recognize the surrounding context value. Fig. 4 shows a configuration for processing an ontology and a data mining method to analyze patterns in the u-SAFE system.

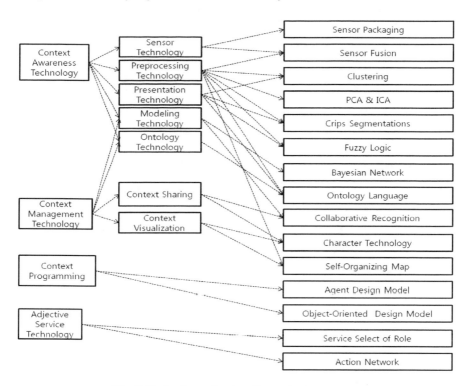

Fig. 3. Technology structure for context-awareness

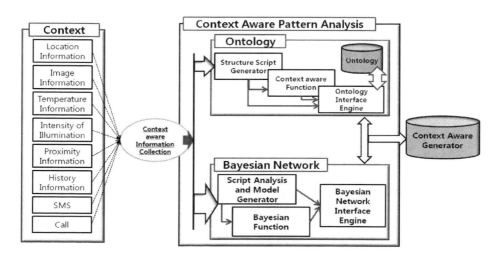

Fig. 4. Structure for context-awareness system

3.2.2 Design of Ontology and Data-Mining

The u-SAFE system receives surrounding context values to check any dangerous situation from sensor, history data and static data and then it checks and infers whether or not there is any data pattern based on ontology and Bayesian networks.

Fig. 5 shows context-awareness agent structure.

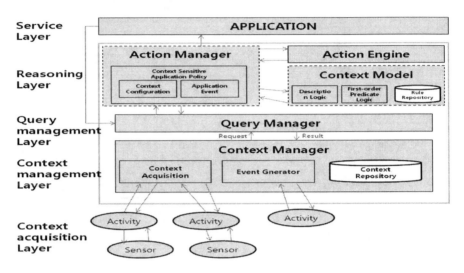

Fig. 5. Context-awareness agent structure using ontology

Fig. 5 shows a configuration for processing an ontology and a data mining method to analyze patterns from our u-SAFE system. Context aware system approach through a Bayesian network for data expresses surrounding context values only with high causality in connection with existing data acquired from emergent situation using methods which compose probable model effectively and are able to infer and learn contexts. Assuming inter-dependency for unexpressed relations, this system uses the Bayesian network structure which expresses probability distribution by defining only conditional probability from direct causality.

Formula 1 represents a combination rule for a basic Bayesian network inference to combine new data coming from historical and sensor data [5].

$$P(X, Y) = P(X|Y)P(Y) = P(Y|X)P(X)$$

$$P(X`|Y`) = P(X`), P(Y`|X`) = P(Y`) \tag{1}$$

Where X` and Y` $\in\in$ independent

Formula 2 represents a binary tree feature to make data inference tree structure using Marginalization feature of conditional probability.

$$P(Y) = \sum P(Y, x_i) = P(Y, x_1) + P(Y, x_2) + \dots + P(Y, x_n) \tag{2}$$

$$= P(Y, x_1) + P(Y, \overline{x_1})$$

3.3 Object-Oriented Modeling Design

We devised the UML-based system considering re-usability and scalability. This system consists of context aware, context analysis, pattern application and command, that suggests a diagram for the context awareness module.

Fig. 6 enumerates system functions extracted from user needs and represents usecase diagram for obtaining context aware.

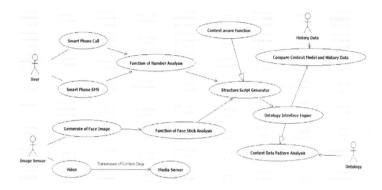

Fig. 6. Usecase diagram for obtaining static situation context-awareness

Fig. 7 is the sequence diagram expressing the flow of user cases that is encapsulated in Fig. 6 as interactions between objects. Fig. 7 captures the dynamics of the exchange of messages between such objects.

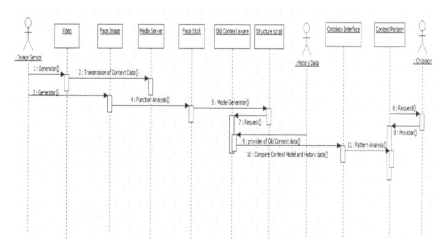

Fig. 7. Sequence diagram for obtaining static situation context

3.4 Ontology Design

Ontology design to implement analysis process of situation recognition system to be proposed by this paper is following.

Fig. 8 shows not a subjective part of the system to be expressed by information, but designing subjective facts.

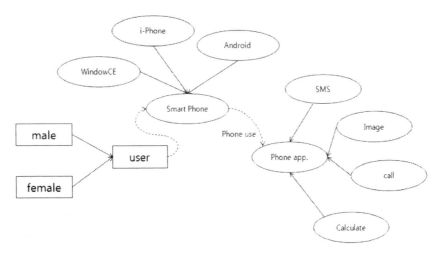

Fig. 8. Express of Information

Fig. 9 shows an expression of situation recognition in which determines system proceeding after comparing surrounding situation data obtained from sensors based on situation recognition pattern analyzed by learning inference with existing situation recognition data.

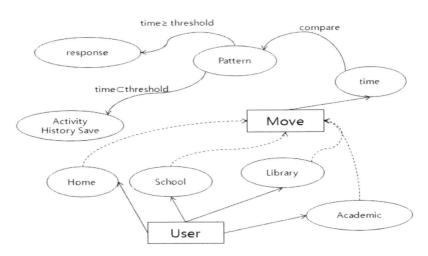

Fig. 9. Express of Context Awareness

4 Conclusion

This paper proposed an Object-oriented UML-based user intelligent situation context aware system. This system intends to postulate functions that enable users to escape from dangerous situations. Invoking the u-SAFE system by compares and analyzes environment data and existing History Activity data that originates from the surrounding situations without any user's physical reaction.

For this purpose, the situation recognition context is defined through pattern analysis using context acquisition of surrounding environments (facial information, voice, image and brightness), data-mining-based Bayesian network inference technology and ontology inference technology. We have postulated a design for our u-SAFE service system for users based on the situation recognition context defined through pattern analysis and from existing history activity data pattern analysis.

And this system is designed based on specification by means of an object oriented UML diagram to minimize maintenance costs depending on stabilized system design and modification and scalability after implementation.

In further thinking about this topic it will became necessary to enhance reliability for reactions to real time-based situations for users.

Acknowledgments. "This research was supported by the MKE(The Ministry of Knowledge Economy), Korea, under the ITRC(Information Technology Research Center) support program supervised by the NIPA(National IT Industry Promotion Agency)" (NIPA-2011-(C1090-1121-0009)).

References

1. Chen, G., et al.: A Survey of Context Aware Mobile Computer Research, Darmouth Computer Science Tech, Report TR2000-381 (2000)
2. Choi, B.-K., Youn, H.-Y.: u-Silvercare based on Context-awareness. Korean Institute of Information Scientists and Engineers 36(3), 200–207 (2009)
3. Benko, C., McFalan, F.W.: Connecting the dots-Aligning Project with Objectives in unpredicatable times, pp. 85–117. Havard Business School Press, Boston (2003)
4. Kwon, S.-H., Park, Y.-T.: Ontology Based Context Aware System for Ubiquitous Environment. Korean Institute of Information Scientists and Engineers 34(1), 281–286 (2007)
5. Shin, D.-J.: Development of Context awareness and service reasoning technique on cellular phone., Master's degree Thesis, Chung-Ang University, Rep. Korea, Appendix: Springer-Author Discount (2009)

An Effective Method to Find Better Data Mining Model Using Inferior Class Oversampling

Hyontai Sug

Division of Computer and Information Engineering, Dongseo University,
Busan, 617-716, Korea
hyontai@yahoo.com

Abstract. Decision trees are known to have very good performance in the task of data mining of classification, and sampling is often used to determine some proper training sets. Among many parameters the accuracy of generated decision trees depends upon training data sets much, so we want to find some better classification models from the given data sets by oversampling the instances that have higher error rates. The resulting decision trees have better accuracy for classes that had lower error rates, but have worse accuracy for classes that have higher error rates. In order to take advantage of the better accuracy and compensate the worse accuracy, we suggest using class association Experiments with real world data sets showed promising results.

Keywords: Decision trees, classification, biased sampling, association rules.

1 Introduction

The task of data mining usually contains incomplete or imperfect data so that there is always the possibility that the trained knowledge model might predict poorly for future or unseen cases. Many data mining algorithms have been suggested to deal with the problem, and among them decision tree algorithms have been studied by many researchers to cope with the problem [1, 2].

Decision tree algorithms have the property that fragment a training data set into many subsets corresponding to terminal nodes. The algorithms split the training data set based on how likely the subsets become purer or are more predictable for a class, so that each training instance becomes to belong to a specific terminal node. But the property of decision tree building might not be good for some data sets, because we often might not have a complete data set, even though the data sets are very large. Since we usually do not have a complete data set for training, some heuristic-based pruning strategies are applied to the generated tree to avoid data over-fitting problem.

Another good point of decision trees is understandability in the condition that the generated trees are not so large, because the structure of decision tree is represented in symbolic form. This good point of understandability of decision trees makes it easy to use the decision trees with some other data mining method. On the other hand, some other representative data mining algorithms like neural networks [3, 4, 5] do not give us much margin for us to use the neural networks with some other data mining

G. Lee, D. Howard, and D. Ślęzak (Eds.): ICHIT 2011, CCIS 206, pp. 576–583, 2011.
© Springer-Verlag Berlin Heidelberg 2011

method, because neural networks have no symbolic representation for their knowledge representation.

We are especially interested in decision trees, because their knowledge structures are represented in symbolic representation, and it is highly possible that they may be used with some other symbolic data mining algorithms like association rules.

In section 2, we provide the related work to our research, and in sections 3 we present our method. Experiments were run to see the effect of the method in section 4. Finally section 5 presents some conclusions.

2   Related Work

Because of the understandable structure of decision trees we may want to use them for the case that we need to understand the structure of trained knowledge models. There have been a lot of efforts to build better decision trees to minimize the error rate with respect to available data sets. Among them C4.5 [6] and CART [7] might be two representative decision tree algorithms, because the two algorithms are based on very different ideas in tree building process and frequently referred in literature [8]. C4.5 uses an entropy-based measure to split branches based on feature values, and the measure selects the most certain split among possible splits of candidate features. So classes that have more certain splits with respect to entropy are preferred. CART uses a purity-based measure, and the algorithm splits the training data set based on how probably the subsets become purer for a class, and it tries to generate more optimized decision trees, so CART generates relatively smaller trees than C4.5. A good point of decision tree algorithms is their scalability because they need relatively shorter computing time than other data mining algorithms like neural networks.

Because decision tree induction method decisively divides the target data sets, the behavior of trained decision tree models are very dependent on the training data set. So, we can infer that the trained knowledge model will be dependent on sample size as well as the composition of data in the samples. Fukunaga and Hayes [9] discussed the effect of sample size for parameter estimates in a family of functions for classifiers. Some interesting fact is that even neural networks are suffered much from imbalanced data sets. In [10] the authors showed that class imbalance in training data has effects in neural network development especially for medical domain. SMOTE [11] used synthetic data for the effect of over-sampling in minority class and showed improved performance in decision trees.

Association rules [12, 13] indicate how often a set of items occur together in transactional databases. These rules give information on association patterns that exist in the databases. Many good algorithms were suggested to find association rules efficiently. Apriori algorithm uses candidate frequent patterns for efficiency [12], and DHP algorithm uses hash-table to find short frequent patterns [14], and in [15] random sampling method was used to find frequent patterns more, in [16] tree structure was used to find association rules more efficiently. Other related work is multidimensional association rules. Multidimensional association rules are basically an application of general association rule algorithms to table-structured databases. If the table has condition attributes and decision attributes, the found association rules are called class association rules (CAR). In papers like [17, 18], class association rules have better accuracy than decision trees for most of example data in small size. They used small data sets, because association algorithms may generate, potentially, a large number of rules and most of the rules are useless. Therefore, some rule selection based on interest measure [19, 20] and generalization is important [21, 22]. Hybrid-dimension association rules are a generalization technique to reduce the number of association rules [23].

3 The Method

Many data sets for data mining have some imbalanced distribution with respect to class values, and this fact can be easily checked if we sort them with respect to classes. Moreover, if we generate decision trees, we can easily check which classes are more accurate in prediction by inspecting confusion matrices.

The method first builds a decision tree. Then, we inspect the number of misclassified objects for each class, and we choose classes that should be over-sampled. Let's call the class inferior class. Over-sampling the instances in the inferior classes might generate worse data set for training. So the decision tree algorithms will try to treat instances in other classes more importantly, and let's call it superior class. Therefore, it is highly possible that the decision tree will be built to cover instances in superior classes more importantly so that the prediction accuracy for the instances will be better than the decision trees from the original training data set. The following is a brief procedure of the method.

After generating decision tree with over-sampled data, we generate conditional association rules for the inferior classes. The parameters for CAR are minimum support and minimum confidence. By inspecting the accuracy of inferior classes of the generated decision tree, we can determine the minimum support. In order to determine an appropriate minimum confidence, we may use domain knowledge.

```
procedure (Output):
  Begin
  /* X, MS, MC: parameters */
  1. Generate a decision tree;
  2. Inspect the accuracy of each class to determine
  over-sampling for some classes; /*inferior classes*/
  3. Do sampling of X % more for the classes
  4. Generate a decision tree for the new sample set;
   /* do discritization first to find CARs */
  5. Generate CARs for inferior classes with MS & MC;
  6. Use the decision tree and CAR to predict;
  End.
```

In the above procedure there are three parameters to be defined, X, MS, and MC. MS represents minimum support to find class association rules. MC represents the minimum confidence of the CAR to find. The values of minimum support and minimum confidence can be dependent on the accuracy of the decision tree for the inferior classes and the domain of the data set.

4   Experimentation

Experiments were run using data sets in UCI machine learning repository [24] called 'adult' [25] and 'statlog(Landsat satellite)' [26] to see the effect of the method. The number of instances in adult data set is 48,842, and the number of instances in statlog data set is 6,435. The data sets were selected, because adult data set may represent business domain and statlog data set may represent scientific domain. The total number of attributes is 14 and 36, and there are two classes and six classes for adult statlog data set respectively. There are six continuous attributes for adult data set, and all attributes are continuous attributes for statlog data set. We used C4.5 [6] to generate decision trees. We did sampling of relatively small sizes for the experiment to reserve relatively large testing data sets. For adult data set sample size of 1,920 is used, and for statlog data set sample size of 480 is used. All the remaining data are used for testing. Before we sample in the above mentioned sample sizes, we sampled the sample size of 1,600 for adult data set, and the sample size of 400 for statlog data set. For each sample size seven random sample data sets were drawn. Table 1 and table 2 show misclassification ratio of each data set for each class for the case that we generate decision trees. All data in the data sets are used to generate the decision trees to determine which class objects should be sampled more.

Table 1. Misclassification ratio for each class of adult data set

Class	Misclassification ratio
>50K	36.9%
≤50K	4.6%

Table 2. Misclassification ratio for each class of statlog data set

Class	Average misclassification ratio
1	0.4%
2	0.4%
3	1.5%
4	6.4%
5	4.1%
6	2.1%

So, 20% more objects were sampled from the object pool of '>50K' class for adult data set, and 10% more objects were sampled for each class 4 and 5 for statlog data set. The following table 3 thru 4 shows average error rate of the decision trees of seven samples for adult and statlog data set. Table 3 shows average error rate of the decision trees with for adult data set.

Table 3. Average error rate for each class of adult data set samples with inferior class over-sampling(ICOS) and conventional sampling

Class	Sampling method	
	ICOS	Conventional
>50K	39.8%	26.4%
≤50K	9.0%	13.2%

If we look at table 3, we can notice that we can find better classification accuracy for superior class and worse classification accuracy for inferior class.

Table 4 shows the result of experiment for statlog data set with inferior class over-sampling. Note that class 4 and 5 has been chosen as inferior classes.

Table 4. Average error rate for each class of statlog data set samples with inferior classes (4, 5) over-sampling(ICOS) and conventional sampling

Class	Sampling method	
	ICOS	Conventional
1	8.1%	9.8%
2	7.7%	9.1%
3	12.1%	15.4%
4	58.6%	54.3%
5	30.1%	28.3%
6	14.0%	22.0%

If we look at table 4 carefully, we can notice that we have some similar result with the result of adult data set.

So we could find some better classification accuracy for superior classes, and some worse classification for inferior classes. The next step is to compensate the classification for inferior classes. We used CAR algorithm to find good classification association rules. For discritization MDL-based method is used [27]. The following rule is an example class association rule for adult data set among 14,426 un-generalized rules.

If marital-status=Married-civ-spouse and capital-gain=$'(5119{\sim}\infty)'$ Then

class= '>50K' confidence:1 (54/54);

The numbers in (54/54) after the confidence means the number of instances that belong to the condition and conclusion. The rules came from the first training data set of adult data set of size 1,600. The given minimum support is 0.5625% that is based on 9 instances in a terminal node of '>50K', and minimum confidence is 90%. The part of corresponding decision tree for the parameters looks as Fig. 1. Similar procedure to generate CARs can be applied to statlog data set.

capital-gain = '(-∞~5119]'
| marital-status = Married-civ-spouse
| | education-num = '(-∞~12.5]'
| | | capital-loss = '(-∞~1881.5]': <=50K (457.0/129.0)
| | | capital-loss = '(1881.5~1938]': >50K (**9.0**)
| | | capital-loss = '(1938~∞)': <=50K (6.0/1.0)
. . . . . .
| | capital-loss = '(1938~∞)': >50K (3.0)
| marital-status = Married-spouse-absent: <=50K (23.0/2.0)
| marital-status = Married-AF-spouse: <=50K (1.0)
capital-gain = '(5119~∞)': >50K (71.0/3.0)

Fig. 1. Decision tree

5 Conclusions

Because data mining tasks usually do not have complete data sets for training, we may want to resort to some heuristics to overcome the problem. Decision trees are considered to be one of the best data mining tools, because we can get some understandable knowledge structures. On the other hand, decision tree algorithms suffer from data fragmentation problem, because a training data set is divided into terminal nodes thru many subtrees and branches. This problem can become worse when the available data sets are well not consisted of good premises for classification. On the other hand, class association rule finding algorithms can find rule exhaustively within given parameters of minimum support and confidence. But it may require a huge computing time and generate a lot of rules unless we give some limitation, so we need some method to make the input of the association rule finding algorithm be more compact.

We propose a method to find better classification models using decision trees and class association rules. We first generate decision trees to determine if there is relatively higher number of errors depending on classes, then we sample more for the classes, and decision trees are generated for the over-sampled data sets. If the trees from the over-sampled data sets have worse accuracy for over-sampled classes and better accuracy for the other classes, we find class association rules for the over-sampled classes, and use the class association rules to compensate the worse accuracy of the decision tree with over-sampling. This can reduce the training data set size for class association rules. Experiments with two real world data sets in business and scientific domain give us the conclusion that we can we can find better data mining models effectively.

References

1. Tan, P., Steinbach, M., Kumar, V.: Introduction to Data Mining. Addison-Wesley, Reading (2006)
2. Russel, S., Novig, P.: Artificial Intelligence: a Modern Approach, 2nd edn. Prentice-Hall, Englewood Cliffs (2002)

3. Bishop, C.M.: Neural networks for pattern recognition. Oxford University Press, Oxford (1995)
4. Heaton, J.: Introduction to Neural Networks for C#, 2nd edn. Heaton Research Inc. (2008)
5. Lippmann, R.P.: An Introduction to Computing with Neural Nets. IEEE ASSP Magazine 3(4), 4–22 (1987)
6. Quinlan, J.R.: C4.5: Programs for Machine Learning. Morgan Kaufmann Publishers, Inc, San Francisco (1993)
7. Breiman, L., Friedman, J., Olshen, R., Stone, C.: Classification and Regression Trees. Wadsworth International Group (1984)
8. Larose, D.T.: Data Mining Methods and Models. Wiley Interscience, Hoboken (2006)
9. Fukunaga, K., Hayes, R.R.: Effects of Sample Size in Classifier Design. IEEE Transactions on Pattern Analysis and Machine Intelligence 11(8), 873–885 (1989)
10. Mazuro, M.A., Habas, P.A., Zurada, J.M., Lo, J.Y., Baker, J.A., Tourassi, G.D.: Training neural network classifiers for medical decision making: The effects of imbalanced datasets on classification performance. Neural Networks 21(2-3), 427–436 (2008)
11. Chawla, N.V., Bowyer, K.W., Hall, L.O., Kegelmeyer, W.P.: SMOTE: Synthetic Minority Over-sampling Technique. Journal of Artificial Intelligence Research 16, 341–378 (2002)
12. Agrawal, R., Mannila, H., Srikant, H.R., Toivonen, H., Verkamo, A.I.: Fast Discovery of Association Rules. In Advances in Knowledge Discovery and Data Mining. In: Fayyad, U.M., Piatetsky-Shapiro, G., Smith, P., Uthurusamy, R. (eds.), pp. 307–328. AAAI Press/The MIT Press (1996)
13. Zaki, M.J.: Scalable algorithms for association mining. IEEE Transactions on Knowledge and Data Engineering 12(3), 372–390 (2000)
14. Park, J.S., Chen, M., Yu, P.S.: Using a Hash-Based Method with Transaction Trimming for Mining Association Rules. IEEE Transactions on Knowledge and Data Engineering 9(5), 813–825 (1997)
15. Toivonen, H.: Discovery of Frequent Patterns in Large Data Collections. phD thesis, Department of Computer Science, University of Helsinki, Finland (1996)
16. Han, J., Pei, J., Yin, Y., Mao, R.: Mining frequent patterns without candidate generation. Data Mining and Knowledge Discovery 8, 53–87 (2004)
17. Li, W., Han, J., Pei, J.: CMAR: Accurate and Efficient Classification Based on Multiple Class-Association Rules. In: Proceedings 2001 Int. Conf. on Data Mining (ICDM 2001), pp. 369–376 (2001)
18. Liu, B., Hsu, W., Ma, Y.: Integrating Classification and Association Rule Mining. In: Proceedings of the Fourth International Conference on Knowledge Discovery and Data Mining (KDD 1998), pp. 80–86 (1998)
19. Toivonen, H., Klemettinen, M., Mannila, H., Rokainen, P., Hatonen, K.: Pruning and Grouping of Discovered Association Rules. In: Workshop Notes of the ECML 1995 Workshop on Statistics, Machine Learning and Knowledge Discovery in Databases, pp. 47–52 (1995)
20. Dimitrijević, M., Bošnjak, Z.: Discovering Interesting Association Rules in the Web Log Usage Data. Interdisciplinary Journal of Information, Knowledge, and Management 5, 191–207 (2010)
21. Klemettinen, M., Mannila, H., Ronkainen, P., Toivonen, H., Verkamo, A.I.: Finding Interesting Rules from Large Set of Discovered Association Rules. In: Proceedings of the Third International Conference on Information and Knowledge Management (CIKM 1994), pp. 401–407 (1994)
22. Perng, C., Wang, H., Ma, S., Hellerstein, J.: Discovery in Multi-attribute Data with User-defined Constraints. ACM SIGKDD Explorations Newsletter 4(1), 56–64 (2002)

23. Chithra, R., Nicklas, S.: A Novel Algorithm for Minng Hybrid-Dimensional Association Rules. International Journal of Computer Applications 1(16), 53–58 (2010)
24. Suncion, A., Newman, D.J.: UCI Machine Learning Repository. University of California, School of Information and Computer Sciences, Irvine, CA (2007), http://www.ics.uci.edu/~mlearn/{MLR}epository.html
25. Kohavi, R.: Scaling up the accuracy of Naive-Bayes classifiers: a decision-tree hybrid. In: Proceedings of the Second International Conference on Knowledge Discovery and Data Mining, pp. 202–207 (1996)
26. Statlog (Landsat Satellite) Data Set, http://archive.ics.uci.edu/ml/datasets/Statlog+%28Landsat+Satellite%29
27. Fayyad, U.M., Irani, K.B.: Multi-interval discretization of continuous valued attributes for classification learning. In: The Proceedings of Thirteenth International Joint Conference on Artificial Intelligence, pp. 1022–1027 (1993)

A New Approach for Calculating Similarity
of Categorical Data

Cheng Hao Jin[1], Xun Li[1], Yang Koo Lee[1], Gouchol Pok[2], and Keun Ho Ryu[1]

[1] Department of Computer Education, Chungbuk National University, Cheongju,
Republic of Korea
{kimsungho,lixun,leeyangkoo,khryu}@dblab.chungbuk.ac.kr
[2] Department of Computer Science, Yanbian University of Science and Technology,
Yanji, China
gcpokyust@gmail.com

Abstract. Similarity measure is very important in data mining techniques such as clustering, nearest-neighbor classification, outlier detection and so on [1][4]. There are many similarity measures have been proposed. For numeric data, there are many Minkowski distance-based similarity measures. However, the similarity measures for categorical data have been studied for a long time, it also has many issues. The main issue is to understand relationship between categorical attribute values. For categorical data, the similarity measure is not clear as well as numeric data. In this paper, we propose a new approach to understand relationship between categorical data. This approach is based on artificial neural network to extract significant features for computing distance between two categorical data objects.

Keywords: Categorical data, Artificial Neural Network, Similarity Measure, Significant Feature.

1 Introduction

Many researchers were interested in studying similarity measure for data objects. Though many similarity measures have been proposed, there are still many problems. The main issue is to understand relationship between attribute values. For numeric data, there are many Minkowski distance-based similarity measures. For categorical data, measuring its similarity between two objects is a very difficult task [6] since the similarity between categorical data is not straightforward as well as numeric data and mainly depends on the specific domain. Therefore, each similarity measure is designed to fits into the characteristic of the specific data set.

In this paper, we propose a novel method to extract the significant features by measuring categorical data similarity. This method is based on artificial neural network (ANN) model. We use this model to extract the significant features from categorical data during artificial neural network testing process and calculate the similarity between two categorical data objects. In this paper, we applied proposed method to k-means clustering to evaluate the goodness of our method.

G. Lee, D. Howard, and D. Ślęzak (Eds.): ICHIT 2011, CCIS 206, pp. 584–590, 2011.
© Springer-Verlag Berlin Heidelberg 2011

2 Related Work

Study of similarity measure has a long history. A lot of similarity measures were proposed for calculating similarity of categorical data. These similarity measures can be divided into two groups. The one is for numeric data and another one is for categorical data.

2.1 Similarity Measurement for Numeric Data

For numeric data, the notion of similarity is clear. The similarity between two numeric data objects is defined as distance of them. The Minkowski distance is one of the most famous similarity measures for calculating distance between two numeric data objects. Especially Euclidean distance and hamming distance are two widely used similarity measure in many data mining techniques such as nearest-neighbor classifier, support vector machine and clustering [1][4][6].

Hamming distance can be used for both numeric and categorical dataset. For numeric data, it also called Manhattan distance. In Manhattan distance, the distance between two points in a grid that is based on a strictly horizontal and vertical path is as opposed to the diagonal or as the crow flies distance. The Manhattan distance is the simple sum of the horizontal and vertical components, whereas the diagonal distance might be computed by applying the Pythagorean Theorem.

Euclidean distance is the straight-line distance between two points which are located in Euclidean space. Because of its efficiency in calculating distance between two vector data objects, it is applied to many data mining techniques; they transform the values of categorical attributes into a series of numbers and calculate their similarity.

2.2 Similarity Measurement for Categorical Data

Categorical data has been studied for a long time in various contexts [1]. Sneath and Sokal discussed categorical similarity measures in detail in their book [7]. The main concerns in the book were biological relevance, and computation efficiency.

There are several measures for similarity of categorical data [6][8]. The chi-square test which was proposed by Pearson is used to test the independence between categorical attributes. Chi-square statistic was modified and extended, leading to several other measures. One of commonly used similarity measure for categorical data is overlap. Overlap similarity measure simply counts the number of attributes that match in the two categorical dataset. The range of the similarity value for overlap is (0, 1). When one of the attribute values match, similarity between two attributes is 1; when no attribute values match, similarity between two attributes is 0.

Overlap measure is the most widely used similarity measure and the popularity of overlap measure is related to its simplicity and ease of use [1]. There are many other similarity measures modified and extended from overlap.

3 Proposed Approach

The proposed approach is based on Artificial Neural Network. For calculating similarity of categorical data, we applies Artificial Neural Network model to extract significant features which are used to compute categorical data similarity.

3.1 Concepts of Artificial Neural Network

The study of artificial neural network is inspired by attempts to simulate biological neural systems [9]. The simplest structure of artificial neural network is called perceptron. This structure consists of two layers which are called input layer and output layer. Each layer includes some nodes which have their own parameters such as sum, value and so on. Usually, the number of input nodes is equal to the number of attributes and the number of output nodes is equal to the number of class labels.

Artificial neural network can have a more complex structure than perceptron model. The artificial neural network model can contain several intermediary layers between input layer and output layer. Such intermediary layers are called hidden layer and the nodes embedded in these layer are called hidden nodes. An example of a multilayer feed-forward artificial neural network is shown in Fig. 1. In a feed-forward neural network, the nodes in one layer are connected only to the nodes in the next layer. These intermediary layers allow model to have more complex relationships between the input layer and output layer.

Commonly, the multilayer feed-forward artificial neural network has one hidden layer [2][3][9]. In multilayer feed-forward artificial neural network, each node in one layer is only connected to the nodes in the next layer. Every node in each layer has its own parameters such as sum, value and so on. The sum is a parameter that is calculated by a formula which is shown as equation, where n is the number of income connections, w and x are the weight and value of node connected by the income connection and k is the index of connections and the values of nodes. Usually the w0x0 is a constant which shows the bias of them.

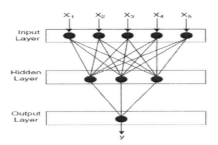

Fig. 1. Example of a Multilayer Feed-forward Artificial Neural Network

3.2 Extracting Significant Features Using Artificial Neural Network

Artificial Neural Network is known as an efficient classification algorithm. Artificial neural network model transmits information from one layer to the next layer. Information is transformed by weighted connection which connects two nodes in different layers.

In fully connected multilayer artificial neural network model, all of information that leaves from one layer is centralized into each node in the next layer. Hence, each node's information is compressed data of the whole information of the previous layer from connections. This information also includes the characteristics of the next layer.

Traditional artificial neural network model has at least one hidden layer. Hidden layer is located at the middle of input layer and output layer. Information of input nodes is centralized into each hidden node and the information of hidden node is transmitted to each output node. Therefore, hidden nodes catch the information of attributes and class label. For calculating similarity between two categorical data, we require some significant features which only take the information of data attributes. We designed artificial neural network structure at which the number of input nodes is the same as that of output nodes. To make the hidden nodes have only information of data attributes, we make the values of output nodes the same as to the input nodes. In this artificial neural network structure, the class label of data set does not need to apply to the ANN model during the step of learning data set.

In order to apply the categorical data into proposed artificial neural network, the values of each categorical attribute should be transformed to numeric format by data preprocessing. There are many transformation functions to give categorical values some identified weights.

The preprocessed data is applied to Artificial Neural Network as training dataset to build a classifier. After that, an Artificial Neural Network model is built from training dataset. We apply the data which is used as training dataset to the obtained model. However, in this method, the Artificial Neural Network model is not used to predict the class label as well as traditional Artificial Neural Network do. The classifier model is used to extract the features from target dataset. So the values of output nodes are not any significance in proposed method. In our method, it catches the value of hidden nodes. The value of each hidden node is come from its sum, and the sum is come from all of the input nodes. The value of each hidden node has the information of all the input nodes. In other words, the value can represent the characteristic of each data instance. Process of extracting significant features is shown as Fig. 2.

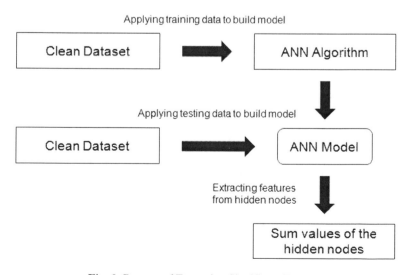

Fig. 2. Process of Extracting Significant Features

4 Experimental Results

In order to evaluate our method, we choose a test data ADULT from UCI data set. ADULT data has 14 attributes that consist of 8 categorical attributes and 6 numeric attributes. The data is grouped by two classes which one is over 50,000 and the other one is less 50,000. Our method is an unsupervised learning so that the class label is not used in the process of extracting features. However, it would be used to get the accuracy and entropy to evaluate our method.

Because the dataset has some numeric attributes which have different domains, this dataset requires an additional data preprocessing task to make these attributes into one of the same domain. For making the dataset to get more efficiency in the ANN model, we make the values of numeric attributes into range of -1 to 1, because the activation function is effective in the range of -1 to 1. The transformation function is shown as equation (1), where the j, i are the index of the attributes and values, min(j) and max(j) are the minimum and maximum value in j^{th} attribute.

$$Nvalue(ji) = \frac{2 \times Ovalue(ji) - (\min(j) + \max(j))}{\min(j) + \max(j)} \tag{1}$$

After finishing this step, we performed the data preprocessing task. In this task, the data instances that have missing values were eliminated and the categorical attributes were assigned a series of numbers from a simple formula is shown as equation (2), where the j, i are the index of the attributes and values, N(j) is the number of values in j^{th} attribute.

$$value(ji) = 2 \times \frac{i}{N(j)} - 1 \tag{2}$$

Obtained dataset was applied to proposed method. At first, the dataset is used in Artificial Neural Network as a training dataset. In this part, we can build an Artificial Neural Network model for extracting the significant features that contain the important information from categorical data. In our experiment, the artificial neural network has three layers - input layer, hidden layer and output layer. The number of hidden nodes is less than number of input nodes. We tried 9 kinds of artificial neural network with hidden nodes changing from 1 to 9 in hidden layer to get the best number of hidden nodes. The values of hidden nodes in each Artificial Neural Network was extracted by our method during the data is applied to ANN model as test dataset. The obtained value is used to calculate the similarity in the k-means algorithm.

To evaluate the efficiency of the method, we evaluated the results using supervised cluster evaluation which is based on entropy. Entropy is a measure that shows the complexity in classification techniques. However clustering is different to classification, it also can be used in cluster evaluation to evaluate the result of clustering. We also got entropy values that are directly applied to k-means cluster algorithm to compare with our method. Graphs shown in Fig. 3 are the entropies in each artificial neural network model that have different number of hidden nodes.

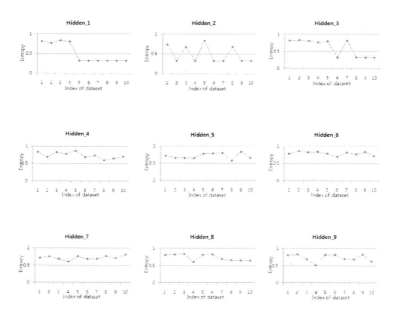

Fig. 3. Entropy for 9 Neural Networks which have Different Number of Hidden Nodes

According to the evaluation results, we can see that entropy graphs whose number of hidden nodes are 1, 2 and 3 are unstable than others. The best number of hidden nodes is 5. But the proposed method can calculate similarity between categorical data.

5 Conclusion

The data mining techniques are used in many fields of studies to understand undiscovered knowledge. The most important part in data mining is to calculate proximity between two data objects.

The notion of similarity for numeric data is well-defined, and there are many useful similarity measures were proposed for numeric data. The most used measure is Euclidean distance. Euclidean distance is used in many clustering techniques as there similarity measure. For categorical data, the similarity between two categorical data objects is not straightforward as well as numeric data. However, several similarity measures for categorical data have been proposed, there are still many issues have to be solved. The main problem of measuring two categorical data objects is to find relationship between there attribute values. Overlap is the most popular simplicity measure for categorical data, because of its simplicity and easy use. However, overlap is cannot represent the relationship of categorical data.

In this paper, we proposed a novel approach for measuring the similarity between two categorical data objects which is based on Artificial Neural Network. We use the Artificial Neural Network whose structure is redesigned for extracting significant features from categorical data to perform unsupervised learning to get similarity between the categorical dataset.

Acknowledge. This research was supported by the National Research Foundation of Korea (NRF) grant funded by the Korea government (MEST) (No.2011-0001044) and a grant (06KLSGB01) from Cutting-edge Urban Development - Korean Land Spatialization Research Project funded by Ministry of Land, Transport and Maritime Affairs.

References

1. Boriah, S., Chandola, V., Kumar, V.: Similarity Measures for Categorical Data: A Comparative Evaluation. In: ACM Computing Surveys (CSUR), pp. 243–254 (2008)
2. Gershenson, C.: Artificial Neural Networks for Beginners (2003)
3. Hornik, K.: Multilayer Feedforward Networks are Niversal Approximators. Neural networks 2, 359–366 (1989)
4. Kelil, A., Wang, S.: SCS: A New Similarity Measure for Categorical Sequences. In: 2008 Eighth IEEE International Conference on Data Mining, pp. 343–352 (2008)
5. Li, X., Hwang, M.Y., Kim, H., Park, K.S., Bae, K.H., Ryu, K.H.: Extracting Method of Significant Features from Categorical Data. In: International Symposium on Remote Sensing (2010)
6. Ahmad, A., Dey, L.: A Method to Compute Distance between two Categorical Values of Same Attribute in Unsupervised Learning for Categorical Data Set. Pattern Recognition Letters 28, 110–118 (2007)
7. Sneat, P.H.A., Sokal, R.R.: Numerical Taxonomy: The Principles and Practice of Numerical Classification (1973)
8. Metzler, D., Dumais, S., Meek, C.: Similarity Measures for Short Segments of Text. LNCS, pp. 16–27 (2007)
9. Yi, W.: Artificial Neural Networks (2005)

Associated Word Extraction System for Search Query Expansion Based on HITS

Jung-Hun Lee[1] and Suh-Hyun Cheon[2]

Department of Computer Science and Engineering, Dongguk University,
26, Pil-don 3-ga jung-gu Seoul 100-715, Korea
leeye123@naver.com, shcheon55@dgu.edu

Abstract. As the utilization of internet becomes generalized, people are able to contact vast information through web. However, as the quantity of information increases rapidly, search engines show the status of limitation in search performance, that they display the information which users do not need. Because of this, it became that users should spend more time and effort to search necessary information. This study suggests a method that a search engine can find out accurate information which users need, and provide it to users swiftly by using query expansion.

Keywords: query expansion, associated word, HITS, personalized search.

1   Introduction

Ordinary information search systems use the method to construct a database with enormous numbered web pages of more than hundreds of millions, and display the ones of higher similarity from the database for the queries input by a user. Because at this time, multiple web pages of diverse themes can be opted, for one query. This phenomenon is due to the ambiguity of the meaning which a word expresses, both in a search query and in a key word. Therefore, it is not easy to find out the result which a user wants, by using only some limited numbered search query. In this case, if more specified queries which are made by adding certain words to the query input by a user, for the first time, is offered to the user, it will enable the user more effective search. That is, it will be possible to make more specified queries by opting associated words in the meaning with the query, among the words occurring in the web pages which a user visited by the search result by the first query which the user input, and offer this more specified queries to the user. For query expansion, a search system can extract words which have high association with the words contained in the input query, or words which fit the search theme, that is, the content of the input query [1]. This study suggests a method for word extraction which uses HITS algorithm and word association relationship. This paper reports on this study and is organized as follows: section 3 describes the procedures required to extract the terms and the subsequent calculation for the weight of each term by means of the weight-value model that we propose; section 4 presents tests that we carried out with the weight-value model and with the TF-IDF weight-value model using NDCG; a discussion together with some conclusion are presented in section 4.

G. Lee, D. Howard, and D. Ślęzak (Eds.): ICHIT 2011, CCIS 206, pp. 591–599, 2011.
© Springer-Verlag Berlin Heidelberg 2011

2   Related Studies

Methods to expand a query can be roughly classified into 2 kinds, a method to construct a word dictionary, and a method to expand a query by extracting related words. First, viewing the method to expand a query by establishing a word dictionary, the method using thesaurus, can be counted. Thesaurus method is one to establish establishing a word dictionary by thesaurus rules for words, and synonyms or related words selected from this dictionary, are offered to a user so that the user might make an additional new query for their original search query [2]. Second method is to use the words which are contained in visiting web pages. According to the performance of algorithms to suggest proper associated words, the performances of each two methods are different. Presently, as methods to expand a query by extracting related words, the method to apply association rule, the method to use TF-IDF, and the method to apply HITS algorithm, exist.

2.2   TF-IDF

TF-IDF(term frequency-inverse document frequency) [3] weight value model is a statistical value model which is used for estimating the importance of a certain word in a certain web page of a linguistic data. TF-IDF model is chiefly used for information search and text mining. The transformed models of TF-IDF weight value model, are also used for determining the priority of similarity between web pages for a user query. By this reason, merely by using this TF-IDF weight value model, search results which a user wants to search may be obtained. Though this method exercises a good performance as much as machine learning method, in this method the association relationship between words is not considered at all, In this method, search is conducted, merely by the way of operation that the weight value of a word which can represent an applied web page is raised.

2.3   HITS

HITS (Hypertext Induced Topic Selection) Algorithm is one which searches hubs and authorities in a web page by using link structure. As in Figure 1, the nodes of hubs and those of authorities are connected by links one another. Accordingly, good authorities can be searched by good hubs, and reversely, good hubs can be searched by good authorities.

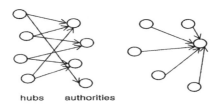

Fig. 1. Connection of Hubs and Authorities by Link

For instance, the more the numbers of connected links, a hub or an authority has, the better one, it is [4]. Though HITS algorithm is used only for calculating the weight value of web pages, initially, Yuanhua Lv applied this algorithm, to weight value measurement too, referring to the associated relation between words and web pages, and make the algorithm to be able to exercise a good performance. [5] However, this algorithm has the fault that it classifies respective words simply on individual level, without considering the association relationship between words.

3   HITS Algorithm Based on Word Link

In this chapter, association relationship of words, and the expression of this association relationship, is elucidated. The words occurring in a same web page have association relationship mutually. This association relationship can be expressed by methods such as a HITS graph.

3.1   Word Directionality

For the words occurring in a same web page, they have the mutual relationship that all of them can be linked, one another, not a unilateral relationship that only a certain words can direct to another word Further, the connected two words have including relationship which makes a link between the words have bi-directionality. D3 of Figure 2, illustrates the bi-directionality between words by bi-directional link of d={A,D,Q,Z} in case of d={t1,t2,t3, ... ,tn}, where d is a web page, and t is a word. Likewise, all the words, occurring in a same web page, are connected by bi-directional links among themselves.

3.2   Method of Graph Composition

As elucidated in Chapter 3.1, all the words occurring in a web page, are linked, bi-directionally. However, if it is regarded that all the words are linked simply, as in Chapter 3.1, the weight value of all the words, which come out as the result value by propagation of weight value of words in HITS. Therefore, it is necessary to remove unnecessary links in order to obtain proper weight value.

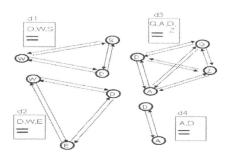

Fig. 2. Link between Words Included in Respective Web Pages

For removing unnecessary links, the words occurred in different web pages, are not connected by links, between themselves. For instance, if the web pages of d1, d2, d3, d4 exist, and the words occurring in the respective web pages are d1={D,W,S}, d2={D,W,E}, d3={Q,A,D,Z}, d4={A,D}, the links between the web pages are like in Figure 2.

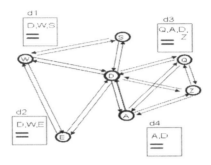

Fig. 3. Between Connectivity of Respective Web Pages by Co-occurring Words

For this status, if the web pages d1, d2, d3, d4 are expressed in one graph, it is like in Figure 3. If the web pages are connected by co-occurring words as the criteria in one graph, the phenomenon that all the words become linked can be avoided, and as the result, unnecessary links can be removed as the criteria by the respective web pages in which a certain word occurs. Because, in this graph, links are removed as the criteria by respective web pages in which a certain word occurs, a word which occurs in all the web pages in common, such as the word D, becomes the connecting ring which connects the individual web pages. In this case, the more similar the themes of the content of respective web pages are, the bigger the weight value of a word which takes the role of connecting ring, becomes, and, on the contrary, when the weight value of the word D, which occurs in all the pages, concurrently, the flow of the weight value which is propagated over all the graph becomes smaller, and in the result, the weight values of all the words in the graph, becomes similar scale.

3.3   Association Relationship Measurement by Using HITS

In HITS Algorithm introduced in 2.3, when nodes are mutually linked, and the links have directionality, the weight value is propagated to the linked node. However, HITS does not consider including relationship between words. At this time, in order to propagate a proper value, a node should not propagate the weight value which the node owns, as it is, but should propagate the value after raising or lowering the weight value, considering the including relationship and association relationship between words.

In this study, a method to adjust propagating value by giving weight value to a link between nodes, as a word is used as a node, is suggested. For instance, in case of A=>B, the association relationship between the two words can be measured, by the ratio between the case that the word B occurs including the word A, and the case that the word B occurs alone by itself. That is, propagating value can be adjusted taking this association relationship as the weight value of a link. This association relationship is called as lift.

$$A_w = \sum_{i=1}^{n} B_{nw} * \log_2 \left(\frac{P(A|B_n)}{P(A|B_n^c)} + 1 \right)$$ (1)

Formula (1) represents one to pursue the weight value which B_n propagates to A by using the weight value of B_n and lift $P(A|B_n)$. Here, A and B_n each indicates a random word. n is the number of words which indicate A, and B_{nw} is the weight value of B_n. A_w is the weight value of the word A. B_n^c is the complementary set of B_n except B_n. $P(A|B_n)$ is the conditional probability that A will occur simultaneously when B_n occurs. In this Formula, because the probability that A will occur when B_n occurs by conditional probability, is pursued, it means, all in all, the probability that B_n occurs including A, is pursued. $P(A|B_n^c)$ is the conditional probability that A will occur simultaneously when the residue different words except B_n occur.

4   Method of Test

4.1   Criteria Words for Performance Estimation

The criteria words for performance estimation are selected among the words contained in the web pages which a search system user visited by the search result which is offered according to the input query by the user. Among the words contained in the web pages visited according to the result of the initial search result, all the words which enable the web pages completely correspondent to the search theme, that is, the intention of a search user, being brought about and placed within the 10 highest positions in retrial search result, when search is retried by attaching the words to the initial query, are selected as criteria words. For the priority of the criteria words for performance estimation, those words which make more respective results which are closer to the search theme, are positioned in the upper level of the search result, are defined to be higher. However, the occurrence frequency of words, that is, how many times a respective word occurred in the visited web pages, is not considered, in the selection of criteria words.

4.2   Performance Estimation Test

For the estimation of respective algorithms, first, the weight value of words contained in the web pages are calculated, by the respective test applied algorithms, and next, when these words are arranged by the priority of the calculated weight value, an algorithm which enable more criteria words selected previously, being contained in the 30 most upper leveled words by the calculated weight value, is determined to be superior. In case that same numbered criteria words are contained in the 30 most higher leveled words in the priority of weight value by respective algorithms, the algorithm which suggests the weight value priority in which higher prior criteria words according to the previously determined priority for the criteria words, will be estimated to be more excellent.

The performance of respective algorithms according to the various modes that a user visits web pages is estimated by the value of Normalized Discounted Cumulative Gain (NDCG) [6]. NDCG uses the method to evaluate the priority of respective units among an applied group by the priority of desirability. NDCG, which is originally a method for estimating web page priority, is used in this test, as it is proper for estimating the result according to the various given conditions [7].

$$DCG_p = rel_1 + \sum_{i=1}^{p} \frac{rel_i}{\log_2 i}, \qquad NDCG_p = \frac{DCG_p}{IDCG_p} \qquad (2)$$

p : A particular rank position rank position.
DCG_p: The DCG accumulated at a particular rank position p.
rel_i: The graded relevance of the result at position i; $rel_i \in \{0,1\}$
$IDCG_p$: This is done by sorting documents of a result list by relevance, producing an ideal DCG at position p.

NDCG is measured using Formula (2). In the experiments, only the Terms ranked in the top 30 were used, and are thus denoted as NDCG@30. For Query Expansion for which the search theme of a search system user should be grasped in the shortest time to suggest associated words for search expansion, the number of web pages for the test is limited and set as 4 pages.

The test is conducted by the assumed 2 types of scenario in the way that users visits web pages. The first scenario is conducting the test by designating the order of visiting web pages by a user, so that the difference between the search theme of a user and the theme of a web page might become lower gradually, though the theme of all the used web pages complies with the search theme of a user. The second scenario is conducting test in the way that, in advance, a web page which does not comply with the search theme, is visited, and next the first scenario is conducted as it is. This scenarios involved the search for Java-related books. From 100 search results, we visited web pages related to "Java network programming", and evaluated performance of each method for term extraction.

For the test, the web pages are used in the status of original text mode without HTML tags and images. The web pages are give a itemset to use in the experiments. This itemset is all texts which except HTML tags and multimedia datas in the web pages.

4.3   Test Result

The value of NDCG algorithm used for examining test result, shows a value which is closer to 1, when the applied sample has high performance, and shows a value which is closer to 0 when the sample has low performance. In the respective tables of test result, below, 'total word number' means the number of total words which occur in the web pages that a user visited. The number of total words is counted by adding the number of words in a newly visited web page to the number of words of previously visited web pages, in accumulation, whenever a user visits a new web page, additionally. But the number of words in a new web page, which occur dually with the words in the previous web pages, is not counted. 'The number of criteria words' is the number of words which are used as criteria words among the entire words. In selecting criteria words, the occurrence frequency of a word is not considered. In this test, the ratio of the words of which the occurrence frequency is higher than the average occurrence frequency is less than 30% of the entire criteria words.

Table 1. Test Result by First Scenario

Number of Visiting Web Pages	1	2	3	4
Simple HITS(S_HITS)	0.21135	0.18704	0.27605	0.13182
Term of Connectivity Based HITS (TCBH)	**0.52238**	**0.48015**	**0.44566**	**0.35544**
TF-IDF	**0.52238**	0.37442	0.35939	0.32730
TF	0.21135	0.18704	0.31822	0.22292
Number of Criteria Words	26	26	35	54
Number of Total Words	280	409	486	798

Table 1 is the test result by the first scenario. In this scenario, as seen in Table 1, when the number of total words increases in high ratio, in case of the existing algorithms, the performance of algorithms to suggest proper associated words for search expansion lowers much. The reason why the performance of the algorithms lowers, is, that the more the number of the total words is, the more words which have no relevance with the search theme of a user, but have high occurrence frequency, are contained in a web page. However, in this scenario, in case of TCBH, suggested in this study, even if the number of the total words increases in high ratio, it does not show the phenomenon that the performance lowers much, and shows the best performance in all the cases by the number of visiting web pages.

Table 2. Test Result by Second Scenario

Number of Visiting Web Pages	1	2	3	4
S_HITS	0.10951	0.10951	0.37114	0.14910
TCBH	**0.52463**	**0.47426**	**0.44324**	**0.35515**
TF-IDF	0.52089	0.42092	0.42490	0.32087
TF	0.10951	0.13630	0.17517	0.19556
Number of Criteria Words	26	26	35	54
Number of Total Words	363	676	1086	1286

This scenario is for estimating the performance of algorithms in case that web pages which have no relevance with the search theme, at all, are included in the web pages for estimation. As seen in Table 2, it is identified that even if web pages which have no relevance with the search theme are contained, this does not lower the performance of TCBH. This phenomenon is the same in case that web pages which have small numbered words having relevance with the search theme, are contained.

Table 3. Result by a Test by Limiting Number of Criteria

Number of Visiting Web Pages	1	2	3	4
TCBH	**0.579398**	**0.669273**	**0.703187**	**0.711706**
TF-IDF	**0.579398**	0.518952	0.561882	0.508262
Number of Criteria Words	10	10	10	10

Table 3 shows the result of a test which is conducted the same as the test by the first scenario, of which the result is described in Table 1, only by decreasing the number of criteria words. In this test, excluding the criteria words of which the occurrence frequency is very low, and using the words of which the occurrence frequency is placed in the first position to the 10th position, TF-IDF, and TCBH, of which the performance appeared to be approximately on similar level, in the result by the first scenario, are compared, once again. The tests of which the results are described in Table 1 and Table 2, are conducted for measuring how much the respective algorithms can embrace the words which have low occurrence frequency by including the words which have high relevance with the search theme, but have low occurrence frequency. In contrast, the test of Table 3 is for estimating the performance of the respective algorithms to extract core associated words which enables web pages fitting to the search theme being placed on upper level in search result. As identified in Table 3, by the estimated result by NDCG, TCBH has higher performance than TF-IDF, and the more web pages a user visits, the larger this difference of performance becomes.

References

1. Lee, J.-H., Cheon, S.-H.: A Term Weight Mensuration based on Popularity for Search Query Expansion. Journal of KIISE: Software and Applications 37(8), 620 (2010) (in Korean)
2. Kristensen, J.: Expanding End-Users' Query Statements for Free-text Searching with a Search-aid Thesaurus. Information Processing and Management 11, 22–33 (1968)
3. Salton, G., McGill, M.J.: Introduction to Modern Information Retrieval. McGraw-Hill, New York (1983) ISBN 0-07-054484-0
4. Kleinberg, J.M.: Authoritative sources in a hyperlinked environment. Journal of the ACM (JACM) 46, 604–632 (1999)

5. Lv, Y., Sun, L., Zhang, J., Nie, J.-Y., Chen, W., Zhang, W.: An iterative implicit feedback approach to personalized search. In: Proceedings of the 21st International Conference on Computational Linguistics and the 44th Annual Meeting of the Association for Computational Linguistics, pp. 585–592 (2006)
6. Jarvelin, K., Kekalainen, J.: Cumulated gain-based evaluation of IR techniques. ACM Transactions on Information Systems 20(4), 422–446 (2002)
7. Liu, T.-Y., Xu, J., Qin, T., Xiong, W.-Y., Li, H.: LETOR: Benchmark dataset for research on learning to rank for information retrieval. In: SIGIR 2007 Workshop on Learning to Rank for Information Retrieval (2007)

MDA Approach for Non-functional Properties of Dependable and Distributed Real-Time Systems

Lichen Zhang and Jifeng He

Software Engineering Institute
East China Normal University
Shanghai 200062, China
Zhanglichen1962@163.com

Abstract. Non-functional properties are highly important for dependable and distributed real-time systems as they are designed to operate in environments where failure to provide functionality or service can have enormous cost both from financial, influential or physical aspects. Therefore it is essential that these properties are calculated as precisely as possible during the design and operation of such systems. In this paper, we propose an aspect-oriented MDA approach for non-functional properties to develop dependable and distributed real-time systems. A case study illustrates the aspect oriented MDA development of dependable and distributed real-time systems.

Keywords: Non-Functional Properties, Aspect-Oriented, MDA.

1 Introduction

Dependable and distributed real-time systems are an enabling technology for the information society. Their economic impact reaches far beyond their immediate market size, since the success of many industrial products depends on the provision of reliable control systems. As the rapidly growing functional and non-functional system requirements cause an enormous increase in system complexity, it is necessary to move into component-based design and aspect-oriented design: to provide pre-validated hardware and software components and an appropriate integration methodology for the design of next generation dependable embedded real-time systems.

Model Driven Architecture (MDA) [1] is based on a series of industry-standard software development frameworks, model drives the software development process, and using support tool model can to achieve automatic conversion among the models, between the model and the code. Its core idea is to establish a Platform Independent Model (PIM) with complete description of system requirements and specific platform implementation technology, through a series of model transformation rule set, the platform independent models to be able to transfer to complete presentation system requirements, and specific implementation techniques related to platform specific model (PSM), finally, using MDA tools will be making platform specific model automatically transferred to code. Aspect-oriented software development methods [2]

G. Lee, D. Howard, and D. Ślęzak (Eds.): ICHIT 2011, CCIS 206, pp. 600–608, 2011.

make up object-oriented software development methods in system development needs of non-functional characteristics of the existing limitations question problem. Use separate technology of concerns separates all the crosscutting concerns of the system, and then analyzed, designed, modeled for each cross-cutting concerns, to address crosscutting concerns in object-oriented software development, the code tangling and scattering problems, enhancing the system's modular degree, lowering coupling between modules.

In this paper, we propose an aspect-oriented MDA approach for non-functional properties to develop dependable and distributed real-time systems.

2 Non-functional Requirements of Dependable and Distributed Real-Time Systems

Dependability is that property of a system that justifies placing one's reliance on it. The dependability of a system is the collective term that describes the availability performance of a system and its influencing factors: reliability, safety, maintainability and maintenance support performance. These non-functional properties are highly important for real-time systems as they are designed to operate in environments where failure to provide functionality or service can have enormous cost both from financial, influential or physical aspects. Therefore it is essential that these properties are calculated as precisely as possible during the design and operation of such systems. Reliability is the ability of a system or component to provide its required functionality or services under given conditions for a specified period of time. Availability is the ratio of total time that a system or a component is functional during a specified period and the length of the period. Maintainability can be specified as the probability that a component or system will be restored to a given condition within a period of time. Safety is described as the absence of serious consequences on the user or environment in case of failure. Safety can be defined as "a property of a system that it will not endanger human life or the environment" .A system is safety-critical if safety cannot be ensured when it fails to provide correct service. Integrity can be specified as the absence of improper alterations on the target system or component. Survivability can be defined as the ability of the system to remain functional after a natural or man-made disturbance. The threats to dependability are faults, errors and failures. There is a relationship between these threats: A fault is a defect in the system which, when activated, leads to an error. An error is an incorrect system state that may affect the external behaviour, thereby causing a failure. A failure occurs when the delivered service deviates from what is considered correct. There exist four general means to achieve dependability: fault prevention, fault tolerance, Fault removal, and fault forecasting. Fault prevention deals with the objective of avoiding to introduce faults during the software development process. There exist four general means to achieve dependability: fault prevention, fault tolerance, fault removal, and fault forecasting. Fault prevention can be considered as an inherent part of it. Fault removal deals with uncovering faults that have happened at any phase of the development process. Fault forecasting is aimed at evaluating the behaviour of the system under the occurrence of faults such that it can be concluded which ones would lead to system failure. Fault tolerance techniques are the means to allow a system to provide correct service even when faults occur. Such techniques use diverse forms of redundancy to detect and recover from faults. The most common approaches use either hardware redundancy, software redundancy, time redundancy or information redundancy to identify erroneous conditions. The subsequent recovery process relies on the remaining fault-free parts of the system to correct the errors and/or prevent them from reappearing.

Timeliness requirements apply to computations in which correctness depends not only on the results produced but also the time at which they become available. Soft real-time requirements are general performance goals, typically expressed via some measure of average response time. Such goals have a probabilistic or statistical flavour which takes them outside the scope of this study. More tangible are hard real-time requirements in which particular events must occur at, or before, certain times. A periodic requirement states that some action must be performed at regular intervals, while a sporadic requirement states that some action must be performed immediately following an external "triggering" event. In real-time systems components do not only have to perform operations correctly, but also have to meet certain timing requirements. General purpose components like graphical user interface frameworks are often not design with a real-time scenario in mind and thus real-time programmers are many times forced to build large parts of their applications from scratch. Building components suitable for real-time applications is a difficult task, as besides the functional requirements attention has also to be paid to the non-functional timing requirements. This additional complexities makes building real-time components more expensive and error prone than general purpose components. Real-time and fault tolerance constraints can impose conflicting requirements on a distributed system. Real-time operation requires an application to be predictable, to have bounded request processing times, and to meet specified task deadlines. This predictability is often the most important characteristic of real-time systems. In contrast, fault tolerant operation requires that an application continue to function, even in the presence of unanticipated, potentially time-consuming events such as faults and fault recovery. Faults are often viewed as asynchronous unpredictable events that can upset a real-time system's scheduled operation. Sustained operation with consistency of application data in the face of faults is often the single most important characteristic of fault-tolerant systems. Thus, there is a fundamental conflict between the philosophies underlying the two system properties of real time and fault tolerance. While real-time performance requires a priori knowledge of the system's temporal operation, fault tolerance is built on the principle that faults can and do occur unexpectedly, and that faults must be handled through some recovery mechanism whose processing time is uncertain. When both real-time and fault tolerant operation are required in the same system today, trade-offs are made at design time, not at run-time.

3   Applying AOP and MDA to Non-functional Requirements

The real-time systems software development process based on aspect-oriented system is divided in five phases [3].

The first phase is a profound analysis of the requirements. The phase includes three steps:

Step one handles the non-functional requirements and then identifies which of those are crosscutting.

Step two performs a traditional specification of functional requirements, in this case, using an UML-like approach where the use case model is the main specification technique.

Step three starts by composing functional requirements with aspects; then it identifies and resolves conflicts that may arise from the composition process.

The concepts of overlapping, overriding and wrapping can be adopted to define the composition part of the model. Overlapping indicates the requirements of the aspect modifies the functional requirements they transverse. In this case, the aspect requirements may be required before the functional ones, or they may be required after them. Overriding indicates the requirements of the aspect superpose the functional requirements they transverse. In this case, the behavior described by the aspect requirements substitutes the functional requirements behavior. Wrapping indicates the requirements of the aspect encapsulate the functional requirements they transverse. In this case, the behavior described by the functional requirements is wrapped by the behavior described by the aspect requirements.

In the design phase, the distributed system will be designed considering both the requirements and the constraints posed by the system and system. Using the MDA approach [4] to produce the platform specific models includes five steps (see Fig. 1):

Step one: Create the PIM for the distributed system.
Step two: Select the target system and create the generic system aspects.
Step three: Transform PIM to enhanced PIM using the application converter.
Step four: Transform the generic aspects to enhanced aspects using the aspect converter.
Step five: Weave the enhanced aspects into the enhanced PIM to produce the PSM.

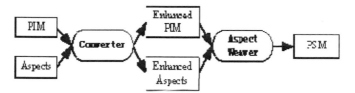

Fig. 1. Process view of the PSM generation

4   Case Study: Intelligent Transportation Systems

Intelligent Transportation systems (ITS)[5] – automotive, aviation, and rail – involve interactions between software controllers, communication networks, and physical devices. These systems are among the most complex cyber physical systems being designed by humans, but added time and cost constraints make their development a significant technical challenge. MDA approach can be used to improve support for design, testing, and code generation. MDA approach is increasingly being recognized as being essential in saving time and money. Transportation systems consist of embedded control systems inside the vehicle and the surrounding infrastructure, as well as, the interaction between vehicles and between vehicle and the infrastructure.

The dynamics of these interactions increases complexity and poses new challenges for analysis and design of such systems.

In real-time systems such as ITS, the passage of time becomes a central feature — in fact, it is this key constraint that distinguishes these systems from distributed computing in general. Time is central to predicting, measuring, and controlling properties of the physical world. The modeling process of non functional requirements time of ITS by aspect–oriented MDA[6][7][8] is shown as Fig.2, Fig.3 and Fig 4.

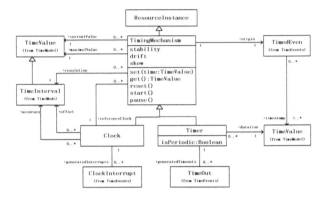

Fig. 2. Time mechanism model-CIM model

```
                    <<aspect>>:TimeAspect
 -isSingleton:bool=true
 -isPrivileged:bool=false
 -mixTime:int
 -maxTime:int

<<pointcut>>::+sensorStartClockPointcut
{<<designator>>=call, <<jointpoint>>=start, <<target>>=Sensor}
<<pointcut>>::+sensorUpdateClockPointcut
{<<designator>>=call, <<jointpoint>>=update,
<<args>>=(a:byte, b:int, c:String), <<target>>=Sensor}
<<pointcut>>::+trafficLightTimerPointcut
{<<designator>>=execution, <<jointpoint>>=setLightState,
<<args>>=(lightstate:int), <<target>>=TrafficLightUI}
<<pointcut>>::+trafficPeccancyTimerPointcut
{<<designator>>=call, <<jointpoint>>=getPeccancy,
<<target>>=VehicleSensor}

<<advice>>::+setClockAdvice
{<<position>>=after, <<pointcut>>=sensorStartClockPointcut,
<<operation>>=setClock, <<target>>=Clock}
<<advice>>::+setClockAdvice
{<<position>>=after, <<pointcut>>=sensorUpdateClockPointcut,
<<operation>>=setClock, <<target>>=Clock}
<<advice>>::+setTimerAdvice
{<<position>>=before, <<pointcut>>=trafficLightTimerPointcut,
<<operation>>=setTimer, <<target>>=new Timer}
<<advice>>::+setTimerAdvice
{<<position>>=before, <<pointcut>>=trafficPeccancyTimerPointcut,
<<operation>>=setTimer, <<target>>=new Timer}

{context TimeAspect:
inv:  the isSingleton is not equal isPrivileged
    if self.isSingleton then self.isPrivileged=false
    else self.isPrivileged=true endif
    self.isSingleton=not self.isPrivileged
    --maxTime always large than mixTime
    self.maxTime>=self.mixTime
    --the Clock has only one instance
    self.setClockAdvice.Clock.allInstances()->size()=1}
```

Fig. 3. Aspect oriented time model: PIM model of time

OCL supplements UML by providing expressions that have neither the ambiguities of natural language nor the inherent difficulty of using complex mathematics. Time aspect is specified by OCL as folows.

Context TimeAspect:

inv: --*the isSingleton is not equal isPrivileged*
    if self.isSingleton then self.isPrivileged=false
    else self.isPrivileged=true endif
    self.isSingleton=not self.isPrivileged
    --*maxTime always large than mixTime*
    self.maxTime>=self.mixTime
    --*the Clock has only one instance*
    self.setClockAdvice.Clock.allInstances()->size()=1

The time constraints of phase is specified by OCL as following:

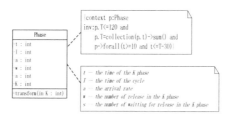

Fig. 4. Time constraints of phase

Now we return to the model transformation, whose essential point is mapping to the special programming language code as shown in Fig.5.

```
package com.aspect.its;
public aspect TimeAspect {
    private boolean isSingleton = true;
    private boolean isPrivileged = true;
    private int mixTime;
    private int maxTime;
    public boolean isSingleton() {
        return isSingleton;
    }
    public void setSingleton(boolean isSingleton) {
        this.isSingleton = isSingleton;
    }
    public boolean isPrivileged() {
        return isPrivileged;
    }
    public void setPrivileged(boolean isPrivileged) {
        this.isPrivileged = isPrivileged;
    }
    public int getMixTime() {
        return mixTime;
    }
    public void setMixTime(int mixTime) {
        this.mixTime = mixTime;
    }
    public void setMixTime(int mixTime) {
        this.mixTime = mixTime;
    }
    public int getMaxTime() {
        return maxTime;
    }
    public void setMaxTime(int maxTime) {
        this.maxTime = maxTime;
    }
    pointcut sensorStartClockPointcut() :
        call(* Sensor.start(..));
    pointcut sensorUpdateClockPointcut(byte a, int b, String c) :
        call(* Sensor.update(byte,int,String))&&args(a,b,c);
    pointcut trafficLightTimerPointcut(int lightstate) :
        execution(* TrafficLightUI.setLight(state(int))&&args(lightstate);
    pointcut trafficOccupancyTimerPointcut() :
        call(* VehicleSensor.getOccupancy(..));
    after() : sensorStartClockPointcut() {
        Clock.setClock();
    }
    after(byte a, int b, String c) : sensorUpdateClockPointcut(a,b,c) {
        Clock.setClock();
    }
    before(int lightstate) : trafficLightTimerPointcut(lightstate) {
        new Timer().setTimer();
    }
    before() : trafficOccupancyTimerPointcut() {
        new Timer().setTimer();
    }
}
```

Fig. 5. Aspect code of time property: PSM model

Considering safety specification, the formal technique can be applied. The train control systems environment consists of train and road[9]. However, since just train is monitored, system environment is specified by MSV variable MSV_{train}, which may have four states(distant, approached, on-crossinng and passed)[9].

$$m_1 = MSV_{train} = \{m_{11} = enter, m_{12} = approached, m_{13} = pull\_in, m_{14} = dis\tan t\}$$

In addition, internal MSV variables, MSV_{timer} is considered to monitor passing of time [9].

$$m_2 = MSV_{time} = \{0, 1, 2, ..., max\}$$

System-controlling component is just the system gate. It is shown by CSV_{gate} variable (called C_{gate}), whose state is set by software as MoveDown, MoveUp and closes/opens the road [9]. Thus,

$$cs_1 = CSV_{gate} = \{c_{11} = open, c_{12} = close\}$$

We have the following safety constraints [9]:

$$approached_{event}(MSV_{train}, f_1) \equiv$$
$$[f_1 = \Delta(k_1, MSV_{train}) \quad and \quad f_2 = \Delta(k_2, MSV_{train})] \quad ,$$
$$k_1 = enter, \quad k_2 = approached$$

$$pull\_in_{event}(MSV_{train}, f_1) \equiv$$
$$[f_1 = \Delta(k_1, MSV_{train}) \quad and \quad f_2 = \Delta(k_2, MSV_{train})] \quad ,$$
$$k_1 = approached, \quad k_2 = pull\_in$$

$$dis\tan t_{event}(MSV_{train}, f_1) \equiv$$
$$[f_1 = \Delta(k_1, MSV_{train}) \quad and \quad f_2 = \Delta(k_2, MSV_{train})] \quad ,$$
$$k_1 = pull\_in, \quad k_2 = dis\tan t$$

The aspect -oriented fault-tolerant model is as shown in Fig.6.

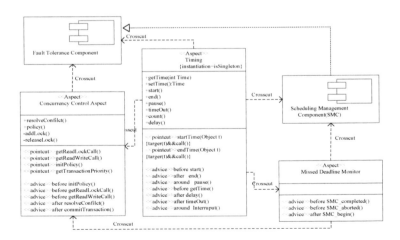

Fig. 6. Aspect-oriented Fault-Tolerant Model: PIM model

```
public aspect Rule {
pointcut TrainArrive (): call(approach (MVS_train, CSV_gate))
after (): approach(){
    if (MVS_train = approached && CSV_gate =fault)
}
}
```

Fig. 7. Aspect code of train arrive: PSM model [9]

5 Conclusion

In this paper, we proposed an aspect-oriented MDA approach for non-functional properties to develop dependable and distributed real-time systems. We illustrated the proposed method by the development of ITS and demonstrated aspect-oriented MDA approach that can be used for modeling non-functional characteristics of complex system, effectively reduce the complexity of software development and coupling between modules to enhance the system's modular.

The further work is devoted to developing tools to support the automatic generation of model and code.

Acknowledgments. This work is supported by National Natural Science Foundation of China under Grant.No.90818008 and No. 61021004), National High Technology Research and Development Program of China (No. 2011AA010101), National Basic Research Program of China (No. 2011CB302904), Doctoral Program Foundation of Institutions of Higher Education of China (No. 200802690018). The authors also gratefully acknowledge support from the Danish National Research Foundation and The National Natural Science Foundation of China (Grant No. 61061130541) for the Danish-Chinese Center for Cyber Physical Systems.

References

1. Object Management Group.OMG MDA guide v1.0.1[EB/OL],
   http://www.omg.org/docs/omg/03-06-01.pdf
2. Wehrmeister, M.A., Freitas, E.P., Pereira, C.E., et al.: An Aspect-Oriented Approach for Dealing with Non-Functional Requirements in a Model-Driven Development of Distributed Embedded Real-Time Systems. In: 10th IEEE International Symposium on Object and Component-Oriented Real-Time Distributed Computing, Greece, May7-9, pp. 428–432 (2007)
3. Liu, J., Yong, Z., Zhang, L., Chen, Y.: Applying AOP and MDA to middleware-based distributed real-time embedded systems software process. In: Asia-Pacific Conference on Information Processing, APCIP 2009, pp. 270–273 (2009)
4. Frankle, D.S.: Model Driven Architecture:Applying MDA to Enterprise Computing. OMG Press
5. Ranjini, K., Kanthimath, A., Yasmine, Y.: Design of Adaptive Road Traffic Control System through Unified Modeling Language. International Journal of Computer Applications 14(7) (February 2011)

6. Clemente, P.J., Sánchez, F., Perez, M.A.: Modelling with UML Component-based and Aspect Oriented Programming Systems. In: Seventh International Workshop on Component-Oriented Programming at European Conference on Object Oriented Programming (ECOOP), Málaga, Spain, pp. 1–7 (2002)
7. Madl, G., Abdelwahed, S.: Model-based Analysis of Distributed Real time Embedded System Composition In: Proceedings of the 5th ACM International Conference on Embedded Software, New Jersey, USA (2005)
8. Lavazza, Quaroni, G., Venturelli, M.: Combining UML and formal notations for modeling real-time systems. In : ACM SIG
9. Babamir, S.M., Jalili, S.: Making real-time systems fault tolerant: a specification-based approach. Journal of Scientific & Industrial Research 69, 501–509 (2010)

Implementation of Smart Car Using Fuzzy Rules

You-Sik Hong[1], Myeong-Bok Choi[2], June-Hyung Lee[3], Cheol-Soo Bae[4],
Jang-Mook Kang[5], Jae-Sang Cha[6], Geuk Lee[7], Seong Jin Cho[8], HyunSoo Jin[9],
Chun-Myoung Park[10], Baek ki Kim[11], Kwang-Deok Han[12], Su Kyun Sun[13],
Chul Jang[14], and S.C. Yu[15]

[1] School of Information and Communication Engineering, Sang JI University, Wonju,
Kangwon, Korea
yshong@sangji.ac.kr
[2] Department of Multimedia Engineering, Gangneung-Wonju National University
[3] Department of Smart Phone, Gang Dong College, ChungCheongbukdo, Korea
[4] Dept. of Information & Communication KwanDong Univ,Gangneung , Korea
[5] Electronic Commerce Research Institute,DonggukUniversity,Gyeongsangbukdo
Korea
[6] Department of Media Engineering, Seoul National University of Science and Technology,
Seoul, Korea
[7] Dept. of Information & Communication, Han-nam University, Daejon, Korea
[8] Department of Computer Engineering, KyungHee University, Korea
[9] Dept.of Information & Communication,BaekSeokUniversity,Cheonan,Korea.
[10] Department of Computer Engineering Chungju National University, Korea
[11] Dept. of Information and Telecom. Eng. Gangneung-Wonju National University
[12] Department of Computer Science, SangJI YoungSeo College,Wonju, Korea
[13] Department of E Business, Tong Won College, Gyong Gi Do, Korea
[14] Director of Public Sales Division, LG N-Sys, Seoul, Korea
[15] Manager of Public Sales Division, LG N-Sys, Seoul, Korea

Abstract. We proposed an algorithm to automatically control the seat, and notify dangerous traffic conditions for smart car system. The system is needed to adapt to both the driver and the circumstance for maximum convenience and safety. In order to solve these problems, we proposed an algorithm for developing smart vehicles.

In this paper, it is simulated to compare the performance of a normal car with that of the algorithm-implemented vehicle in a potential accident. Through the computer simulation, we proved that the algorithm-implemented vehicle automatically adjusted the mirrors and seat for the maximum comfort and driver's awareness of the circumstance. We used fuzzy neural network algorithm for design of optimal car speed, and proposed the system to prevent intelligently against traffic accidents.

1 Introduction

Smrat cars require a human-friendly technology. Ubiquitous technologies are certainly needed that vehicles become human –friendly. People could get pleasures to

G. Lee, D. Howard, and D. Ślęzak (Eds.): ICHIT 2011, CCIS 206, pp. 609–616, 2011.

be able to manage vehicles. We must be able to provide status information of a vehicle to a driver to realize ubiquitous functions on a vehicle. If we can transfer the information of vehicles effectively via cellular phones and PDAs, people will know easily the fact that their vehicles are stolen or broken [1-2]. The technology that we propose should be implemented in future cars. We are going to implement intelligence in vehicles. Making it as mentioned above, we think that future-oriented vehicles may be completed. Sensor network combines a number of disciplines of various fields. Sensor network is mixed technology of a sensor technology, communications technology, SoC, and MEMS technology and so on. This technology is a key technology of network and semiconductor industry for next generation. The purpose of this paper is to develop intelligent devices and improve driver's convenience and safety on vehicles [3-6]. We are going to develop a car with an intelligent algorithm which has an electronic control system. This device is connected to TPEG terminal with a wireless Internet. This terminal provides the status of the vehicles to the driver. The driver can know engine temperatures, conditions of exhaust gas, tire and gasoline [7-10]. Traffic accidents break out on the road. Drivers and a road do not hold a close relationship. Specially, they occur due to the causes such as insufficient safe facilities, erroneous road design, and uncomfortable mind of drivers. This paper is organized as follows. Section 2 explains automatic control algorithm which can automatically adjust car seat depending on the driver's height. Section 3 explains the fuzzy rule algorithms for smart car. Section 4 shows that it can automatically adjust car speed when it finds an obstacle, or receives the traffic message broadcasted via a wireless data channel, and describes traffic accident simulation results. Finally, Section 5 will present our conclusions.

2 Fuzzy Rule Algorithm for Smart Car

For safe driving, a driver must be able to see the road well. Hence, a vehicle has a rearview mirror. Even though there are a number of deluxe cars, there is no vehicle that controls rearview mirror automatically. Drivers need a car to automatically control the rearview mirror, because of the need for a vehicle's safety. The principles for controlling optimum seat and the rearview mirror for driver are as follows.

$e = R - Y$
$Ce = e2 - e1$

Constrained condition

    Y: Optimum safety seat control

    R: Reference input (driver average physical condition)

    E: Error (reference input - driver's height, weight)

    Ce: Change amount of error

    E 2: Present error (upper body and lower body)

    E 1: The present sampling previous error

Table 1. Safety seat control to consider physical condition

Quantization step	Range of quantization step to consider driver's physical condition
-6	x<=-60 meters
-5	-60<x<=-40
-4	-40<x<=-30
-3	-30<x<=-20
-2	-20<x<=-10
-1	-10<x<=0
0	0<x<=10
+1	10<x<=20
+2	20<x<=30
+3	30<x<=40
+4	40<x<=50
+5	50<x<=60
+6	x<60 meters
OVER	NONE

Table 2. Fuzzy rule for automatic seat control

	NB	NM	NS	Z	PS	PM	PB
NB	NB	NS	NB	NB	NB	NS	NB
NM				NM	NM		
NS			NM	NS			
Z		NS	NS	Z	PS		
PS			PS	PS	PM		
PM			PM	PM			
PB	PS	PM	PB	PB	PB	PM	PM

Table 1 shows the most suitable value for safety seat control to consider driver's height condition. Also, physical condition changes the input value of fuzzy control.

Table 2 is the fuzzy rule to revise for seat coordination.

If the height is the same as the weight and the upper body is longer than the lower body, it will make an error. Also, when the driver's inclination differs from the result of the automatic seat control device, a variation occurs.

In this paper, fuzzy safety seat control rule is as follows.

(RULE 1)   IF DPSV IS PB
           AND USPC IS NS
         THEN OPRG IS PB
(RULE 2)   IF DPSV IS PB
           AND USPC IS NM
         THEN OPRG IS PM
(RULE 3)   IF DPSV IS PS
           AND USPC IS NS
         THEN OPRG IS PS

In above routine, the constrained condition is as follows.

> DPSV: Upper body condition and lower body condition error (E)
> USPC: Driver's visual height
> Driver's inclination: Error change amount (CE)
> OPRG: The most suitable seat control

To consider fuzzy control rule and driver's physical condition, it produces the most suitable seat control output (OPRG).

(Rule 1)

[0.3/4, 0.5/5, 1/6] | ^ [0.7/-3, 0.6/-2, 0.8/-1, 0.4/0, 0.1/1]
  ^ [0.3/4, 0.5/5, 1/6]
= 0.3 ^ 0.7^ [0.3/4, 0.5/5, 1/6]
= [0.3/4, 0.5/5, 1/6]

(Rule 2)

[0.3/4, 0.5/5, 1/6] | ^ [0.3/-6, 0.2/-5, 0.8/-4, 0.5/-3, 0.4/-2,0.2/-1]
  ^ [0.1/2, 0.5/3, 1.0/4], 0.5/5,0.2/6]
= 0.3 ^0.5^[0.1/2, 0.5/3, 1.0/4, 0.5/5, 0.2/6]
= 0.1/2, 0.3/3, 0.3/5, 0.3/5, 0.2/6

(Rule 3)

[0.3/1, 0.9/2, 0.7/3, 0.3/4] | ^ [0.7/-3, 0.6/-2, 0.8/-1, 0.4/0, 0.1/1] | ^ [0.3/1, 0.9/2, 0.7/3, 0.3/4]
= 0.3, 07^[0.3/1, 0.9/2, 0.7/3, 0.3/4]
= 0.3/1, 0.3/2, 0.3/3, 0.3/4

Non-fuzzification method:

$$U = \frac{\sum (\text{Big set that have membership value of function} \times \text{Its value of function})}{\text{Value of membership function}}$$

3 Design of Optimal Car Speed

The traditional Max-Min CRI method has a considerable error region. To overcome those problems, we adopt the new Max-Min CRI method [10]. The new Max-Min CRI method has an additional process which computes a similarity degree by a new similarity measure, and applies it to the inference process of the Max-Min CRI method. Most papers have discussed the above similarity measure in general point of view. A similarity measure needs not to have generality but to have specialty according to domain area and inference method. The above mentioned similarity measures can not apply to the Max-Min CRI method, because they can not be guaranteed except estimates of language similarities.

$$SM(A, A') = \tau = \frac{f(\mu_A(u) \wedge \mu_A'(u))du}{f(\mu_A(u) \vee \mu_A'(u))du}, \text{for} \quad all \quad u \in U \qquad (4)$$

For all universes of discourse ($u \in U$), the shadowed part is the intersection part of two fuzzy sets and the white part is the union part of two fuzzy sets.

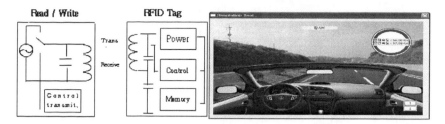

Fig. 1. Optimal car speed depending on road condtions

ID	도로명	날씨	노면상태	도로경사	안전속도
guriic	구리IC	비/눈	연화포장	-5%	55.83
namyangjuic	남양주IC	비/눈	콘크리트포장	-10%	57.99
topyungic	토평IC	폭우/폭설/빙...	아스팔트포장	0%	62.0
gangilic	강일IC	습도	아스팔트포장	0%	57.99
toigewonic	퇴계원IC	건조	아스팔트포장	0%	62.0
amsaic	암사IC	건조	콘크리트포장	-10%	57.99
sangiljc	상일JC	건조	아스팔트포장	0%	63.9
hanamjc	하남JC	비/눈	연화포장	+15	63.9
shinnaeic	신내IC	건조	아스팔트포장	0%	62.0
mookdongic	묵동IC	건조	아스팔트포장	0%	62.0
wolengic	월릉IC	건조	콘크리트포장	0%	62.0
nokchenbr	녹천교	비/눈	아스팔트포장	0%	62.0
sangkyebr	상계교	건조	아스팔트포장	0%	62.0
sooraku	수락지하차도	비/눈	아스팔트포장	0%	62.0
nowonbr	노원교	건조	아스팔트포장	0%	62.0
changdongbr	창동교	건조	아스팔트포장	0%	62.0
wolkye1br	월계1교	비/눈	아스팔트포장	+15	57.99
hawolgokr	하월곡 램프	건조	아스팔트포장	0%	62.0
jungiengr	정릉 램프	건조	콘크리트포장	0%	49.01
wolgok	월곡	건조	아스팔트포장	0%	62.0
jungrangbr	중랑교	비/눈	아스팔트포장	0%	62.0
janganbr	장안교	비/눈	연화포장	0%	57.99
gunjabr	군자교	건조	아스팔트포장	0%	62.0
majang	마장	습도	아스팔트포장	0%	57.99
sagn	사근	건조	아스팔트포장	0%	62.0
sungdongjc	성동JC	습도	아스팔트포장	0%	57.99
sungsoojc	성수JC	건조	아스팔트포장	0%	62.0
kilemr	길음 램프	건조	연화포장	0%	57.99
jungingt	정릉 터널	건조	아스팔트포장	0%	62.0
hngjimunt	홍지문 터널	건조	아스팔트포장	0%	62.0
hongeenr	홍은램프	폭우/폭설/빙...	아스팔트포장	0%	62.0

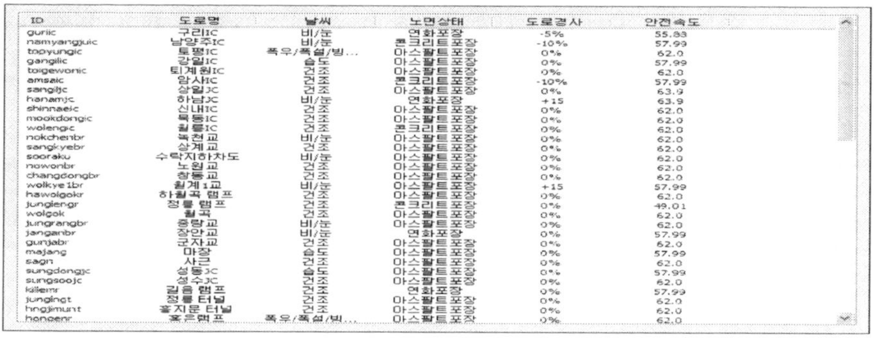

Fig. 2. Simulation for optimal car speed

Fig. 1 and Fig. 2 show the simulation of optimal car speed depending on different road conditions.

(1) Fuzzification procedure

The fuzzification membership functions in a fuzzy rule base are triangular types def ined by equation (5) with $a, b, u \in U$. The fuzzy rule base is composed of MISO (M ulti-Input Single-Output) typed rule base. Each fuzzy membership function in a fuzzy rule base has a membership value area $[0,1]$, and should be normalized in this area.

$$y = \begin{cases} \dfrac{2}{b-a}(x-a), & a \le u \le b, \quad u \in U \\ 0, & otherwise \end{cases} \tag{5}$$

This interval includes all possible values for the variable in universe of discourse (U). All fuzzy sets in a fuzzy rule base have the same support interval $[a,b]$. The equation can be represented all types of fuzzy membership functions including both fuzzy and n on-fuzzy membership functions.

(2) Inference Procedure

The main process of new MAX-MIN CRI (NCRI) is as the following equation. Comparing to the Max-Min CRI method, NCRI makes use of the proposed similarity measure between input facts and fuzzy sets in a condition part (antecedent) of a rule.

The similarity degree which is estimated by the similarity measure is represented as τ in (6). For all rule i, the following steps are performed for inference.

$$
\begin{aligned}
\mu_{B'}(v) &= \underset{u \in U, v \in V}{Max} \ Min(\mu_{A'}(u), \mu_R(u,v)) \\
&= \underset{u \in U, v \in V}{Min} \ (Max \ Min(\mu_{A'}(u), \mu_A(u)) \times \tau, \mu_B(v)) \qquad (6) \\
&= \underset{u \in U, v \in V}{Min} \ (\alpha \times \tau, \mu_B(v))
\end{aligned}
$$

It is very difficult to produce the proper periodic signal and minimum car waiting time. Because the length of intersecting roads, the speed of car, the number of road lanes and straight or rotating car lanes must be changed, it is difficult to produce the proper periodic signal to allow easier flow of traffic as shown in Fig.1. and Fig. 2. Optimal green time using fuzzy neural network algorithm is given as Table 1.

4   Intelligent Prevention Technology against Traffic Accidents

Wireless communication is applied to various application fields. One of them is wireless data channel. It can offer traffic information to a driver. Such application could contribute to reduce the traffic accident. Communication between a vehicle and road side is basically achieved between a RSE (Road Side Equipment) on the road side and OBE (On Board Equipment) on the vehicle.

Specially, the algorithm is developed to take advantages of information that is transmitted from the road and controls vehicle speed. Speed control of vehicles uses a throttle. The following is the vehicles speed control algorithm.

O Entry alarm to a curve in the road
O Alarm for the speed limit
O Accident alarm and control for invisible area

Fig. 3 could be explained as follows. It loads the speed limit, to make various controls and alarm information, received by the vehicles at a curve in the road. We propose the system that the sensors on RSE detect the vehicle's speed violation, and

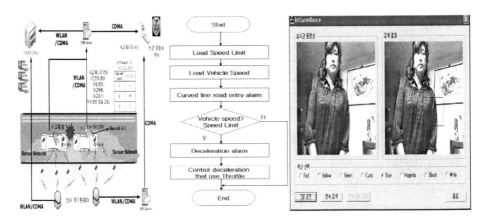

Fig. 3. Communication Flow for Prevention against traffic accident

the RSE sends to the car's ECU (Electronic Control Unit) the commands to reduce its speed in rainy day or when a car passes the accident sector. The driver's risk can be reduced very much if they can predict the road conditions in advance, such that the road is wet, icy or in the fog. Fig. 3 shows a flowchart to prevent traffic accidents.

It could be explained as follows. When an accident occurs or an obstacle is found while a vehicle is driven, the RSE transfers such information. The RSE transfers the information to successive RSE through DSRC (Data Short Range Channel) network. The driver pushes a button to signal when an accident occurs. It could be explained as follows. Vehicles behind can obtain the existence of vehicles ahead by the RSE. Moreover If vehicles in front are in an accident, vehicles are notified of the accident occurrence information.

This paper simulated the system to primarily search ex-convicts (dangerous people group DB) stored in the database for searching the criminal when a crime was arisen around a railway station. Furthermore, the simulation was developed to search the person with the same clothes color of the criminal at the scene of a crime around the station if the clothes color of the criminal was entered at the site. Fig.3 is the screens executing the program, and the real time video area shows the real time video from the USB camera. The color is selected from 8 colors to search for the real time video, and a screen is divided into 16 areas for each frame of the video to check which color is mostly existed in each divided area, and the result is shown in the search result window. Since there is an option of the one time search and the continuous search, it could be decided whether searching the area where the suspect with the color once or continuously after specifying a color. Moreover, this paper proposed an intelligent traffic system which can adjust the speed limit adaptively rather than the fixed, according to weather condition. Fuzzy algorithm is used to improve the existing system to provide real-time traffic information. The traffic controller based on fuzzy logic computes the optimal traffic signal and vehicle speed according to sensor and traffic information. Simulation is carried out to show the validity of our approach, which is based on fuzzy traffic control under intelligent traffic system environment. The method utilizing test car which is used for collecting traffic data at present is still exploited despite of many weak points. This paper assumed that temperature and humidity sensor on the road perceived weather conditions and performed trial examination for automatic speed limit system that changed speed limit into over 1/3 as adjusting speed limit to the weather conditions of rain or snow.

5   Conclusion

In this paper, we proposed an algorithm to automatically control the seat, and notify dangerous traffic conditions for smart car system. The system is needed to adapt to both the driver and the circumstance for maximum convenience and safety. In order to solve these problems, we proposed an algorithm for developing smart vehicles.

Future traffic must be safe. It could be explained as follows. Vehicles behind can know the existence of vehicles ahead by the RSE. If vehicles in front are in an accident, vehicles are notified of the accident occurrence information. Also, people should be joyful to drive a car. Recently, people are interested in smart cars. Every smart car has some electronic equipment so that they can communicate each other. As a result, every car can share information in all places. It will become more convenient for vehicles if they can control smart vehicles using fuzzy rules. This paper proposed an algorithm to automatically control the seat, and notify dangerous traffic conditions. Future vehicle must be both driver-friendly and able to adapt to changes in the road conditions. Through the computer simulation, we proved that the algorithm-implemented vehicle automatically adjusted the mirrors and seat for the maximum comfort and driver's awareness of the circumstance. We used fuzzy neural network algorithm for design of optimal car speed, and proposed the system to prevent intelligently against traffic accidents.

Moreover, this paper simulated the system to primarily search ex-convicts (dangerous people group DB) stored in the database for searching the criminal when a crime was arisen around a railway station. Furthermore, secondly the simulation was developed to search the person with the same clothes color of the criminal at the scene of a crime around the station if the clothes color of the criminal was entered at the site.

Acknowledgment. This work was supported by the Security Engineering Research Center, granted by the Korea Ministry of Knowledge Economy.

References

[1] Hwang, S.: Considerations on Preventing and Solving Hit-and-Run Accidents. Journal of Korea Police Academy (2004)

[2] Kim, W.S., Park, T.U., Lee, S.K.: Ubiquitous Computing's concept and industry same native place,The 1035th ETRI, Week, technology same native place (February 27, 2002)

[3] Kim, B., Park, K., Han, M.: Detecting Traffic Accidents on the Crossroad by Recognizing the Sound from Accidents. In: Joint Spring Academic Conference, Korea Management Academy/Korea Industrial Engineering Academy (2001)

[4] Park, B.H.: A Study of traffic accident estimate model development of Chungchong Province. Journal of Korean Society of Transportation, The 10th book 1, 81 (1995)

[5] Lim, S., Cho, K.: A Study on GSIS to Manage Current Status of Traffic Accidents and to Build DB, Collection of Papers, Korea Civil Engineering Academy (2004)

[6] Gibby, A. R., Reed, T. C, Washington, S. P.: Ferrara., Evaluation of High-Speed Isolated Signalized Intersections in California. Transportation Research Record 1376

[7] Zador, P., Stein, H., Shapiro, S., Ternoff, P.: Effect of Signal Timing on Traffic Flow and Crashes at Signalized Intersections. In: Transportation Research Record, vol. 1010, pp. 1–15. Transportation Research Board, Washington (1985)

[8] Hall, J.W., de Hurtado, M.P.: Effect of Intersection Congestion on Accident Rates. In: Transportation Research Record, vol. 1376, pp. 71–77. Transportation Research Board, Washington (1992)

[9] Barbaresso, James, C.: Flashing Signal Accident Evaluation. In: Transportation Research Record, vol. 956, pp. 25–29. Transportation Research Board, Washington (1984)

[10] Ha, W.K., Kim, D.W., Cheo, N.H.: Ubiquitous collection of books, Ubiquitous IT revolution and the third space. Electron Newspaper Publishing Company (November 2002)

A Fuzzy Framework for Software Libraries Matching

Nicolás Marín, Clara Sáez-Árcija*, and M. Amparo Vila

Intelligent Databases and Information Systems Research Group,
Department of Computer Science and A.I.,
University of Granada, 18071, Granada, Spain
{nicm,clarasa,vila}@decsai.ugr.es
http://idbis.ugr.es

Abstract. The use of intelligent computational techniques to improve the software engineering process is a problem of increasing interest in the literature. This paper is focused on the use of these techniques with the aim to assist in the process of software libraries matching. Most integration systems only take into account the information about the structural component of the concepts. However, many conventional data models consider both structural and behavioral knowledge. In this paper, a novel approach that uses both kinds of knowledge is presented. To do that, multiple criteria that can be taken into account to determine class matches are considered, including those that are related to this *new* behavioral information. These criteria are then integrated into a hierarchical fuzzy multicriteria decision making model.

Keywords: data integration, schema matching, ontology, software libraries, fuzzy decision making.

1 Introduction

Information systems integration is a problem of increasing interest in many areas, as Business Intelligence, Customer Relationship Management, Enterprise Information Portals, E-Commerce, or E-Business. Often, software libraries are part of the information systems, and for this reason, software libraries integration is a topic of increasing interest.

There are many definitions of Information Integration among the literature. Those definitions change depending on the research area (see, for example, [10,12]). In [7], a compendium of some of the referenced definitions can be found:

Definition 1. Information Integration *is the task that aims at building a global system which provides an unified access to the information from many information sources. Those information systems can be distributed (placed in different places), autonomous (independently managed) and heterogeneous (with different software, hardware, data model, etc.).*

---

* Corresponding author.

G. Lee, D. Howard, and D. Ślęzak (Eds.): ICHIT 2011, CCIS 206, pp. 617–624, 2011.

This information integration process has three main steps [7]: *Semantics Extraction*, capturing the semantics from each information source and translating into a same conceptual model; *Schema Integration*, looking for matchings between the conceptual models of the different information sources; *Data Consolidation and Manipulation*, looking for matchings between data instances from the different information sources, and dealing with irregular objects that represent instance matches.

There is much previous work about information integration. Nevertheless, this research does not take into account software libraries as part of the information systems. The proposal introduced in this work intend to aid in the second step of the information integration process, when we deal with java software libraries as data sources.

Current research in information integration aims to automate the schema integration process. This is a topic of increasing interest in the context of schema translation and integration, knowledge representation, machine learning, and information retrieval. Good surveys about the automatic schema matching approaches and their classification are in [6,8,10].

Most of tools for conceptual modeling allow to express information about the structural component of the concepts, and current integration systems use this type of knowledge in the matching process. In our case, the conceptual model is the Java data model, where the concepts are the library classes and its structural component refers to the classes name and fields, as well as to the class hierarchy.

Nevertheless, in some conceptual models (e.g., object-oriented data model), the semantics consists of two types of knowledge: structural and behavioral knowledge, that are related to the structural component and the behavioral component of the model, respectively. In our case, the behavioral component refers to the information about the methods of the classes. The use of this behavioral information can enrich the matching process thanks to additional criteria concerning to the behavioral component.

We use Jar2Ongology ([5]), a tool to semantically models the structure and the behavior of the classes embedded in a jar file. Jar2Ontology can obtain as output an ontology that includes a comprehensive set of information, because it models both the structural and the behavioral knowledge from the java library. With this tool we perform the semantic extraction step in the software libraries integration process. It permits the matching process be afforded from a more complete point of view, using a more comprehensive set of criteria.

In order to find a good matching, not only an appropriate set of criteria (as wide as possible) has to be considered, but also the partial matching degrees (related to each criterion) must be suitably aggregated into a global matching opinion. As we will see, this problem can been formulated as a hierarchical fuzzy multicriteria decision making process. The parametric approach of the model lets users choose the importance of the diverse criteria, as well as the operators that will be used to calculate the criteria aggregation.

The paper is structured as follows. Initially, section 2 is devoted to the description of the criteria that will be used as basic information in the matching

Table 1. Matching Criteria

Matching Criteria		
Class Matching Criteria	Element Level Criteria	Class Name Matching Degree
		Package Matching Degree
		Modifiers Matching Degree
		Interfaces Matching Degree
		Structural Component Matching Degree
		Behavioral Component Matching Degree
	Structure Level Criteria	Graph Based Matching Degree
		Taxonomies Based Matching Degree
Field Matching Criteria	Field Name Matching Degree	
	Modifiers Matching Degree	
	Constraint Matching Degree	
Method Matching Criteria	Method Name Matching Degree	
	Modifiers Matching Degree	
	Return Type Matching Degree	
	Parameters Type Matching Degree	
	Local Variables Matching Degree	
	Invoked Methods Matching Degree	
	Exceptions Matching Degree	
	Code Matching Degree	

process. Then, section 3 shows how to aggregate the partial matching degrees by means of a hierarchical fuzzy multicriteria decision making problem. Finally, in section 4, some concluding remarks and guidelines for future work end the paper.

2   Matching Criteria

As before mentioned, we use Jar2Ontology ([5]) to represent the structure and the behavior of software libraries in OWL (Ontology Web Language). Once the ontologies that model input libraries have been obtained, the next step is to find matchings between the elements of the libraries. To accomplish this task, the first thing to do is to select the criteria to be used in order to find matchings. This section explains the semantic information that can be taken into account and the diverse techniques that are necessary to determine matchings for each criterion.

Table 1 depicts the diverse criteria that can be taken into account in the software libraries matching process.

The criteria that can be used to calculate the partial matching degrees between classes (table 1, section Class Matching Criteria) can be classified as follows:

- *Element Level Criteria*: They compute matching degrees between two classes by analyzing them in isolation, ignoring their relations with other classes. These criteria obtain as a result the matching degrees between the two classes, depending on the classes names, packages, modifiers, interfaces, structural components and behavioral components.

– *Structural Level Criteria*: They analyze how classes appear together in a hierarchy in order to find the matching. Graph based matching degrees and taxonomies based matching degrees are some of this kind of criteria.

When comparing the structural component of both classes (i.e., the set of fields), two mechanisms are needed: a basic mechanism to match fields and a mechanism to compare the sets themselves. The criteria to match fields are based on the metadata about the fields that are modeled in the ontologies, i.e., the field name, type, and modifiers.

A mechanism to compare sets is also needed when comparing the behavioral component of both classes (i.e., the set of methods). In this case, the sets to compare are the sets of methods of the classes. Method matching criteria depend on the metadata about the methods that are modeled in the ontologies, i. e., method name, modifiers, return type, parameters type, local variables, invoked methods, exceptions, and code (this kind of information is usually not taken into account when looking for correspondences, though it is available in many data models - like the object oriented one - and can be of help in the matching process).

In order to obtain some of the partial matching degrees, some matching techniques have to be used (for example, name matching techniques, constraint matching techniques and structural level matching techniques). Interested readers can find good surveys of these techniques in [6,8].

3   Fuzzy Matching Criteria Aggregation

Section 2 has explained a survey of the criteria that can be used to determine the matching degree between two classes. These criteria can be applied in the matching process, obtaining as a result a set of partial matching degrees that have to been aggregated to obtain a global one.

This section shows how they can be aggregated, expressing this problem as a hierarchical fuzzy multicriteria decision making model. First, a hierarchical approach to aggregate the partial degrees is presented. Then, a fuzzy multicriteria decision making model needed to aggregate the partial degrees in each level of the criteria aggregation hierarchy is described.

3.1   Criteria Aggregation Hierarchy

A global matching degree between two clases is an aggregation of the partial matching degrees related to element and structure level criteria, i.e., class name, package, modifiers, interfaces, structural component, behavioral component, graph matching, and taxonomies matching (see table 1).

Some of these partial matching degrees are obtained as an aggregation of another partial degrees. For example, to calculate field matching (table 1), partial degrees are obtained, depending on the used matching criteria: field name, modifiers, and constraints.

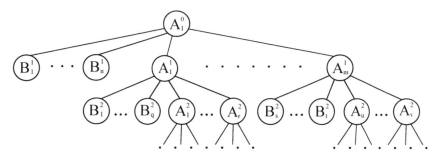

Fig. 1. Criteria Hierarchical Aggregation Tree

Furthermore, there are some criteria that entail comparison of sets of elements, for example, when computing the matching degree according to the interfaces, modifiers, structural component, or behavioral component of the compared clases. There exist remarkable works that take advantage of fuzzy set theory in order of accomplish this comparison problem in the literature, as for example, [1,4].

The methodology that we use is based on the theoretical hierarchical approach of works like [4,2]. Following the before mentioned ideas, a hierarchical model can be built to aggregate all the partial matching degrees and obtain a global one.

Let us formalize this *Model for the Hierarchical Aggregation of Criteria (MHAC)*, that is used to compare two software classes.

Definition 2. *We define a* Model for the Hierarchical Aggregation of Criteria *as a tree that organizes the diverse matching criteria to compare two classes. A graphic representation of this tree is shown in figure 1 and it is described as follows:*

- *The root node A_1^0 represents the global matching degree that determines to which extent two classes match.*
- *Each internal node is calculated as the aggregation of the criteria represented by its children, that are grouped in two types:*
    - Basic nodes: *A basic node B_p^n represents a basic matching technique that provides a partial matching degree in [0,1].*
    - Aggregated nodes: *An aggregated node A_p^n represents a criterion that is the result of the aggregation of many criteria that are represented by its children. This aggregation produces a matching degree in [0,1].*
- *For each internal node A_p^n, there exists an aggregation procedure that combines the decisions of its children.*

This approach to build a MHAC can be used in any integration system, with any kinds of criteria, and aggregation procedure. In this context, a matcher will be a particular tree built following the above definition.

Figure 2 represents a particular example tree generated with the criteria before mentioned in section 2. Nodes in dark background are basic nodes, and nodes in white background are aggregated nodes.

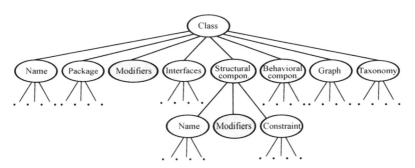

Fig. 2. Criteria Hierarchical Aggregation Tree Instance

3.2   Fuzzy Multicriteria Decision Making Model

Once the criteria are hierarchically organized, an aggregation procedure should be executed. This aggregation procedure can be considered as a hierarchical decision making problem, where soft computing techniques are pretty useful (see [3,9]).

Our proposal is based on two mainstays. On the one hand, we use a parametric approach, which lets the user set up the configuration of the matcher according to the desired comparison semantics. On the other hand, we want the aggregation process to simulate a group of human experts looking for a consensus.

In order to solve this problem, quantified sentences of Zadeh [11] are used. Quantified sentences are usually classified into two classes, namely, type I sentences and type II sentences. A type I sentence is a sentence of the form: *"Q of X are A"*, where $X = \{x_1, ..., x_m\}$ is a finite set, Q a linguistic quantifier, and A a fuzzy property defined over X. A type II sentence can be described in general as: *"Q of D are A"*, where D is also a fuzzy property over X.

We use type II sentences because we want to use importance degrees to weight children opinions, as well as to adjust the weight of such opinions in the aggregation process using importance degrees. Thus, the accomplishment degree of the matching property provided by a given node is computed depending on the accomplishment degree of the quantified sentence

"Q of its important children think that the classes match"

This approach let us work in the nodes in a flexible way, because the user can decide:

– What criteria are more important: providing an importance degree $(w_{X_p^n})$ to each criterion associated with a basic or aggregated node X_p^n. This importance degree indicates the importance of this node when computing the aggregated criterion represented in its parent.
– The used quantifier. A quantifier $(Q_{A_p^n})$ is provided to each criterion associated with an aggregated node A_p^n.

Figure 3 depicts the extension of the tree shown in figure 1, according to the proposed fuzzy model approach. Let us formalize it.

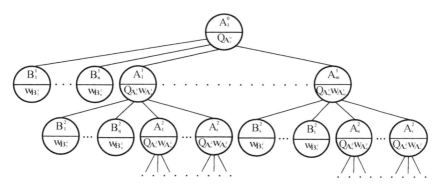

Fig. 3. Fuzzy Criteria Hierarchical Model

Definition 3. *We define a Fuzzy Model for the Hierarchical Aggregation of Criteria as an extension of the MHAC. Let O_1 and O_2 be two sets of classes, X_p^n a node in the tree, the matching degree between $c_i \in O_1, c_j \in O_2$ associated to the node X_p^n, $matching_{X_p^n}(c_i, c_j)$, is computed as follows:*

- *If X_p^n is a basic node (B_p^n): $matching_{B_p^n}(c_i, c_j)$ is the matching degree between c_i and c_j produced by the basic matching technique associated to the node.*
- *If X_p^n is an aggregated node (A_p^n): Let $Children_{A_p^n}$ be the set of children of the node A_p^n in the tree, $matching_{A_p^n}(c_i, c_j)$ is the accomplishment degree of the quantified sentence "$Q_{A_p^n}$ of the criteria that are $W_{A_p^n}$ are $M_{A_p^n}$", where $W_{A_p^n}$ is the fuzzy set induced by the importance degrees $w_{X_k^{n+1}}$ in $Children_{A_p^n}$ and $M_{A_p^n}$ is the fuzzy set induced by the matching opinions $matching_{X_k^{n+1}}(c_i, c_j)$ in $Children_{A_p^n}$.*

4   Conclusion

In this work we have presented a generic model for software libraries matching. The approach takes as input two Java libraries (sets of object code files), and uses two ontologies that model both the structural and behavioral components of the classes. Then, a set of matching criteria are taken into account to determine class matches.

Our proposal innovates mainly in two aspects: On the one hand, the set of matching criteria taken into account is more comprehensive than in current integration systems, specially because we are the only that use both the structural and behavioral component of the concepts. On the other hand, the use of computational intelligence makes our proposal be a parametric approach that lets users choose the importance of the diverse criteria, as well as the operators that will be used to calculate the criteria aggregation.

This matching model lets the user set up the importance of each criterion in fuzzy terms, and choose the fuzzy operators and quantifiers to use in the aggregation of the partial matching degrees (related to each criterion).

Acknowledgment. Part of the research reported in this paper is supported by the Andalusian Government (Junta de Andalucía, Consejería de Economía, Innovación y Ciencia) under project P07-TIC-03175 "Representación y Manipulación de Objetos Imperfectos en Problemas de Integración de Datos: Una Aplicación a los Almacenes de Objetos de Aprendizaje".

References

1. Hallez, A., Bronselaer, A., De Tré, G.: Comparison of Sets and Multisets. International Journal of Uncertainty, Fuziness and Knowledge-based Systems 17, 153–172 (2009)
2. Hallez, A., De Tré, G.: A hierarchical approach to object comparison. In: Melin, P., Castillo, O., Aguilar, L.T., Kacprzyk, J., Pedrycz, W. (eds.) IFSA 2007. LNCS (LNAI), vol. 4529, pp. 191–198. Springer, Heidelberg (2007)
3. Lu, J., Zhang, G., Ruan, D.: Intelligent Multi-Criteria Fuzzy Group Decision-Making for Situation Assessments. Soft Computing 12, 289–299 (2008)
4. Marín, N., Medina, J.M., Pons, O., Sánchez, D., Vila, M. A.: Complex object comparison in a fuzzy context. Information and Software Technology 45, 431–444 (2003)
5. Marín, N., Sáez-Árcija, C., Vila, M.A.: Jar2Ontology: A Tool for Automatic Extraction of Semantic Information from Java Object Code. In: Proceedings of the 9th International Conference on Enterprise Information Systems, ICEIS (2011)
6. Rahm, E., Bernstein, P.A.: A Survey of Approaches to Automatic Schema Matching. The VLDB Journal (2001)
7. Sáez-Árcija, C., Marín, N., Vila, M.A.: A Lazy-Typing Based Architecture for a Data Integration System. In: Workshop on New Trends on Intelligent Systems and Soft Computing, vol. 2, pp. 1–18 (2009)
8. Shvaiko, P., Euzenat, J.: A Survey of Schema-based Matching Approaches. Journal on Data Semantics(JoDS) (2005)
9. Vaníček, J., Vrana, I., Aly, S.: Fuzzy Aggregation and Averaging for Group Decision Making: A Generalization and Survey. Knowledge-Based Systems 22, 79–84 (2009)
10. Wache, H., Vögele, T., Visser, U., Stuckenschmidt, H., Schuster, G., Neumann, H., Hübner, S.: Ontology-Based Integration of Information - A Survey of Existing Approaches. In: Workshop on Ontologies and Information Sharing at the International Joint Conference on Artificial Intelligence (IJCAI), pp. 108–117 (2001)
11. Zadeh, L.A.: A Computational Approach to Fuzzy quantifiers in Natural Languages. Comp. and Maths. with Appls. 9(1), 149–184 (1983)
12. Ziegler, P., Dittrich, K.R.: Three Decades of Data Integration: All Problems Solved? In: 18th IFIP World Computer Congress (WCC 2004), IFIP International Federation for Information Processing, pp. 3–12 (2004)

Simulation of Non-linear Singular System Using RK-Butcher Algorithm

V. Murugesh[1], K. Murugesan[2], and Kyung Tae Kim[3]

[1] Department of Microsoft Information Technology, Keimyung Adam College,
Keimyung University, 2800 Dalgubeoldaero Dalseo-Gu, Daegu 704-70, Republic of Korea
murugesh72@gmail.com
[2] Department of Mathematics, National Institute of Technology, Tiruchirappalli – 620 015,
Tamilnadu, India
[3] Department of Information and Communications Engineering, Hannam University,
133 Ojung-dong, Daejeon 306-791, Republic of Korea
ktkim@hnu.kr

Abstract. In this paper, a new method of study on non-linear singular systems from fluid dynamics using the RK-Butcher algorithm is presented. To illustrate the effectiveness of the RK-Butcher algorithm, four cases in non-linear singular systems from fluid dynamics have been considered and compared with the classical fourth order Runge-Kutta, and are found to be very accurate. Local truncation error graphs for the non-linear singular system based nuclear reactor core problem are presented in a graphical form to show the efficiency of this RK-Butcher method. This RK-Butcher algorithm can be easily implemented in a digital computer and the solution can be obtained for any length of time.

Keywords: Runge-Kutta method fourth order, RK-Butcher Algorithm, Non-linear singular systems and Ordinary differential equations.

1 Introduction

Runge –Kutta (RK) methods have been used by many researchers [1-15] to determine numerical solutions for the problems, which are modeled as Initial Value Problems (IVPs) involving differential equations that arise in the fields of Science and Engineering. Although the RK method was introduced at the beginning of the twentieth century, research in this area is still very active and its applications are enormous because of its nature of extending accuracy in the determination of approximate solutions and its flexibility.

Butcher [16] derived the best RK pair, together with an error estimate and this is known as the RK-Butcher algorithm. It is nominally considered sixth order; since it requires six function evaluations (it looks like a sixth-order method, but in fact is fifth-order method). In practice the "working order" is closer to 5 (fifth order), but accuracy of the results obtained exceeds all the other algorithms examined, including RK-Fehlberg, RK-Centroidal mean and RK-Arithmetic mean methods.

G. Lee, D. Howard, and D. Ślęzak (Eds.): ICHIT 2011, CCIS 206, pp. 625–632, 2011.
© Springer-Verlag Berlin Heidelberg 2011

Bader [17-18] introduced the RK-Butcher algorithm for finding the truncation error estimates and intrinsic accuracies and the early detection of stiffness in coupled differential equations that arises in Theoretical Chemistry Problems. Murugesan et al. [2] discussed the non-linear singular systems from fluid dynamics using the RK-methods based on variety of means. In this paper, we consider the same non-linear singular systems from fluid dynamics (discussed by Murugesan et al [2]) but presenting a different approach by the RK-Butcher algorithm with more accuracy.

2 Representation of Equations of Flow as a Non-linear Systems

The simplified model consists of two connected sub channels filled with a steadily flowing fluid. Control volumes and flow variables for the system. Here, m_i , represents the axial mass flow rate in sub channel i and w represents the cross-flow rate per unit length, assumed positive if the flow is from sub channel 1 to sub channel 2.

An application of the principles of conservation of mass, momentum and energy to the control volumes yields the following set of equations for sub channel 1.

Continuity   : $\dfrac{dm_i}{dx} = -w$ (3)

Axial momentum: $\dfrac{d}{dx}(m_1 u_1) + w.[H(w)u_1 + H(-w)u_2] = -F_1 - A_1 \dfrac{dp_1}{dx}$ (4)

Energy   : $\dfrac{d}{dx}(m_1 h_1) = q_1 - w.[H(w)h_1 + H(-w)h_2]$ (5)

Analogous equations for sub channel 2 can be obtained from these by substituting $-w$ for w and by interchanging subscripts 1 and 2. In this equation set, H is the Heaviside unit step function. F represents pressure loss per unit length due to friction, A is the cross-sectional area, q represents the heat energy added per unit length, and the variables, u, p and h stand for particle velocity, pressure and enthalpy respectively.

In analogy with the pressure drop due to friction in a long pipe, a lateral momentum balance may be taken as $p_1 - p_2 = Cw|W|$, where C is a cross-flow friction factor.

To simplify the above equation, the following assumptions are made. Cross-sectional area is constant; the coolant is incompressible; there is no enthalpy change; and the frictional pressure loss function is of the form $F_1 = m_1 u_1 F$, where F is a constant.

With these assumptions, the equations may be combined and written in the following form:

$$\frac{dm_1}{dx} = -w$$

$$\frac{d}{dx}(w|w|) = \epsilon^{-1}\left\{\frac{1 - 2m_1}{2} + 2w[1 - H(w)m_1 + H(-w)(m_1 - 1)]\right\}$$ (6)

To make the above system (6) into the symmetric form, take $x = m_1 - \frac{1}{2}, y = \frac{w}{2}$ and $t = x$.

Hence we get $\frac{dx}{dt} = -2y$

$$\frac{d}{dt}(y|y|) = (4\varepsilon)^{-1}[-x + 2(y - 2x|y|)] \tag{7}$$

Replacing x by x_1 and y by x_2, we have

$$\dot{x}_1 = -2x_2$$

$$\frac{d}{dt}(x_2|x_2|) = (4\varepsilon)^{-1}[-x_1 + 2(x_2 - 2x_1|x_2|)] \tag{8}$$

An analysis is carried out in four different ways depending upon the values of x_2 and ε as given below :

$(i)\, x_2 > 0 \ \text{and}\ \varepsilon \neq 0$

$(ii)\, x_2 < 0 \ \text{and}\ \varepsilon \neq 0$

$(iii)\, x_2 > 0 \ \text{and}\ \varepsilon = 0$

$(iv)\, x_2 < 0 \ \text{and}\ \varepsilon = 0$

In the first two cases the parameter ε has been varied from $10^0, 10^1, 10^2, \ldots, 10^7$ and in the last two cases, ε has been set to zero.

Case (i) : When $x_2 > 0$ and $\varepsilon \neq 0$
In this case eq. (8) becomes

$$\dot{x}_1 = -2x_2$$

$$8\varepsilon x_2 \dot{x}_2 = -x_1 + 2x_2 - 4x_1 x_2 \tag{9}$$

The above two equations can be considered as a system of equations of the form

$$\begin{bmatrix} 1 & 0 \\ 0 & 8\varepsilon x_2 \end{bmatrix}\begin{bmatrix} \dot{x}_1 \\ \dot{x}_2 \end{bmatrix} = \begin{bmatrix} 0 & -2 \\ -1 & 2 \end{bmatrix}\begin{bmatrix} x_1 \\ x_2 \end{bmatrix} + \begin{bmatrix} 0 \\ -4x_1 x_2 \end{bmatrix} \tag{10}$$

This is of the form

$$K(x(t))\dot{x}(t) = Ax(t) + f(x(t)) \tag{11}$$

The first order non-linear system (10), representing the highly simplified two channel model of a nuclear reactor core from fluid dynamics, when $x_2 > 0$ and $\varepsilon \neq 0$, can be converted into a second order equation in order to reduce the number of equations, as well as the number of unknowns, and is given as

$$\ddot{x}_1 = \frac{1}{2\varepsilon}\left[\frac{-x_1}{\dot{x}_1} - 1 + 2x_1\right], \tag{12}$$

where $x_2 = \frac{-\dot{x}_1}{2}$

Hence eq. (12) is of the form $\ddot{x}_2 = \phi(\varepsilon) f(t, x_1, \dot{x}_1)$,

where $\phi(\varepsilon) = \dfrac{1}{2\varepsilon}$

Case (ii) : When $x_2 < 0$ and $\varepsilon \neq 0$

In this case eq. (8) becomes

$$\dot{x}_1 = 2 x_2$$
$$8\varepsilon x_2 \dot{x}_2 = x_1 + 2x_2 + 4x_1 x_2 \tag{13}$$

i.e.,

$$\begin{bmatrix} 1 & 0 \\ 0 & 8\varepsilon x_2 \end{bmatrix} \begin{bmatrix} \dot{x}_1 \\ \dot{x}_2 \end{bmatrix} = \begin{bmatrix} 0 & 2 \\ 1 & 2 \end{bmatrix} \begin{bmatrix} x_1 \\ x_2 \end{bmatrix} + \begin{bmatrix} 0 \\ 4 x_1 x_2 \end{bmatrix} \tag{14}$$

This is also of the form (11)

The first order non-linear system (14), representing the highly simplified two channel model of a nuclear reactor core from fluid dynamics, when $x_2 < 0$ and $\varepsilon \neq 0$, can be converted into a second order equation in order to reduce the number of equations, as well as the number of unknowns, and is given as

$$\ddot{x}_1 = \frac{1}{2\varepsilon} \left[\frac{x_1}{\dot{x}_1} + 1 + 2 x_1 \right], \tag{15}$$

where $x_2 = \dfrac{\dot{x}_1}{2}$

Hence eq (15) is of the form $\ddot{x}_1 = \phi(\varepsilon) f(t, x_1, \dot{x}_1)$,

where $\phi(\varepsilon) = \dfrac{1}{2\varepsilon}$

Case (iii) : When $x_2 > 0$ and $\varepsilon = 0$

In this case eq.(8) becomes

$$\dot{x}_1 = -2 x_2$$
$$0 = -x_1 + 2 x_2 - 4 x_1 x_2 \tag{16}$$

i.e.,

$$\begin{bmatrix} 1 & 0 \\ 0 & 0 \end{bmatrix} \begin{bmatrix} \dot{x}_1 \\ \dot{x}_2 \end{bmatrix} = \begin{bmatrix} 0 & -2 \\ -1 & 2 \end{bmatrix} \begin{bmatrix} x_1 \\ x_2 \end{bmatrix} + \begin{bmatrix} 0 \\ -4 x_1 x_2 \end{bmatrix} \tag{17}$$

The system (17) is a singular non-linear system and it is of the form (11). This system can be written as

$$\dot{x}_1 = \frac{x_1}{2 x_1 - 1} \quad \text{and} \quad \dot{x}_2 = \frac{-\dot{x}_1}{2}$$

The above equations has been converted into a second order equations as

$$\ddot{x}_1 = \frac{-\dot{x}_1}{(2 x_1 - 1)^2}, \tag{18}$$

where $x_2 = \dfrac{-\dot{x}_1}{2}$

Hence eq. (18) is of the form $\ddot{x}_1 = f(t, x_1, \dot{x}_1)$,

Case (iv) : When $x_2 < 0$ and $\varepsilon = 0$

In this case eq. (8) becomes

$$\dot{x}_1 = 2x_2$$
$$0 = -x_1 - 2x_2 - 4x_1x_2 \tag{19}$$

i.e.
$$\begin{bmatrix} 1 & 0 \\ 0 & 0 \end{bmatrix}\begin{bmatrix} \dot{x}_1 \\ \dot{x}_2 \end{bmatrix} = \begin{bmatrix} 0 & 2 \\ -1 & -2 \end{bmatrix}\begin{bmatrix} x_1 \\ x_2 \end{bmatrix} + \begin{bmatrix} 0 \\ -4x_1x_2 \end{bmatrix} \tag{20}$$

The system (20) is a singular non-linear system and it is of the form (11). This system can be written as

$$\dot{x}_1 = \frac{-x_1}{1+2x_1} \quad \text{and} \quad \dot{x}_2 = \frac{\dot{x}_1}{2}$$

Further, the above equations has been converted into a second order equations as

$$\ddot{x}_1 = \frac{-\dot{x}_1}{(2x_1-1)^2}, \tag{21}$$

where $x_2 = \dfrac{-\dot{x}_1}{2}$

Hence eq. (21) is of the form $\ddot{x}_1 = f(t, x_1, \dot{x}_1)$,

3 Numerical Methods and Results in Graphical Form

The objective of this chapter is to find discrete solutions to the simplified two channel model of a nuclear reactor core from fluid dynamics under all four cases discussed in the section 2. In the case of (i) and (ii), the system has been reduced to a singular system, which is further converted into a second order equation.

It is very difficult to obtain the exact solution of this non-linear equation. Hence it has been analyzed by the following numerical methods by the way of determining the discrete solutions at different time intervals:

   (i) Classical fourth order Runge-Kutta method (RK(4)).
   (ii) RK-Butcher Algorithm (RK-B) discussed by various authors [7-8, 19-21].

The discrete solutions of a two channel model of a nuclear reactor core problem for the cases   (i) when $x_2 > 0$ and $\varepsilon \neq 0$ (ii) when $x_2 < 0$ and $\varepsilon \neq 0$ [i.e., equations(12) and (15)] have been determined using the methods RK(4) and RK-Butcher by varying the parameter ε from 10^0 to 10^7 with $x_1(0) = 1$, $\dot{x}_1(0) = 1$ and the discrete solution for the cases (iii) when $x_2 > 0$ and $\varepsilon = 0$ (iv) when $x_2 < 0$ and $\varepsilon = 0$, [i.e., singular systems] have been determined using RK(4) with $x_1(0) = 1$, $\dot{x}_1(0) = 1$

To exhibit the efficiency of the discussed methods, an error graph is presented for the observer variable $x_1(t)$ and $x_2(t)$ in figures 1 to 4 at various time intervals up to the time $t = 5.0$ and from this it is observed that the RK–Butcher algorithm gives more accurate results when compared with the STWS-I and STWS-II methods discussed by Murugesan *et al.* [2].

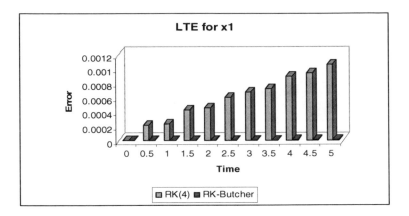

Fig. 1. Local Truncation Error graph for x_1 case (i) at $\varepsilon = 10^7$

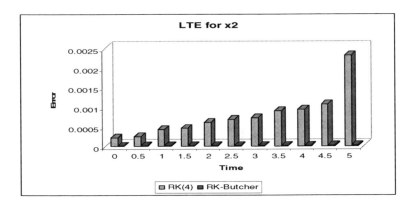

Fig. 2. Local Truncation Error graph for x_2 case (ii) at $\varepsilon = 10^7$

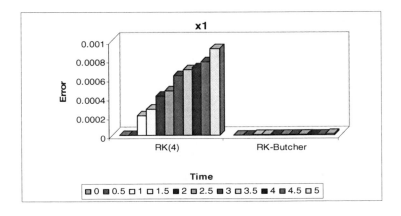

Fig. 3. Local Truncation Error graph for x_1 case (iii)

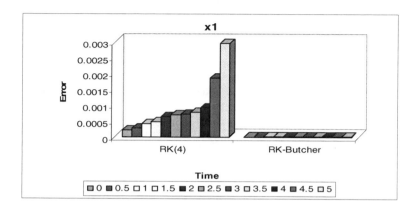

Fig. 4. Local Truncation Error graph for x_1 case (iv)

4 Conclusions

The nuclear reactor core problem has been studied under four different cases (specified in section 2) by way of determining the discrete solutions for different time 't' using the classical fourth order Runge-Kutta method and RK-Butcher algorithm. In [2], for the same problem, the approximate solution was determined using STWS method and it was mentioned that the classical RK method failed to obtained approximate solutions when the parameter $\varepsilon \geq 10^{-3}$. But in this paper, it has been established that the Runge-Kutta methods are adequate enough to determine approximate solutions for all values of ε (i.e., $\varepsilon = 0, 10^{0}, 10^{-1}, \ldots, 10^{-7}$). In cases (iii) and (iv), when $\varepsilon = 0$, the system reduces to a singular system for both $x_2 > 0$ and $x_2 < 0$. It is observed that, for a singular system, the discrete solutions obtained by the classical Runge-Kutta method and RK-Butcher algorithms are found to be similar. By observing the graphs in the Figures-1 to 4 one could also easily visualize the Local truncation errors obtained by classical fourth order Runge-Kutta and RK-Butcher algorithm. Hence, by comparing the results obtained for the nuclear reactor core problem discussed under four cases; the RK-Butcher Algorithm is more suitable for studying the nuclear reactor core problem.

References

1. Butcher, J.C.: Numerical Method for Ordinary Differential Equations. JohnWiley & Sons, Chichester (2003)
2. Murugesan, K., Paul Dhayabaran, D., Evans, D.J.: Analysis of non-linear singular system from fluid dynamics using extended Runge-Kutta methods. Int. J. Comput. Math. 76, 239–266 (2000)
3. Murugesan, K., et al.: A comparison of extended Runge-Kutta formulae based on variety of means to solve system of IVPs. Int. J. Comput. Math. 78, 225–252 (2001)

4. Murugesan, K., et al.: A fourth order embedded Runge-KuttaRKACeM(4,4) method based on arithmetic and centrodial means with error control. Int. J. Comput. Math. 79(2), 247–269 (2002)
5. Murugesan, K., et al.: Numerical solution of an industrial robot arm control problem using the RK–Butcher algorithm. Int. J. Comput. Appl. Technol. 19(2), 132–138 (2004)
6. Murugesan, K., Murugesh, V.: RK-Butcher algorithms for singular system-based electronic circuit. Int. J. Comput. Math. 86(3), 523–536 (2009)
7. Murugesh, V.: Raster cellular neural network simulator for image processing applications with numerical integration algorithms. Int. J. Comput. Math. 86(7), 1215–1221 (2009)
8. Murugesh, V.: Image Processing Applications via Time-Multiplexing Cellular Neural Network Simulator with Numerical Integration Algorithms. Int. J. Comput. Math. 87(4), 840–848 (2010)
9. Alexander, R.K., Coyle, J.J.: Runge-Kutta methods for differential-algebric systems. SIAM J. Numer. Anal. 27(3), 736–752 (1990)
10. Evans, D.J.: A new 4th order Runge-Kutta method for initial value problems with error control. Int. J. Comput. Math. 139, 217–227 (1991)
11. Evans, D.J., Yaakub, A.R.: A new fifth order weighted Runge-Kutta formula. Int. J. Comput. Math. 59, 227–243 (1996)
12. Evans, D.J., Yaakub, A.R.: Weighted fifth order Runge-Kutta formulas for second order differential equations. Int. J. Comput. Math. 70, 233–239 (1998)
13. Lambert, J.D.: Numerical Methods for Ordinary Differential Systems. The Initial Value Problem. JohnWiley & Sons, Chichester (1991)
14. Yaakub, A.R., Evans, D.J.: A fourth order Runge-Kutta RK(4, 4) method with error control. Int. J. Comput. Math. 71, 383–411 (1999a)
15. Yaakub, A.R., Evans, D.J.: New Runge-Kutta starters of multi-step methods. Int. J. Comput. Math. 71, 99–104 (1999)
16. Butcher, J.C.: The Numerical Analysis of Ordinary Differential Equations: Runge-Kutta and General Linear Methods. JohnWiley & Sons, Chichester (1987)
17. Bader, M.: A comparative study of new truncation error estimates and intrinsic accuracies of some higher orderRunge-Kutta algorithms. Computational Chem. 11, 121–124 (1987)
18. Bader, M.: A new technique for the early detection of stiffness in coupled differential equations and application to standard Runge-Kutta algorithms. Theor. Chem. Accounts 99, 215–219 (1998)
19. Park, J.Y., et al.: Optimal control of singular systems using the RK–Butcher algorithm. Int. J. Comput. Math. 81(2), 239–249 (2004)
20. Shampine, L.F.: Numerical Solution of Ordinary Differential Equations. Chapman and Hall, NewYork (1994)
21. Shampine, L.F., Gordon, M.K.: Computer Solution of Ordinary Differential Equations-The Initial Value Problem. W.H. Freeman, San Francisco (1975)

A Memory Access Pattern Based Data Distribution Technique for Array Processors

Doosan Cho

Department of Electronic Engineering, Sunchon National University
dscho@sunchon.ac.kr

Abstract. The successful use of array-type multiprocessors strongly depends on how to efficiently distribute data and code in target applications. However, producing efficient code for such multiprocessors is a difficult task. Hand-coding efficient parallel programs for these processors can be extremely difficult, time consuming and error-prone, so developers have turned to automatic compiler techniques to ease the task.

To efficiently execute an multimedia application on array processors, a good data partitioning is required. Default partitioning such as row-wise or column-wise may not produce a good data distribution and, thus, will prolong the total execution time of an application.

The two main approaches to this are using compilers that 1) generate data transfer code, and/or 2) generate code for data distribution. Neither has been completely successful for all types of programs.

In this paper, we propose efficient methods for finding the data distribution for array processors while minimizing the total execution time. The experimental results indicate that our noble technique would be able to support many user programs efficiently.

Keywords: Compiler, Memory Hierarchy, Performance, Data Partitioning, Optimization.

1 Introduction

Array processors are a popular choice for image processing. The main components of array processors include the PE (Processing Element) array and the local memory. The PE array is a 2D array of possibly heterogeneous PEs connected with a mesh interconnect, though the exact topology and the interconnects are architecture dependent.

The local memory is typically a high speed, high bandwidth, highly predictable random access memory that provides temporary storage space for array data, which are often input/ output of loops that are mapped to array PEs. To provide high bandwidth, local memories are often distributed due to their relatively low cost and potentially high performance. The main drawback of the distributed local memory system is that the most natural and efficient way to program these architectures is via explicit data transfer codes, which is well known to be a difficult programming model and can make the cost of developing and debugging complex programs intolerably high. All data is explicitly moved between PEs via operations in the user program.

G. Lee, D. Howard, and D. Ślęzak (Eds.): ICHIT 2011, CCIS 206, pp. 633–640, 2011.
© Springer-Verlag Berlin Heidelberg 2011

We are primarily addressing two areas. The both are the automatic distribution of data and generation of data transfer codes. These areas have been studied extensively over the last decade. Much work has been done on clever ways to optimize the data transfer codes generated by a compiler.

Automatic data distribution and data transfer code generation tend to be successful in situations where the data access pattern in a program is primarily simple and completely understood by the compiler. It tends to have excessive unnecessary data transfer overhead even for simple access patterns. Thus, we tackle the problem of inefficiently generated data transfer codes in this work.

The remaining parts of this paper are organized as follows. Section 2 discusses the existing work for array processor architectures. Section 3 describes overall workflow of this work. Section 4 gives the detailed algorithm that analyzes data access pattern and generate data distribution. Section 5 provides some experimental results. Section 6 talks about some future work and concludes this work.

2 Related Work

A compiler of an array processor must take responsibility for collecting and analyzing the data access pattern information from a sequential code, determining the distribution of data, and then inserting data transfer codes in the proper places to exchange data between PEs.

Research on data transfer generation of compilers began in the early 1990s. People soon found out that the efficiency of generated code relied heavily on its data distribution. Improper data distribution decisions could degrade the performance significantly. Since then, many efforts have been made, and are continuously being made, for developing automatic data distribution algorithms, which play an important role in parallelizing compilers [9,1,4,2,6]. Many algorithms try to pick a near-optimal solution from the possible distribution candidates, while finding the optimal solution is an NP-complete problem [7]. But in the real world, applications are so varied that most of these algorithms can only succeed in limited cases. This is not only because of the program's complexity, but also because often run-time input information that is necessary for the data distribution algorithm is unavailable at compile time.

However, the overhead for maintaining the runtime information will exist. Some researchers [5] have proposed simple schemes to handle limited special cases, but in general, it is quite difficult to reduce the extra overhead. Research [3], mostly done on small benchmark codes or application kernels, indicates that the compiler targeting data transfer code is more effective on programs with regular access patterns.

The difference between them and ours is that ours can easily take advantage of any new improvement in the underlying data transfer codes without changing the interface, as well as the improvements in compiler techniques in the field of explicit data transfer.

3   Overview of the Code Generating Framework

We give a brief outline of the proposed algorithm framework here. To describe the outline, we firstly classified data by its usage patterns. Data accesses can be classified into two classes: private data and distributed (shared) data.

Private data consists of variables that are only accessed by a single PE. No transfer is needed for this data. Distributed data consists of variables, with simple usage patterns, which might need to be moved between PEs. Movement of these variables is handled by the transfer codes. Based on the classification, the algorithm consists of three phases, preceded by the parallelization of the program targeting to PE array.

- Parallelization

This phase produces a parallel program for a certain array processor. It may be done by a parallelizing compiler, or by coding the program in an explicitly parallel fashion by hand. In either case, the result of this phase is a parallel program with variables privatized as necessary, reductions indicated, and parallel regions marked. In this step, we assume that data will be privatized based on a single parallel loop nest.

- Data classification and distribution

In this phase, we classify unprivatizable user data as distributed data. Distributed data is divided among the PEs. The movement of distributed data will be handled by the explicit data transfer codes. Data access patterns are analyzed in this phase. Detailed algorithms of data access pattern analysis and pattern classification are represented in Section 4.1 and Section 4.2 respectively.

Data having simple-enough access patterns is considered as distributed data. All the other data in the program is considered as non-distributable data, except the privatizable data, which was identified in the preliminary parallelization. The definition of simple-enough depends on the data transfer capabilities of the compiler.

- Data transfer code generation

Codes for data transfers are generated and inserted in this phase. We generate explicit data transfer codes for the distributed data. In the following sections, we will discuss the second phase in more detail.

4   The Proposed Approach

4.1   Summarizing Data Access Patterns

Prior techniques for array access analysis sometimes fail because they are unable to recognize some hidden simple access regions. To overcome this limitation, we use a linear array access descriptor, which is developed from [11], generated from memory access footprints. To create the footprint, we perform profiling several times. Using the profiled information, access descriptors represent the access pattern precisely and enable analysis techniques to expose the simplicity of array access patterns.

A linear array access descriptor is described by the triple $start + [stride, span]$. The *start* is the offset, from the first element of the array, for the first location accessed. A dimension is a movement through memory with a consistent stride and a computable number of steps. The *stride* gives the distance between two consecutive array accesses in one dimension. The *span* is the distance (in memory units) between the offsets of the first and last elements that are accessed in one dimension.

For example, an access footprint of 0,8,16, ..., 80 will be described by a linear array access descriptor as 0+[8,80]. If the array is accessed in a two level nested loop, which the outer loop has step 1 and span 5, the descriptor might be a multi-dimension pattern like 0+[1,5][8,80].

4.2   Data Classification

In this section we describe data access pattern classification procedure by intersecting both linear array access descriptors. Intersecting two arbitrary array access descriptors is very complex and probably intractable. But if two array descriptors have the same strides, or the strides of one are a subset of the strides of the other, which has been quite often true in our experiments, then they are similar enough to make the intersection algorithm tractable.

To illustrate intersection with two array access pattern descriptors, we present in Figure 1 one simple intersection algorithm, which accepts two descriptors $A1$ and $A2$, and produces a set of array access pattern descriptors that summarize the array regions represented by $A1 \cap A2$. The output is a linear array access pattern descriptor set **Overlap**. An array descriptor set called **Non-overlap** represents the area of $A1$ that does not overlap the area of $A2$.

```
Input: linear array access descriptors A1=start1+[s1, span1], A2=start2+[s2, span2]

Procedure Intersection:
  if  start2 − start1 > 0 then
    if  start2 − start1 − span1 > 0 then return {A}; fi
```

$$l' = 0; \; l = \left\lceil \frac{start2 - start1 - 1}{s1} \right\rceil \cdot s1 + s1 \; ;$$

```
    Non-overlap = {start1+[s1, l − s1]};
  else
    if start1 − start2 − span2 > 0 then return {A}; fi
```

$$l = 0; \; l' = \left\lceil \frac{start1 - start2 - 1}{s2} \right\rceil \cdot s2 + s2 \; ;$$

```
  fi

  if start2 + span2 − start1 − span1 ≥ 0 then
```

$$u = span1; \; u' = span2 - s2 - \left\lceil \frac{start2 - span2 - start1 - span1 - 1}{s2} \right\rceil \cdot s2 \; ;$$

```
    Non-overlap = Non-overlap ∪ {start1 + u + s1 + [s1, span1 − s1 − u]};
  else
```

$$u' = span2; \; u = span1 - s1 - \left\lceil \frac{start1 - span1 - start2 - span2 - 1}{s1} \right\rceil \cdot s1 \; ;$$

```
  fi

  if u − l < 0 or u' − l' < 0 then return Non-overlap; fi
```
S_{LC} = Least Common Multiple (s1, s2)

Sub1 = sub-region descriptors of start1 + l + $[S_{LC}$, u − l];

Sub2 = sub-region descriptors of start2 + l' + $[S_{LC}$, u' − l'];

Overlap = Common_descriptors(Sub1, Sub2);

return Overlap;

Fig. 1. The algorithm for finding the intersection of both linear array descriptors

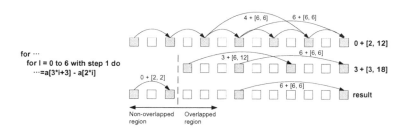

Fig. 2. Intersection of the two references

To explain this algorithm with an example, consider the code in Figure 2. In order to partition the array a within the loop of Figure 2, we would have to identify what region of the array a is overlapped. For the code in Figure 2, we would need to perform $0+[2,12] \cap 3+[3,18]$, then determine whether the result is empty. According to the intersection algorithm, the nonoverlapping area in $0+[2,12]$ should be first found, as shown in Figure 2:

Non-overlap $= 0 + [2, 2]$.

In the overlapping area, we intersect $4+[2,12]$, which is the subregion of $0+[2,12]$ in the area, with $3+[3,18]$. For this, we first find the least common multiple (LCM) of the strides of both accesses, LCM(2,3)=6. Then, we calculate a set of sub-region descriptors (which have θ as a stride) for $4+[2,12]$, which is
$S1 = \{4+[6,6], 6+[6,6]\}$,
and a set of sub-region descriptors, which have stride=θ, for $3+[3,12]$, which is
$S2 = \{3+[6,12], 6+[6,6]\}$.
By intersecting S2 with S1, we can obtain the results in the overlapping area. The operation S1 $\cap$ S2 is straightforward since they have a common stride 6; that is, it can be performed by simply comparing the elements of the sets. This results in:
$Overlap = S1 \cap S2 = \{6+[6,6]\}$.
This process continues until it can either be determined that no overlap occurs, or until the inner-most dimension is reached where it can make the final determination as to whether there is an intersection between the two.

4.3   Data Distribution

In the first phase (parallelization), privatizable data is recognized and parallelizable loops are marked. In the next phase, we further classify the remaining user data, which is involved in parallel loops, as distributed and non-distributed data. Notice that the algorithms used in our framework are quite simple as shown in Figure 3.

The input to the algorithm is the set of references to a given array, found in a program, and the loop information for loops surrounding each reference. The output of the algorithm is the distribution decision for the array. The distribution of each dimension of array is considered separately in this algorithm. The function of Step 1 computes the access pattern for every dimension of each data reference with respect to the loop

```
                    Algorithm of data distribution
Input:
    Each Array,
    The reference set of array and involved loop information.
Output:
    The distribution for each array.

// the first step creates array access pattern descriptors for each array
GenerateAccessPatterns(Array* array){
    For(every reference of array)
        For(every dimension of the reference)
            ComputeAccessPattern(ref(array))
        EndFor
    EndFor
}

// step 2 Initilization
    Initialize all the access pattern descriptors of the arrays as "*" (Undecided)

Main Procedure(array)
    GenerateAccessPatterns(array)

//step 3 determines distributability of arrays by using their descriptors
    for(each descriptors of array)
        Analyze(access pattern descriptor)
        if(the corresponding loop of this dimension is parallelizable)
            if(access pattern is separated or simply overlapped)
                DistributableSet(thisDescriptor)
            else
                nonDistributableSet(thisDescriptor)
            endif
        else
            nonDistributableSet(thisDescriptor)
        endif
    endFor

//the step 4 makes decision to distribute array
    for(each element of DistriburableSet)
        if(access pattern is separated)
            each accessed elements of array can be placed on each PE
        else//then they are simply overlapped
            each overlapped elements of array can be replicated on PEs which access
them
        endif
    endfor
endprocedure
```

Fig. 3. Data distribution algorithm

index. The access pattern could be non-overlapped, simply overlapped, or irregular, as described in Section 4.2. This is the basic step for the following analysis. In the next step of the algorithm, the distribution decision for each descriptor is initialized as *, meaning the distribution is undecided in this dimension. In Step 3, we decide the possible distribution for each array reference. Notice that in this step, we only consider the distribution of distributable data in the scope of an individual reference. Thus, the distributions are local decisions. We analyze the array access pattern in each loop with the procedure described in Section 4.1. If a loop is parallelizable, and the access pattern of the corresponding dimension where the loop index is involved is non-overlapped or simply overlapped, then we determine that the involved array in the loop can be distributable.

If the access pattern is too complicated, or cannot be precisely described at compile time, we determine that such array is non-distributable. If a loop is sequential, we also make decision the array to be involved in the non-distributable set, because in a parallel program, a sequential loop implies the data is not distributed among PEs.

The reference-based local distribution decisions in step 3 may conflict with each other for a given array dimension. The conflict is resolved based on the precedence of the types of distributions. non-Distributable has higher precedence than Distributable, which has higher precedence than *. By using the precedence rule, we can solve the conflict of distribution in the scope of the whole program.

In this algorithm, for simplicity, we only consider two distribution patterns: independently distributed array data and replicated array data, based on which can be used most efficiently. In Step 4, we decide which distribution pattern we should choose for each distributable array. For each parallel loop, we classify accessing patterns of elements being accessed in each distributable reference to overlapped or separated. If the patterns are involved in separated group, then they can be placed on PEs independently. Otherwise, accessed elements of array should be replicated to each PE to distribute such array.

5   Experiments

We have designed the framework of our technique for ICD-C [10], a source-to-source translator developed at the University of Dortmund. ICD-C analyzes sequential codes and provides the information necessary to generate data transfer codes. The actual experiments were conducted on a coarse grained reconfigurable array processor, called RSPA [8]. RSPA consists of 16 (4x4) processing elements (PEs) in which each PE is connected to 4 neighboring PEs and 4 diagonal ones.

The benchmark set includes Jacobi, Gauss, Swim, Tomcatv, and SpMatMul. Jacobi is a partial differential equation solver, using an iterative method. Gauss is Gaussian elimination with partial pivoting, for solving a set of linear equations. Swim is a benchmark for weather prediction from SPEC, solving difference equations on a two dimensional grid. Tomcatv is a SPEC benchmark. Most of the access patterns in these codes are regular. SpMatMul is a code for multiplying a sparse matrix by a vector. The sparse matrix is stored in the compressed row storage(CRS) format. This program has both regular and irregular access patterns. The results of the experiments are shown in Table 1. With the first four benchmarks, Jacobi, Gauss, Swim and Tomcatv, in which regular accesses dominate the access patterns, the original codes show that it executed sequentially. By introducing explicit data transfer, the transformed code with data transfer improved performance by efficiently using memory bandwidth. SpMatMul has both regular and irregular accesses. Irregular access patterns are hard to distribute into array PEs. Thus, that yields less speedup than regular patterns.

Table 1. Experimental results

	Speedup (%)	Code size increase (%)
Jacobi	42.85	84
Gauss	15.68	59
Swim	238.88	104
Tomcatv	1060	78
SpMatMul	33.33	22

As we can see from Table 1, the array processor executes applications 15% to 1060% faster than the original code. We also estimate the amount of code size increase. The overall code size increase from the proposed technique ranges from 22% to 104%, and the average increase is 69.4%. However, the applications are critical loop kernels which typically account for 5% - 10% of entire application code size. By carefully applying the proposed technique to mission critical loops with profiling, overall code size increase can be moderated.

6 Conclusion

We have presented a data distribution algorithm for array PEs which minimize the total execution time. Array access descriptors can be collected before hand by the profiling method or the earlier run of an application. From these data access descriptors, distributability of data can be determined. The proposed approach can effectively be applied to a wide range of applications. The experimental results show significant improvement compared with the original code.

Acknowledgment. This work was supported in part by Basic Science Research Program through the National Research Foundation of Korea(NRF) funded by the Ministry of Education, Science and Technology (No.2010-0024529) and Sunchon National University Research Fund in 2011.

References

1. Anderson, J.M., Lam, M.S.: Global optimizations for parallelism and locality on scalable parallel machines. In: Proceedings of the ACM SIGPLAN 1993 conference on Programming language design and implementation, PLDI 1993, pp. 112–125 (1993)
2. Bixby, R.E., Kennedy, K., Kremer, U.: Automatic data layout using 0-1 integer programming. In: Proceedings of the IFIP WG10.3 Working Conference on Parallel Architectures and Compilation Techniques
3. Coxt, A.L., Dwarkadast, H., Lut, H., Zwaenepoelt, W.: Evaluating the performance of software distributed shared memory as a target for parallelizing compilers (1997)
4. Garcia, J., Ayguade, E., Labarta, J.: Dynamic data distribution with control flow analysis. In: Proceedings of the 1996 ACM/IEEE Conference on Supercomputing (CDROM), Supercomputing 1996 (1996)
5. Han, H., Tseng, C.-W.: Compile-time synchronization optimizations for software dsms (1998)
6. Huang, C.-H., Sadayappan, P.: Communication-free hyperplane partitioning of nested loops. J. Parallel Distrib. Comput. 19, 90–102 (1993)
7. Kremer, U.: Np-completeness of dynamic remapping. In: 4th International Workshop on Compilers of Parallel Computers (1993)
8. Lee, J.-e., Choi, K., Dutt., N.D.: Compilation approach for coarse-grained reconfigurable architectures. IEEE Des. Test 20 (2003)
9. Navarro, A.G., Asenjo, R., Zapata, E.L., Padua, D.: Access descriptor based locality analysis for distributed-shared memory multiprocessors. In: Proceedings of the 1999 International Conference on Parallel Processing, ICPP 1999, p. 86. IEEE Computer Society, Los Alamitos (1999)
10. University of Dortmund. Icd-c compiler framework
11. Paek, Y., Hoeflinger, J., Padua, D.: Simplification of array access patterns for compiler optimizations. In: PLDI 1998: Proceedings of the ACM SIGPLAN 1998 Conference on Programming Language Design and Implementation (1998)

Design of Efficient Reconfigurable Digital Compensator for Advanced Wireless Transmitter Using Computational Digital Signal Processing

Jonggyun Lim and Hyunchul Ku

Electronic Engineering, Konkuk University
1-Hwayang-dong, Kwangjin-gu, Seoul, Korea
{rain0820,hcku}@konkuk.ac.kr

Abstract. This paper presents an efficient digital compensator that can be used for adjustment of gain and phase imbalances in linear amplification with nonlinear components (LINC) transmitters. The proposed compensator is designed in signal component separator (SCS) block using computational digital signal processing (DSP). Complex baseband equivalent signals and systems are considered to analyze error vector magnitude (EVM) and adjacent channel power ratio (ACPR) of the transmitter. Look up tables (LUT) based on the analyzed EVM and ACPR are used to compensate gain/phase imbalances of RF paths in the LINC transmitter. The suggested method is efficient to implement using computational software and DSP, and thus the compensator is reconfigurable for various input signals and RF transmitters. The LINC transmitter with the proposed compensator is implemented, and the suggested scheme is verified by showing the improvement of EVM and ACPR performance for the transmitted signals.

Keywords: LINC, DSP, Transmitter, ACPR, EVM, DPD.

1 Introduction

Spectral efficiency and power efficiency are the most important factors in modern wireless communication transmitter systems. The spectrum is limited resource, but the number of wireless data and users has been rapidly increased. In result, it is required to send more data on the limited bandwidth. Complex modulation scheme such as 16-, 64-, and 256- quadrature amplitude modulation (QAM) is used to achieve high spectral efficiency. However, the complex modulation scheme causes high peak-to-average power ratio (PAPR) and wide dynamic range of the transmitted signal. The large output back-off (OBO) is required for the high PAPR signal to reduce nonlinear distortion of the transmitted signal. The large OBO decreases power efficiency, which is measured by the ratio of the output RF power to the supplied DC power in the transmitter, significantly.

Considering this problem, many techniques have been proposed and developed to increase power efficiency and linearity for high PAPR signal [1]. Such advanced wireless transmitter technologies are digital pre-distortion (DPD) scheme, linear

G. Lee, D. Howard, and D. Ślęzak (Eds.): ICHIT 2011, CCIS 206, pp. 641–648, 2011.
© Springer-Verlag Berlin Heidelberg 2011

amplification with nonlinear component (LINC) scheme *etc.* DPD is a scheme to insert a digital block which has inverse nonlinear characteristics of RF power amplifier (PA). DPD can linearize the output signal of RF PA, reduce OBO of the transmitted signal, and improve power efficiency of the wireless transmitter system. LINC is a method to make two constant envelop signals using out-phasing scheme [2,3]. A signal component separator (SCS) is used to make two constant envelop signals in the LINC system. The LINC technique is a highlighted method in recent wireless communication system to improve efficiency and linearity of the transmitter [4,5]. However, the LINC system uses multiple RF paths, and gain / phase imbalances of two RF paths cause the significant output signal distortion [5-8]. To compensate the imbalance in LINC system, the pre-compensation scheme is required.

The digital technology enables to implement the advanced wireless transmitter with low cost and efficiently. It has already been noted that DPD has become the most active area for PA linearization development [1]. The nonlinear behavior of the power amplifier can be characterized with the digitized data, and compensated by a digital system.

The digital technology also enables to implement the efficient compensation system for the LINC transmitter. In this paper, we analyze a RF path imbalance problem, and suggest a digital compensator to improve error of the transmitted signal occurring from path imbalances. A digital compensator is designed effectively with look-up-table (LUT) using computational digital signal processing (DSP). The computational DSP enables effective and reconfigurable design for various RF transmitters with various input signals. The suggested scheme can be used in software defined radio (SDR) system to improve performance of wireless transmitter.

This paper is organized as follows. In section 2, we analyse the signal distortion occurring from RF path mismatch in the LINC system and provide a design of compensator. In section 3, an advanced wireless transmitter using LINC scheme and the suggested digital compensator is implemented. The proposed system is verified by measuring error vector magnitude (EVM) and adjacent channel power ratio (ACPR) of the digitally modulated signals. The experimental results are summarized in section 3.

2   Design of Reconfigurable Digital Compensator

2.1   Analysis of RF Path Imbalance Problem

LINC technique enables the wireless transmitter to improve efficiency and linearity. Fig.1 shows the architecture of the LINC system. In Fig. 1, the complex envelop of the input signal is as follows:

$$\tilde{x}_0(t) = r_0(t)e^{j\theta(t)} \tag{1}$$

where $r_0(t)$ denotes the magnitude, and $\theta(t)$ is the phase of the input complex envelop signal. The SCS block converts the input signal $\tilde{x}_0(t)$ to constant envelope signals, $\tilde{x}_1(t)$ and $\tilde{x}_2(t)$.

$$\tilde{x}_{1,2}(t) = r_{1,2}(t)e^{j(\theta(t)\pm\alpha)} \tag{2}$$

The modulated signals by up-converted are

$$x_{1,2}(t) = \mathrm{Re}\left[r_{1,2}(t)e^{j(\omega_c t + \phi_{1,2})}\right] \tag{3}$$

where $\omega_c(t)$ is the carrier frequency. For the LINC transmitter, the magnitude and phase of the separated signals are calculated such as

$$r_{1,2}(t) = c = \frac{\max(r(t))}{2}, \quad \phi_{1,2} = \theta(t)\pm\alpha(t) = \theta(t)\pm\cos^{-1}\left(\frac{r_o(t)}{2c}\right) \tag{4}$$

The up-converted signals are amplified by RF PA with gain G_1 and G_2 for each path

$$y_1(t) = G_1 \cdot x_1(t), \quad y_2(t) = G_2 \cdot x_2(t) \tag{5}$$

The output signal combined by power combiner is

$$y_0(t) = y_1(t) + y_2(t) \tag{6}$$

The complex envelope signals in the LINC transmitter are shown in Fig .1.

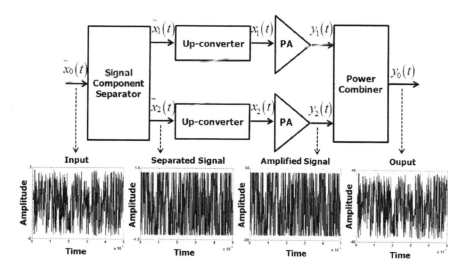

Fig. 1. Block diagram of LINC system and its Signals

Parallel branches that form the LINC system have unavoidable path imbalance on their gain and phase difference. In order to exactly recover the original signal, the gain and phase imbalances from the each path have to be perfectly compensated. Any gain and phase mismatch will cause the significant deterioration in linearity of the system. Thus it degrades ACPR and EVM performance of the wireless system.

ACPR arises from spectrum regeneration when a band-limited digitally modulated signal is applied to nonlinear system. In general, ACPR is defined as the ratio of integral of power spectral density (PSD) function for the specific two frequency bands

[9] (usually in signal band from f_1 to f_2, in adjacent channel band from f_3 to f_4) as shown in Fig.1 (a).

$$ACPR_{UPPER} = \frac{\int_{f_3}^{f_4} PSD \cdot df}{\int_{f_1}^{f_2} PSD \cdot df} \tag{7}$$

EVM is defined as the magnitude from the ideal constellation to the actual transmitting constellation caused by path imbalance as shown in Fig.2 (b). And the deterioration of effects for path imbalance are described in Fig.2 (c) and (d). To model this effect, assuming that the ratio of gain of path1 to path2 is ΔG ($\Delta G = G_2/G_1$, and $G=G_1$) the difference between phase delay of path1 and path2 is $\Delta\varphi$. Then, the output signal of LINC transmitter is

$$\begin{aligned} y_o(t) &= y_1(t) + y_2(t) \\ &= Ge^{j\phi}(\tilde{x}_1(t) + \Delta G e^{j\Delta\phi}\tilde{x}_2(t)) \end{aligned} \tag{8}$$

The simulation data are described in Fig. 2 (c) and (d) by using the Matlab program for a given gain and phase mismatch. The 16-QAM signal is considered as an input signal of the system in this case. The results show that the LINC transmitter has considerable error degradation because of path imbalances.

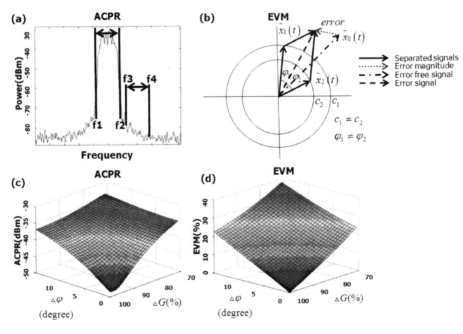

Fig. 2. (a) ACPR of the transmitted signal, (b) EVM of the transmitted signals, (c) Deterioration of ACPR versus gain and path imbalance of two paths, (d) Deterioration of EVM versus gain and path imbalance of two paths

2.2 Design of Digital Compensator

The digital baseband signal is separated and pre-compensated in DSP part as shown in Fig. 3. The transmitter structure is similar with DPD system which compensates the nonlinear distortion of PA in the digital baseband domain.

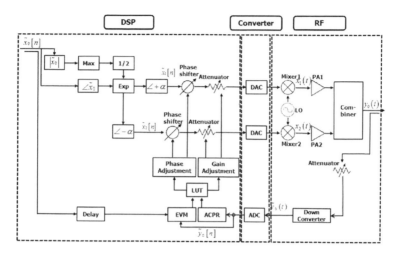

Fig. 3. The block diagram of digital compensator for RF path imbalance

For the input signal, LUTs for the EVM and ACPR versus gain and phase imbalances are extracted analytically using the equations in [9]. The LUTs are two-dimensional mapping for gain imbalance and phase imbalance. The EVM and ACPR are calculated from the input and output complex envelop signals, and the extracted values are used to get gain and phase imbalance values from LUTs as shown in Fig. 3. The gain and phase compensator uses phase shifter and attenuator. The compensator block can be implanted in DSP part or in analog part.

3 Reconfigurable Wireless Transmitter Using Digital Compensator Using Computation DSP

3.1 Implementation of System

The advanced wireless transmitter using LINC scheme and digital path imbalance compensator is implemented as shown in Fig. 4. The Matlab program is used to create two outphased signals in the computer. The separated signals are converted into analog signal by IFRIO (NI-PCI5640R) [10]. The separated signals go into mixer (RF2051) and are up-converted to RF signals with 2.3 GHz center frequency. The signals are amplified at RF PAs located at each RF path. Both outputs of PAs are combined by the Wilkinson combiner. To get digital complex envelop signal of the output of the combiner, the RF signal analyzer (SA) is used in the measurement. The compensator block in Fig. 4 is implemented using computer and IFRIO in Fig. 4.

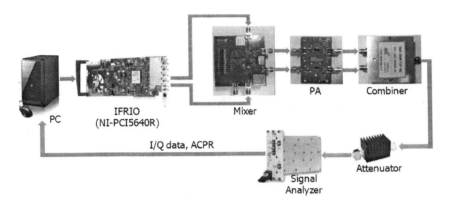

Fig. 4. Block Diagram of overall system

The digital part of the system, which works as SCS and compensator, is shown in Fig. 5. The PCI-5640R is a 2-channel IF input and 2-channel IF output PCI board with a Xilinx field programmable gate array (FPGA). The PCI interface is provided by the NI DAQ, which has four direct memory access (DMA) channels able to transport streams between the host CPU and the Xilinx FPGA. Digital up-conversion and digital down-conversion are done within the analog digital converters (ADC)/ digital analog converters (DAC), thus offloading the Xilinx FPGA. The Matlab in the computer generates LUTs for EVM and ACPR using the input data from PCI-5640R and the digitized measured output data from the signal analyzer in Fig. 5. The generated LUTs are uploaded in the FPGA and the input signal is compensated using the LUT. The pre-compensated signal is converted into analog signal at the DAC and sent to the mixer through output ports in PCI-5640R.

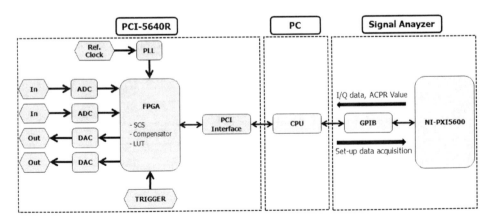

Fig. 5. Digital Compensator using Digital Computational Digital Signal

The compensator based on the architecture shown in Fig. 5 can be applied adaptively to various RF transmitters and various input signals because LUTs are generated by computational software and uploaded on FPGA. The proposed compensator is reconfigurable using Matlab program and computational DSP.

3.2 Experimental Results

ACPR and EVM are measured without/with the proposed compensator. ACPRs are plotted and compared in Fig. 6 for the 16-QAM digitally modulated input signal. For various input signals, ACPR and EVM performances are summarized in Table. 1. The results show that the proposed scheme can efficiently reduce EVM and ACPR.

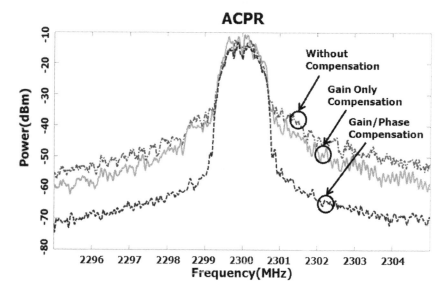

Fig. 6. Measurement ACPR Results

Table 1. Summary of Performance

	Case1 (Tx HW 1, 16-QAM signal)		Case2 (Tx HW 2, 64-QAM Signal)		Case3 (Tx HW 2, 256-QAM signal)	
	Without Compensation	With Compensation	Without Compensation	With Compensation	Without Compensation	With Compensation
Gain Ratio (%)	82	99	67	99	67	99
Phase Delay (°)	14.5	0.5	101	0.5	101	0.5
ACPR (dBc)	-24.93	-51.95	-25.57	-49.06	-25.14	-47.17
EVM (%)	28	2	99	2	99	2

4 Conclusion

In this paper, path imbalance of the LINC transmitter has been discussed analytically. Based on the analytical results, the mismatch of gain and phase can be estimated from EVM and ACPR of the output signal, providing a useful guide for compensation circuit design. Compensation block has been designed based on LUT structure. The block is implemented efficiently using software and computational DSP, and thus it is reconfigurable for various input signals and various transmitters. The proposed reconfigurable compensation block is demonstrated by showing the improvement of EVM and ACPR for the various digitally modulated signals such as 16- and 64- QAM signals. The proposed architecture and design method can be applied to SDR system in the advanced wireless communication system.

Acknowledgments. This work was supported by the National Research Foundation of Korea (NRF) grant funded by the Korea government (MEST) (No. 2011-0003299).

References

1. Cripps, S.C.: RF Power Amplifiers for Wireless Communications, pp. 285–425 (2006)
2. Chireix, H.: High Power Outphasing Modulation, vol. In: Proc. IRE, vol. 23, pp. 1370–1392 (1935)
3. Cox, D.C.: Linear Amplification with Nonlinear Components. IEEE Trans. 22, 1942–1945 (1974)
4. Jheng, K.Y., Chen, Y.J., Wu, A.Y.: Multilevel LINC System Designs for Power Efficiency Enhancement of Transmitters. IEEE Journal of Selected Topics in Signal Processing 3, 523–532 (2009)
5. Myoung, S., Lee, I., Yook, J., Laskar, J.: Mismatch Detection and Compensation Method for the LINC System Using a Closed-Form Expression. IEEE Trans. 56, 3050–3057 (2008)
6. Zhang, X., Larson, L.E., Asbeck, P.M., Nanawa, P.: Gain/Phase Imbalance Minimization Techniques for LINC Transmitters. IEEE Trans. 49, 2507–2516 (2001)
7. Lim, J., Kang, W., Ku, H.: Compensation of Path Imbalance in LINC Transmitters using EVM and ACPR Look up Tables, TH3G-08. In: APMC, Yokohama, pp. 1296–1299 (2010)
8. Gard, K.G., Gutierrez, M., Steer, M.B.: Characterization of Spectral Regrowth in Microwave Amplifiers Based on the Nonlinear Transformation of a Complex Gaussian Process. IEEE Trans. 47, 1059–1069 (1999)
9. National Instruments, http://www.ni.com

Joint Rate Control Algorithm with Look-Ahead in Multi-channel IPTV Services

Young-Il Yoo, Kyeong-Hoon Jung, Ki-Doo Kim, and Dong-Wook Kang

School of Electrical Eng., Kookmin University, Jeongneung-Gil 77,
Jeongneung-Dong Seongbuk-Gu Seoul 136-702, Korea
{yiyoo0125,khjung,kdk,dwkang}@kookmin.ac.kr

Abstract. This paper presents a joint rate control method for the multi-channel video services. Each encoder analyzes the spatio-temporal complexities of its video frames and reports the result to the look-ahead statistical multiplexer. And then the multiplexer assigns a QP level to each encoder. The proposed joint QP control scheme regulates the quality of the video sequences and minimizes decoder buffering delay, and in consequence improves the overall QoS level in multi-channel IPTV system.

Keywords: H.264, AVC, Joint Rate-Control, QoS.

1 Introduction

As multimedia coding and bandwidth efficiency have been greatly improved, the multiple channel service became common in most conventional broadcasting networks such as terrestrial, satellite and cable. Especially the birth of IPTV accelerated this trend and the problem of how to jointly deal with multiple video sequences became very important. Many studies have shown that joint rate control is an effective way in maintaining a more uniform picture within a sequence and among sequences by allocating dynamically the available bandwidth to each sequence according to its relative complexity [1]-[5]. Most approaches are based on bit allocation strategy in which joint rate controller assigns target bitrates for each sequences and then encoder individually controls according to the assigned target bitrates [1]-[4]. In this approach, there can be some delay in responding time since the rate controller cannot instantaneously change the target bitrates and the previously assigned target bitrates are maintained at least during an intra-period.

Meanwhile, the joint rate control scheme which is based on QP control strategy is proposed [5]. It introduces the concept of super frame and joint buffer. The encoding bits for each channel can be allocated by frame level in this approach. But it is not easy to manage the scene change in a single channel since every frame should be encoded as same frame type. Also some degradation over every channel is likely to occur when super frame is encoded as the I-frame.

In this paper, we propose an efficient joint rate control algorithm with look-ahead capability. The statistical multiplexer controls the QP of each channel by frame level

G. Lee, D. Howard, and D. Ślęzak (Eds.): ICHIT 2011, CCIS 206, pp. 649–656, 2011.

and each frame can be encoded as different type so as to provide more consistent visual quality while making full use of total available bandwidth.

2 Proposed Joint QP Control Scheme

Fig. 1 depicts the joint QP control scheme. Each encoder analyzes the spatial and temporal complexity of the input frame and estimate the bits required for encoding the frame based on the complexities in advance of actual encoding. And then it sends the information to the statistical multiplexer. The joint QP controller decides the unified QP level based on the information of the overall future-required bits of the encoders and the joint buffer constraint. After that, the QP is announced to the encoders in order to apply actual encoding.

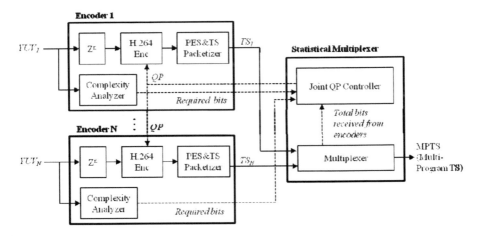

Fig. 1. The proposed joint QP control scheme with a look-ahead multiplexer

2.1 Complexity Analysis

The complexity analyzer computes the macroblock-based spatial and temporal complexity of the input frame and estimates the bits required for encoding a frame. Fig 2 shows the frame analysis scheme.

The spatial complexity of the i-th macroblock of the n-th frame is calculated by (1):

$$c_s(n, i) = \sum_{(x,y) \in B_{n,i}} [|G_v(x, y)| + |G_h(x, y)|] . \tag{1}$$

where G_v and G_h are the outputs of vertical and horizontal gradients, and $B_{n,i}$ is the set of pixels in the i-th macroblock of the n-th frame.

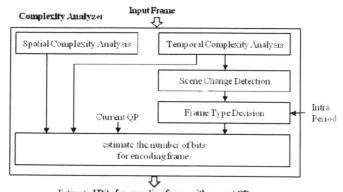

Fig. 2. The bits estimation procedure in complexity analyzer

The temporal complexity of the same macroblock is computed by (2):

$$C_t(n,i) = \sum_{(x,y)\in B_{n,i}} \left| f_n(x,y) - \hat{f}_n(x,y) \right| + \lambda \cdot R(MVD_i) .$$

(2)

where f_n is a input frame and $\hat{f}_n$ is its open-loop motion compensated prediction and $R(MVD)$ is the bits required for encoding the motion information and λ is Lagrangian multiplier [6].

After that, the number of bits required for encoding the macro-block is estimated by (3) for intra mode and (4) for inter mode.

$$b_{intra}'(n,i) = \alpha(QP_{n-N}) \cdot c_s(n,i) .$$

(3)

$$b_{inter}'(n,i) = \begin{cases} 0, & skip\ mode . \\ \beta(QP) \cdot c_t(n,i), & otherwise . \end{cases}$$

(4)

where α and β are the functions of QP which are to be experimentally determined. The skip mode is selected when $MVD(i) = 0$ and $\beta(QP) \cdot c_t(n,i) < T_{skip}(QP)$.

In order to estimate the total required bits for encoding the frame, we should consider with the frame type and the characteristic of a macroblock. The frame type can be determined by scene change information and intra period in analysis phase. Scene change can be easily determined by comparing predetermined threshold and the temporal complexity for a frame. Finally, Total bits required for encoding the frame is estimated differently depending on the frame type by (5):

$$b'(n) = \begin{cases} \sum_{i\in B_n} b_{intra}'(n,i), & I - frame . \\ \sum_{i\in B_n} Min\left(b_{intra}'(n,i), b_{inter}'(n,i)\right), & P - frame . \end{cases}$$

(5)

where B_n is the set of macro-blocks in the n-th frame.

2.2 Joint QP Control

The bits rate for encoding the frame is controlled by the above-mentioned estimated bits of the future frames and leaky-bucket joint buffer model. The number of bits required by the overall encoders for the succeeding N frames in the future is predicted by the joint QP controller with (6):

$$G'(n) = \sum_{j=0}^{L-1} \sum_{x=n+1}^{n+N} b_j'(x).$$ (6)

where N is the size of the look-ahead window and L is the number of programs.

The actual buffer level after n-th frame encoding is updated by (7):

$$B(n) = B(n-1) + \sum_{j=0}^{L-1} b_j(n) - C.$$ (7)

where b is the number of bits actually generated for encoding frame and C is the target transmitted bits per frames for the overall channel.

Then, we can expect the future joint buffer level when we are to hold the current QP for the next N frames:

$$R'(n+N) = \frac{B(n) + G'(n) - C \times N}{S}.$$ (8)

where S is the total buffer size (i.e. $S = total\ bitrate \times delay$).

In consequence, we can control the joint buffer level by adjusting the QP according to $R'(n+N)$:

$$QP_{n+1} = \begin{cases} QP_n + 2, & 0.9 < R'(n+N). \\ QP_n + 1, & 0.7 < R'(n+N) \le 0.9. \\ QP_n, & 0.5 < R'(n+N) \le 0.7. \\ QP_n - 1, & 0.3 < R'(n+N) \le 0.5. \\ QP_n - 2, & R'(n+N) \le 0.3. \end{cases}$$ (9)

3 Experimental Results

3.1 Performance of the Required Bits Prediction

Experiments are performed for the 2 type QVGA@15fps sequences, 'Soccer' is high motion sequence and 'Game' is low motion sequence. The JM 10.2[7] is used for encoding.

Fig. 3 and 4 show the comparison between the estimated and the actual bits for encoding with the options that the QP is 35 and the intra period is 2 second. They show the prediction accuracy of the proposed method is reliable in both high and low motion sequence and in both I-frame and P-frame type.

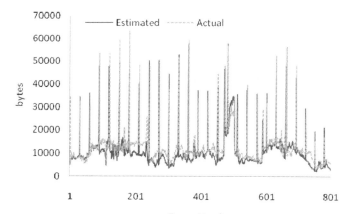

Fig. 3. Estimated bits and actual bits 'Soccer' sequence

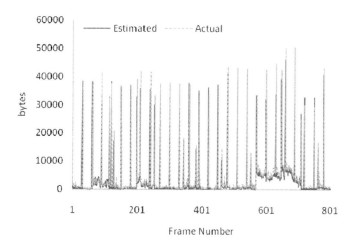

Fig. 4. Estimated bits and actual bits for 'Game' sequence

3.2   Performance of the Joint Rate Control

Experiments are performed for the service of concurrent 8 QVGA@15fps sequences. Target bits-rate for overall channels set to 1360 kbps (170kbps x 8 for independent individual encoding). Fig. 5 shows that the unified QP level yields the variable bit-allocation effects to each encoder.

The buffer regulation property of the proposed scheme is shown in Fig. 6. The buffer fullness varies from 50% to 70%, which means that decoder buffering can be controlled by setting the desired buffer level in Eq. (9).

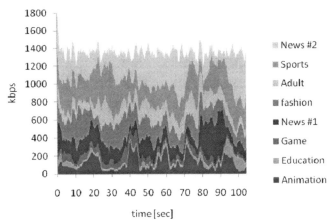

Fig. 5. The variation of bit-allocation for each channel

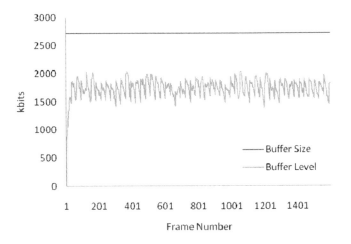

Fig. 6. Joint buffer fullness of the proposed joint QP control algorithm

Table 1. Comparison between JM and the joint QP-control

Sequence Name	Independent Rate-Control		Joint Rate-Control	
	Bitrate[kbps]	PSNR[dB]	Bitrate[kbps]	PSNR [dB]
Animation	169.73	35.73	122.94	35.73
Education	169.45	41.89	42.59	38.35
Game	169.66	35.76	68.57	35.15
News #1	169.32	32.21	193.13	35.30
fashion	168.69	35.86	202.55	37.40
Adult	169.22	36.15	136.37	36.05
Sports	169.27	32.13	308.93	34.93
News #2	170.24	31.47	301.08	34.13

Table 1 and Fig. 7 also show the performance of the proposed algorithm. The output video quality (per-frame PSNR) of the proposed joint QP control algorithm is much less varying than that of the independent individual rate control. Fig. 7 shows that the proposed joint QP control algorithm improves 99.9% interval in the frame PSNR distribution by about 6 dB, compared with the individual rate-control algorithm, which means that the QoS level of the proposed scheme will set to be higher by the same amount.

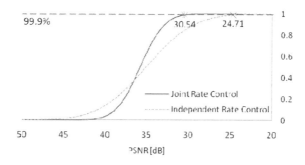

Fig. 7. Cumulative distribution of the per-frame PSNR

4 Conclusion

We proposed a joint rate control method for the multi-channel video service. In the proposed scheme, the statistical multiplexer controls jointly QP levels of the multiple encoders, which results in improvement of coding efficiency and good balance of visual quality among the sequences as well as within a sequence. In addition, the proposed method maintains buffer to the desired level and hence minimizes the decoder buffering delay, which decreases the waiting time for the channel change. The proposed scheme can be applied to all types of multi-channel video service platforms such as terrestrial, satellite, cable and IPTV systems.

References

1. Boroczky, L., Ngai, A.Y., Westermann, E.F.: Joint Rate Control with Look-Ahead for Multi-Program Video Coding. IEEE Trans. Circuits Syst. Video Technol. 10(7), 1159–1163 (2000)
2. Yang, J., Fang, X., Xiong, H.: A Joint Rate Control Scheme for H. 264 Encoding of Multiple Video Sequences. IEEE Trans. Consumer Electron 51(2), 617–623 (2005)
3. He, Z., Wu, D.O.: Linear Rate Control and Optimum Statistical Multiplexing for H.264 Video Broadcast. IEEE Trans. Multimedia 10(7) (November 2008)
4. Hsu, C.-Y., Ortega, A., Reibman, A.R.: Joint Selection of Source and Channel Rate for VBR Video Transmission under ATM Policing Constraints. IEEE J. Sel. Areas Commun., Special Issue on Real-Time Video Services in Multimedia Networks 15(6), 1016–1028 (1997)

5. Rezaei, M., Bouazizi, I., Gabbouj, M.: Joint Video Coding and Statistical Multiplexing for Broadcasting Over DVB-H Channels. IEEE Trans. Multimedia 10(8), 1455–1464 (2008)
6. Wiegand, T., Sullivan, G.J., Bjntegaard, G., Luthra, A.: Overview of the H.264/AVC Video Coding Standard. IEEE Trans. Circuits Syst. Video Technol. 13(7), 560–576 (2003)
7. JVT/H.264 JM v10.2 Software, http://bs.hhi.de/suehring/tml/

SERLOG: Generating and Analysis Server Push Workloads

Dong-Il Cho and Sung-Yul Rhew

Soongsil University, Seoul, Korea
chodongil@yahoo.co.kr,
syrhew@ssu.ac.kr

Abstract. In the WWW, server push, which allows bidirectional communication between the server and the web browser, is an essential part of modern web application. In order for the server push to ensure high levels of interaction with the browser, the performance of the server application is of utmost importance. The server push application has a different communication mechanism from the typical web application, hence making it hard for the currently-existing load test tools to run an accurate performance test. In this paper, in order to run an accurate performance test of the server push application, we analyzed the workload characteristics of server push, while designing and implementing SERLOG, a load test tool that can run accurate tests on the performance. In addition, through comparisons with the other web load testing tools, we have proved how SERLOG accurately represents the workload of server push and provides the necessary performance data.

Keywords: Server Push, Performance measurement, Load testing.

1 Introduction

Server push is a web-based bidirectional communication architecture that allows the server to deliver information to the browser without its request[1]. In the WWW, server push is an essential part of modern web applications because of its ability to allow high-quality interaction between the server and the browser[2].

Server push must be able to provide active, supporting services to masses of simultaneously-accessing users, hence making the server application's performance very important. The server application's performance is quantified through the load test[3,4,5], but the complicated and unique communication mechanism of server push, differing from the previous web-based load test tools, makes it impossible for correct representation of server push workload. Hence, in order for a correct quantification analysis and performance predictions of server push application to be established, there is a need for a load test tool that can handle and gather data through the server push application's environment, such as response time and throughput.

In this paper, in order to lay hands on accurate quantified performance analysis and estimation, we have analyzed the characteristics of the server push workload and defined its representation model. In addition, we took our defined model and

G. Lee, D. Howard, and D. Ślęzak (Eds.): ICHIT 2011, CCIS 206, pp. 657–664, 2011.

implemented SERLOG(SERver Push WorkLOad Generator), a load test tool. SERLOG, in comparison with the previously-existing web based load test tool JMeter, showed that it could better represent the server push workload and recorded a higher server resource-usage. In addition, it provided accurate performance data on the response time, service success rates and throughput that matched the server push environment.

2 Server Push Characteristics

In order to make an accurate performance prediction of the server application, we focus on the working sections of the server business application.

2.1 Comparison between Traditional Web Application and Server Push Application

In typical web application, response time is the time elapsed from the user click until the arrival of the last object composing a web resource[6,7]. If we only focus on processing the server application, the response time will be defined as the time duration between the moment one HTTP connection requests the service and the moment they received response data. Hence, a transaction starts when HTTP connection's request is sent to the server, and ends when the server's response is received. The matter of success/failure of a transaction depends on whether the HTTP connection receives a legitimate response; if yes, it is a success, and if not, it is a failure. The HTTP request parameter is usually sent in a *key=value* format.

On the other hand, server push differs from the browser's HTTP connection in having a different HTTP connection to process the server's response. The server's response is sent through a long-lasting, continuous receive-only connection called long-polling[1]. Long-polling receives a server's response only if the server's event or the other browser's event satisfies the sending conditions in server application. In other words, request and response can be caused by unidentified server events, unlike the typical web application[1,2]. Such request/response pattern makes it difficult for the previously-existing tool server to run performance tests. Server push's response time can be defined as the time duration between the moment of the request of a browser event from an HTTP connection and the moment of the long-polling response received from another HTTP connection[8,9]. Also, the transaction starts when this event request is delivered to the server and ends when the long-polling's response is received. If the response to the request is well-sent as the long-polling's response, this transaction succeeds, and if not (for example, failing to deliver message, timeout, server error...etc.), it fails. Also, the long-polling's data request format that gets sent from the browser to the server, follows the constructive data format defined by the server push application, often using formats such as JSON and XML[1,10].

2.2 Server Push Workload Characteristics

A main-stream server push technology, COMET, as the communication model such as Fig.1.

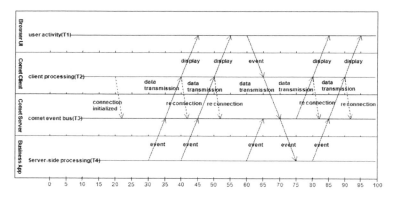

Fig. 1. Comet Web Application Interaction Model[11]

In Fig. 1, the *Comet Client* and the *Browser UI* is a Javascript application that works in a browser. *Comet Client* is a module in charge of COMET communications. *Comet Server* is a server module that processes the COMET request and browser event request received from the *Comet Client*. *Business App* is a module that communicates with the *Comet Server*, processing business logic.

When the browser first connects to a *Comet Server*, its connection is initialized by the *Comet Client (T2(20))*. From here, when an event takes place in the server, a push-pattern of message sending from the server to the browser is made possible. After that, like *T1(60)*, when a specific event takes place on the browser, the server processes that event and sends it back to the browser *(T4(80))*.

SERLOG records the COMET request and browser event request of *Comet Client* and sends it to the server after mimicking this data as many as the number of the VUser(virtual user) during load test; when this happens, the performance data such as the response time and throughput is gathered.

2.3 Method for Server Push Performance Analysis

In Fig. 1, the web transaction can be defined as the *T2(65)* section that delivers the browser's event and the *T3(85)* section that delivers the processed message back to the browser *T2(65)* → *T3(85)*. The response time is defined as the time difference between the start of transaction at the *Comet Client* and the end of it *T2(90) - T2(65)*. The matter of success in a transaction is when, upon *T2(65)*'s request, the intended *T2(90)* response is received. In all other cases, it is defined as a failure.

Server push uses a long-lasting and continuous long-polling requests, hence allowing timeouts to occur, as the browser and server's processes are delayed. The client timeout occurs when *T2(40, 50, 80) + Client Timeout Duration) < T2(50, 80, 90)*. Server timeout occurs when server-side processing time of browser event request exceeds the server timeout, and in Fig.1, it occurs when *T3(70) + Server Timeout Duration < T3(85)*. The hit count is the number of times a transaction succeeded in a testing period, and can be defined as the *Count(success)*. Throughput can be defined as the *Hit Count / Test Duration*.

3   Obstacles and Solution to Creating Server Push Workload

Server push has different communication pattern from the typical web applications, and hence there are many obstacles in the way of correctly representing the workload. We have divided these obstacles into three kinds(*Response Event Obstacle, Transaction Obstacle and Transmission Data Model Obstacle*).

3.1   Response Event Obstacle and Assumption

In a server push application, a transaction is defined as long-polling's response following the browser's event request. Here, the long-polling's response from the server depends on the different business terms. Such terms show that an event request cannot be met by a response, and this proves to be an obstacle in measuring the response time or the transaction's matter of success. Hence we assume, for the sake of measurement feasibility, that one event request must be met by one response.

3.2   Transaction Obstacle and Solution

From the assumption of section 3.1, we can simplify the transaction into a browser's event request and long-polling's response sectors, because the browser's request is always met with a response. However, a problem occurs that the two HTTP connections synchronize to provide the results for the request and response. In order to solve this transaction obstacle, we have designed an algorithm where one VUser can process two connections.

Algorithm : Generation of a User Behavior

```
run() {
  while(scenarioChain.hasNext()) {
    try {
      step = scenarioChain.next();
      longpollingRequest =
createLongpollingRequest(step);
      sendRequest(longpollingRequest);
      waitForRequestComplete(longpollingRequest);
      waitForTime(serviceInterval);
      messagePushRequest =
createMessagePushRequest(step);
      sendRequest(messagePushRequest);
      waitForAllComplete();
      reportResult();
    } catch(e) {
      reportError();
}}}
```

The test scenario consists of multiple tasks where a browser communicates with the server. Each task records information about one server push transaction. In the algorithm, the test scenario is a *scenarioChain* object that gets delivered to the VUser thread. Each task, as a part of the *scenarioChain*, is objectified as objects called *step*, and they record long-polling's request to receive server response, the corresponding *longpollingRequest* information, and the *messagepushRequest* information which responds to the browser's event request.

Each VUser chronologically processes each of the steps which are a part of the *scenarioChain*, and collects data. At each *step*, a VUser first creates a *longpollingRequest* and sends it to the server. When the *longpollingRequest* is successfully delivered, it will wait for a preset interval before sending *messagePushRequest* to the server. A VUser will wait until the response to both requests are received, then record the results, and continue the following *steps* using the same pattern until the chain ends.

3.3 Transmission Data Model Obstacle and Solution

In a server push, a browser and a server exchange constructed data. Generally, JSON and XML are frequently used. This format is in the *key=value* structure that is hard to formalize, and hence a request parameter recording method is required to run each step of the load test tool.

SERLOG does not limit the HTTP Body domain's data to *key=value*, and allows it to be edited in order to support such structured request parameter.

4 Implementation and Validation of SERLOG

We have implemented SERLOG based on the solution we used to represent the server push workload that we analyzed. We also ran performance tests of the frequently-used web-based load test tool Apache JMeter and SERLOG under the same server push environments, comparing the load of work they send to the server and the offered performance test data.

4.1 Implementation

SERLOG was constructed using Java in order to avoid influences from the OS. JDK used Sun Java SDK 1.6.0 for the sake of the native performance and better thread scheduling. UI component used SWT 3.6.2 for the sake of fast UI rendering speeds, OS-friendly screens, and various UI components. Finally, we used Jetty 7.3 to represent the workload of server push. Jetty 7.3 provides API that processes HTTP connections in a non-blocking IO. This API allows the server push transaction's two connections to be synchronized at the thread without delay. We were able to use Jetty 7.3's API to create an algorithm that, in a thread, sent browser event requests to the server when a long-polling's request was successfully received by the server.

4.2 Experimental Validation

We have run comparison tests between SERLOG and Apache JMeter. Because, in the comparison test, JMeter and SERLOG run tests about the same server application program in the same environment, let we analyzed the performance data gathering of message latency without considering the influence from network, hardware and running environment.

Experimental Setup. The hardware and software environment where the test took place is as follows in Table.1.

Table 1. Test Environment

	Server	Test Client
CPU	Intel Xeon Quard Core 2G x 2	Intel Core 2 DUO 2.4G
Memory	8 GB	3 GB
OS	Windows Server 2008 64 bit	Windows 7 32 bit
Language/Version	Sun Java SDK 1.6.0	Sun Java SDK 1.6.0
Tool	Tomcat 7.0 Springframework 3.0 CXF 2.3.3	SERLOG JMeter 2.4
Monitoring	JConsole 1.6.0	SERLOG JMeter 2.4
Network	1 Gbps LAN	

We have implemented a simple server push software for the test. The implemented server application used Tomcat 7.0 as its server, Springframework 3.0 and CXF 2.3.3 as its foundation for processing long-polling and browser event requests.

JMeter can only process one HTTP connection from one task, and hence we ran two applications simultaneously to separately process the long-polling request and the message push. SERLOG, in one program instance, follows each step of the test scenario chronologically.

Result. Each test was done repeating a scenario 10 times with 150 VUser. The interval between each step was set to 100ms. From the test and JConsole monitoring, we collected server resource usage as follows in Fig.2.

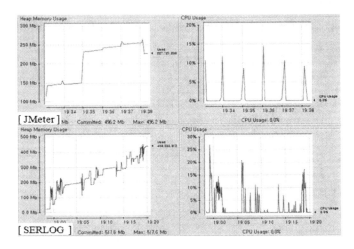

Fig. 2. Push Server CPU, Memory Usage : The two graphs on the top are the test results using JMeter, and the lower two graphs are the test results using SERLOG.

As a result of the test, we found out that JMeter used 14% of Server CPU and 260 MB of memory while SERLOG using 28% Server CPU and 460 MB of memory. As visible from Fig.2, under the same VUser situation, SERLOG was able to better mimic the server push browser behavior, hence resulting in higher CPU and memory usage and more frequent usage. Also, when compared to the server log, the response time provided by SERLOG had approximately 50ms of error per transaction, and this can be described as accurate data considering the delays due to the processing by each network and tier. On the other hand, JMeter required for human to analyze and estimate from the JMeter log and results in order to collect server push performance data.

5 Conclusion and Future Research

In this paper, we described the load test tools for the server push application. We first compared the communication characteristics of browsers and servers in server push applications and typical web applications. In the comparison data, the server push application sent its requests and responses through different HTTP connections, and they were sent in uncommon structural formats such as XML and JSON. This characteristic served as the reason why current web-based load test tools could not run performance tests on the server push applications. In order to overcome this problem, we have classified the obstacles into three categories and provided solutions for each. The proposed solution was realized in the form of SERLOG.

In order to verify SERLOG, we ran comparison tests with the commonly-used web-based load test tool JMeter. The test was done under the same server push application environment, with the JMeter and SERLOG using the same number of VUsers to run load tests; then we collected and compared the amount of work delivered to the server. As a result of the test, we could find the twice usage of CPU and the memory in server in the case of SERLOG, compared to JMeter. In addition, SERLOG provided test result information needed for accurate estimation of the server push performance, and when compared to the server log, we discovered that the values were quite close.

In this paper, we need to make an assumption before SERLOG can solve a problem proposed in section 3.1. There are plenty of limitations in accommodating various server push event environments. Also, there is need for further research in order for the recording and the more automatic performance analysis of the server push workload.

Acknowledgments. This research was supported by the Soongsil University Research Fund.

References

1. Cane, D., McCarthy, P.: COMET and Reverse AJAX: The Next-Generatioon Ajax 2.0. APress (2009)
2. Pohja, M.: Server push with instant messaging. In: Proceedings of the 2009 ACM symposium on Applied Computing, pp. 653–658. ACM, New York (2009)

3. Ruffi, G., Schifanella, R., Sereno, M., Politi, R.: WALTy: A User Behavior Tailored Tool for Evaluating Web Application Performance. In: Third IEEE International Symposium on Network Computing and Applications, pp. 77–86. IEEE Press, Boston (2004)
4. Jiang, Z.M., Hassan, A.E., Hamann, G., Flora, P.: Automated performance analysis of load tests. In: IEEE International Conference on Software Maintenance, pp. 125–134. IEEE Press, Canada (2009)
5. Subraya, B.M., Subrahmanya, S.V.: Object driven performance testing of Web applications. In: Proceedings of First Asia-Pacific Conference on Quality Software, pp. 17–26. IEEE Press, China (2000)
6. Andreolini, M., Colajanni, M., Valente, P.: Design and Testing of Scaleable Web-Based Systems with Performance Constraints. In: 2005 Workshop on Techniques, Methodologies and Tools for Performance Evaluation of Complex Systems, pp. 15–25. IEEE Press, Italy (2005)
7. Barford, P., Crovella, M.: Generating representative Web workloads for network and server performance evaluation. In: The 1998 ACM SIGMETRICS Joint International Conference on Measurement and Modeling of Computer Systems, pp. 151–160. ACM, New York (1998)
8. Ying, G., Zhenxing, D.: A research of the instant messaging system architecture based on Comet and Message Queue. In: 2nd International Conference on Education Technology and Computer (ICETC), pp. 390–394. IEEE Press, Los Alamitos (2010)
9. Bozdag, E., Mesbah, A., van Deursen, A.: A Comparison of Push and Pull Rechniques for AJAX. In: 9th IEEE International Workshop on Web Site Evolution, pp. 15–22. IEEE Press, Los Alamitos (2007)
10. Bayeux - a JSON protocol for publish/subscribe event delivery protocol 0.1draft3, http://svn.cometd.org/trunk/bayeux/bayeux.html
11. Comet: Low Latency Data for the Browser, http://infrequently.org/2006/03/comet-low-latency-data-for-the-browser/

Hierarchical Views for Distributed Databases of Semantically Condensed Data through Web Links to Sensor Data

MinHwan Ok

Korea Railroad Research Institute, Woulam, Uiwang, Gyeonggi, Korea
panflute@informatics.krri.re.kr

Abstract. A number of sensor networks may produce a huge amount of data, and there has been a necessity the data are processed in a single system. However the viewing or navigating toward data interested has not been evolved suitably because of the overwhelming volumes of sensor data. This work introduces a data browsing technique from distributed databases of sensor data condensed according to semantics shared among servers. Web links of a colored presentation could enhance the view of sensor data indexed along a hierarchy from distributed databases in the geographic regions.

Keywords: Sensor Databases, Data Browsing, Semantic Condensing.

1 Introduction

Sensor devices are being embedded in public facilities for pervasive computing these days. In many cases, however, much of data collected for a long time are nearly meaningless. For the volume reduction of sensor data, most of the data managements refine and filter raw data. In the process there is hardly any condition to filter off data definitely useless. Further data captured by diverse kinds of sensor devices have diverse filtering or capturing conditions different from each other. As an example, a number of sensor nodes are capturing environmental factors such as carbon monoxide(CO), dust density and smoke in public facilities. As such public facilities are spread in a region, the volume of aggregate data should be huge in the sensor database of the region, constituted with organization databases of the public facilities. If a central management of those sensor data is necessitated for the regional or national coverage, reducing volumes of the aggregate is required.

The motivating application in this work is monitoring environmental factors for habitat concerns. A condensing method is introduced for reducing the volume of aggregate data and it constructs hierarchical views in the form of Web links to the sensor data the user is interested. Data lost in condensing the raw data is supplemented by building alternative multiple hierarchies. An early system designed to provide a worldwide sensor database system is IrisNet[1]. It supports distributed XML processing over a worldwide collection of multimedia sensor nodes, and addresses a number of fault-tolerance and resource-sharing issues[2].

G. Lee, D. Howard, and D. Ślęzak (Eds.): ICHIT 2011, CCIS 206, pp. 665–668, 2011.

2 A Motivating Application and Semantic Condensing

Among the public facilities transportation facilities are often crowded with people. Rare events such as fire or toxic gas however could develop into a catastrophe. In the subway stations sensor nodes with several kinds of sensor devices are being deployed against several kinds of environmental factors including air temperature, carbon monoxide, dust density, smoke, and carbon dioxide(CO_2). For the case of fire, rapid escalation of air temperature might implicate imminent fire. However this single property is not sufficient to indicate imminent fire. Properties such as current temperature and average temperature for one hour could be associated for higher probability. A forest for the navigation toward sensor data of the environmental factor, air temperature, is shown in Fig. 1. An organization(subway station) operates a few sensor networks with several kinds of sensor devices, and maintains a database of raw data gathered. The data of each organization is marked as a circle. These data are condensed and supplied to a region database that represents organization databases in the region. The condensed data of each region database is marked as a rhombus. Among these condensed data only a little above a threshold are supplied to the central database, as they are data might indicate a phenomenon evolving into significant situation. The condensed data of the nation database is also marked as a rhombus.

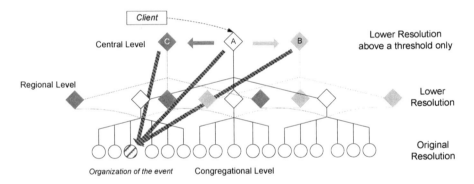

Fig. 1. Spawn queries to the interested sensor data in the trees of condensed data

In Fig.1, the trees of root A, B, and C consist of temperature variations for a preset duration, current temperatures, and average temperatures for another preset duration, respectively. The sample scenario is as follows; in the tree of root A, the temperature variations greater than 2.5 degrees centigrade for 1 minute are supplied to the regional level. Variations greater than 2.5 degrees centigrade above 35 degrees centigrade are supplied to the central level on its occasion. On this update, the event is reported to the user connected to the central server of the center database. The server automatically queries to current temperature of uncondensed data, average temperature of uncondensed data, and temperature variation of uncondensed data. Those properties of air temperature are displayed to the client and the client makes a decision on whether to notify the regional command of the fire brigades. Although that event is also notified by the regional server and the organizational server, one central server covers the occasion of fire in regions and the regional servers aid co-operation of regional fire brigades. The clients could be connected to each of regional servers directly. Additionally, querying to the forest of CO and that of smoke should be helpful in detecting imminent fire if the forests are built in the databases. Reducing rates by the semantic condensing is derived in our previous work[3] with a similar sample scenario. A similar work except condensing the sensor data is dealt in [4][5]. The work is to integrate real-world devices to the Web, for easy combination with other virtual and physical resources. The proposed architecture is implemented on the Sun *SPOT* platform and on the *Ploggs* wireless energy monitors, with a resource oriented approach. Our work deals with upper layers in the system covering large area.

3   Web Links to Sensor Data in the Hierarchical Views

The semantic condensing described is vertical decomposition of one environmental factor into properties and filtering off insignificant data according to predefined conditions. The central server and regional server are able to manage all the events concentrated from the lower level, with the reduced volumes of condensed aggregate data.

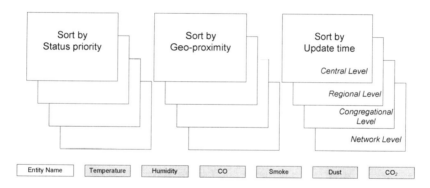

Fig. 2. Status browsing at each level along the hierarchy and a tuple listed in a screen

The user interface is designed basically in semi-CUI, as shown in Fig. 2. Every entity is sorted and listed at each level along one of three screen modes. The entity name is the region, organization, network, and node(place) names respectively, from the central level. The attributes of environmental factors are displayed in colors at each level except the *Network level*. For the attributes of temperature, i.e., from the regional level one of 8 statuses, one of 4 statuses(ORDINARY, NOTICE, WARNING, DANGER), one of 2 statuses(NORMAL, ABNORMAL) is displayed in colors, respectively. Colors of congregational and network level are the same to that of regional level. Each attribute of the environmental factor is a link to the lower level. When the user click on a rectangle of one level, a query on the attribute is requested to the server of that level, and the data in that level is displayed soon. Note that, the highest status among the attribute values determines the attribute value of the

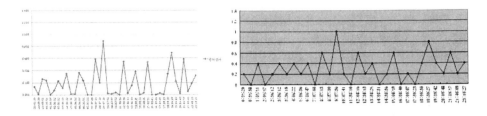

Fig. 3. Captured and stored data in the congregational level(left) and in the regional level(right)

entity. In the central level, one region's temperature value is ABNORMAL as soon as one organization has the temperature status corresponds to ABNORMAL in the region.

Those data captured and supplied are stored as records for future references. Fig. 3. shows the records of Dust, for example, in the congregational level and the regional level. The data in the original resolution in the level are condensed into data in a lower resolution (1/100). There were 32 value changes at the resolution of condensing thus only 32 tuples are sent to the regional server for this factor. At the time one occurrence above a threshold (1.000) is reported from a region database although the corresponding original data is below the threshold in the organization database. This round-off error, by the condensing into the lower resolution, however makes the center database record one tuple from the region database. The CUI is planned to evolve toward a Google Map-style GUI.

References

1. Campbell, J., Gibbons, P.B., Nath, S.: IrisNet: An Internet-Scale Architecture for Multimedia Sensors. In: Annual ACM International Conference on Multimedia, pp. 81–88. ACM, New York (2005)
2. Deshpande, A., Nath, S., Gibbons, P.B., Seshan, S.: Cache-and-query for wide area sensor databases. In: ACM SIGMOD International Conference, pp. 503–514. ACM, New York (2003)
3. Ok, M.: A Hierarchical Representation for Recording Semantically Condensed Data from Physically Massive Data Out of Sensor Networks Geographically Dispersed. In: Meersman, R., Herrero, P., Dillon, T. (eds.) OTM 2009 Workshops. LNCS, vol. 5872, pp. 69–76. Springer, Heidelberg (2009)
4. Guinard, D., Trifa, V.: Towards the Web of Things: Web Mashups for Embedded Devices. In: Workshop on Mashups, Enterprise Mashups and Lightweight Composition on the Web. ACM, New York (2009)
5. Guinard, D., Trifa, V., Pham, T., Liechti, O.: Towards physical mashups in the Web of Things. In: Int. Conf. Networked Sensing Systems, pp. 1–4. IEEE, Los Alamitos (2009)

An Efficient Nested Query Processing
for Distributed Database Systems

Yu-Jin Kang[1], Chi-Hawn Choi[2], Kyung-En Yang[1],
Hun-Gi Kim[2], and Wan-Sup Cho[1]

[1] Dept. of Management Information Systems, u-BIZ BK21,
Chungbuk National University, Korea
yingying77@chungbuk.ac.kr,
hbbs01@hanmail.net
wscho@chungbuk.ac.kr
[2] Dept. of Bio-Information Technology,
Chungbuk National University, Korea
Charisma0629@gmail.com,
bluehist@nate.com

Abstract. Performance of OLAP queries becomes a critical issue as the amount of data in the data warehouses increases rapidly. To solve this performance issue, we proposed a high performance database cluster system called HyperDB in which many PCs can be mobilized for excellent performance. In HyperDB, an OLAP query can be decomposed into sub-queries, and each of the sub-queries can be processed independently on a PC in a short time. But if an OLAP query has nested form (i.e., nested SQL), it could not be decomposed into sub-queries. In this paper, we propose a parallel distributed query processing algorithm for nested queries in HyperDB system. Traditionally, parallel distributed processing of nested queries is known as a difficult problem in the database area.

Keywords: Distributed Database System, Data Warehouse, OLAP, Query Parallel Distributed Processing, Nested Query.

1 Introduction

DW (data warehouse) is a collection of information for business intelligence systems [2]. DW maintains subject-oriented, integrative, time-variant, and non-volatile organizational data[3], and supports decision-making in an organization[3]. *OLAP (On-Line Analytical Processing)* is a tool which quickly provides answers for decision-making using DW[12]. OLAP queries are usually more complex than conventional *OLTP (On-Line Transaction Process)* queries[2] because complex logic for decision making should be represented in the OLAP queries. For this reason, OLAP queries' performance has been a critical issue in DW area.

To solve performance issue, we have proposed a high performance distributed database system called *HyperDB*[4,5,6,10]. HyperDB consists of multiple PCs

G. Lee, D. Howard, and D. Ślęzak (Eds.): ICHIT 2011, CCIS 206, pp. 669–676, 2011.
© Springer-Verlag Berlin Heidelberg 2011

connected by high speed network. Each node participates in the computation of complex OLAP queries via *NFS (Network File System)*. A specialized chip called *LanLinux* (http://www.Lanlinux.com) is used for the seamless connection of the PCs.

In this paper, we propose an algorithm for nested query decomposition and parallel query evaluation. Typical OLAP query can be decomposed easily and its decomposed queries are able to be processed independently on each PC node[10]. But if OLAP query has nested form (i.e., nested SQL), it could not be decomposed into sub-queries[11]. Therefore, nested queries can't exploit the merit of multiple PCs in HyperDB[4].

To process nested query on HyperDB, we use Kim[9] and Ganski[7]'s efficient technique to rewrite the initial nested query form to its canonical equivalent non-nested form. But, temporary tables generated in the processing of the nested queries are major obstacle in the parallel distributed processing of the nested queries in HyperDB. We propose a temporary table allocation method in HyperDB. Performance analysis for TPC-H database benchmark (http://www.tpc.org/) has been done and the result shows good performance for the nested queries.

The paper is organized as follows. In Section 2, we introduce HyperDB system. In Section 3, we describe nested query decomposition technique and how to deal with temporary tables in HyperDB. In Section 4, we present the experimental results for TPC-H benchmark. In Section 5, we conclude the paper.

2 Related Work - HyperDB

In this section, we introduce HyperDB[4,5,6,10] developed by Chungbuk National University. HyperDB consists of several PCs. Fig 1 shows HyperDB architecture. PC-nodes are booted from Boot Server, not from their local disks. After booting, each

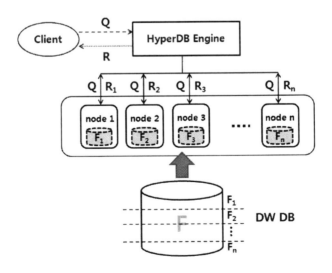

Fig. 1. HyerpDB Architecture [4,5,6.10]

node mounts a DW DB stored in a Server and caches a specific partition of the database via NFS. This means that DW DB, physically stored in a server, is shared by all the PC nodes. HyperDB provides parallel distributed query processing like intra-query parallelism. Intra-query parallelism consists of having multiple nodes processing, at the same time, different operations of the same query[1].

Fig 2 (a) shows the original star schema in a data warehouse: the fact table F and 4 the dimension tables D_1, D_2, D_3, D_4. If we have n-nodes in HyperDB, each node has dimensional tables (replication) as shown in Fig 2 (b). Note that the dimension tables are very small compared with the fact table. However, the fact table is stored in the server system, and the node PC_i is able to access predefined part (denoted by F_i, $_i$=1,2,...,n) of the fact table via NFS. A field in the fact table can be designated as the partition key whose values are used to form F_i[4,5,6,10].

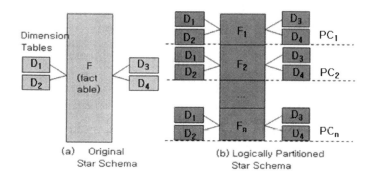

(a) Original
Star Schema

(b) Logically Partitioned
Star Schema

Fig. 2. HyperDB Storage Model [4,5,6.10]

3 Nested Query Processing on HyperDB

In this section, we propose a nested query processing on HyperDB. Since nested queries can be transformed into flat queries[7,9], decomposition algorithm[10] can be applied for the transformed flat queries. Some types of nested queries can be transformed into flat queries by using temporary tables[7,9], and these temporary tables must be treated in an appropriate way for the performance of HyperDB.

If we decompose nested query Q into sub-query Q_i and process it on node PC_i independently ($i = 1,2,...,n$), we could not get the correct results. When the SELECT clause in the inner query block includes an aggregate function (SUM, AVG, MAX, MIN, COUNT) and the WHERE clause of the inner query block contains a join predicate which references the relation of the outer query block, the inner query block have to scan the entire fact table. Since each node has only a part of fact table in our storage model, logical full replication, inner block need to access remote fact table which causes severe performance delay. For these reasons, we adopt well-known nested query transformation techniques[7,9] instead of query decomposition[4,5,6] for nested queries.

Kim[9] first introduced an efficient technique to rewrite the initial nested query form to its canonical equivalent non-nested form. First, he identified five types of nested queries. Table 1 shows the query types and corresponding conditions. For example, if a query satisfies C2 condition (i.e., SELECT clause includes an aggregate function), the query is type A.

Table 1. Nested Queries classification (Won Kim, 1982)

Query Type / Conditions	Type A	Type N	Type J	Type JA	Type D
C1: Inner query has a join predicate that references the outer query block	X	X	O	O	O
C2: SELECT clause includes an aggregate function	O	X	X	O	X
C3: Inner query block contains a division predicate	X	X	X	X	O

Kim[9] showed that nested queries could be transformed to logically equivalent flat queries using a join predicate. He proposed several transformation algorithms called *algorithm NEST-N-J*[9], *algorithm NEST-JA*[9] and *algorithm NEST-D*[9]. Ganski and Wong[7] revised the algorithms later.

In Kim[9]'s transformation algorithm for *Type N, J* nested queries, we can rewrite nested queries with simple join condition. Then we can process it easily on HyperDB system with the query decomposition algorithm[4,5,6].

However in cases of *Type-J, Type-JA and Type-D*, to transform a nested query into the non-nested form, a temporary table Ts is created to compute aggregation and join condition of the inner query block. Ts is made of the table on From clause in the inner query block. Then, we can transform nested query to non-nested form through the join predicate between the temporary table Ts and the outer query block. Finally, new query can be decomposed to sub-queries Q_i, and the sub-queries can be processed independently on the PC_i in HyperDB.

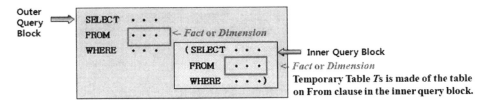

Fig. 3. Nested Query's Structure and Temporary Table

At this point, the size and divisibility of the temporary table are important in the query performance. Temporary table Ts may be small because of the GROUP BY clauses and aggregation functions. But, in some cases, that may not be true. We classify three cases of the nested query based on the size and divisibility of Ts. (1) If Ts is very small, Ts can be duplicated in every node like dimension tables. (2) If Ts is large like the fact table and it is divisible, the temporary table Ts is stored in the server system, and the node PC_i is able to access and cache predefined part Ts_i $(i=1,2,...,n)$ of the temporary table via NFS. (3) If Ts is large and it is indivisible, Ts is stored in the DW server and each node PC_i accesses the remote Ts table at the sacrifice of performance for correct result[10].

By the analysis of nested queries, we can estimate the size and divisibility of the query and construct a query evaluation plan. There are four types of join between outer query block and temporary tables Ts from inner query block:

1) Type-1: Dimension ⋈ Ts from Dimension
In this case, both inner and outer query blocks use the dimension tables. Ts is generally very small compare with the dimension table of the inner query block. Since each node has copied dimension tables as shown in Fig 2 (b), original query Q can be processed on a single PC without partitioning Ts.

2) Type-2: Fact ⋈ Ts from Dimension
The outer query block uses the fact table and the inner query block uses the dimension tables. In this case Ts is very small and each node PC_i has copied temporary table Ts. Since the node PC_i has full copy of Ts and can access predefined part of F_i, we can decompose query Q into sub query Q_i and process Q_i on the node PC_i.

3) Type-3: Dimension ⋈ Ts from Fact
Outer query block uses the dimension tables and inner query block uses the fact table. When the inner query block uses the fact table, Ts can be large like the fact table occasionally. In this case, the Ts is stored in the DW server, and the node PC_i is able to access and cache the predefined part of the temporary table (:Ts_i) via NFS. Since the dimension table used by outer query block is not divided, the temporary table can be split logically, and query Q can be decomposed into sub-query Q_i and process Q_i on the node PC_i.

4) Type-4: Fact ⋈ Ts from Fact
Both outer and inner query blocks use the fact table. The FROM clauses of the inner query block and the outer query block use the fact table together. In this case, the temporary table Ts is indivisible even though Ts is large because PC_i has already partitioned fact table (F_i). If the records which should be compared with Ts_1 on PC_1 is stored in fact table F_2 on PC_2, we could not obtain the correct result (Fig 4). In this case, Ts is stored on DW sever and each node PC_i has to access full temporary table Ts and predefined part of F_i to search correct result. However, note that, since most of all Ts are small, this case is not often occurred.

Fact	T from Fact		Example)
F_1	T_1	PC$_1$	Select sum(F_price)
F_2	T_2	PC$_2$	From Fact, T Where F_num = T_num and F_quantity < T_AVQ:
F_3	T_3	PC$_3$	To find the records satisfying where condition,
⋮	⋮		F_1~F_n have to scan the entire table T. But, if table T is divided up,
F_n	T_n	PC$_n$	F_n on PC_n can use T_n only.

Fig. 4. Partitioned Temporary Table Problem on Type 4 Query

4 Experiment

We use *TPC-H benchmark* data for OLAP query performance evaluations[8]. TPC-H benchmark consists of nine nested queries and related database. Among nine queries, 4 queries use only dimension tables. Therefore we select 5 typical nested OLAP queries which include a huge fact table (*LineItem*): Q4, Q18, Q21(Type-2), Q20(Type -3), Q17(Type-4). The size of the database is about 1GB. The *LineItem* table of TPC-H database is the largest table, usually called the fact table in DW. We define an *L_OrderKey* attribute of a *LineItem* table as a partition key. Table 2 shows our experiment environment.

Table 2. Experiment environment

Category	Description
DBMS	MySQL 5.0
Number of nodes	16
OS	Red Hat Linux Release 9
PC specification	CPU: Pentium Ⅳ 2.4 GHz
	Main Memory: 512MB
Network Bandwidth	100Mbps

In HyperDB, 16 PCs are used in this database system. All PCs participating in a distributed database system do not use their local disk. Instead, they are booted from the remote Linux boot server[8]. We configured the key buffer, sort buffer, and record buffer of MySQL as 8MB, 2MB, and 1KB respectively.

Fig 5 shows performance result for nested queries. Query performance improves in proportion to the number of nodes. Note that performance upgrade is significant until the number of PCs becomes 8.

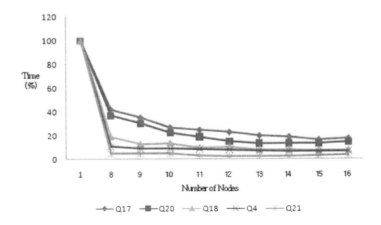

Fig. 5. Correlation between Number of Nodes and Nested Query Performance

5 Conclusion

HyperDB is a novel distributed database system for high performance OLAP query processing. We just need the *LanLinux* chip additionally for each PC to participate in the this system. Since the cluster nodes are loosely-coupled, we can get higher extensibility and minimized management cost [4,5,6,10].

However HyperDB system has a limitation in the query evaluation of nested queries[10]. Traditional nested query processing techniques, such as Kim[9]'s, and Ganski and Wong[7]'s, are helpful to solve the problem when the size of the temporary table in the query processing is small. For large temporary table, we need another solution to keep the performance of HyperDB.

We proposed a temporary table allocation method to the PCs in HyperDB. We classified 4 types of nested queries, and proposed appropriate solutions for each query type. Performance analysis for TPC-H database benchmark (http://www.tpc.org/) has shown that the query response time for the nested queries is decreased significantly as the number of PCs increases.

Acknowledgement. This work was supported by the grant of the Korean Ministry of Education, Science and Technology" (The Regional Core Research Program/Chungbuk BIT Research-Oriented University Consortium) and the Ministry of Education, Science Technology (MEST) and Korea Industrial Technology Foundation (KOTEF) through the Human Resource training Project for Regional Innovation.

References

[1] Alexandre, A.B., et al.: Parallel OLAP Query Processing in Database Clusters with Data Replication. Distributed and Parallel Databases 25(1-2), 97–123 (2009)

[2] Akal, F., Böhm, K., Schek, H.-J.: OLAP query evaluation in a database cluster: A performance study on intra-query parallelism. In: Manolopoulos, Y., Návrat, P. (eds.) ADBIS 2002. LNCS, vol. 2435, pp. 218–231. Springer, Heidelberg (2002)

[3] Inmon, W.H.: Building the data warehouse, 2nd edn. John Wiley&Sons, Inc, West Sussex (1996)

[4] Kim, T.K., et al.: HyperDB: A PC-Based Database Cluster System for efficient OLAP Query Processing. In: Proc. of 19th IASTED International Conference on Parallel and Distributed Computing and Systems (PDCS 2007), Cambridge, USA (November 2007)

[5] Kim, T.K., et al.: A LanLinux-Based Grid System for Bioinformatics Applications. In: Proc. of International Conference on Advanced Communication Technology (ICACT), vol. 3, pp. 2187–2192 (February 2006)

[6] Kim, T.K., et al.: A Hybrid Grid and Its Application to Clustering Orthologous Groups for Multiple Genomes. In: Proc. International Symposium on Computational Life Science, pp. 11–20 (2006)

[7] Ganski, R.A., Wong, H.K.T.: Optimization of Nested SQL Queries Revisited. ACM Transaction on Database System, 23–33 (1987)

[8] TPC (Transaction Processing Performance Councile) Benchmark,
http://www.tpc.org/tpcd/default.asp

[9] Kim, W.: On Optimizing and SQL-like Nested Query. ACM Transaction on Database System 7(3), 443–469 (1982)

[10] Kang, Y.J., et al.: An OLAP Query Decomposition Technique for PC-based Database Cluster Systems. In: Proc. of the 18th IASTED International Conference on Parallel and Distributed Computing and Systems (PDCS 2009), Innsbruck, Austria (February 2009)

[11] Kang, Y.J., et al.: A Query Decomposition Technique for Nested Query on Database Cluster System Environment. Korea Institute of Information Technology Architecture 7(1), 89–96 (2010)

[12] Colliat, G.: OLAP, Relational and Multidimensional Database System. In: Proc. ACM SIGMOD Record, vol. 25(3) (1996)

Author Index